KT-142-894

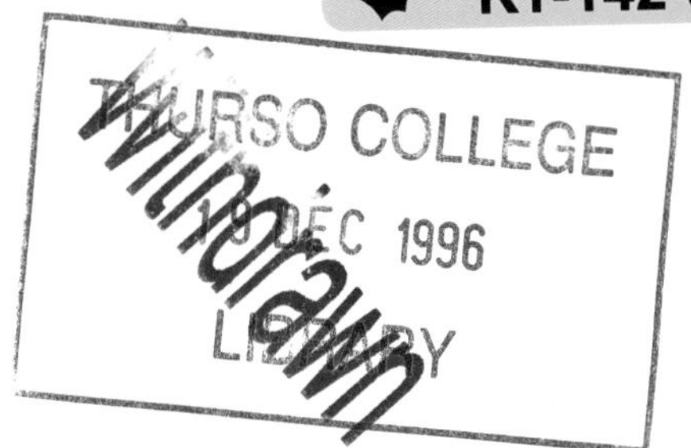
THURSO COLLEGE
10 DEC 1996
LIBRARY
Withdrawn

PRACTICAL HPLC METHODOLOGY AND APPLICATIONS

PRACTICAL HPLC METHODOLOGY AND APPLICATIONS

BRIAN A. BIDLINGMEYER

A Wiley-Interscience Publication

JOHN WILEY & SONS, INC.

New York / Chichester / Brisbane / Toronto / Singapore

This text is printed on acid-free paper.

Library of Congress Cataloging in Publication Data:

Bidlingmeyer, Brian A.
Practical HPLC methodology and applications / Brian A. Bidlingmeyer.
p. cm.
"A Wiley-Interscience publication."
Includes bibliographical references and index.
ISBN 0-471-57246-2
1. High performance liquid chromatography. I. Title.
QD79.C454B53 1992
543′.0894—dc20 92-4680

Printed in the United States of America

10 9 8 7 6 5 4 3

To
Art
and
Buck

CONTENTS

PREFACE ix

1 OVERVIEW OF MODERN LIQUID CHROMATOGRAPHY 1

2 LIQUID CHROMATOGRAPHY AS A RESEARCH TOOL 27

3 THE CHROMATOGRAM AND WHAT CONTRIBUTES TO IT 68

4 APPROACHING THE PROBLEM 104

5 DEVELOPING THE SEPARATION 131

6 CONSIDERATIONS FOR PROPER OPERATION OF A LIQUID CHROMATOGRAPH 207

7 GRADIENT ELUTION CHROMATOGRAPHY 284

8 EXPERIMENT 1: DEMONSTRATING THE FUNDAMENTALS 318

9 EXPERIMENT 2: NORMAL-PHASE CHROMATOGRAPHY 332

10 EXPERIMENT 3: EFFECT OF COLUMN LENGTH AND RECYCLE 345

11 EXPERIMENT 4: GEL PERMEATION CHROMATOGRAPHY USING DUAL DETECTORS (UV AND RI) 358

12 EXPERIMENT 5: DEVELOPING A REVERSE PHASE CHROMATOGRAPHIC SEPARATION 370

13 EXPERIMENT 6: QUANTITATION 384

14 EXPERIMENT 7: MONITORING KINETICS 405

15 EXPERIMENT 8: PREPARATIVE LIQUID CHROMATOGRAPHY 415

16 EXPERIMENT 9: ANALYSIS OF ESSENTIAL OILS (STEAM DISTILLATES) 425

17 EXPERIMENT 10: GRADIENT ELUTION 435

INDEX 449

PREFACE

The first commercial introduction of high performance liquid chromatography (HPLC) was in 1969, and since that time it has become one of the most important and fastest-growing techniques in the modern laboratory. Although HPLC is only beginning its third decade of use, modern HPLC analyses are routinely fast and efficient with detection of as little as 200 pg of material. Over the years HPLC has progressed from a difficult "art" into a straightforward separation science used to solve a host of problems in every area of chemical analysis. Opportunities to apply HPLC are almost unlimited, with the result that HPLC instruments have become indispensable tools for a variety of scientists and industries. This widespread popularity underscores the importance of "getting started" in HPLC at an early stage in an individual's scientific education.

The purpose of this book is to provide the novice with sufficient practical information necessary to begin developing useful separations. The "science" of HPLC is application focused, and it should be no surprise that most successful chromatographers are very application driven. By this it is meant that an understanding of how the chromatography is used to solve problems provides insight into which technique(s) will accomplish the separation of interest. Understanding how HPLC is being applied to solve problems puts the "problem-solving strategy" into the proper context. In other words, knowing how to develop a separation is more valuable if the individual also understands where and how to apply HPLC. Although numerous HPLC texts exist, until now no text has been totally focused on "getting started."

This book will also provide the intermediate practitioner with a useful bridge between the strictly introductory texts and the more advanced treatises. The pragmatic discussions of separation describe the rationale that is used to develop a procedure. This detail gives the intermediate reader a broader foundation in chromatography. It leads to a better understanding of the benefits of a systematic study of the interactions of the sample with the stationary and mobile phases and sets the stage for more advanced discussions and work.

Although modern HPLC is more than 20 years old, not everyone has access to those 20 years of experience. Therefore, this book includes subjects that were actively discussed in the formative years of HPLC. As a result, some topics are from manufacturers' literature which is no longer in print and some information is from "dated" references. However, this information is as crucial today to proceed along the learning curve as it was in the early days.

Also included in this book are nine tested experiments to teach the basics with "hands-on" investigations. All of the experiments need not be run; however, relevant ones should be to instill needed experimental and operational skills. Each experiment contains a brief discussion to put the subject matter into perspective and, in keeping with the theme of "getting started," some steps include comments on why they are necessary and/or how they contribute to the experiment. Example results are actually shown and discussed. Each section is designed to minimize preparation time of the instrument and sample(s) so that the hands-on experience with the equipment is maximized. For advanced investigations, supplementary experiments are often suggested.

It is hoped that after reading this book and applying the information contained here in, the reader can develop successful separations and, thereby, maximize the performance output of HPLC equipment in solving practical problems.

BRIAN A. BIDLINGMEYER

Hopkinton, MA
March 1993

CHAPTER 1

OVERVIEW OF MODERN LIQUID CHROMATOGRAPHY

Introduction
- Selectivity
- Efficiency
- Using the chromatogram

Historical perspective

Application growth
- Survey of general LC application areas
- Gel permeation chromatography
- Ion chromatography
- Preparative HPLC
- Compound synthesis

The future

Glossary

References

INTRODUCTION

Column liquid chromatography (LC) is one of the fastest-growing segments of analytical instrumentation, owing in large part to the excellent quantitative and preparative capabilities offered. Liquid chromatography can be applied to a greater variety of samples than any other separation technique and is applicable to the extremely complex mixtures that scientists encounter in chemical and biological systems. Training in LC separation technology is essential for modern chemists and biochemists as LC separations will continue to play a complementary and vital role in all areas of chemical analysis despite additional advances in selective chemical reagents and improvements in physical techniques for measurement.

As a result of technological developments during the past decade which have brought significant improvements to instrumentation and column packings, high-performance liquid chromatography (HPLC) has emerged as the preferred method for the separation and quantitative analysis of a wide range of samples. Modern HPLC analyses are fast and efficient, and detection of as little as 200 pg of material is routinely achievable. Opportunities for applying HPLC are almost unlimited, with the result that HPLC instruments have become indispensable tools for a variety of scientists and industries. This widespread popularity underscores the importance of introducing future scientists to the principles of LC at an early stage in their training.

Liquid chromatography is based upon the phenomenon that, under the same conditions, each component in a mixture ordinarily interacts with its environment differently from all other components. When used as an analytical tool, LC can determine the number of components in a mixture, how much of each is present, and the degree of purity of each component. When employed in conjunction with other analytical techniques, it assists in identifying components in a mixture.

Separation is required when (1) a mixture is too complex for a direct analytical measurement (e.g., spectroscopy), (2) the materials to be analyzed are very similar, such as isomers, (3) it is necessary to prepare highly purified materials, and (4) a measurement of the amount of a particular material is needed. Filtration, open-column chromatography, and thin-layer chromatography are used for relatively easy separations. Modern HPLC is a technique for making precision separations of complex mixtures and offers high-resolution separating capability to solve problems faster and better.

However, HPLC is not a typical analytical instrument in which you bring the sample to the instrument and insert it, the instrument preforms the measurement, and then gives a signal (analogue or digital) which represents the amount present. In HPLC many parameters must be determined and set before the sample is placed into the device. Hence to be successful with HPLC, sufficient time is needed to "set up" the instrument for each analysis. This can be hours or days depending upon the problem (application), the knowledge of the operator, and the type of experimentation which is done. With experience and a working knowledge of the literature, HPLC methods can be developed effectively in a reasonable time.

Figure 1-1 shows the components of a simple HPLC instrument. The mixture to be analyzed is dissolved in a suitable solvent, introduced ("injected") at one end of the "column," and carried through the column by a continuous flow of the same solvent ("mobile phase") in which the mixture was dissolved. The separation takes place in this column, which contains "sorptive" particles of large surface area. These particles are referred to as the "stationary phase." The device for applying a precise volume of sample onto the column is the injector. Sample components that are injected revers-

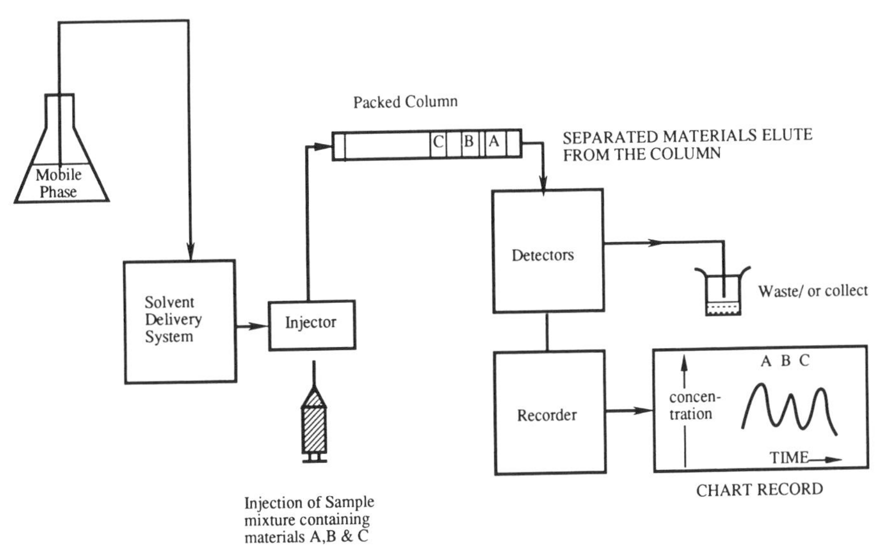

FIGURE 1-1. Block diagram showing the components of an isocratic HPLC instrument.

ibly interact with the stationary phase in a continuous manner. The mobile phase (also called the eluent) is pumped through the column bed of the tightly packed chromatographic particles using a solvent delivery system ("pump"). With the selection of the proper mobile phase and column packing material, some components of the mixture will travel through the column more slowly than others. As the sample components emerge from the column, a suitable detector is used to monitor and transmit a signal to a recording device. The "chromatogram" is a record of the detector response as a function of time and indicates the presence of the components as "peaks." An example of a chromatogram (1) is shown in Figure 1-2. Sometimes the instrumentation also contains a solvent-mixing device and another pump for gradient elution. The technique of gradient elution is discussed in Chapter 7.

The chromatographic process is depicted in Figure 1-3*a*. The three basic steps which take place are injection, separation, and elution. During injection components are "sorbed" onto the stationary phase. Migration of the sample components through the stationary phase is essentially the result of two forces—movement driven by the mobile phase and retardation resulting from the stationary phase. Thus the sample molecules are held by the stationary phase and transported by the mobile phase. These two opposing forces cause an "on again, off again" activity which is mathematically described by an equilibrium distribution between the two phases for each compound. When a sample is first injected onto an LC column, it forms a narrow band at the head of the column (Fig. 1-3*b*1). If the equilibrium distributions

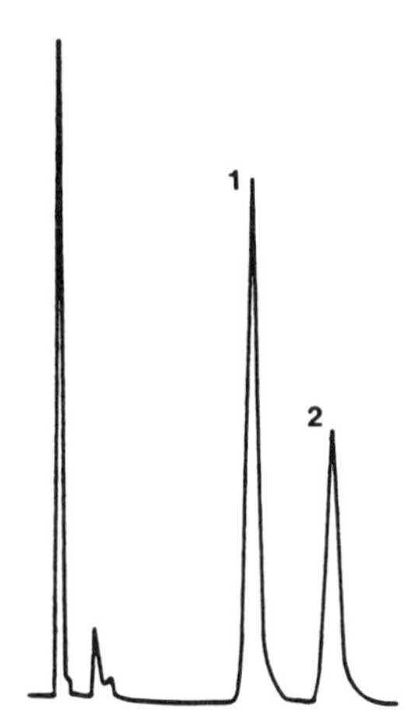

FIGURE 1-2. A typical chromatogram. Separation of Mountain Dew® soft drink sample: peak 1 is caffeine and peak 2 is benzoic acid. The two unmarked peaks are unidentified components. (Reproduced from reference 1 with permission.)

for two compounds differ, the chromatography will result in a different rate of migration for each compound. As the mobile phase passes through the column, the initial band separates into individual solute bands (Fig. 1-3*b*2), each of which migrates at a rate governed by the equilibrium distribution of solute between the mobile phase and the surface of the column packing (stationary phase). Finally, the segmentation of the compounds is complete (Fig. 1-3*b*3) and the separated components continue to elute from the column as zones or bands that pass through a detector. The amount of

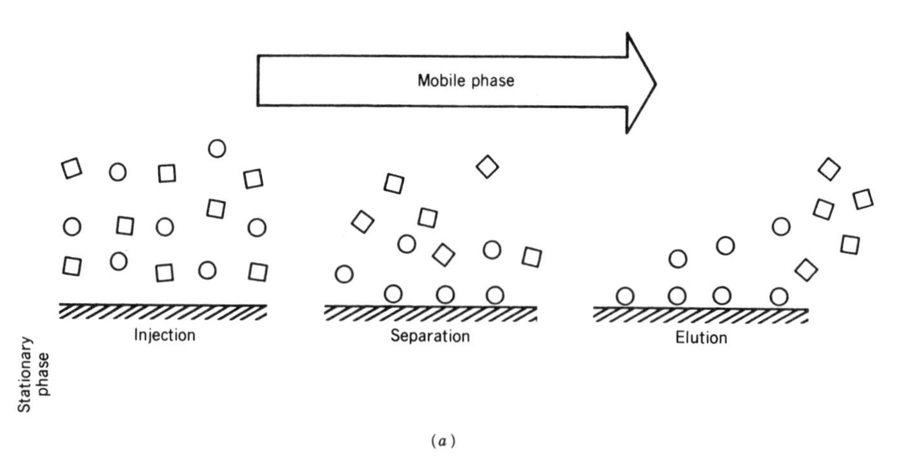

FIGURE 1-3. The liquid chromatographic process. (*a*) Segmentation at the surface. (*b*) Migration through the column.

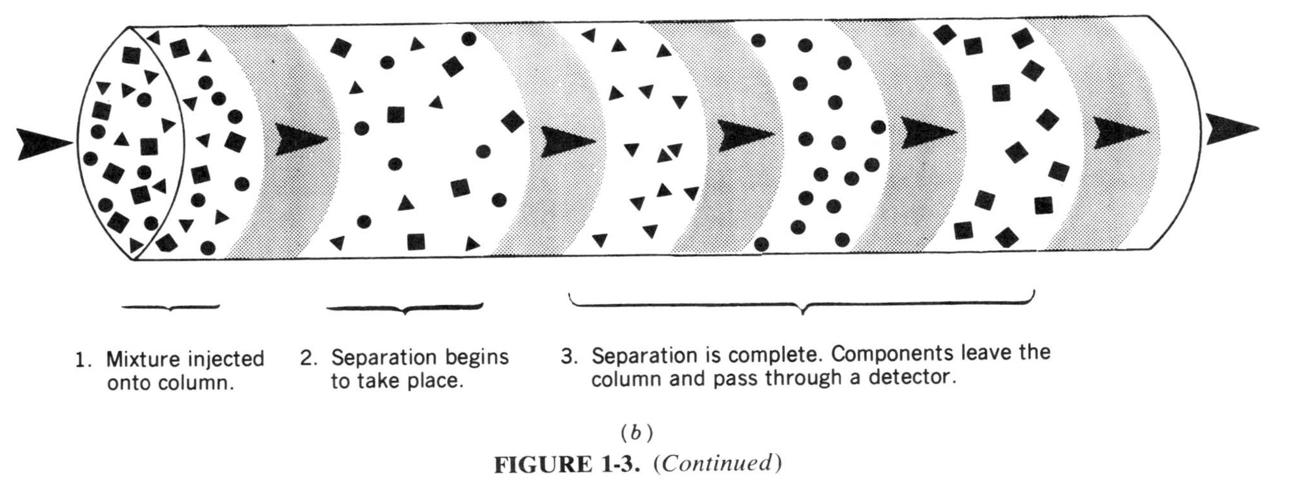
1. Mixture injected onto column.
2. Separation begins to take place.
3. Separation is complete. Components leave the column and pass through a detector.

(*b*)

FIGURE 1-3. (*Continued*)

component present in an eluted zone is sensed by the detector and is proportional to the peak height recorded on the chromatogram (Fig. 1-2).

By adjusting the composition of the mobile phase, a wide variety of solutes of differing polarity can be separated easily. Even closely related molecules can be resolved on efficient HPLC columns. Mobile phases may need to be complex—depending on the nature of the sample components—and it may be necessary to use several solvents and additives such as acids, bases, salts, or surfactants. The stationary phase can be changed to aid in achieving a separation, but this necessitates owning an assortment of packed LC columns, which is expensive. In addition, the diversity of mobile phases that can be prepared is far greater than the variety of commercially available stationary phases.

Selectivity

To develop and achieve the separation, it is necessary to manipulate the experimental variables that have the greatest influence on the equilibrium distribution—the composition of the mobile phase, the nature of the stationary phase, and, less importantly, the temperature. The forces responsible for solute interactions between the two phases are those that explain solubility: electrostatic, dipole, and dispersive (Van der Waals) forces. For maximum flexibility in developing a separation, the mobile and stationary phases are usually chosen to have contrasting polarities. In normal-phase LC, the stationary phase is polar while the mobile phase is composed of nonpolar solvents. The situation is opposite for reverse-phase LC, in which the mobile phase is more polar than the stationary phase. In reverse-phase LC, polar solutes are more strongly attracted to the mobile phase than to the stationary phase and, therefore, elute more quickly from the column. Less-polar solutes spend more time in interactions with the stationary phase and are retained longer by the column. The interested reader is directed to several texts (2–8) and review articles (9–12) for additional information. Sometimes it may not be possible to achieve the separation of all components in a sample in a reasonable time when a single mobile phase is used. Therefore, the technique of gradient elution, which results in additional mobile phase control of separations, is required. Gradient elution is the use of two or more solvents forming a mobile phase whose concentration ratio is varied with time. The gradient elution approach is often used for samples containing solutes of widely differing polarities for which a single mobile phase (an isocratic separation) is not appropriate. Gradient elution is discussed in detail in Chapter 7.

Efficiency

In addition to the differential migration of solutes, the chromatographic separation of a mixture is characterized by the subsequent spreading of the molecules of each solute as they move along the column. This band broadening is a measure of how wide the solute band is when it elutes from the column relative to how long it is retained by the column. Band broadening is closely related to the efficiency of the column and is the result of physical processes that occur in the LC column. Highly efficient HPLC columns contain small particles of less than 10-μm diameter tightly packed into steel, glass, or plastic tubes and the result is narrow peaks that elute from the column. High-performance solvent delivery systems are required to move the mobile phase across these columns at pressures up to 6000 psi. Preparative columns are often less efficient owing to the use of large (50–100 μm) particles. A more complete discussion of column efficiency and band broadening may be found elsewhere (2–8).

Using the Chromatogram

Once a separation is developed, several pieces of information about the sample can be ascertained from the chromatogram. First, by counting the peaks, one can estimate how many components are present in the mixture. Second, by the use of standards, both the *identity* and *concentration* of each compound present can be obtained. Lastly, if the mixture is totally unknown, the peaks can be collected and the identity confirmed by other instrumental methods of chemical analysis (e.g., infrared, nuclear magnetic resonance, or mass spectroscopy).

To perform qualitative and quantitative analyses correctly, it is necessary to fully separate the peaks of interest in the chromatogram. The "goodness" of separation is referred to as the resolution between two peaks and is defined as the distance between peak centers (ΔV) compared with the average width of the two peaks $\frac{1}{2}(W_A + W_B)$, as shown in Figure 1-4. Also included in Figure 1-4 is the analytical expression of the calculation. Generally, for good qualitative and quantitative analysis, a resolution of 1.0 or greater is desired.

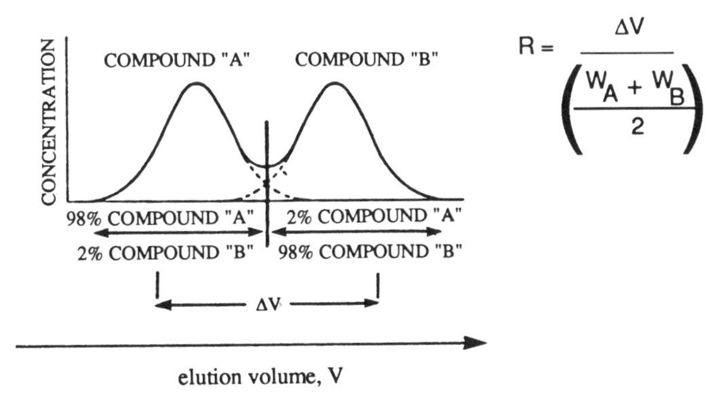

$$R = \frac{\Delta V}{\left(\frac{W_A + W_B}{2}\right)}$$

FIGURE 1-4. Resolution of two chromatographic peaks at a resolution value of 1.

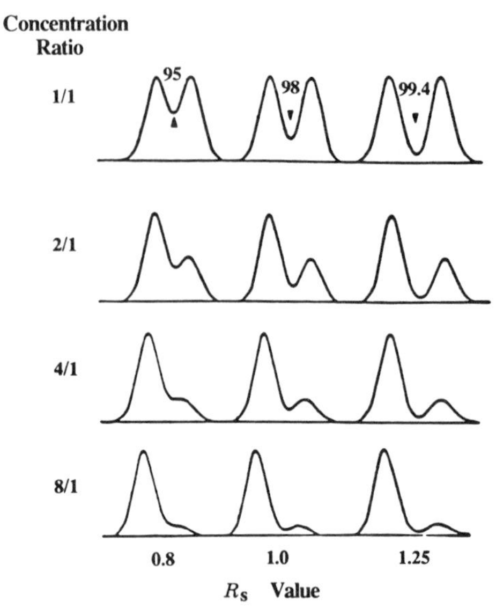

FIGURE 1-5. Comparison of overlapped peaks at several resolution values.

Figure 1-5 shows a visual comparison of overlapped peaks for several resolution values and several ratios of sample concentration of the first peak to the second peak. At a resolution value of 1.25, each peak is 99.4% pure and only 0.6% of each peak lies beneath the other peak. At a resolution value of 0.8 there is 5% overlap of the first peak into the second. It is also apparent that the peak heights of the resolved peaks are independent of one another at a resolution value greater than 1; therefore, as long as $R \geq 1$, peak heights are an adequate measure of concentration in the chromatogram. It is also clear from the diagrams that the ability to discriminate two peaks from one another depends mainly upon the resolution value and secondarily upon the concentration with respect to the neighboring peak.

In addition to providing the capability to do good qualitative and quantitative analyses, another characteristic of LC is that the compounds being studied are easily collected after passage through the system since they are not destroyed in the detector. This capability of LC is an important advantage when it is necessary to do rigorous identification of the separated compounds. The collection of LC fractions is referred to as preparative chromatography.

Preparative chromatography is useful for obtaining a highly purified compound of interest (or fraction of it) from quantities of a few milligrams to multigrams. The scope of preparative LC includes: (1) preliminary cleanup by removing extraneous materials prior to gas chromatography (GC) or analytical LC analysis; (2) analysis support by preparing pure standards; (3) synthesis support by isolating products for identification, determining the purity of starting materials, studying reaction mechanisms and kinetics, and optimizing yields; (4) biological support by preparing ultrapure compounds for biological testing; (5) natural product isolation and purification; and (6) commercial preparation of special chemicals and biologicals.

Remember that HPLC is not one scientific discipline but several, ranging from fluid mechanics to surface chemistry to the physics of detection. In a sense, HPLC is simply a very sophisticated sample preparation technique combined with an instrumental measurement (e.g., spectroscopy). High performance liquid chromatography is a unique analytical tool in that one does not simply set a dial and let the instrument do the measurement. Often a new HPLC method needs to be developed for each sample or sample type. Furthermore, it is not uncommon that each method is significantly different in terms of which objectives are being met and how the parameters of the instrument (the method) are determined. The performance objectives and operational parameters are chosen by the "application world" in which the problem resides. As shown in Figure 1-6 the application is *the* issue all individuals address and developing a successful HPLC separation requires meeting the many objectives by making key decisions in choosing the operating parameters. Making these decisions and meeting the objectives are

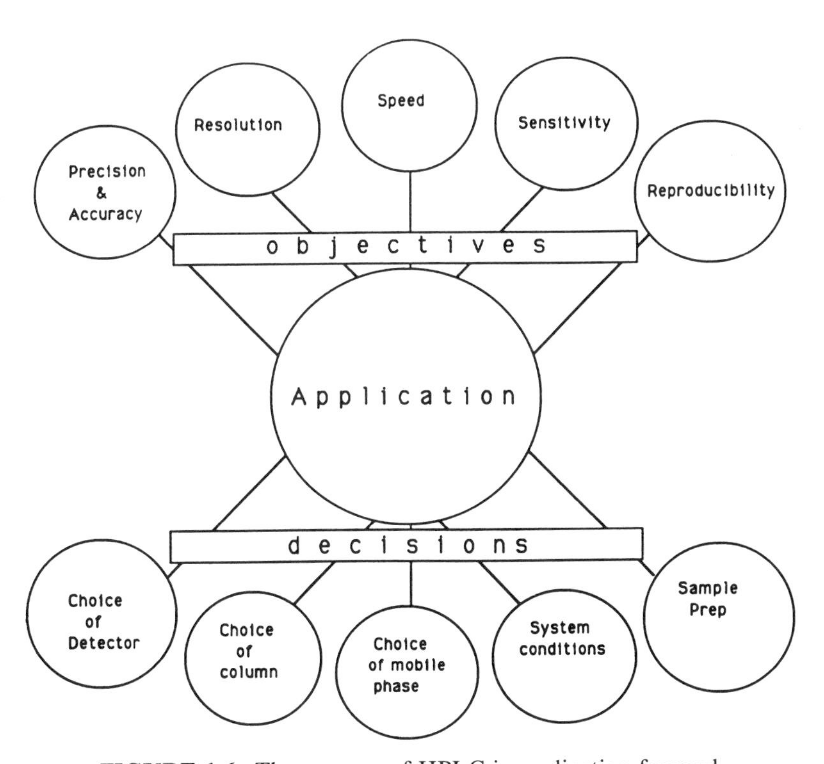

FIGURE 1-6. The success of HPLC is application focused.

always compromised based upon the application. Understanding how to compromise the decisions and trade off the objectives is the "science" of HPLC. This book is directed toward fostering an understanding of the application focus of HPLC so that successful HPLC separations can be accomplished.

HISTORICAL PERSPECTIVE

Although many are just now enjoying the benefits of modern HPLC, liquid chromatography is an old separation technique. Table 1-1 lists many of the milestones in the development of modern HPLC. Tswett is considered to have "invented" the chromatographic technique when he reported separations of chloroplast pigments into a series of colored bands on a packed column. He named this technique "chromatography." In the 1940s, Martin and Synge developed the theory of partition chromatography and used mathematics to describe the separation process resulting from the use of a liquid-coated solid phase and a moving liquid phase. As a result Martin was awarded the Nobel Price in chemistry. In his acceptance speech, he commented that a gas could be used as an eluting phase instead of a liquid.

Later Martin followed his own idea and published the first application of gas chromatography with James. "Gas–liquid" chromatography provided much better efficiencies than "liquid–liquid" or "liquid–solid" chromatography, supplying impetus for many applications of this new separation tool. As a result, many fundamental studies working toward the optimization of high-speed gas separations took place during the 1950s. This work, coupled with the practical uses reported, resulted in the commercialization of an instrument for GC.

TABLE 1-1. Some Historical Milestones in Liquid Chromatography

1903	Tswett	Development of chromatography
1938	Reichstein	Flowing chromatograph
1941	Martin and Synge	Partition chromatography
1944	Consden, Gordon, and Martin	Paper chromatography
1952	Martin and James	Gas chromatography
1952	Alm	Gradient elution
1953	Wheaton and Bauman	Exclusion principle observed
1959	Porath and Flodin	Cross-linked polydextrans—Gel filtration
1962	Moore	Cross-linked polystyrene gels—Gel permeation
1965	Giddings	"Unifying theory" and comparison of GC to LC
1969		First commercial instruments

During the 1960s when Giddings (13) reported the use of "reduced" parameters, all chromatographic (gas and liquid mobile phases) systems were able to be compared and contrasted. This provided a shift of attention to LC, which was being used under operating conditions that were far from optimum. By the late 1960s these concepts of LC were well established, and the stage was set for the commercial development of HPLC. Many of the problems and decisions faced by the early designers of LC hardware were essentially the same as the problems faced by designers today—control of fluid flow, column size, packing material separation capability, and detection. Bringing together certain precision mechanical components—precision pumps for accurately controlling flow rates and volume; columns with smooth, precisely machined internal surfaces containing efficient packing materials; and detectors capable of sensing low levels of components in the chromatography eluent—enabled the first commercially available liquid chromatograph to be built in 1969.

Gas and liquid chromatography are similar in that both separation processes are due to a difference in the equilibrium distribution of sample components between two different phases. Separation results from different velocities of migration as a consequence of the difference in equilibrium distributions. Gas chromatography separations are based primarily on the effective vapor pressure (boiling point) differences of the compounds. Liquid chromatography separations are based mainly on solubility differences. The choice of the carrier gas used as the mobile phase in GC is dictated predominantly by the type of detector or sensor used to monitor the effluent from the column. The carrier gas is inert and the separation will be almost independent of the type of carrier gas. With LC, the composition of the mobile phase is of prime importance in the separation. It makes a great deal of difference whether hexane, acetonitrile, or water is used as the moving phase. Another difference between the two chromatographic techniques is that in LC diffusion in the mobile phase is extremely low compared to that in GC. Because of the low diffusion in the mobile phase, high flow rates can be used and high-speed LC is possible. In addition, the effects of temperature are only of secondary importance in LC.

Because of these differences, these two separation techniques may be compared on the basis of the molecular weights of the compounds they can separate. This comparison is shown in Figure 1-7. It has been estimated that because the sample must be volatile or be made volatile through an appropriate chemical reaction, GC is useful for only approximately 15% of known compounds while the remaining 85% are potential candidates for LC.

APPLICATION GROWTH

High performance liquid chromatography is one of the most widely utilized analytical techniques because it has a large diversity of applications, as

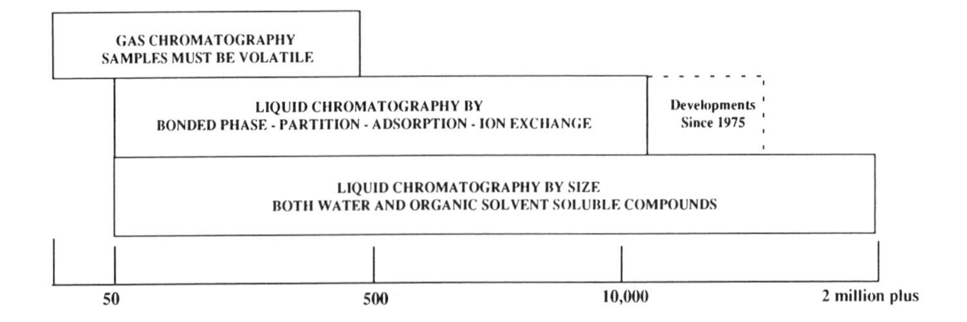

FIGURE 1-7. Comparison of chromatography by molecular weight.

shown in Table 1-2. Inspection of the scientific literature supports this premise. From essentially no references to HPLC in 1969, through the explosive growth of references recorded in the 1970s, to the present, a number of highly significant applications areas are apparent.

Survey of General LC Application Areas

Scientists working in the pharmaceutical industry were among the first to recognize the efficacy of HPLC. Early classes of compounds analyzed by HPLC included antibiotics, analgesics, over-the-counter pharmaceutical preparations, vitamins, steroids, and tranquilizers. With the development of reverse-phase columns and the use of ion pairing as an alternative to ion-exchange methods, the number of pharmaceutical compounds analyzed increased dramatically. Figure 1-8 plots only two areas of journal references of many in pharmaceutical literature as an indicator of the rapid rise in usage. Closely related to pharmaceuticals was the effective analysis of cosmetic and toiletry products, fatty acids, FD&C dyes, illegal drugs and their metabolites (cannabinol), and forensic-related compounds.

The first applications of HPLC in the analysis of food materials began in the late 1960s. Its use soared with the evolution of column packing materials that would separate sugars and spurred the development of a host of different applications (see Fig. 1-9). Using HPLC to analyze sugars also was justified economically as a result of sugar price increases in the mid 1970s which motivated the soft-drink manufacturers to add high-fructose corn syrup as a substitute for sugar. Monitoring the sweetener content by HPLC assured a good-quality product.

Other early food applications included the analysis of pesticide residues in fruits and vegetables, organic acids, lipids, amino acids, toxins (e.g., aflatoxins in peanuts, ergot in rye), and contaminants. As with pharmaceutical analysis, HPLC provides the ability to analyze for vitamin content in food

TABLE 1-2. Typical Uses for HPLC

Chemistry and biochemistry research—University and industry

- Analyzing complex mixtures
- Purifying chemical compounds
- Developing processes for synthesizing chemical compounds
- Isolating natural products that have beneficial biological characteristics
- Predicting physical properties (e.g., molecular-weight distributions of polymers)

Quality control

- Ensuring purity of raw materials
- In-process testing to control and improve process yields
- Quantitative assay of final products to assure conformance to specification
- Evaluating product stability and monitoring degradation

Environmental control

- Analyzing air and water pollutants
- Monitoring for materials that may jeopardize occupational safety or health
- Monitoring pesticide levels in the environment

Federal and state regulatory agencies

- Surveillance of food and drug products
- Identifying confiscated narcotics
- Adherence to label claims

preparations rapidly and simultaneously, without interferences from other sample constituents. Associated food-product analyses include the nutritional content of animal feed products and drugs in feeds.

One of the first applications of HPLC in the clinical field was the quantitation of theophylline in asthmatic infants. This highly accurate measurement was an important test because of the very low amount of sample required and the accuracy of the determination (see Fig. 1-10). More recent clinically related HPLC separations include drugs and drug metabolites, neurochemicals and their metabolites, histamines, thyroid hormones, and enkephalins. The earliest bioresearch applications of HPLC included the determination of peptides, proteins, and amino acids. Application of HPLC to the analysis of these compounds remains important, as indicated by the rapid growth in references (Fig. 1-11). Bioresearch remains one of the most rapidly expanding growth areas of LC.

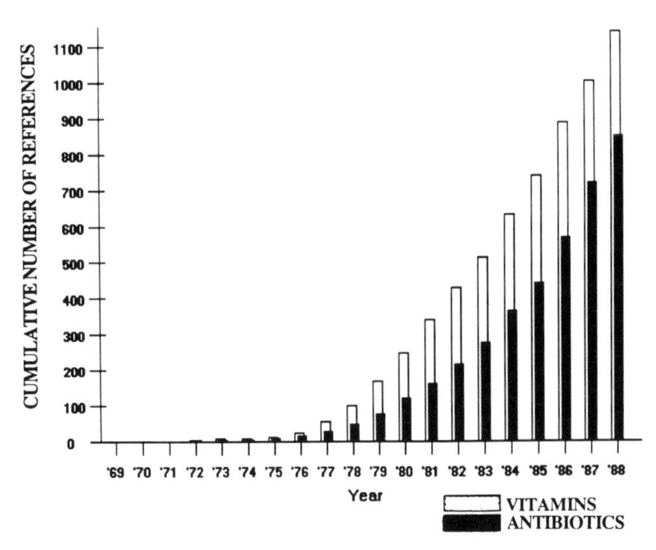

FIGURE 1-8. Journal references to the HPLC analysis of antibiotics and vitamins, 1969–1988. (Adapted from reference 14 with permission.)

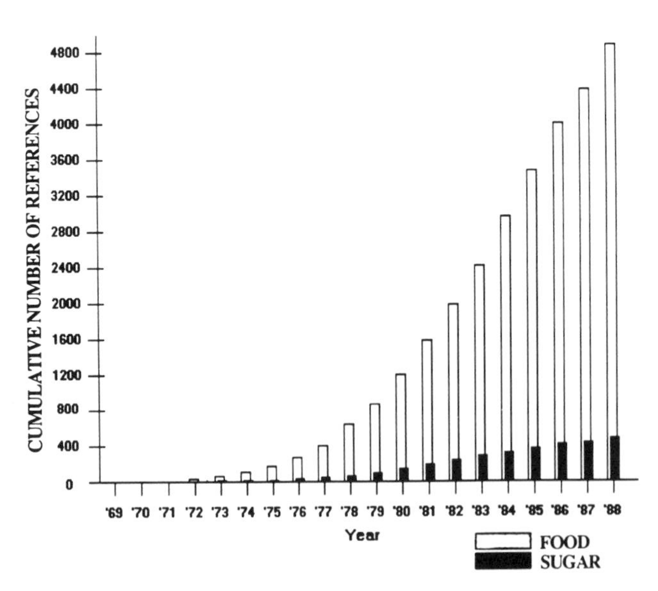

FIGURE 1-9. Journal references to HPLC analysis of food and sugar, 1969–1988. (Adapted from reference 14 with permission.)

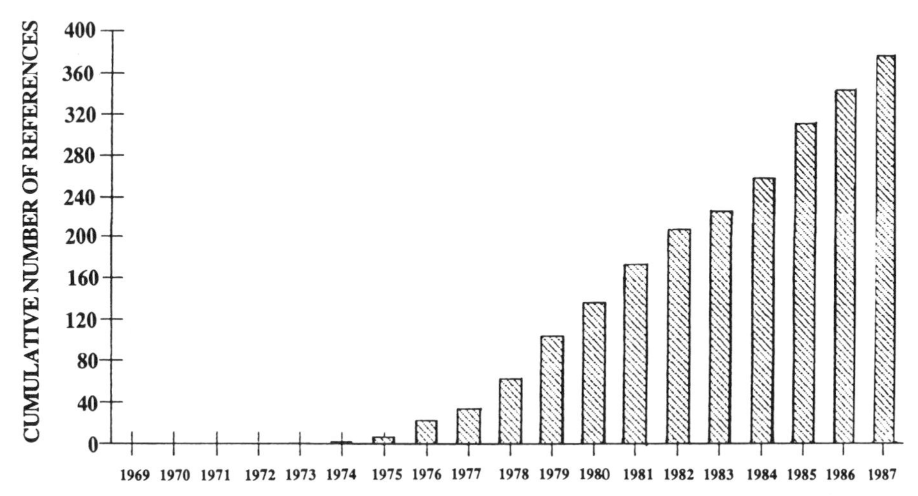

FIGURE 1-10. Journal references to the analysis of theophylline by HPLC, 1969–1987. (Reprinted from reference 14 with permission.)

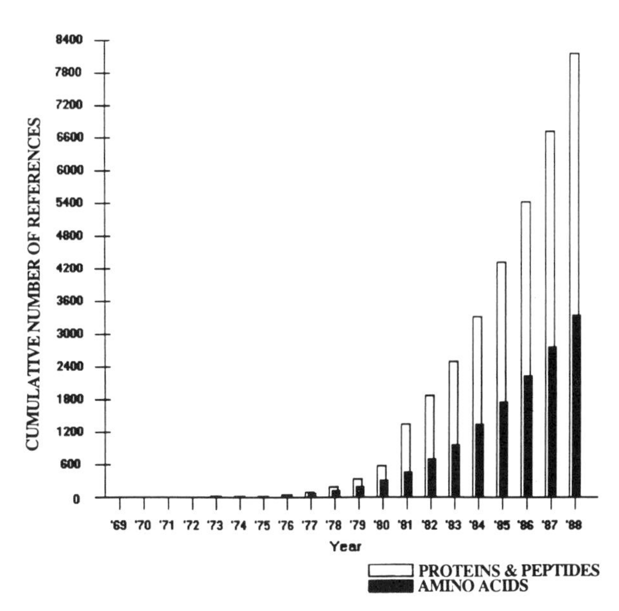

FIGURE 1-11. Journal references to peptides, proteins, and amino acids, 1969–1988.

In the late 1970s HPLC provided an ideal tool for the analysis of pollutants and other environmental contaminants. Techniques were developed for analyzing chlorophenols, pesticide residues, and metabolites in drinking water and soil (parts per trillion) and trace organics in river water and marine sediments, and for monitoring industrial waste water and polynuclear aromatics in air. Techniques were also developed for determining fungicides and their decomposition products and herbicide metabolites in plants and animals.

Gel Permeation Chromatography

The development of HPLC was paralleled by that of gel permeation chromatography (GPC), which is the separation of molecular species according to size. The first applications for GPC were for the characterization of the polymers polyethylene and polystyrene. Any polymer that can be dissolved in a solvent can potentially be characterized by GPC. Early GPC use was largely in the area of understanding or modifying polymer reactions. As such, the development and growth of GPC are closely associated with the development and growth of the plastics industry. As the plastics industry has become more sophisticated, GPC has been used increasingly in the design of polymers with specific end-use characteristics. The many applications of GPC include: (1) determination of the quality of resins to help processors discover problems before they occur on the production line; (2) determination of the biocompatibility of plastic devices that are used in the body; (3) lowering of the rejection rates of printed circuit boards in the electronics industry; and (4) improvement of paints and plastic coatings.

Ion Chromatography

Ion exchange chromatography (IEC) is the separation of ionic compounds through their interaction (exchange) with fixed ionic sites on the resin (stationary phase). It was in use as early as 1915 but its development as an analytical tool was slow, primarily because of the limitations of appropriate ion exchange resins. With the development of silica and rigid polymer-based packing materials, IEC has become widely used in the separations of biologicals, especially proteins, peptides, and nucleic acids.

Currently, ion chromatography (IC) is a specialized field of IEC. The development of suitable ion exchange columns and utilization of the conductivity detector has led to the use of IC for the analysis of inorganic ions such as metal ions, F^-, Cl^-, SO_4^{2-}, and PO_4^{3-}. References to IC applications have increased significantly each year since the mid-1970s (see Fig. 1-12). As the capabilities of HPLC expand, the sharp distinction between HPLC and IC becomes increasingly blurred.

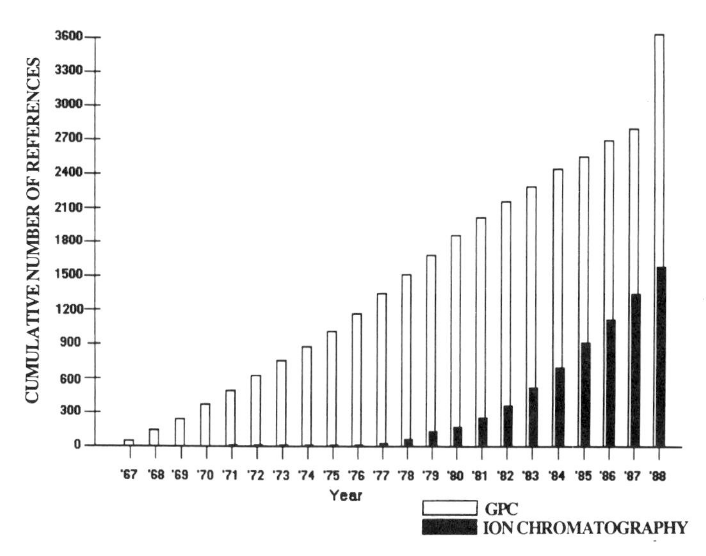

FIGURE 1-12. Journal references to IC and GPC, 1967–1988. (Adapted from reference 14 with permission.)

Preparative HPLC

An obvious outgrowth of the success of analytical HPLC is the development of modern preparative HPLC, particularly as it applies to separations of organic and biological compounds. Preparative LC first evolved in Germany in the 1920s with the use of open-column chromatography, which was based upon the ideas of Tswett. Open-column chromatography was used extensively in the dye industry for testing the purity of compounds, characterizing structures, and determining physical properties.

Preparative LC involves more than just a scale-up of instrumentation and column size (3). Sample loads must be far greater to achieve a sufficient amount of separated materials. Analytical LC developed the concepts of high resolution and high speeds. Today, specialized instruments, columns, and column packing materials attempt to imitate these characteristics for scale-up of analytical separations to preparative separations on the laboratory and pilot-plant scale.

Compound Synthesis

Intimately connected with preparative LC is the development of synthetic compounds that are either new or identical to those found naturally. The capability that HPLC added to synthesis procedures—the ability to verify compound purity, separate stereoisomers, and confirm compound identity—was utilized in the syntheses of vitamin B_{12} and oligonucleotides, and in natural product isolation. These areas are discussed in Chapter 2.

THE FUTURE

The only limitation for LC is that the sample must be soluble in the mobile phase. This means that LC will grow in applications as our knowledge grows—those that at some point seem impossible may become possible the next year Some observers think that the growth of HPLC is reaching a plateau. At present, although the growth rate of HPLC is slowing, separations are continually being published which only five years ago were believed to be impossible. An example is the portion (dotted area) of Figure 1-7 that shows the expanded use of LC in the protein separation arena from a limit of molecular weight 10,000 in 1975 up to several hundred thousand today. New instrumentation, techniques, data processing, and data interpretation continue to contribute to the growth of the separations market. Hence the growth of HPLC applications may be boundless. It is no wonder, then, that with the tremendously wide range of applicability and the seemingly unlimited potential for application, HPLC is an essential research tool for all laboratories.

GLOSSARY

Adsorption. The noncovalent attachment of one substance to the surface of another. *See* Liquid–solid chromatography.

Analyte. The substance that is being analyzed; sometimes referred to as the sample.

Analytical chromatography. Operation at any sample loading level in which the objective is a chromatogram representing the components present. Identification of each component is made by comparing elution volumes with those of known standards. Quantitation is obtained by plotting detector response versus concentration for known standards and reading directly from this graph the concentration corresponding to an observed response.

Alpha (α). A measure of the separation of two components at their peak apexes. It does not take into account the width of the two peaks.

AUFS. An acronym denoting the output from a spectrophotometer in Absorbance Units (per) Full Scale (deflection).

Band. The volume of mobile phase that contains the sample as it travels along the column. Occasionally this term is expressed as the distance of the column which the band (volume) occupies. *See* Peak width.

Band broadening. The dilution of the sample with mobile phase during the chromatographic process. For example, a 10-μL volume is injected, and by the time it reaches the detector, the sample is spread throughout a 200-μL volume. Band broadening in some cases can affect resolution of peaks within the chromatogram and, as such, should be kept as low as possible.

Baseline resolution. Separation of components at the peak base, that is, no overlap of any peak area.

Bed. The fixed structure of stationary phase particles contained in the column. *See* Stationary phase and Column.

Binary eluent. A two-solvent blend used as the mobile phase.

Bonded-phase packings. Column packing materials made by chemically bonding a ligand (an organic moiety) to a solid particle. Bonded-phase packing materials have largely replaced liquid–liquid systems, where the particles are mechanically coated with a suitable liquid. The popularity of bonded-phase packing materials results from their greater stability and ease of use.

Buffer. A substance capable of maintaining a specified acidity or basicity of a solution.

Capacity factor, k'. A measure of the interaction of a sample component with a specific packing material and solvent combination. The k' term is a measure of the solvent volume required to elute a component from the column, expressed as multiples of the column void volume.

Chromatogram. A permanent record of the separation, the chromatogram shows the detector signal as sample components emerge from the column and pass through the detector cell. This printed or displayed plot is the representation of units of time on the "*x* axis (abscissa) and the "*y* axis" (ordinate) is in units of the detector's response which is proportional to (and representing) the relative amount of the sample component which elutes as a "peak."

Column. A steel, glass, or plastic tube containing the stationary phase.

Column volume. A volume of solvent equal to the volume of the column occupied by the mobile phase. A column volume is also referred to as the void volume of the column or the mobile phase volume.

Counter ion. A mobile-phase ion of opposite electrical charge to either a charged stationary phase (packing material) or an ionic sample. *See also* Ion exchange and Paired-ion chromatography.

Detector. A device for monitoring the effluent from the HPLC by sensing a chemical or physical property of the sample.

Differential refractometer. An LC detector that measures changes in the refractive index (RI) of the mobile phase as components emerge from the column. Since all compounds have a refractive index, this detector is "universal."

Efficiency. A measure of column performance reflecting the amount of peak spreading (dilution) that occurs as the separation takes place.

Eluent. The moving solvent in a chromatographic column (sometimes spelled eluant). *See* Mobile phase.

Elute. To travel through and emerge from the column.

Elution volume. The amount of solvent required to elute a certain component from the column. *See* Capacity factor and Retention volume.

Eluotropic series. A ranking of relative solvent polarities.

Exclusion chromatography. *See* Size exclusion chromatography and Gel permeation chromatography.

Exclusion limit. A rating of the size-sorting capability of GPC packings. The exclusion limit is the nominal diameter of the largest molecular size which is excluded from the pores of the size packing material. Molecules larger than this size pass directly through the column being excluded from the pore volume. Smaller molecules can enter the packing pores and, therefore, these molecules travel more slowly through the column. *See* Gel permeation chromatography.

Flow programming. Increasing the flow rate in a predetermined manner. Normally, this technique is used when low flow rates are needed to obtain sufficient resolution for early eluting materials and late eluting materials have an unnecessary amount of resolution. Flow rates can be continuously increased or changed as a step increment.

Flow rate. The rate at which a volume of mobile phase is moving (flowing) per unit of time.

Fronting. A condition in which the front of a peak is less steep than the rear relative to the baseline. This condition results from nonideal equilibria in the chromatographic process.

GC. Gas chromatography.

Gel filtration chromatography. A mode of LC in which molecules are separated according to their size using an aqueous mobile phase. Gel filtration chromatography most often is used to separate proteins and peptides. See Gel permeation chromatography for a discussion of the mechanism. Gel filtration and gel permeation chromatography are sometimes referred to as size exclusion chromatography (SEC).

Gel permeation. *See* Gel permeation chromatography

Gel permeation chromatography (GPC). A mode of LC in which samples are separated according to molecular size using an organic solvent as a mobile phase. The packing consists of porous particles with controlled pore sizes. Small sample components can enter these pores; larger molecules, however, cannot penetrate all of the interior portions of the stationary phase. Because their size excludes them from the porous packing, very large molecules travel only through the interstitial spaces in the column. Thus, large molecules travel through the column faster than small molecules and always elute first. Gel permeation chromatography is used to characterize polymers and to determine their molecular weight distribution. The technique is often abbreviated GPC. *See also* Size exclusion chromatography.

GPC. Gel permeation chromatography.

Gradient. Changing the percentage composition of two or more discrete solvents in the mobile phase over time. *See* Gradient elution.

Gradient Elution. A technique for decreasing analysis time by reproducibly changing solvent composition during a separation. The technique varies the composition of the moving solvent in a predetermined manner to ensure the elution of all peaks within a reasonable period of time. Gradient elution is generally used when some components in the sample elute within a reasonable time (capacity factor, k' value, between 2 and 10), while other components remain on the column much longer. The solvent composition can vary continuously during the course of the run or can be changed as a step increment.

HETP (Height equivalent to a theoretical plate). A measure of band spreading which compensates for the effect of column length.

HPLC. High performance liquid chromatography.

IC. Ion chromatography.

IEC. Ion exchange chromatography.

Injection. The act of introducing sample into the solvent (mobile phase) stream and onto the column.

Injector. The device used to introduce the sample into the mobile phase.

Ion exchange. A retention mode in which sample components are separated based on differences in their charge and in their ionization constants. Only ionized sample components can be separated by this technique. *See* Ion suppression and Paired-ion chromatography.

Ion-pair chromatography. See *Paired-ion chromatography.*

Ion suppression. A buffer is added to the mobile phase to adjust the pH so that sample components will be present in their nonionized forms. In this way, ionic sample components can be separated by a reverse-phase mode on a bonded-phase packing, a much easier technique than traditional ion exchange.

Isocratic elution. A technique in which solvent composition and flow rate are held constant throughout the separation.

Isoeluotropic. Two eluents with the same eluting strength. The peaks for a particular sample in two chromatograms generated with isoeluotropic eluents have the same k' value.

k'. *See* Capacity factor.

LC. Liquid chromatography.

Linear velocity. The actual speed, in units such as millimeters per second, with which a liquid moves through the column. Linear velocity is proportional to flow rate for a given column.

Liquid–liquid chromatography. A retention mode in which the sample components are separated based on their relative solubilities in the mobile phase and a second, immiscible, phase coated onto a solid support. An advance in LC technology was the introduction of packings with the coating chemically (covalently) bonded to the substrate to avoid loss of the coating. *See also* Bonded-phase packings.

Liquid–liquid extraction. Sample components are partitioned between two liquid phases. The desired component(s) remains in one of the liquid phases while the others are discarded with the other liquid phase.

Liquid–solid chromatography. A retention mode of LC in which separation of the sample components is based on differences in adsorption and desorption rates of the sample components on the surface of porous particles. *See also* Reverse phase and normal phase.

Matrix. A term describing the material in which a sample is present.

Mobile phase. The flowing solvent (also known as the eluent). The liquid phase involved in the chromatographic separation containing the dissolved sample as it travels past and through the stationary phase.

Mobile phase volume. The volume of solvent in a packed column, V_M, given by the amount of mobile phase required to elute a sample component which does not interact with the packing material (V_M also known as the void volume, V_0).

Normal phase. A form of chromatography in which the stationary phase is relatively polar (e.g., alkylamine modified silica or unbonded silica) and the mobile phase is relatively nonpolar (e.g., hexane). In normal phase separations, nonpolar components emerge from the column first.

Packing. A term commonly used to refer to the stationary phase. Also, a term used to describe the act of placing the stationary phase into a column.

Paired-ion chromatography. An alternative to ion exchange using a reverse-phase column. A large organic counter ion is included in the mobile phase and separation of an ionic compound is accomplished by reverse-phase techniques.

Partition. *See* Liquid–liquid extraction.

Partition chromatography. *See* Liquid–liquid chromatography.

Peak. On the recorded trace, each emerging sample component is seen as a deviation from baseline, a "peak." For a specific compound, peak height or area can be directly related to concentration. *See* Chromatogram.

Peak area. In a chromatogram the area, in relative units, within the outline of a peak and the associated baseline.

Peak height. The distance in a chromatogram from the baseline of the peak to a point at the maximum of the peak envelope.

Peak maximum. The highest point on the peak from the baseline. The highest peak height.

Peak width. For a peak in a chromatogram, the width of the base (time duration) from a baseline start point to a baseline return point; used to determine area and characterize the peak.

Pellicular packing. A solid core surrounded by a thin porous crust of smaller particles. These particles are an economical means of obtaining moderately high efficiencies for analytical work. Also, they are popular as packings in guard columns to protect analytical columns.

Plate height. *See* HETP.

Plates. A measure of the efficiency of the column based on the elution volume and the peak width at the baseline. The narrower the peak, the more efficient the column and the higher the plate number (N) will be.

Polarity. A general characteristic of materials caused by the presence of electronegative and electropositive groups in the compound. The number, strength, and distance between these groups in the molecule contribute to the overall polarity of the molecule. Polarity influences molecular interactions, including solvent–packing, solvent–sample, and sample–packing attraction.

Porous packing. A stationary phase which has tiny holes (usually between 100 and 500 Å) throughout the entire structure of the particle.

Preparative liquid chromatography. Any scale of operation in which the objective is the collection of sample components for subsequent identification or use. Preparative LC may involve submilligram or multigram quantities.

Recycle. An inexpensive technique for increasing the efficiency (separation power) of the LC. An unresolved or partially resolved sample is directed from the detector(s) back into the pumping system for as many additional passes through the column as necessary to achieve satisfactory resolution.

Resolution. A measure of how well any two components have been separated. Resolution, R, takes into account the separation at peak maxima and the width of the peaks. Components with R equal to 1.5 are baseline separated.

Retention chromatography. A general term for all modes of separation that rely on the relative attractions between the surface of the packing material, the mobile phase, and the components in solution.

Retention time. Time elapsed from the moment of injection to the moment when the maximum concentration of the component elutes. Retention time is a function of retention volume and flow rate.

Retention volume. The volume of mobile phase required to elute a particular component from the column. Retention volume is a measure of the attraction of a sample component for the packing material when a specific mobile phase and set of operating conditions is used. V_R is given by the volume of solvent that passes through the column from the time of injection to the time where the maximum concentration of the component emerges. Also known as elution volume.

Reverse phase. A form of chromatography where the packing material surface is relatively nonpolar, and the solvent relatively polar. In a reversed-phase separation, the most polar compounds elute first. This order of elution reverses that found with older, normal-phase separations; hence reverse phase. See also Bonded-phase packings.

RI. Refractive index.

Run. The chromatographic analysis.

Sample. The component or components to be analyzed.

Sample clean-up. All the steps necessary to prepare the sample for final LC analysis. These steps generally are taken to remove extraneous material from the sample.

Selectivity. *See* Alpha.

Separation. The goal of the chromatographic process whereby the sample components are isolated and no longer mixed.

Size exclusion chromatography (SEC). A mode of LC in which molecules are separated according to size. Small molecules penetrate all of the pores within the packing while larger molecules only partially penetrate the pores. The large molecules cannot penetrate into all of the pores and elute before the smaller molecules. See Gel permeation chromatography for a more complete discussion of the mechanism. Size exclusion chromatography is a general term used to describe either the gel filtration (aqueous mobile phase) or gel permeation (organic mobile phase) chromatographic process. Sometimes the term *exclusion chromatography* is used.

Size separation. *See* Gel permeation chromatography.

Solid phase extraction. A sample preparation technique using liquid–solid "sorption" to remove contaminants from the sample or isolate the sought for compounds from the sample.

Solute. A single chemical substance. Often it is used to refer to sample components, but it is also used to refer to additives to the mobile phase; *see* Analyte.

Solvent. A substance capable of dissolving another to form a uniformly dispersed solution. In chromatography, solvents are used to make up the mobile phase.

Solvent delivery system. A device for pumping the mobile phase at precise flow rates and at high pressure.

Solvent programming. *See* Gradient and Flow programming.

Sorption. The noncovalent attachment of one substance to the surface of a packing. The term is generally used to indicate any or all types of attractive modes of LC that may be involved, for example, ion exchange, liquid–solid, and liquid–liquid partition.

Stationary phase. The material that is contained in the column and does not move during the chromatographic process. The sample components are selectively attracted to the surface of this material and this results in selective retardation and eventual separation.

Tailing. A condition where the front of the peak is steeper than the rear relative to the baseline. This condition results from nonideal equilibria in the separation process.

Trace enrichment. A technique used to concentrate samples with low concentrations by passing a large volume of dilute sample across an LC column or solid-phase extraction device such that the trace impurities are concentrated (enriched) on the top of the column.

UV/Visible detector. A photometer used in LC. This detector measures changes in the absorbance of ultraviolet (UV) or visible (VIS) light resulting as the components pass through the detector. This is a sensitive and specific detection technique for compounds that contain chromaphoric groups. The wavelength of absorbance can be chosen to enhance specificity and/or sensitivity for a specific compound.

Van Deemter plot. A graph of column efficiency, expressed as HETP versus linear velocity of the mobile phase. This plot indicates the optimum linear velocity (and, thus, flow rate) for a particular column.

Void. An area in the LC column containing no packing. A void is the result of the stationary phase particles settling (collapsing). A void usually occurs at the top of a column and causes a large decrease in the efficiency.

Void time. The time equivalent to the void volume (void time = void volume ÷ flow rate).

Void volume. The total mobile phase volume in a packed column. The volume between the packing particles (interstitial volume) and the volume within the packing pores added together equal the void volume, V_0. Void volumes are typically 40–80% of the empty column volume and are determined by injecting a nonretained component, for example, heptane on a silica column, using chloroform as the mobile phase. The void volume is also referred to as the mobile phase volume, V_M.

Zone. See *Band.*

REFERENCES

1. M. F. Delaney, K. M. Pasko, D. M. Mauro, D. S. Gsell, P. C. Korologus, J. Morawski, L. J. Krolikowski, and F. V. Warren, Jr., *J. Chem Educ.,* **62,** 618 (1985).
2. B. A. Bidlingmeyer, Ed., *Preparative Liquid Chromatography,* Elsevier, Amsterdam, 1987.
3. P. R. Brown, *High Pressure Liquid Chromatography, Biochemical and Biomedical Applications,* Academic, New York, 1973.

4. B. L. Karger, L. R. Snyder, and C. Horvath, *An Introduction to Separation Science,* Wiley, New York, 1973.
5. A. M. Krstulovic and P. R. Brown, *Reversed-Phase High Performance Liquid Chromatography: Theory, Practice and Biomedical Applications,* Wiley, New York, 1982.
6. D. A. Skoog and D. M. West, *Analytical Chemistry, An Introduction,* Holt Rinehart and Winston, New York, 1974.
7. L. R. Snyder and J. J. Kirkland, *Introduction to Modern Liquid Chromatography,* Wiley Interscience, New York, 1974.
8. H. Engelhardt, *High Performance Liquid Chromatography,* Springer-Verlag, New York, 1979.
9. D. H. Freeman, *Science,* **218,** 235 (1982).
10. P. T. Kissinger, L. J. Felice, W. P. King, L. A. Pachla, R. M. Riggin, R. E. Shoup, *J. Chem. Educ.,* **54,** 50 (1977).
11. Bulletin N65, Waters Associates, Milford, MA.
12. J. Korpi and B. A. Bidlingmeyer, *Am Lab.,* **13**(6), 110 (1981).
13. J. C. Giddings, *Anal. Chem.,* **37,** 60 (1960).
14. K. Conroe and B. A. Bidlingmeyer, *Am. Lab.,* **20**(10), 82 (1988).

CHAPTER 2

LIQUID CHROMATOGRAPHY AS A RESEARCH TOOL

Contributions to the biotechnology industry
- Commercialization of human insulin
- Interferon isolation
- Gene synthesis
- Protein analysis
- Peptide isolation
- Peptide mapping
- Amino-acid sequencing
- Total amino-acid composition
- Summary of LC benefits to biotechnology

Contributions to environment research
- Air quality
- Water quality

Distinguishing between good and bad plastics
- The concept of molecular weight and molecular-weight distribution
- Using GPC to determine molecular weight

An aid to organic synthesis
- Rapid optimization of reaction yields
- Purification of reaction mixtures—An indispensable aid in the synthesis of vitamin B12
- HPLC is an indispensable tool in the laboratory

Therapeutic drug monitoring
- Monitoring of asthma treatment
- Management of epilepsy treatment

A qualitative analysis tool
- Testing incoming raw materials

Testing for product integrity
Determining probable causes
Summary
References
Acknowledgment

As we discussed in Chapter 1, modern LC is applicable to compounds of molecular weight up to 20 million and because of this, HPLC is a rapidly growing separation technique used in the modern analytical laboratory. Liquid chromatography is capable of separating very similar compounds from each other and of providing interference-free analytical information and/or quantities of pure material from complex mixtures. Once an LC separation has been developed, typical methods take from 5 to 20 min. Sample preparation is minimal, usually consisting only of a dissolution and filtration. Accuracy and reproducibility for LC methods are normally equal or superior to those of the traditional methods they replace. The ability of LC to separate vast arrays of compounds by a variety of separation mechanisms, simply and with a minimum amount of sample, makes it an almost indispensable research technique. Industry and academia are making extensive use of the capabilities of LC; as a result, HPLC methods are in daily use in thousands of laboratories throughout the world.

When a newcomer first hears about HPLC, the most commonly asked question is "What can HPLC do for me?" This chapter is intended to answer that question. However, since it is impossible to survey all the areas of chemical composition management to which LC is bringing benefits, this chapter is intended to highlight the wide versatility and utility of the technique in many of the evolving areas of science and technology. By emphasizing the benefits of using LC, the reader should glean a "feel" for the power of this separation technique and be able to "see" the relevant place for it in solving his/her research problems.

CONTRIBUTIONS TO THE BIOTECHNOLOGY INDUSTRY

Commercialization of Human Insulin

Human insulin was the first commercial health care product produced by recombinant DNA technology. Eli Lilly, the producer of this synthetic insulin, often relied on HPLC to confirm the structure and to determine the potency of synthetic human insulin. The story behind Lilly's recombinant DNA-produced insulin is described in an article in *Science* (1): "High performance liquid chromatography (HPLC) techniques developed at Lilly can detect proteins that differ by a single amino acid, and HPLC tests show human insulin (recombinant DNA) is identical to pancreatic human insulin . . . A chromatogram of human insulin (recombinant DNA), pancreatic human insulin, and a mixture of the two, showed that they were superimposable and identical." Furthermore, LC ". . . was useful for ensuring that we had the appropriate disulfide bonds and lacked other types of protein or peptide contaminants of a specifically degraded sample. In addition, the peptide maps following enzymatic digestion are identical for human insulin (recombinant DNA) and pancreatic human insulin."

The article concluded: "HPLC has now become an important analytical tool to determine structure and purity and is now considered to be a more precise measurement of potency than the rabbit assay, although most government regulatory agencies around the world still emphasize the rabbit potency assay. . . . In the end, we employed 12 different tests to establish that what we had produced was human insulin. We believe that correlation among three of the tests was particularly important—the radioreceptor assay, radioimmunoassay, and HPLC."

Interferon Isolation

In an article on the development interferon (2), Dr. Sidney Pestka of the Roche Institute of Molecular Biology said, "Just a few years ago, it was an enormous struggle to obtain a few micrograms of any interferon in pure form. With the rapid application of new technologies, many interferons have been purified and produced in relatively large amounts. These new technologies consist of three general areas: high performance liquid chromatography, [production of] monoclonal antibodies, and gene splicing. These three areas have merged to provide major achievements in interferon research." Dr. Pestka concluded that HPLC was "instrumental in achieving the first purification of human leukocyte interferon and in providing the initial surprising result that the leukocyte interferons were a family of closely related proteins, not a single species. HPLC has been used to purify other interferons as well."

The scientists at Alpha Therapeutics, a Los Angeles biotechnology company, who were looking for new subspecies of naturally occurring leukocyte interferon are using size exclusion and reverse phase HPLC "with great success" (3). Dr. Steve Herring, a Senior Research Scientist at Alpha Therapeutics stated: ". . . it was found that several classes of human interferon exist (leukocyte, fibroblast, immune), and that numerous interferon 'subspecies' may be present within each of these classes. Because these different interferon species can be quite similar, showing between 70 to 95 percent homology in amino acid sequence, it has been difficult until recently to separate them from each other." (3) Dr. Herring concludes the article by saying, "HPLC continues to be used in the purification of bacterially produced interferons" and that "columns and equipment are now available which would allow researchers to carry out the same separations to produce gram quantities of the interferon in a matter of minutes."

Gene Synthesis

In addition to playing a major role in commercializing the recombinate DNA industry, HPLC has been a significant contributor in the synthesis of genes. Nobel laureate Har Gobind Khorana directed the research that completed the synthesis of the first man-made gene that was fully functional in a living cell. The synthesized material was the tyrosine suppressor transfer RNA gene of the *Escherichia coli* bacterium (see Fig. 2-1). To build the large molecule (gene) whose subunits must be linked in a unique order, Khorana's group made short stretches of oligonucleotides that later were joined together. To do this process you need pure "building blocks," which is where LC played a significant role.

Khorana's research team constructed the gene in several steps, using the principles of subassembly. Forty such segments were synthesized, an effort spanning nine years and involving the participation of 24 postdoctoral fellows. After each segment was made, it was necessary to purify it. Dr. Khorana credited the particularly dramatic progress in this area of purification to a rapid, high-performance liquid chromatograph. Furthermore, Dr. Khorana stated that with HPLC separations were obtained in minutes with protecting groups intact, a minimum of solvent consumed, and no need for subsequent characterization by other chromatographic means. In addition

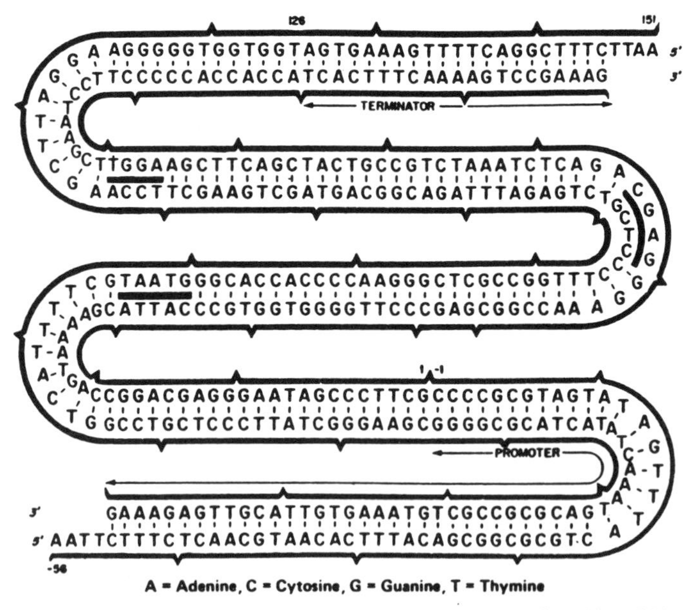

FIGURE 2-1. The tyrosine suppressor RNA gene of *E. coli*. This RNA gene consists of 39 sequence-specific 10–12 unit oligonucleotides. Once these 39 oligonucleotides have been synthesized, they are enzymatically joined to form the DNA duplex.

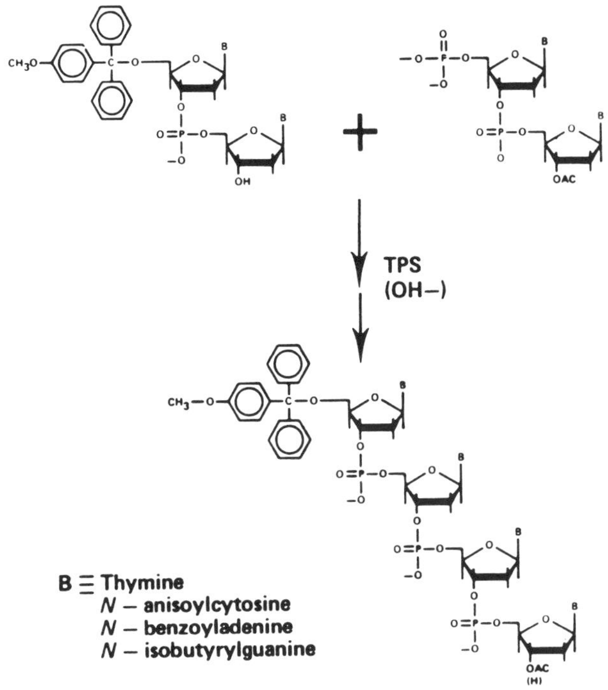

FIGURE 2-2. General method for the synthesis of oligonucleotides.

". . . LC proved to be very powerful on both analytical and preparative levels in terms of resolution, sensitivity, speed, recovery, and capacity." (4)

As shown in Figure 2-1, a gene is essentially a double-stranded DNA molecule composed of a long series of nucleotides joined in a specific sequence. The synthesis of a gene is performed in multiple steps. First, mononucleotides are condensed to form di-, tri-, and tetranucleotides (shown in Fig. 2-2) which are in turn condensed to form 10- to 12-unit oligonucleotides. At each condensation step, a set of "protecting groups" is used to permit only formation of the desired covalent linkages.

Examples of the capabilities of HPLC are shown in Figure 2-3 for analytical analyses and Figure 2-4 for preparative isolation. The purification was first optimized on an analytical scale with a C18 column. Each constituent of the reaction mixture was then purified on a preparative scale with a large particle C18 column. These chromatograms show the purification of the reaction mixture from the condensation of the dinucleotide pTpT–OAc, and the hexanucleotide, $MMTrG^{ib}pG^{ib}pA^{bz}pG^{ib}pC^{an}$–OH.

The starting material $MMTrG^{ib}pG^{ib}pA^{bz}pG^{ib}pC^{an}$-OH (chromatogram A in Fig. 2-3) elutes from the analytical column after about 6 min at 35% acetonitrile concentration. This compound was condensed with a large excess of

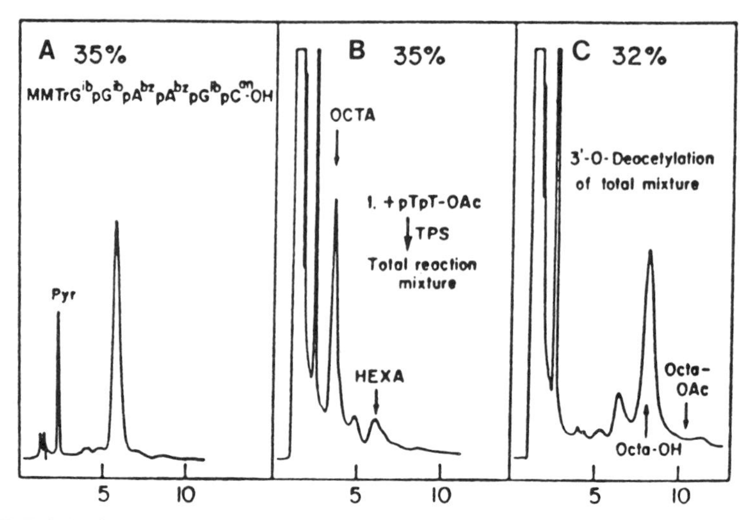

FIGURE 2-3. Separations of oligonucleotides. Chromatogram A is the starting material, chromatogram B is the reaction mixture, chromatogram C is the deacetylated reaction mixture. (Reprinted from reference 4 with permission.)

pTpT–OAc to ensure a high turnover of the starting hexanucleotide. The excess reagent, 2,4,6-triisopropylbenzene sulfonyl chloride (TPS), was removed by extraction and the products in the reaction mixture were precipitated from ether. Liquid chromatographic analysis of this mixture (chromatogram B in Fig. 2-3) showed that some hexanucleotide was still present. The large peak corresponds to the desired octanucleotide. The material eluting very fast is the excess of the dinucleotide, its pyrophosphate and pyridine. since thymine in the dinucleotide block shown above makes no significant lipophilic contribution, the elongated chain elutes faster than the hexanucleotide starting material, due mostly to the presence of two more phosphates. This shift to shorter retention time can be further increased by deacetylation (performed on the total reaction mixture). Even with an octanucleotide, an acetyl group has a significant influence as shown in chromatogram C in Figure 2-3. The main peak at longer retention time now is the octanucleotide with a free 3′-hydroxyl group.

A typical preparative separation is shown in Figure 2-4. For the preparative run (Fig. 2-4*a*), 250 mg of the reaction mixture was injected onto a C18 column and gradient elution (profile is dotted line) was performed over 60 min. The small peak occurring at approximately 43 min was the starting hexanucleotide material. The main peak at approximately 38 min was collected and pooled as indicated, and was analyzed for purity by injecting into the analytical HPLC (Fig. 2-4*b*). There is a trace of pyridine showing up in the pooled material and marked as such in the chromatogram. As an extra

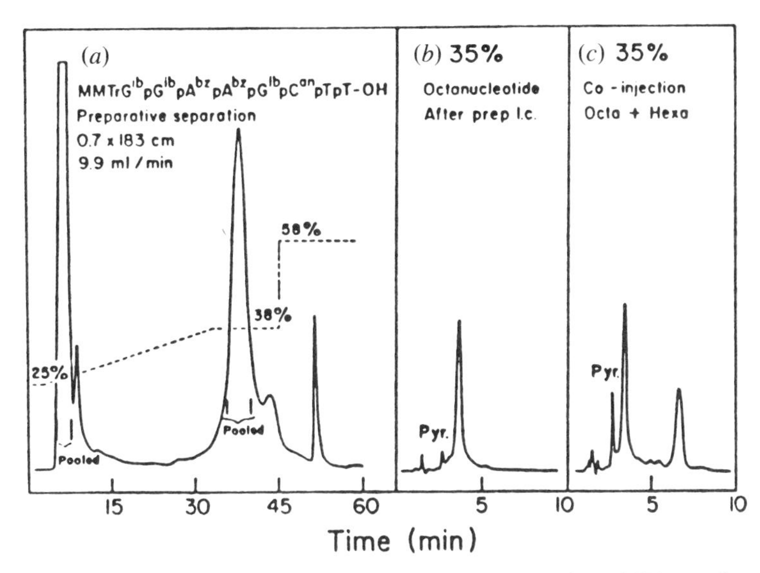

FIGURE 2-4. A typical preparative separation. (*a*) Preparation of 250 mg of reaction mixture. The small peak appearing at about 43 min in this separation contains the starting hexanucleotide. The main peak containing the desired octanucleotide was pooled as indicated. (*b*) The product checked for purity by analytical LC. Absence of the starting material was then confirmed by coinjection of the hexa- and isolated octanucleotides shown in chromatogram (*c*). (Reprinted from reference 4 with permission.)

check for purity the absence of the starting material was confirmed by coinjection of the hexa- and the isolated octanucleotide as shown in Figure 2-4*c*.

The importance of Dr. Khorana's work is that man-made (chemically synthesized) genes can now be made available for systematic studies of how the structure of a gene influences the function of the gene. With regard to the importance of HPLC, it was stated at a meeting of the American Chemical Society: ". . . even at this stage of development, a saving of at least 50% of the time has been effected in the synthesis of oligonucleotides." (4) Often the previous methods would take 2 wk or more, so the time savings resulting from using HPLC was substantial.

Protein Analysis

Life scientists are also involved with protein analysis and purification, and it should be no surprise that HPLC also is a significant contributor in this area. In addition to the fact that LC often is able to do in 30 min that which would require overnight by more conventional techniques, LC is able to accomplish some separations not possible by other separation approaches. For example, a variant β-globin was discovered and subsequently purified using HPLC (5) when a number of other commonly used analytical separation techniques revealed no abnormal hemoglobins. The other analytical tools included:

1. Standard electrophoresis on cellulose acetate and citrate agar.
2. Isoelectric focusing in pH gradient from 6 to 8.
3. Anion exchange LC.
4. Cation exchange LC.

Only the HPLC using a bonded phase in the reverse-phase mode accomplished the separation of the hemoglobins. The chromatogram is shown in Figure 2-5.

This work uncovered the fact that a substitution of the amino acid alanine for valine at position 126 in the β-chain of hemoglobin occurred in a hematologically normal adult of Lebanese extraction. The variation β-globin was initially observed and subsequently purified by reverse-phase liquid chromatography (see Fig. 2-5). Also LC was used to isolate the variant tryptic peptide of β-T13 that had alanine replacing the valine at amino-acid residue position 126. This is shown in Figure 2-6. (Peptide mapping is discussed later in this chapter.)

This discovery of "hemoglobin Beirut" illustrates the usefulness and sig-

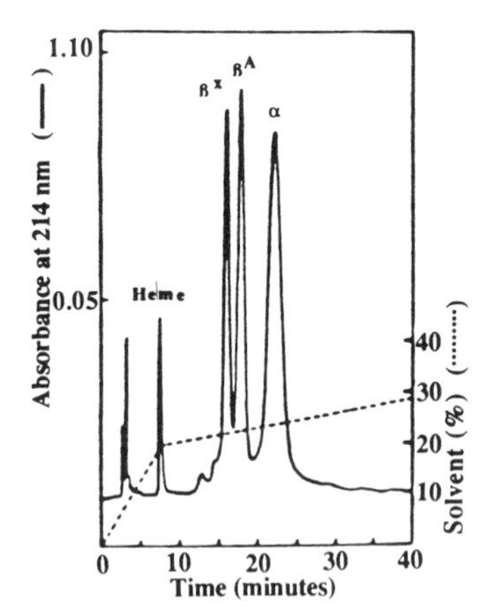

FIGURE 2-5. Detection of a β-globin variant (β^X). Column: μBondapak C_{18}, 3.9 mm ID × 30 cm. Mobile phase: linear gradient segments from solvent A consisting of acetonitrile and water (40:60) with 0.2% trifluoroacid acid (TFA) (weight to volume) to solvent B consisting of acetonitrile and water (60:40) with 0.2% TFA (weight to volume). (Reprinted from reference 5 with permission.)

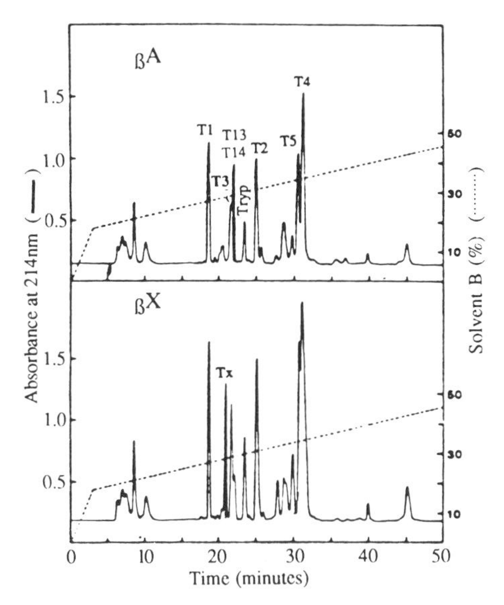

FIGURE 2-6. Separation of tryptic peptides. Column: same as Figure 2-5. Mobile phase: linear gradient segments from solvent A consisting of 0.1% TFA in H_2O (weight to volume) to solvent B consisting of 0.05% TFA in acetonitrile (weight to volume). (Reprinted from reference 5 with permission.)

nificance of using HPLC for the detection of neutral amino-acid substitutions in proteins. Furthermore, this work has shown that the ability of HPLC to detect neutral substitutions in undigested proteins may be pertinent to the monitoring of genetic variation in humans.

Peptide Isolation

When biochemists are identifying new proteins, the first step is to do a "peptide map" on a tryptic digest. After the peptides are isolated, each needs to be sequenced. By essentially breaking up the protein into smaller subunits of peptides and then by breaking up the peptide into smaller subunits of amino acids, knowledge about the original protein structure can be determined. Reassembling the structure of each peptide and properly coupling the order of each peptide linking can result in the complete structure of the original protein.

Before a sequencing operation can begin, peptides from a tryptic digest must be separated from the parent protein and from each other. The goal of this separation is to obtain as pure a peptide sample as possible in order to eliminate or minimize secondary (minor) peptide cosequencing. An example of this is shown in Figure 2-7 which is the tryptic digest of bovine calcium binding protein (BCBP) (6).

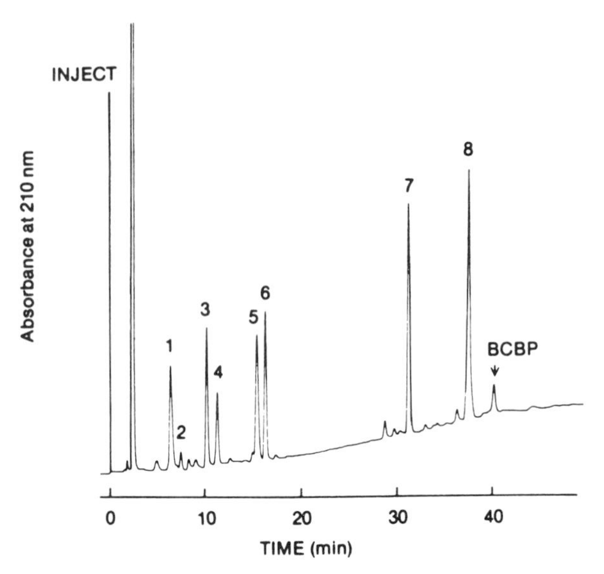

FIGURE 2-7. This chromatogram indicates the peptide profile of BCBP after one hour of tryptic digestion. (Reprinted from reference 6 with permission.)

Peptide Mapping

In addition to the ability to isolate the peptides from a peptide map, using HPLC is also useful in monitoring the digestion reaction (Figs. 2-8 and 2-9) by observing small changes in the peptide profile. Figure 2-8 shows a chromatogram of the peptide profile of bovine calcium binding protein (BCBP)

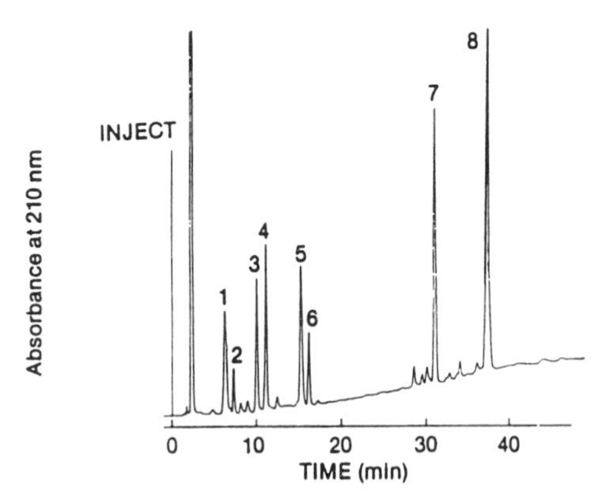

FIGURE 2-8. This chromatogram represents the peptide profile of BCBP after four hours of tryptic digestion. The profile shows eight major peaks with apparent minor impurities. Note the impurities surrounding peak 7 and the relative heights of peaks 2, 4, and 6 compared with the peptide profile in Figure 2-7. (Reprinted from reference 6 with permission.)

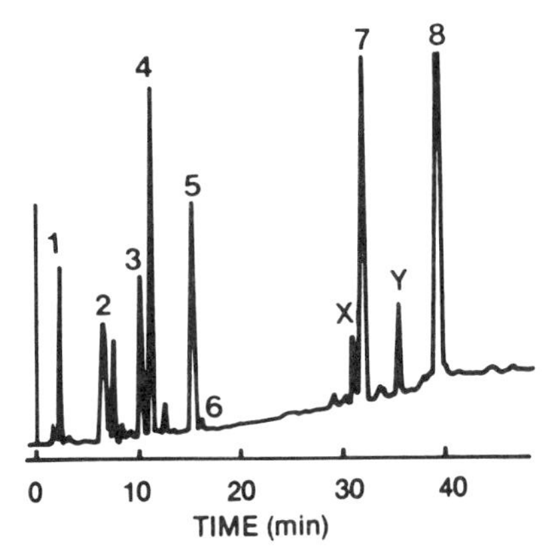

FIGURE 2-9. This peptide profile represents 16 hours of tryptic digestion. Note that peak 6 has decreased in height while 2 and 4 have increased. The peptide corresponding to peak 6 contains an internal lysine (K) which is slowly cleaved by trypsin. Peptide 8 also contains an internal lysine (K) which is well-protected from cleavage. As a result peaks X and Y (cleavage products of peptide 8) are only produced after long exposure to trypsin. (Reprinted from reference 6 with permission.)

after four hours of tryptic digestion. There are eight major peaks with some minor impurities showing up as small peaks. After 16 h of tryptic digestion, the peptide profile changes as shown in Figure 2-9. Note that the impurities surrounding peak 7 in Figure 2-8 and peaks 2, 4, and 6 are significantly changed in Figure 2-9. Peak 6 has decreased in height while peaks 2 and 4 have increased. The peptide corresponding to peak 6 contains an internal lysine which is slowly cleaved by trypsin. Peak 8 also contains an internal lysine which is well protected from cleavage. As a result, peaks X and Y (Fig. 2-9) are cleavage products of peptide 8 and are only produced after long exposure to trypsin. Thus, it is obvious that using LC for the observation of minor reaction products and/or components and/or the presence of impurities can give valuable information about the nature of the amino acids surrounding a cleavage site as well as knowledge about the kinetics of the cleavage reaction.

Amino-Acid Sequencing

After the isolation of the peptides from the peptide map, the structure of individually isolated peptides needs to be determined. The "building blocks" of a peptide are amino acids and the order of occurrence in a peptide is done by a process called "sequencing." Sequencing is the ability to cleave only one amino acid at a time from the peptide backbone. Total amino-acid sequencing can be accomplished by the Edman method, followed by identification of each "cleaved" amino-acid group by HPLC. After LC purification

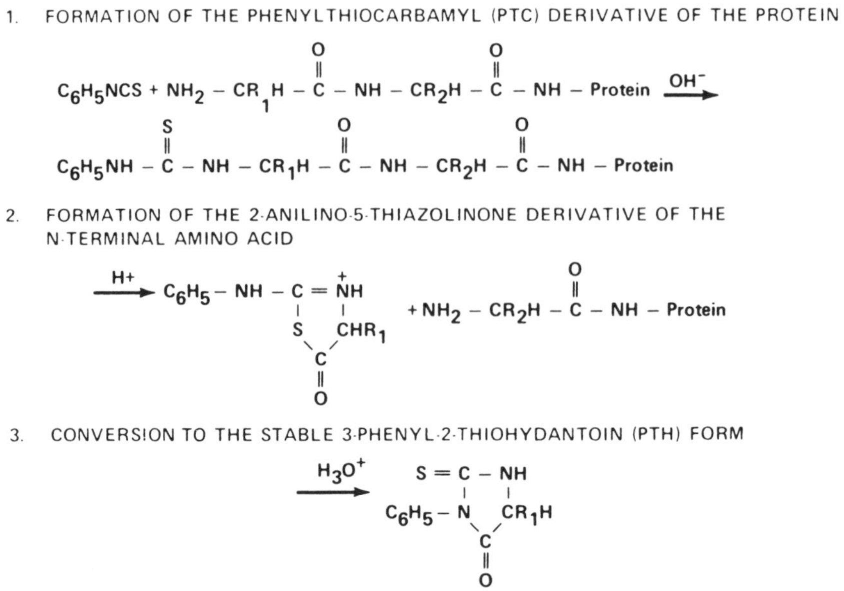

1. FORMATION OF THE PHENYLTHIOCARBAMYL (PTC) DERIVATIVE OF THE PROTEIN

$$C_6H_5NCS + NH_2 - CR_1H - \overset{\overset{O}{\|}}{C} - NH - CR_2H - \overset{\overset{O}{\|}}{C} - NH - \text{Protein} \xrightarrow{OH^-}$$

$$C_6H_5NH - \overset{\overset{S}{\|}}{C} - NH - CR_1H - \overset{\overset{O}{\|}}{C} - NH - CR_2H - \overset{\overset{O}{\|}}{C} - NH - \text{Protein}$$

2. FORMATION OF THE 2-ANILINO-5-THIAZOLINONE DERIVATIVE OF THE N-TERMINAL AMINO ACID

$$\xrightarrow{H+} C_6H_5 - NH - C = \overset{+}{N}H \;(\text{ring: } C - S - C(=O) - CHR_1 - NH) \quad + NH_2 - CR_2H - \overset{\overset{O}{\|}}{C} - NH - \text{Protein}$$

3. CONVERSION TO THE STABLE 3-PHENYL-2-THIOHYDANTOIN (PTH) FORM

$$\xrightarrow{H_3O^+} S = C - NH \;(\text{ring: } C_6H_5 - N - C(=O) - CR_1H)$$

FIGURE 2-10. Edman degradation reaction. The first step of the three-step Edman method forms the PTC-peptide, resulting from phenylisothiocyanate to reacting with the N-terminal amino group of a peptide. The PTC tag enhances the chromatographic qualities of the peptide and provides a UV-absorbing group which is maintained during conversion to the PTH-tagged amino acid. This also enables the use of a UV detector.

of the desired peptide, a reaction is performed to form a "PTC-peptide" (Fig. 2-10). This reaction may be done in solution (solution phase) or after the peptide has been attached to a small solid bead (solid phase). The N-terminal group can be removed via Edman hydrolysis and the PTH derivative formed as shown in Figure 2-10. This PTH-amino acid can then be chromatographed (Fig. 2-11) and identified. The entire process is repeated until the complete amino-acid attachment sequence is known.

Each cycle creates additional chemical "noise." In TLC or GC, this noise build-up creates a substantial problem. In LC, however, the very polar, noise-causing compounds elute before the PTH-amino acids and, therefore, do not interfere with the analysis. The sequencing of a protein or peptide can be done manually or on a commercial device called a "sequencer." The HPLC is the analysis tool which aids in the identification of the amino acid by matching the retention time of the unknown to a known, specific amino acid. An example of the separation and identification of the output of a "sequencer" is shown in Figure 2-11.

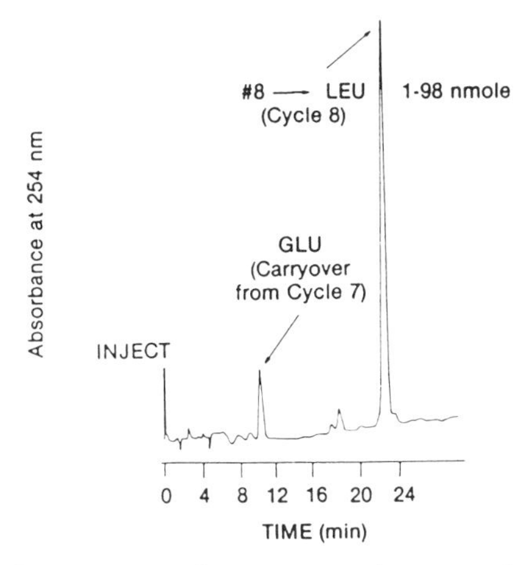

FIGURE 2-11. This chromatogram demonstrates the separation and identification of sequencer products (PTH–amino acids) from peak 8 in Figure 2-9. Detection of the PTH–amino acids is possible at low nanomole levels with HPLC. (Reprinted from reference 6 with permission.)

Total Amino-Acid Composition

In addition to knowing the sequence of the amino-acid attachment, the total amino-acid content of a peptide or protein aids in the structural composition analysis. Total amino-acid composition gives information against which the sequence information can be referenced. Also the total amino-acid composition is an easier analysis to run routinely to verify protein–peptide composition. To do amino-acid analysis, the first step is to break the linkages between the amino acids in the protein (or the peptide) by hydrolysis of the bonds, often using hydrochloric acid at elevated temperatures. The resulting mixture of amino acids is then separated and quantitated using HPLC.

Amino-acid analysis was pioneered by Moore et al. (7) who developed an ion exchange separation of the amino acids on a cation ion exchanger using a series of buffers as the eluent. Detection of the separated amino acids is accomplished by passing the mobile phase through a post column reaction device where a colorimetric reaction of the amino acids with ninhydrin occurs. This approach has been the mainstay in the protein laboratory for over 20 years. With the rapid development of HPLC, many variations of the classical analysis have been configured to refine and automate the original approach of Moore to a point where the analysis takes approximately one hour.

In concert with the improved HPLC techniques, many chemistries have been applied to derivatizing the amino acid prior to separation so that the faster reverse-phase mode of LC may be used. The most significant of these

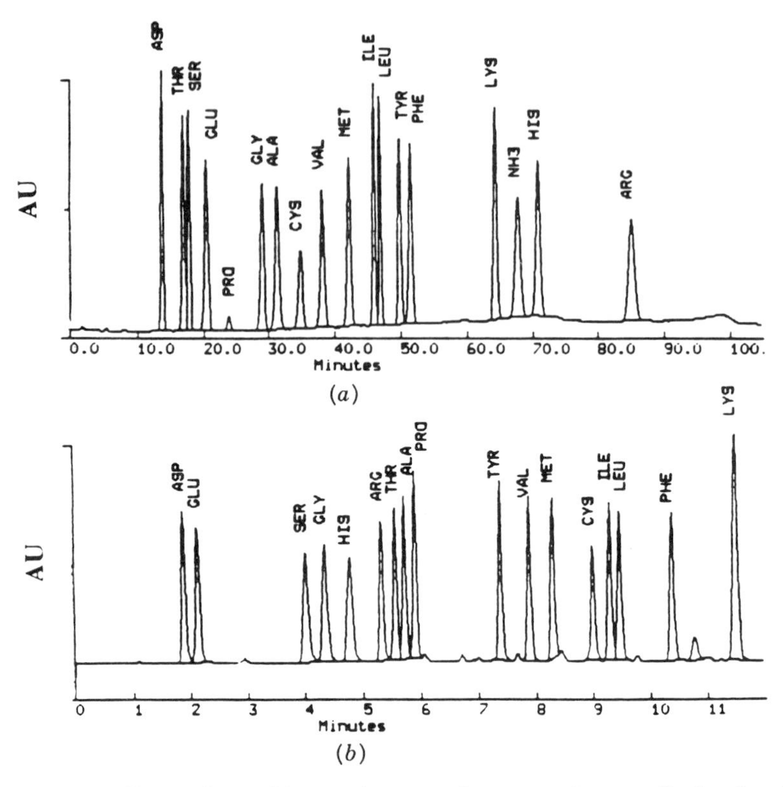

FIGURE 2-12. Comparison of ion-exchange and reverse-phase methods of amino-acid analysis. (*a*) The ion-exchange separation. (*b*) The method of reference 8 (commercialized by Waters under the name of "Pico-Tag™ Amino Acid Analysis").

"precolumn" derivatization approaches was introduced by Bidlingmeyer et al. (8) and was the first major development in amino-acid analysis since Moore's original approach (9). This method takes the sample and reacts ("tags") it with a reagent (phenylisothiocyanate) which forms a chemical compound which strongly absorbs ultraviolet light. The tagged amino acids are phenylthiocarbamyl derivatives and can be analyzed using reverse-phase LC. The separation process takes 10 minutes.

Both the postcolumn derivatization and precolumn derivatization approaches are shown in Figure 2-12. Because of its speed and accuracy at low levels (1–250 picomoles), the precolumn derivatization followed by reverse-phase LC is becoming a popular approach. Most current research which is being done requires high sensitivity at low levels, which can only be done by the precolumn approach. One research group, referring to the precolumn derivatization method of reference 8, stated: "It is really much better than we had before. It has lowered our routine sensitivity by a factor of 50 and has reduced analysis time from 2 hours to 20 minutes." (9)

Summary of LC Benefits to Biotechnology

The use of LC for the analysis of biological materials has many advantages over other classical methods of analysis and purification. Biological molecules range from simple small molecules to structurally complex biopolymers. They also may be highly polar (e.g., nucleic acids), nonpolar (e.g., lipids) or both in the same molecule (proteins). They may be linear linkages (e.g., peptides) or highly branched materials (e.g., polysaccharides). They may have high molecular weights (e.g., 1×10^6 for nucleic acids), intermediate molecular weights (e.g., 60,000 for proteins) or low molecular weights (e.g., 120 for amino acids). Liquid chromatography is uniquely suited to separate all of these biological materials as a result of the wide availability of selectivity of the packing–mobile phase–analyte interactions.

CONTRIBUTIONS TO ENVIRONMENT RESEARCH

Air Quality

In our industrialized society, pollutants are being released into our atmosphere in the form of gases and chemical compounds adsorbed onto tiny particles. For example, one such class of compounds which is of concern is polynuclear aromatic hydrocarbons (PAH) since some of these compounds are reported to be carcinogenic. Three PAHs which can be formed during incomplete combustion have been identified in steel coking operations, automobile exhaust, and cigarette smoke.

In addition to the need to monitor known problematic compounds, newer compounds are being identified as potential threats to humans and as such need to be monitored in the atmosphere. For example, researchers reported (10) that several chemical and instrumental analyses of HPLC fractions provided evidence for the presence of *N*-nitroso compounds in extracts of airborne particles in New York City. The levels of these compounds were found to be approximately equivalent to the total concentrations of polycyclic aromatic hydrocarbons in the air. Since 90% of the N-nitroso compounds that have been tested are carcinogens (10), the newly discovered but untested materials may represent a significant environmental hazard. The procedure involved collecting samples of breathable, particulate matter from the air in New York City. These samples were extracted with dichloromethane. Potential interferences were removed by sequential extractions with 0.2 N NaOH (removal of acids, phenols, nitrates, and nitrites) and 0.2 N H_2SO_4 (removal of amines and bases). The samples were then subjected to a fractional distillation and other treatments. Readers interested in the total details should consult the original article (10). Both thin-layer chromatography (TLC) and HPLC were used to separate the compounds present in the methanolic extract.

Thin-layer chromatography was used to separate the N-nitroso com-

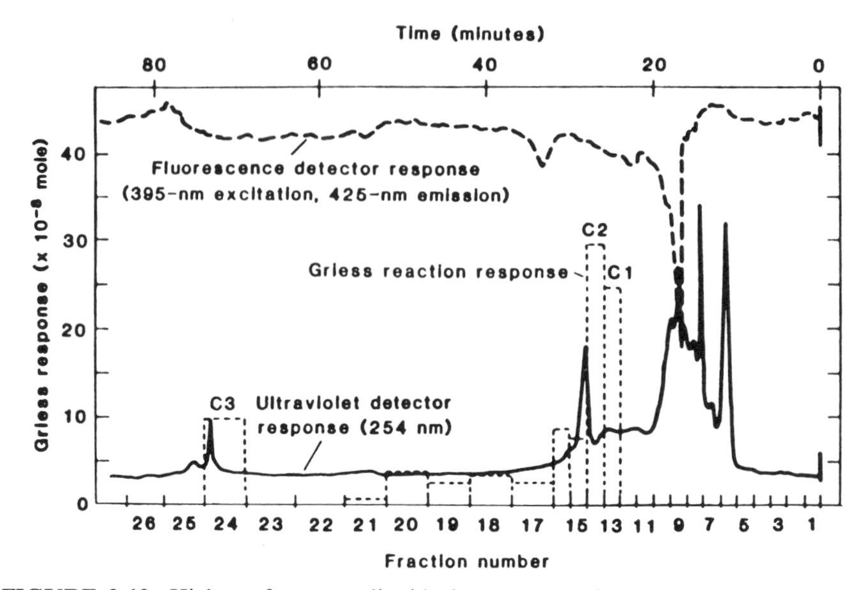

FIGURE 2-13. High performance liquid chromatographic separation of unknown *N*-nitroso compounds. The colorimetric, photometric, and fluorometric detection responses are shown. Column: μBondapak C_{18} 3.9 mm ID × 30 cm. Mobile-phase composition was programmed from 20 to 80% acetonitrile in water (pH = 4, phosphoric acid) over 20 min, was held isocratic (constant mobile phase composition) for 22 min, and then was programmed from 80 to 100% acetonitrile in 1 min and held at 100% acetonitrile for 20 min. The flow rate was 1.0 mL/min. (Reprinted from reference 10 with permission.

pounds in the methanol solution from aliphatic and PAH classes by development of the silica-gel TLC plates with cyclohexane. Separations with HPLC (shown in Fig. 2-13) then gave three fractions that were consistently positive by the Griess reaction. The authors state that if the compounds found in this study are as potent as the other nitroso compounds evaluated in animal studies, then they may be more hazardous then benzo [α] pyrene. They state, "This estimate that the hazard from this class of compounds may equal that for total PAH indicates the need to determine the structures of the *N*-nitroso materials and to test their biological activity." (10) Clearly, LC has a significant role to play in this type of research.

Water Quality

As was the case in air-pollution research, in water pollution work the largest concern is the monitoring of the environment to determine if contamination is occurring and to follow the subsequent reduction of the level of problematic substances. As was mentioned earlier, PAHs are of environmental concern and, as such, are being monitored in air and water. Figure 2-14 shows the HPLC analysis of a National Bureau of Standards (NBS) sample of

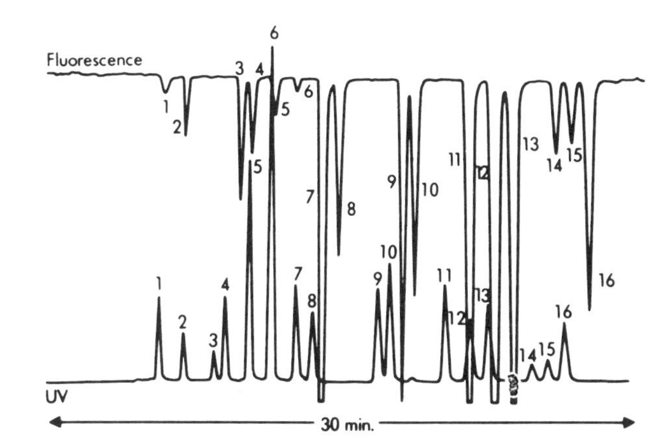

FIGURE 2-14. Analysis of PAHs. Sample: NBS Standard No. 1647; Column: PAH Analysis (Waters) 5 mm ID × 10 cm; injection volume 10 μL. Mobile phase: linear gradient from 35% acetonitrile in water to 100% acetonitrile over 30 min. Flow rate: 2 mL/min. Detection: UV at 254 nm and fluorescence with Ex = 254 nm and Em = 375 nm.

PAHs using both spectrophotometric and fluorometric detection. When a water (or air) sample is analyzed, the absence of a peak implies that it is not present. The presence of a peak at the same retention time suggests that a PAH presence is highly possible. In some cases another confirmative analysis is needed.

Another area of interest in the water-pollution area is the presence of pesticides in agricultural run-off water or possible drinking water supplies. Widespread use of pesticides has caused environmental researchers to focus on monitoring analysis such as the one shown in Figure 2-15. This method is for screening four classes of pesticides: (1) organophosphates, (2) triazine herbicides, (3) carbamates, and (4) chlorophenoxy acids and esters. This chromatogram shows a water sample spiked with 20 ppb of each of 20 pesticides. Three of these pesticides, 2, 4D, Silivex (2,4,5-TP) and methoxychlor, are presently regulated under the Safe Drinking Water Act.

Regulatory agencies are now establishing LC methodologies for monitoring and testing for specific compounds. In an effort to protect sources of drinking water, regulatory agencies need to monitor new compounds as soon as they are determined to be a possible contaminant. One example is the analysis of carbamate pesticides. These compounds are commercially available, highly effective pesticides with a broad spectrum of activity. Carbamates are used worldwide and are considered generally less toxic to humans than the traditional organohalogen pesticides. As carbamate use increased, environmental scientists began to find the parent compounds and their degradation products in more and more water resources. Most probably the carbamates have found their way into the water supplies from agricultural run-off following an application to food crops. High performance liquid

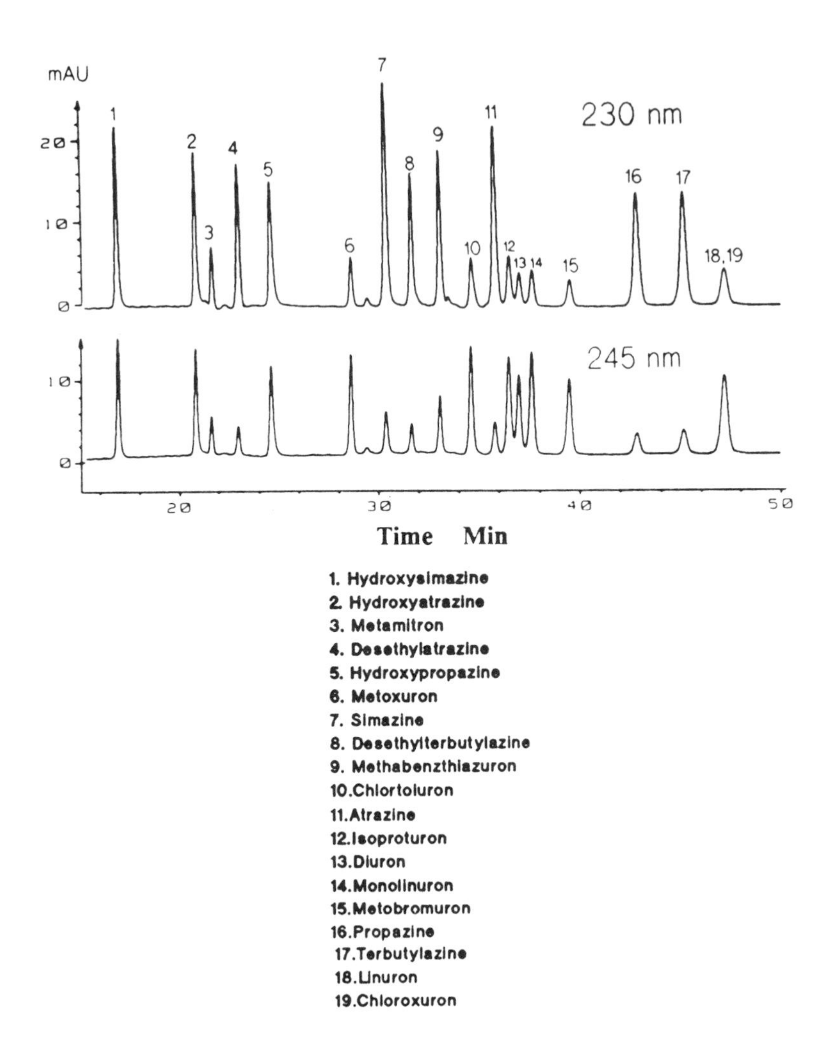

FIGURE 2-15. Pesticide screening using HPLC. Sample: 1 μg/L, 50 ul injection. Column: Zorbax C8 (5 μm), 4 mm ID × 30 cm (Knauer). Detector: Photodiode array at 230 nm and 245 nm. Mobile Phase: multiple step gradient from 95% water (2 mM sodium acetate, pH = 6.5) to 95% acetonitrile in 60 min. (Reprinted with permission from C. Schlett, *Am. Env. Lab., 12/90,* 12(1990).)

chromatography is an effective method for separating and analyzing the various carbamate pesticides.

As a further substantiation of the role of LC in analyzing and monitoring contaminants in the environment, Tables 2-1 to 2-3 list the HPLC approved methods for the EPA, NIOSH, and OSHA agencies. This list will probably increase in size as the need to do monitoring increases.

TABLE 2-1. HPLC Analyses Published by EPA[a]

Analyte(s)	Method	Matrix	Analyte(s)	Method	Matrix
Bendiocarb	639 (Draft)	WW	Glyphosate	TBA (Draft)	DW
Benomyl and carbendazim	631	WW	Hexachlorophene and dichlorophen	604.1 (Draft)	WW
Bensulide	636 (Draft)	WW	MBTS and TCMTB	637 (Draft)	WW
Bentazon (basagran)	643	WW	Mercaptobenzothiazole	640 (Draft)	WW
Benzidines	605	WW	*N*-Methyl carbamamoyloximes and *N*-methyl carbamates	531	DW
Biphenyl and orthophenyl-phenyl	642	WW	Oryzalin	638 (Draft)	WW
Carbamate and amide pesticides	632.1	WW	Picloram	644	WW
Carbamate and urea pesticides cleanup	632	WW	Polynuclear aromatic hydrocarbon	8310	HW
			Polynuclear aromatic hydrocarbons	510 (Draft)	DW
Diquat, paraquat	TBA (Draft)	DW	Polynuclear aromatic hydrocarbons	610	WW
Formaldehyde	TO-11	Air	Rotenone	635 (Draft)	WW
Gel permeation cleanup	3640	HW	Thiabendazole	641	WW

[a] DW, drinking water; WW, wastewater; HW, hazardous waste.

TABLE 2-2. HPLC Analyses Published by NIOSH*

Analyte(s)	Method	Analyte(s)	Method
Acetaldehyde	3507	Isocyanate group	5505
Ammonia	6701	Orano arsenic	5002
Anisidine	2514	Paraquat	5003
Benzidine and 3.3-dichloro-benzidine	5509	Polynuclear aromatic hydro-carbons (list of 17)	5506
Benzoyl peroxide	5009	Pyrethrum (6 active constitu-ents)	5008
Bromoxynil octanoate	5010		
Chloroacetic acid	2008	Rotenone	5007
Chloroacetate ion		Strychnine	5016
Dyes: benzidine	5013	Sulfur dioxide	6004
Q-Dianisidine		Sulfate ion	
Q-Tolidine		Sulfite ion	
Hippuric acid	8301	Thiram	5505
Methyl hippuric acid		Toluene-2.4	2535
Hydroquinone in urine	5004	Diisocyanate	
Inorganic acids	7903	2,4,7-Trinitrofluoren-9-one (TNF)	5018
HBr, HNO_3, H_2SO_4			
HF, HCl, H_3PO_4		Warfarin	5002
Iodine	6005	2,4 D; 2,4,5, T	5001
Iodide ion			

Source: NIOSAH, *Manual of Analytical Methods,* 3rd ed., 1984; 1st supplement, 1985; 2nd supplement, 1987.

DISTINGUISHING BETWEEN GOOD AND BAD PLASTICS

A common problem in the polymer industry is that when two reactions are run to make the same polymer, one of the resulting polymer batches will be formulated into a "good" product and the second batch will form a "bad" product. The question then is how to tell "good" from "bad" polymer. If only we could see a "polymer" molecule.

The Concept of Molecular Weight and Molecular-Weight Distribution

If one had a microscope with a magnification of several million, the micron-long polymer molecules would be "seen" as strands several inches long. These lengths would be made up of repeating links referred to as repeating monomer units, M. If we are viewing a polystyrene polymer the repeating monomer would be styrene. If we use our hypothetical microscope and look at each batch of polymer we would see that viewed as a bulk mass, both batches look like a mass of intertwined, tangled "spaghetti." To understand how the two batches of polymer differ, we must pull apart the pieces of spaghetti into the individual strands and compare them to one another. To

TABLE 2-3. Analyses Published by OSHA*

Analyte(s)	Method	Analyte(s)	Method
Acetic acid	ID-119	MOCA	24
Acrylic acid	28	4,4′ Methylene	
Benzene	12	bis (*O*-chloroaniline)	
Bromine	ID-108	*N*-Nitrosodiphynelamine	23
Bromide, bromate		*N*-Nitrosodiethanol	31
Diethylamine	41	NDELA, bulk	
Diisocyanates	18	samples	
2,4-TDI, MDI		Nitric acid	ID-127
Diisocyanates	42	Pentachlorophenol	39
Dimethylamine	34	Phenol/cresol	32
Diphenylamine	22	Phosphoric acid	ID-111
Ethylamine	36	Sulfur dioxide	ID-104
Ethylene glycol	43	Sulfate	
Dinitrate (EGON)		Sulfite	
and Nitroglycerin (NG)		Sulfur dioxide	ID-107
Formic acid	ID-112	Solid sorbent	
Hydrazine	20	Sulfuric acid	ID-113
Maleic anhydride	25	2,3,4,6 Tetrachlorophenol	45
Methyl isocyanate (MIC)	54	Urea derivatives	33
Methylamine	40	(of MDI & TDI)	
Methylene bisphenyl isocya-		for confirmation of	
nate (MDI)	47		

Source: OSHA, *Analytical Methods Manual,* 1985.

simplify the discussion, assume that each batch is unraveled and each contains three pieces. Viewing the pieces through our microscope would reveal a situation as shown in Figure 2-16.

In the "bad" batch all of the molecular chains are 4 in. long and all weigh the same (Fig. 2-16*a*). This is the molecular size and molecular weight. The average molecular size can be determined statistically: 4 in. + 4 in. + 4 in. = 12 in. divided by 3 molecules = 4 in. which is the average molecular size of the polymer chains.

After untangling the polymer strands of the other sample which was formulated into a "good" product, it was shown to consist of a different group of three molecules (see Fig. 2-16*b*). The first was 2 in., the second was 4 in., and the third was 6 in. in length. Statistically 2 in. + 4 in. + 6 in. = 12 in. divided by 3 molecules is a molecular size average of 4 in., which was the same average size as the "bad" batch.

Remember that the assumption was that the 6-in. molecule weighs 3 times more than the 2-in. molecule. Therefore, it can be concluded that both polymer batches are made up of the same number and types of atoms and statistically both have an average length of 4 in. Thus, the weight-average

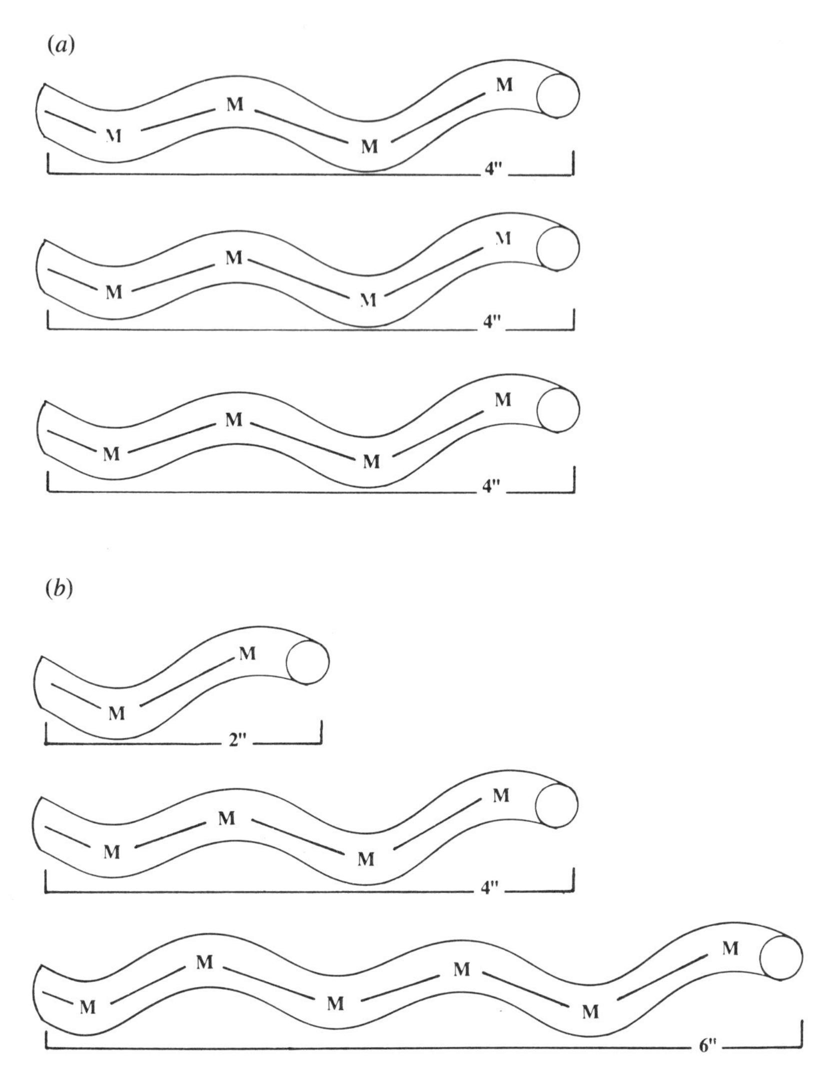

FIGURE 2-16. Two polymers with the same molecular weight average but different number-average molecular weight. (*a*) The polymer chains are all the same length. (*b*) The polymer chains are of varying lengths.

molecular weights are equal because weight and size are proportional. Therefore, it is clear that the difference between the samples is in the individual lengths. The "bad"-batch polymer has a narrow molecular weight distribution (all polymer strands are 4 in.) and the "good" sample has a broad molecular distribution (polymer strands range from 2 to 6 in.).

Our observation of the performance of "good" and "bad" batches can be rationalized and supported by polymer chemistry. In polymers the large strands impart strength and the smaller strands result in flexibility. Often, a balance of sizes is needed to result in a "good" polymer which can be processed correctly. Too many long chains yield a brittle product while too many small chains make a soft product. Thus, there is a way of determining "good" and "bad" polymer batches using a microscope with several million magnification. Unfortunately, no such microscope exists.

What does exist, however, is GPC, which separates the molecules in a polymer on the basis of molecular size (weight). The large molecules elute from the column earlier than small molecules and from this elution profile both the *weight-average molecular weight (size)* and the *molecular weight (size) distribution* can be calculated.

Using GPC to Determine Molecular Weight

By examining the weight-average molecular weight and the molecular weight distribution of the molecules in a polymer, the scientist can better monitor and control the polymerization to produce a better product. Also, by using GPC purchased polymers can be selected to reduce variability and to better control the quality of the final product.

An example of how GPC can aid in controlling the quality of a polymer is the analysis of polyethylene coffee can lids. Polyethylene resins are used in the injection molding of coffee can lids. A polyethylene with much high-molecular-weight material is brittle and hard. A lid made of this material will crack easily. A polyethylene with much low-molecular-weight material is very soft. A lid made of this material will not hold its shape. Pretesting the polymer using GPC analysis to determine the molecular weight distribution of the polyethylene resin prior to injection molding can prevent losses by eliminating the manufacture of faulty lids and lessen the possibility of equipment damage. Using the GPC mode of LC to test the polyethylene provides a "fingerprint" of the resin, as shown in Figure 2-17. By using the chromatogram, the lid manufacturer can determine whether the polymer has too much high- or low-weight material from the shape of the peak profile. Thus, by using GPC, the manufacturer will know whether the resin will perform satisfactorily.

Gel permeation chromatography can also be used to profile lower-molecular-weight materials. For instance, by monitoring the condition of a lubricating oil using GPC, both unnecessary shut down of production machinery and the possibility of machinery damage can be avoided because the oil is changed only when necessary. With prolonged use, lubricating oils break down and can also recombine to form high-molecular-weight compounds that do not flow freely. This increases wear on the machinery. Changing the oil when this breakdown–recombination process starts protects the machinery. Unfortunately, changing the oil too early leads to unnecessary loss of machine use time.

A GPC analysis of the oil can be performed on a single drop of oil—avoiding machine shut down. Figure 2-18 shows a profile of the molecular

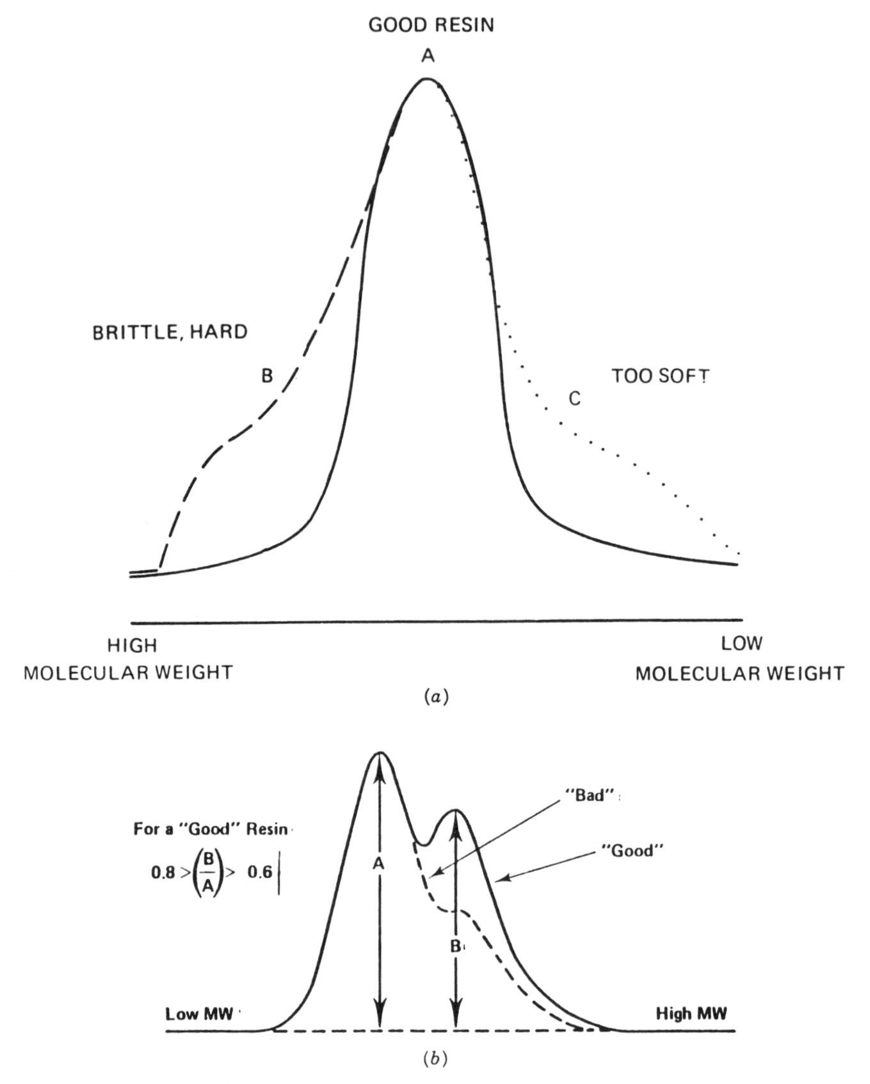

FIGURE 2-17. Contribution of molecular weight to polymer properties. (*a*) Molecular weight distribution profile of a typical polymer. (*b*) Quality control of an alkyd resin. It was determined that a good paint resulted only when the ratio of the peak heights was between 0.6 and 0.8. Detector: refractive index.

weight distribution of oil. Without GPC, monitoring material changes with time and/or use can be very difficult, especially when there is only a qualitative indicator such as a change in color or smell. In addition, with products such as lubricating oils, paints, or adhesives, relying on a qualitative "best guess" of useful life can lead to waste or even failure. Monitoring changes in the GPC profile, as shown in Figure 2-18, can give the necessary information quickly and easily.

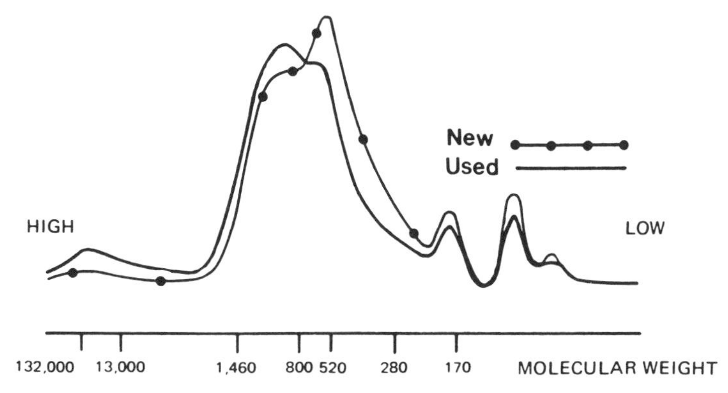

FIGURE 2-18. Loss of lubricating properties results from an increase in the high molecular weight material in the oil. This increase is seen in the profile as a "bump" in the very high-molecular-weight region and as a general shift in the profile toward the high-molecular-weight region.

AN AID TO ORGANIC SYNTHESIS

Rapid Optimization of Reaction Yields

A valuable application of LC is to rapidly establish optimum chemical reaction conditions. After developing the separation, the first step is to collect and identify the individual reaction products. Most detectors have a linear response of peak height to amount present; therefore, once the characteristic retention time and detector response versus concentration are known, it is easy to establish the optimum reaction conditions. In chromatography the retention time establishes the absence or presence for each compound of interest. The calibration curves indicate the amount of each component. Comparison of the chromatograms of reaction mixtures is a simple and rapid method of determining the preferred operating reaction conditions and concentrations. Since HPLC is much faster than TLC and is carried out in a closed system, a minor benefit is that it is easier to prevent oxidative degradation in HPLC than in TLC.

To illustrate the usefulness of HPLC in optimizing reaction yields, it is instructive to look at a real example. At one stage in the synthesis of vitamin B_{12} by Professor R. B. Woodward and his group at Harvard University, it was necessary to convert the methyl propionate groups of dicyanoheptamethylcobrinate (cobester) to amides without destroying any of the nine asymmetric centers in the molecule. However, treatment of cobester with ammonia in ethylene glycol gave two products (reaction should in Fig. 2-19). In the major product, a lactam had formed in the B ring. Although formation of this lactam did not affect the asymmetric centers of the molecule, the lactam could not be converted to the amide form without disruption of some of the asymmetric centers. Unfortunately, the desired product was only 3% of the reaction mixture.

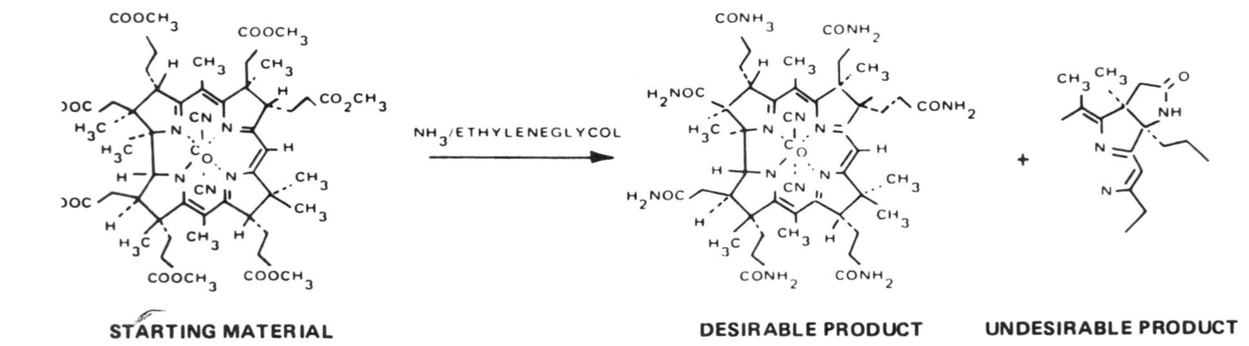

FIGURE 2-19. Reaction of cobester to amides and lactam. The chromatography is shown in Figure 2-20.

Using analytical HPLC it was possible to vary the reaction conditions systematically and follow the resulting changes in product composition without isolating any of the reaction products. The results of some of these runs are shown in Figure 2-20. In addition to changing the duration of the reaction (chromatograms 2-20-2, 2-20-3, 2-20-5) and the temperature at which the reaction occurred (chromatograms 2-20-3 and 2-20-4), final solution of the problem required the addition of ammonium chloride (chromatograms 2-20-6 and 2-20-7). This made the solution slightly acidic and slowed the rate of lactam formation sufficiently to allow Professor Woodward's group to eventually achieve 80–85% yield of the desired aminolysis product (chromatogram 2-20-7). Because the reactions were run in parallel and the LC analysis time was so short, optimum reaction conditions were established much more rapidly than possible by other methods.

Another example of LC in this area is the optimizing of photochemical reactions. Dr. H. H. Wasserman and his group at Yale University had a problem similar to that of Dr. Woodward. As shown in Figure 2-21, compound I is converted primarily to compound II in the presence of UV radiation and oxygen. However, concentration and examination of the impurities by LC revealed that compound IV, which was of interest in their research, was present as 0.1% of the reaction mixture. They, therefore, wanted to determine conditions that would increase the yield of compound IV.

Following a method similar to that of Woodward, the Yale group systematically varied reaction conditions and used LC to follow quantitative and qualitative changes in the reaction products. They quickly determined that eliminating oxygen from the reaction vessel and increasing the strength of the UV radiation gave a 50–55% yield of compound IV.

Purification of Reaction Mixtures—An Indispensable Aid in the Synthesis of Vitamin B_{12}

In addition to aiding in determining the optimum parameters under which to run a chemical reaction, HPLC is quite useful in purifying the final product(s) during a complex synthesis route. For example, Woodward and his research group used HPLC as a significant time saver during the research necessary to complete the total synthesis of vitamin B_{12}. At one stage in the synthesis it became necessary to prepare the individual monopropionamides of dicyanoheptamethylcobyrinate (cobester). To accomplish this, cobester was reacted with ammonia, resulting in a mixture of four positional monoamide isomers, di- and tri-amides, and unreacted cobester, as shown in Figure 2-22. The reaction yield is approximately: cobester, 56%; b-amide, 5%; d-amide, 17%; e-amide, 8%; f-amide, 8%; and di- and tri-amides, 6%.

Separation of the individual monoamides from each other proved to be a major obstacle to the completion of the project. This separation was not possible using TLC or open-column chromatography methods. With TLC, only the d-amide was separated from the b-, e-, and f-amides. Using HPLC,

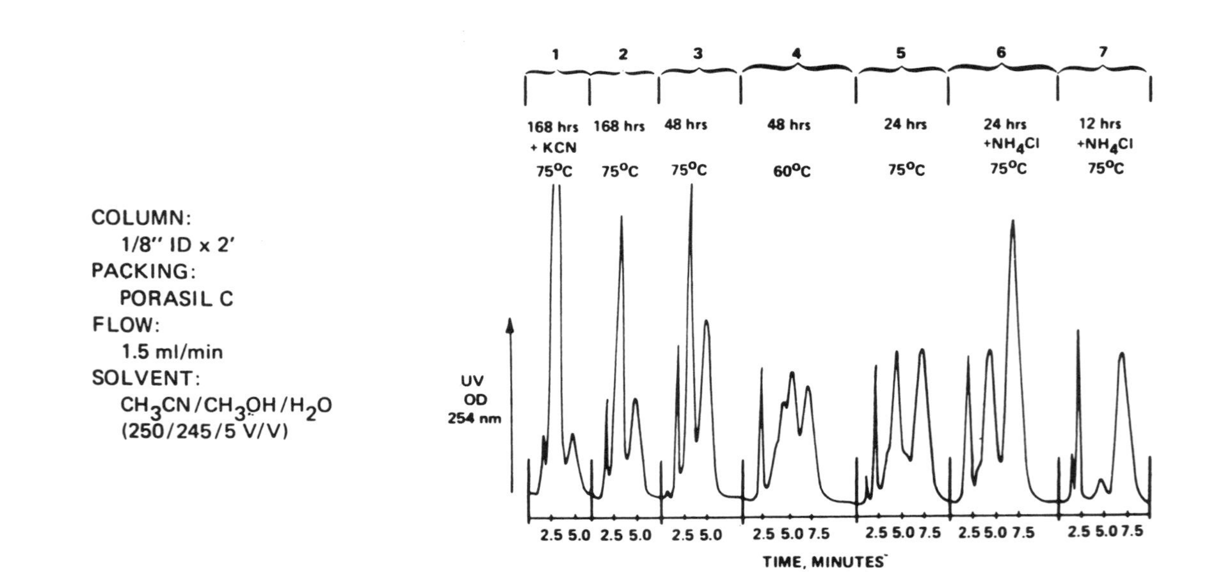

FIGURE 2-20. Optimization of reaction conditions using LC as a monitor. The retention times and peak heights were used to evaluate the completion of the reaction. The unreacted starting material eluted at approximately 2.5 min, the undesired lactam eluted at approximately 5 min, and the desired aminolysis product had a retention time of approximately 7.5 min. Column: Porasil C (silica, 37–75 μm), 2 mm ID × 60 cm. Mobile phase: acetonitrile, methanol, water (250/245/5, v/v). Flow rate: 1.5 mL/min. Detector: UV at 254.

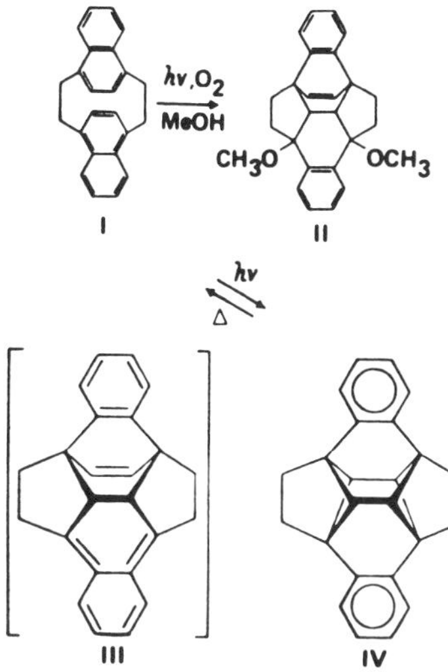

FIGURE 2-21. Photochemical reaction scheme.

a greatly improved separation was obtained. The cobester peak, the b-, c-, d-, and f-amide peaks, were well separated as shown in Figure 2-23. The analytical separation of the individual b-, e-, and f-amides were accomplished using a different mobile phase and is shown in Figure 2-24.

This satisfactory analytical separation provided the data for scale up to a preparative separation. It was decided to obtain 200 mg of each amide from a 600-mg mixture of the b-, e-, and f-amides previously separated by thick-layer chromatography. When the cobester reaction mixture was separated using LC, seven other compounds were found, each comprising 2% or less of the reaction mixture.

At this point in the program to synthesize vitamin B_{12}, it was decided that 1 g of each of the four pure amides was required. As shown in the diagram in

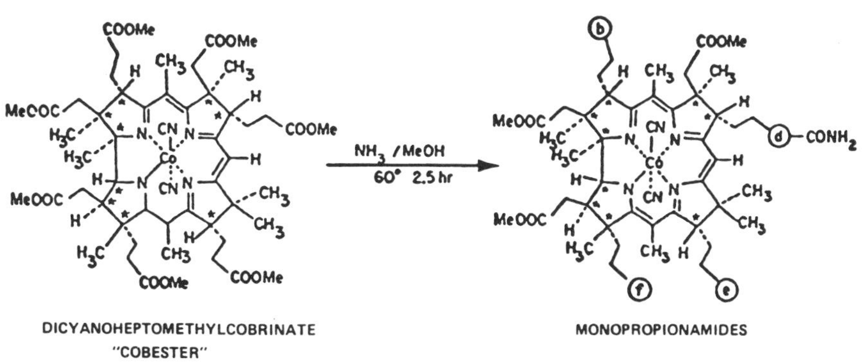

FIGURE 2-22. Reaction of cobester.

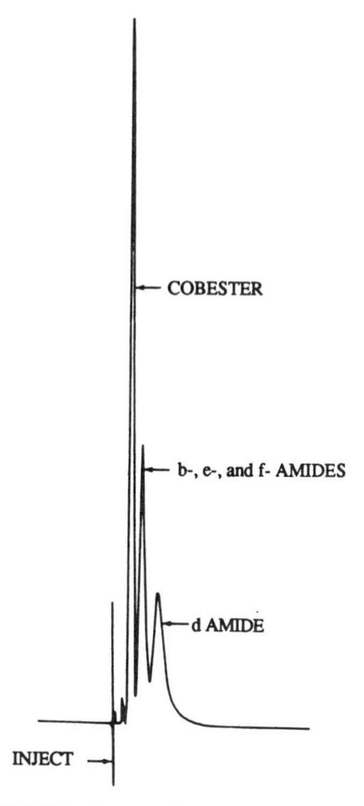

FIGURE 2-23. Separation of reaction products.

Figure 2-25, to obtain 1 g of each amide required the reaction and separation of 18 g of cobester. The cobester was first reacted to obtain the amide mix. The preparative route shown on the left side of Figure 2-25 was first considered since the research group had extensive experience with preparative TLC. It involved a thick-layer plate separation followed by a $\frac{3}{8}$-in. diameter LC column separation. The raw product is processed on thick-layer plates to separate the b-, e-, and f-fraction from the d-fraction. To separate 18 g of the reaction product by TLC, it is necessary to develop 240 thick-layer plates. Since each plate requires five sequences of elution and drying, the total operation requires 1200 elution and drying steps. The b-, e-, and f-amide is then separated using the high performance preparative column described in the center box in the lower third of Figure 2-25. To inject a total of 3 g of material to obtain 1 g each of the b-, e-, and f-amides, 20 preparative runs with 150-mg injections would be required.

Because the time needed for purification by following the route on the left side of Figure 2-25 is long, it was decided to scale up the HPLC separation and thereby reduce the man-hours of effort required. The route on the right involves three separations, the two shown on the chart are HPLC types. The three steps are:

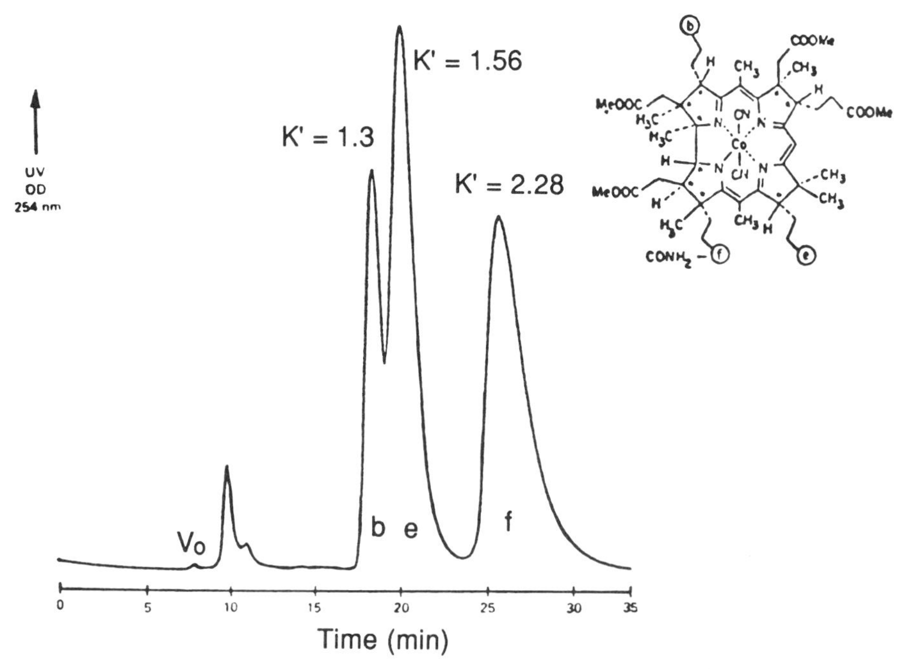

FIGURE 2-24. Separation of b-, e-, and f-amides. Column: Corasil II (silica, 37–50 μm), 2 mm ID × 180 cm. Mobile phase: methylene chloride, acetonitrile, methanol saturated with hydrogen cyanide (175/120/5). Flow rate: 0.45 mL/min. Detector: UV at 254 nm.

1. A liquid–liquid separation using a two-phase system consisting of benzene and water to separate the cobester, the b-, d-, e-, and f-amides into the benzene phase while the more polar di- and triamides remained predominantly in the water phase.
2. Five grams of the mixture (cobester and b-, d-, e-, and f-amides) were then injected on a 0.9-in. ID × 8 ft silica gel column, and eluted with a mixture of methanol/isopropanol/hexane. The 5-g mixture was injected three times at a flow rate of 34 mL/min. The result of this separation was the separation of 2.3 g of d-amide, 8.5 g of cobester, 3.1 g of mixed b-, e-, and f-amides, and 0.8 g of di- and triamides. The chromatogram in Figure 2-26 shows this separation.
3. The HPLC separation previously made using 0.3-in. ID diameter columns was scaled up to 0.9-in. ID columns. The scaleup was straightforward and permitted 1.0 g to be injected on the 24 ft of 0.9-in. ID columns. The amount of sample injected was increased by the ratio of the cross-sectional areas of the two column systems. The flow rate was increased by the ratio of the cross-sectional areas of the 0.3- to 0.9-in. ID columns to 30 mL/min. The linear velocity and hence the separation time remained essentially the same.

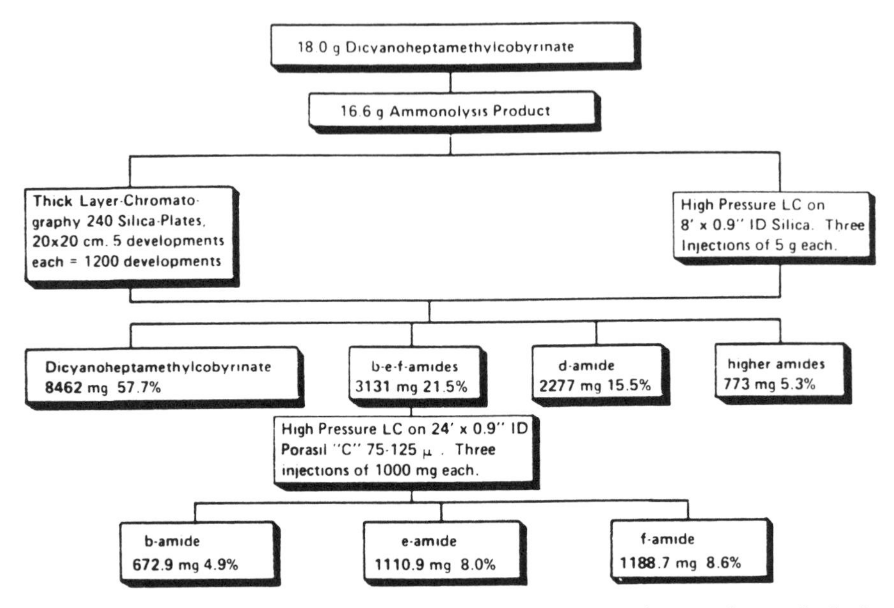

FIGURE 2-25. Liquid chromatography separation flow diagram for aminolysis products of dicyanoheptamethylcobyrinate (cobester).

With the scaled-up separation (the right route), the 1200 thick-layer developments were essentially replaced by three 5-g preparative runs on 8 ft of 0.9-in. ID silica gel columns. The 20 injections of 150 mg each on 0.3-in. diameter columns were replaced by 3 injections of 1.0 g each on 0.9-in. ID columns.

HPLC is an Indispensable Tool in the Laboratory

Liquid chromatography enabled Professor Woodward to obtain gram quantities of the four isomer monoamides and proceed with the synthesis scheme. On more than one occasion during the course of synthesizing vitamin B_{12}, Dr. Woodward's research group faced the problems of the separation of very closely related molecules. The resolution of these separation problems were made possible using HPLC. In Dr. Woodward's words:

> Here I should say that of absolutely crucial importance to all of our further work has been our taking up the use of high pressure liquid or liquid–liquid chromatography to effect the very difficult separations with which we were faced from this point onwards. *The power of these high pressure liquid chromatographic methods hardly can be imagined by the chemist who has not had experience with them;* they represent relatively simple instrumentation and I am certain that they will be indispensable in the laboratory of every organic chemist in the very near future (11).

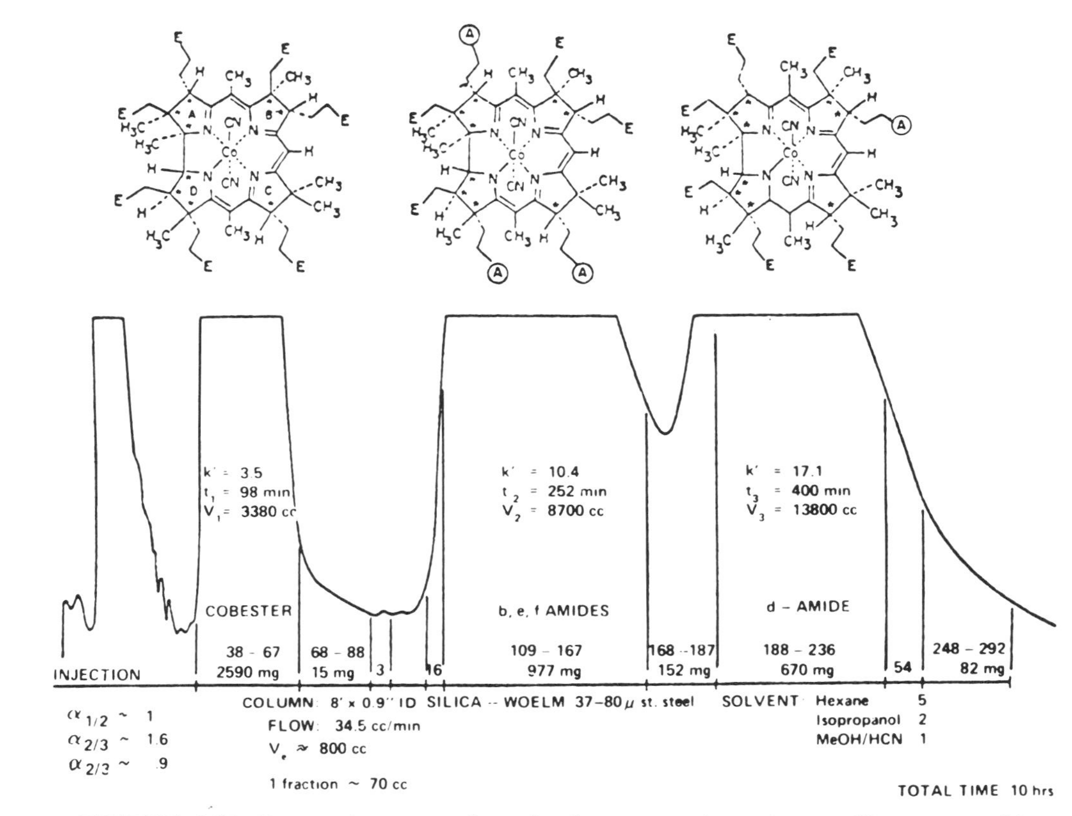

FIGURE 2-26. Preparative separation of cobester reaction mixture. Chromatographic conditions are shown on the figure.

Dr. Woodward became a rapid convert to LC and the degree to which he found it an absolute necessity to be depended upon was summarized by his statement:

> The cobyric acid was crystalline, and *it was identical in all respects, most particularly in liquid chromatographic behavior, with cobyric acid derived from natural sources*. (Italics are Woodward's.)

Dr. Woodward obtained his first HPLC in 1971 and eventually his group had three. Very quickly, his group benefited from using LC and two short years later at the 1973 IUPAC Symposium (11), Dr. Woodward announced the synthesis of vitamin B_{12} and emphasized the power of HPLC. At that time he became the strongest spokesman for the importance of the technique to the organic chemist and many followed his lead and quickly incorporated LC into their laboratories. Thus, Dr. Woodward played a key role in the tremendously rapid growth of HPLC.

Dr. Woodward's colleague, Dr. A. Eschenmoser of Eidgenossische Technische Hochschule (ETH) (Zurich, Switzerland), also praised LC in his account of his part in the synthesis of vitamin B_{12}:

> . . . this presented difficulties not of principle but rather of an experimental nature. That these difficulties were successfully overcome with the aid of high-pressure liquid chromatography, which appeared at precisely the right moment, proved to be one of the first illustrations of the efficacy of this new chromatographic separation procedure in organic synthesis. (12)

It is interesting to note that prior to Woodward's and Eschenmoser's successful applications of HPLC, the technique was almost universally viewed as a tool for the analytical chemist. The synthesis and natural product chemists used TLC and open column chromatography routinely but the tremendous benefits of HPLC were essentially unknown.

THERAPEUTIC DRUG MONITORING

The rate of drug metabolism by a patient on long-term chemical therapy can change radically over the period of time during which the drug is administered. This change of metabolism can cause corresponding changes in the concentration of the drug in the patient's blood, resulting, among other things, in a toxic condition or in an ineffective dosage level. To improve this situation, serum or plasma levels of a drug should be measured periodically. The dosage can then be adjusted to compensate for changes in the metabolic breakdown of the drug in the body and, therefore, to keep the drug level in the therapeutic range.

Commonly used analytical methods of measuring drug serum levels include IR and UV spectrophotometry, GC, TLC, and HPLC. High-performance liquid chromatography, the "new kid on the block," provides a fast and simple technique for excellent quantitative separations of drugs from serum plasma. It also usually overcomes the limitations imposed on serum drug analysis by some of the other commonly used analytical techniques.

Monitoring of Asthma Treatment

One of the first successful examples of the use of LC for drug monitoring was the clinical management of serum theophylline levels. Theophylline is widely accepted as an effective bronchodilater in the treatment of asthma, apnea (13,14), and obstructive lung disease. Theophylline (1,3-dimethylxanthine), shown in Figure 2-27, is used as the major active ingredient in a variety of preparations that can be administered orally or intravenously. To be effective, however, theophylline serum concentrations must be within a therapeutic range (15–18). with serum levels in this range, theophylline has been found to give relief of bronchospasm to most patients. Below the therapeutic level, most patients show little benefit from the drug; above the level, there is a high incidence of toxicity which usually takes the form of nausea, vomiting, or headache (19,20).

There are very marked differences in the rate of metabolism of theophylline among individuals (15,21); therefore, dosage suggestions in milligrams per kilogram of body weight might be insufficient to ensure maintenance of therapeutic dosage levels. Liquid chromatography methods have fulfilled this need to rapidly, reliably, and accurately determine theophylline levels in blood with the result that, when appropriate, therapy might be individualized. At the time of its development (22,23) LC could do the analysis job much better than UV spectrophotometry and GC, which required larger sample amounts and longer waiting times for the results.

With HPLC, serum concentrations of this important bronchodilator can be determined in about 10 min on approximately 5 μL of serum, allowing quick dosage adjustment to ensure therapeutic levels while avoiding toxicity. In addition, LC eliminates the need for extraction procedures and prevents interference from metabolites, other medications, and dietary xanthines such as caffeine. Using an internal standard, the technician can obtain

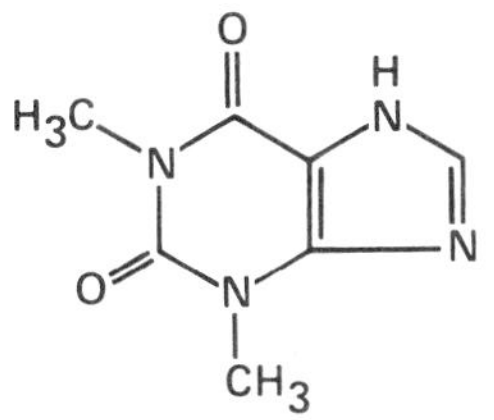

FIGURE 2-27. Structure of theophylline (1,3-dimethylxanthine).

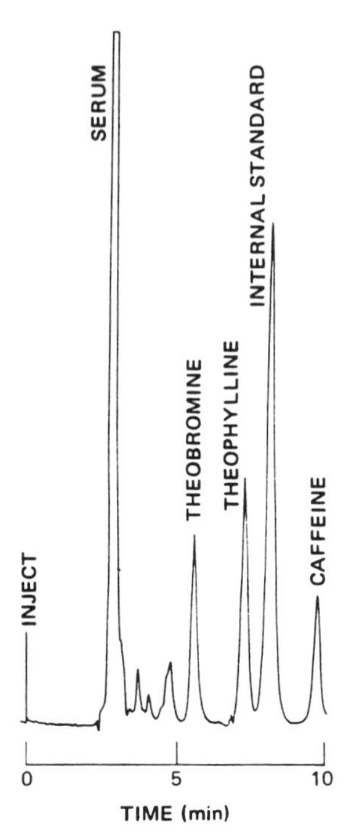

FIGURE 2-28. Separation of theophylline and other xanthines.

reproducible quantitative data with ease. An example of the type of results obtained using LC is shown in Figure 2-28. Notice that the peak corresponding to theophylline is clearly separated from the other materials in the serum sample (24).

Management of Epilepsy Treatment

Another early example of the use of LC was in the analysis of diphenylhydantoin and phenobarbital levels in the plasma of a patient receiving these drugs for the management of epilepsy (25,26). As shown in Figure 2-29, by using LC the plasma was found to contain 8.5 μg of phenobarbital and 9.2 μg of diphenylhydantoin per milliliter. The LC method, additionally, has several advantages over other methods. These advantages include faster separation, lack of derivatization, better sample stability, and smaller sample size (25,26).

The advantages of HPLC mentioned above apply to the monitoring of many drugs in plasma and other body fluids. In most cases, the body metabolizes a drug by converting it into successively more polar compounds, which are easier for the body to excrete. With LC you can develop a separation that allows you to use these polarity changes to your advantage. When

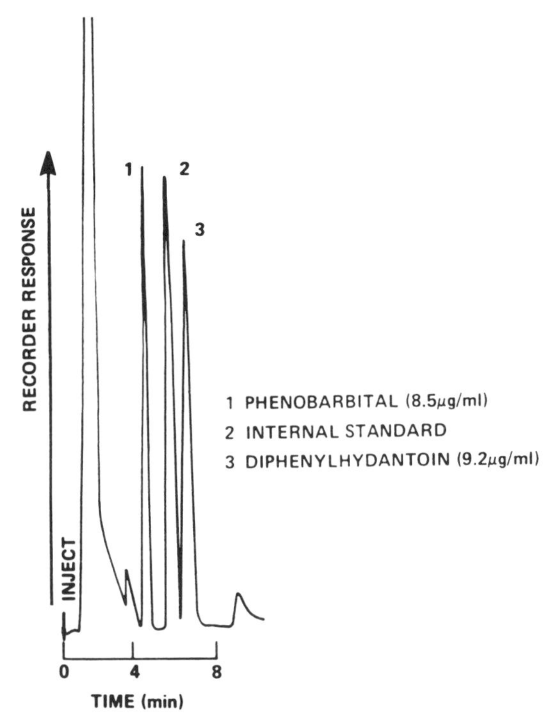

FIGURE 2-29. Separation of the extract of a patient plasma. (Adapted with permission from reference 25.)

you are trying to identify the metabolites of a certain drug, knowing that the metabolites have eluted by the time the drug itself appears on the chromatogram can be extremely useful. In addition, LC permits rapid determinations of many endogenous compounds. Such rapid and reliable analyses are playing an increasingly important role in clinical research.

A QUALITATIVE ANALYSIS TOOL

Because HPLC has such a broad capability for separating widely differing types of compounds, it can be helpful in analyzing and comparing complex mixtures. A quick check of composition and comparative concentrations by LC can give a distinctive chromatogram for each material which represents a quick and informative comparison of sample composition and relative concentrations. Thus, when you have to monitor changes in a material with time and/or use, identify the source of a sample, or control the quality of raw materials, LC can help. Comparing the two materials by other analytical techniques could be very difficult since each component might require a different technique and/or procedure and, therefore, result in a great deal of time and effort.

Because LC is a separation technique, it allows you to monitor the individual constituents in a complex raw material. Once this is done for known "good" (i.e., acceptable) raw materials, chromatographic specifications can be set for accepting or rejecting new materials. This technique is particularly valuable for very complex materials. For such materials, averaged chemical and physical properties such as pH, specific gravity, color, odor, and solid contents may not adequately predict processing of final characteristics; but chromatograms may make such predictions.

However, the chromatograms should be used cautiously and with good "analytical sense." When substances are chromatographically separated, all that can be said is that this is positive proof that the two substances are *not* identical. If there is no peak in a chromatogram which occurs at a retention time characteristic of a particular compound, the conclusion is that the particular compound which would elute at this retention time is not present in the sample. Also, if a peak does occur at the particular retention time characteristic of that particular compound, the LC data "suggests" that the particular compound is present. However, the LC data is not "proof positive." Proof "beyond a shadow of a doubt" must come from another analysis performed on the collected material from the HPLC, for example, mass spectrometry. Thus, chromatography combined with another chemical or instrumental analytical method is necessary to identify chemical species.

Testing Incoming Raw Materials

One example of using the chromatogram for inspecting incoming materials can be found in the beverage industry. The quality of some beverages depends, in large part, upon the quality of the lemon oil used. Lemon oils contain both volatile and nonvolatile materials. Comparing the volatile portions of lemon oils is easily done by sniffing or by GC. The nonvolatile materials also play an important role in the final taste of a beverage, but cannot be checked by these methods. Differences in the composition and relative concentrations of the nonvolatiles, however, show up very clearly on the LC chromatograms of different lemon oil batches. This is shown in Figure 2-30.

Testing for Product Integrity

An example of testing product integrity is the analysis of aspirin. Because salicylic acid, a degradation product of aspirin, is believed to be responsible for some toxic side effects associated with taking aspirin, FDA regulations strictly regulate the amount of salicylic acid that can be present in an aspirin tablet when it is sold. Analysis of this trace degradation product in aspirin tablets takes only minutes by LC. Because of this, changes in the chemical

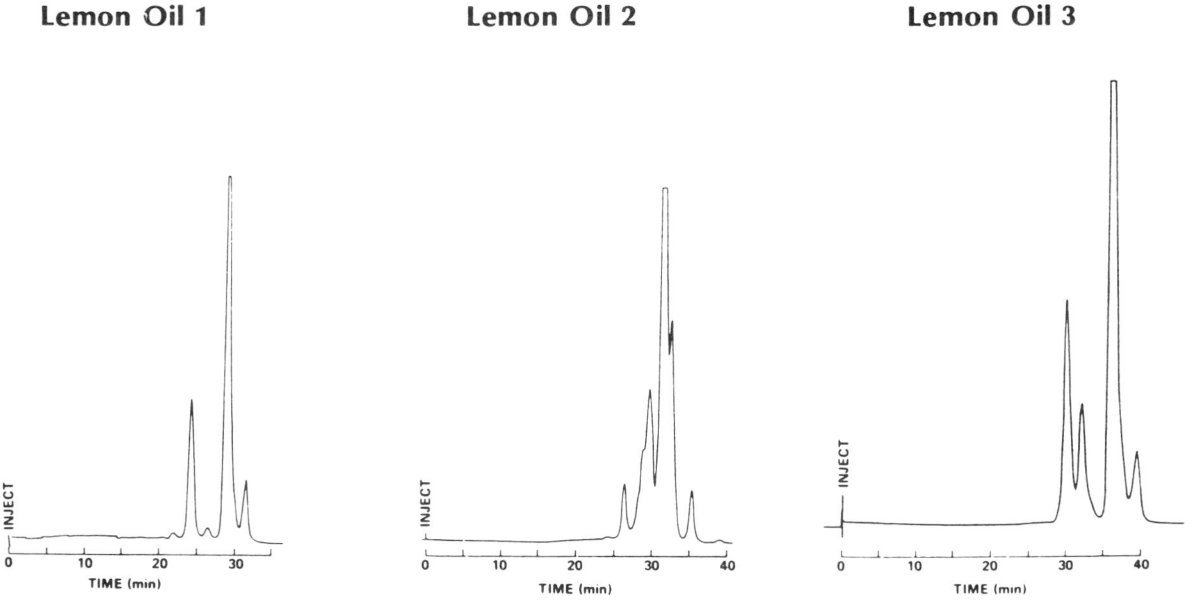

FIGURE 2–30. Comparison of lemon oils. Use of synthetic lemon oils gives lemon-flavored beverages an "off" taste. Natural lemon oils contain high molecular weight, nonvolatile components which are normally absent from synthetics. These compounds do not show up on the survey print of lemon oil No. 3, indicated by the arrow. This sample is probably synthetic. All the samples were identical by GC.

composition of aspirin can easily be monitored as a function of time. In this way, information can be gained that is helpful in determining the shelf life of products.

Determining Probable Causes

In certain post-mortem examinations where the search for a probable cause of death may depend upon the presence or absence of a specific compound. With LC, you can rapidly screen samples for that compound. For instance, when morphine is suspected, a quick (5 min.) screening test can be done. If morphine is present, it will show up as a "peak" in a characteristic position in the chromatogram and confirmatory tests can be run. If morphine peak is absent, the investigation will take another direction.

In another example, an investigation depends on identifying the source of a complex chemical product or a chemical product in a stain. In this example, comparing chromatograms of the chemical mixture or extracts of the stain and the suspected product can confirm identity. In these cases, physical and chemical comparisons (e.g., color or smell) alone are not sufficient to prove that the stain was traceable to a particular product brand. Since LC is a qualitative and quantitative tool, it can only demonstrate that a particular product is not the culprit. If a peak occurs and implicates a product as the culprit, other chemical characterizations are required after the initial LC screening to confirm that the product matches as the most probable source.

SUMMARY

As the wide range of LC applications discussed in this chapter demonstrate, it should be apparent that HPLC is an essential tool for modern scientists. The HPLC technique is utilized in a wide range of scientific disciplines and is often faster than the techniques which it replaces. It is not sample limited when used for most analyses, and yet large amounts of sample may be purified if materials are collected for identification. While HPLC is an indispensable separation tool for many situations, it is not a "cut-and-dried" technique. A separation must be developed for each type of sample. Different samples probably will require different analysis conditions (eluent, column, etc.). Therefore, the remainder of this book is intended to help you understand what is needed and how to develop a successful separation.

REFERENCES

1. I. S. Johnson, *Science,* **210,** 632 (1983).
2. S. Pestka, *Genetic Engineering News,* (Sept./Oct.) 1982.
3. S. Herring, *Genetic Engineering News,* (Mar./Apr.) 1983.

4. R. E. Harmon, Ed., *Chemistry of Nucleosides and Nucleotides*, Academic, New York, 1977.
5. J. R. Strahler, B. B. Rosenbloom, and S. M. Hanash, *Science*, **221,** 860–862 (1983).
6. C. S. Fulmer and R. H. Wasserman, *J. Biol Chem.*, **15,** 254 (1979).
7. S. Moore, D. H. Spackman, and W. H. Stein, *Anal. Chem.*, **30,** 1185 (1958).
8. (a) B. A. Bidlingmeyer, S. A. Cohen, and T. L. Tarvin, *J. Chromatogr. (Biomed. Appl.)*, **336,** 93 (1984). (b) B. A. Bidlingmeyer, S. A. Cohen, and T. L. Tarvin, *International Symposium on HPLC in the Biological Sciences*, paper presented at Melbourne, Australia, February 20–22, 1984.
9. T. H. Maugh, *Science*, **225,** 6296 (1984).
10. T. J. Kneip, J. M. Daisey, J. J. Solomon, and R. J. Hershman, *Science*, **221,** 1045 (1983).
11. R. B. Woodward, *Pure Appl. Chem.*, **33,** 145 (1973).
12. A. Eschenmoser and C. E. Wintner, *Science*, **196,** 1410–1420 (1977).
13. J. F. Lucey, *Pediatrics*, **55,** 584 (1975).
14. D. C. Shannon, F. Gotay, I. M. Stein, M. C. Rogers, J. D. Todres, F. M. Moylin, *Pediatrics*, **55,** 589 (1975).
15. J. w. Jenne, R. Wyze, F. S. Rood, and F. M. MacDonald, *Clin. Pharmacol. Ther.*, **13,** 349 (1972).
16. F. R. Jackson, R. Garrido, H. I. Silverman, and H. Salem, *Ann. Aller.*, **34,** 413 (1973).
17. M. M. Weinberger and E. A. Bronsky, *J. Pediat.*, **34,** 421 (1974).
18. M. M. Weinberger and S. Riegelman, *N. Engl. J. Med.*, **291,** 151.
19. E. Bresnick, W. K. Woodward, and E. B. Sagemen, *JAMA*, **136,** 397 (1948).
20. W. E. Zwillich, et al., *Ann. Int. Med.*, **82,** 784 (1975).
21. F. E. R. Simons, W. E. Pierson, and E. W. Bierman, *Pediatrics*, **55,** 35 (1975).
22. M. M. Weinberger, R. A. Matthay, E. J. Ginchansky, C. A. Chidsey, and T. L. Petty, *JAMA*, **225,** 2110 (1976).
23. M. Weinberger and C. Chidsey, *Clin. Chem.*, **21,** 834 (1975).
24. L. C. Franconi, G. L. Hawk, B. J. Sandmann, and W. G. Haney, *Anal. Chem.*, **48,** 372 (1976).
25. S. Atwell, V. Green, and W. Haney, *J. Pharm. Sci.*, **64,** 806 (1975)
26. J. C. Roger, G. Rodgers, Jr., and A. Soo, *Clin. Chem.*, **19,** 590 (1973).

ACKNOWLEDGMENT

Figures 2-1, 2-2, 2-12, 2-17 through 2-26, 2-28, and 2-30 are courtesy of the Waters Chromatography Division of Millipore.

CHAPTER 3

THE CHROMATOGRAM AND WHAT CONTRIBUTES TO IT

Characteristics of the chromatogram
A properly operating system
- Reservoir
- Solvent delivery system
- Injector
- Precolumn filter and guard column
- General considerations in system configuration
- Diagnostics

"Seeing" the compounds
- Wide scope vs. high sensitivity
- Time constant

The separation
- Capacity factor
- Selectivity
- Efficiency
- Resolution
- Selectivity vs. efficiency

Molecular distribution between moving and stationary phases
- Normal and reverse phase
- Liquid–solid chromatography (LSC)
- Bonded-phase chromatography (BPC)
- Ion-pairing chromatography (IPC)
- Ion exchange chromatography (IEC)
- Gel permeation chromatography (GPC)
- Summary

Summary
References
Acknowledgment

The chromatogram is a representation of the response of the LC detector as a function of time. It is a record of what happened during the chromatography. All decisions that are made after the use of the HPLC technique are based upon the chromatogram. Among other things, the chromatogram can tell you the number of compounds present, the precise retention times for the compounds, qualitative assignments of identity, the concentration of the different compounds, and the purity of separated fractions. The chromatogram will represent "accurate" information if the chromatograph is functioning correctly. However, if it is not, decisions will be based on inaccurate information obtained from the chromatogram. It is indeed unfortunate, but nevertheless true, that a chromatogram will result even when the chromatograph is physically or chemically malfunctioning; and, in this situation, bad decisions may result. Therefore, it is important to understand the characteristics of the chromatogram and what contributes to it in order to insure that decisions are based only upon chromatograms that result from a correctly operating liquid chromatograph.

CHARACTERISTICS OF THE CHROMATOGRAM

An example of a chromatogram is shown in Figure 3-1. The peaks are a result of the detector's sensing of the sample components as they emerge from the column. The components that are not adsorbed by the packing will not be retained by the column. Therefore, these components will directly pass through the column and elute first. The total volume of mobile phase eluting from the column between the time of injection and the appearance of the nonretained species equals the volume of the column not occupied by the packing. This is one "column volume," often referred to as the "void volume," V_0, of the column and its occurrence in the chromatogram is illustrated as the V_0 peak in Figure 3-1. The "retention volume" or the "elution

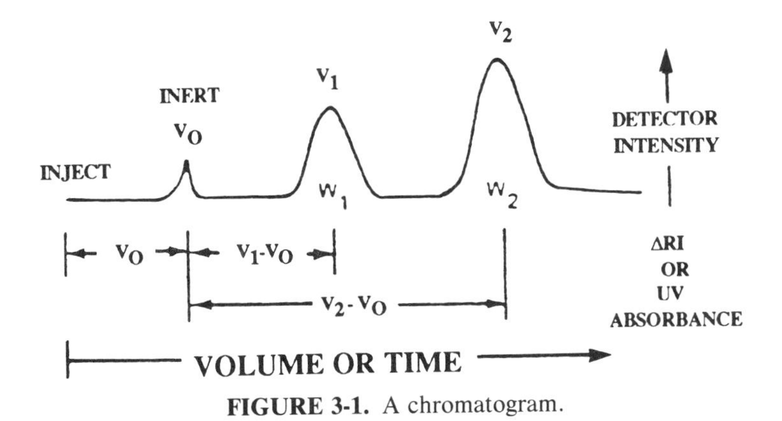

FIGURE 3-1. A chromatogram.

volume" of a retained species is always larger than the void volume. The retention volume is specific for each sample component under a given mobile phase, column, and operating conditions.

Components 1 and 2 in Figure 3-1 have elution volumes of V_1 and V_2, respectively. Because an HPLC uses a constant flow rate, the retention (elution) volumes are often measured in units of time and are, therefore, called the retention times or elution times of a peak. The retention time of a peak in a chromatogram is unique for a particular compound in a given mobile phase. For this reason, qualitative analysis is performed by matching the retention time of a peak from a known, standard compound to that of a peak in the mixture.

When the chromatograph is operating correctly, quantitative analysis may be accomplished because there is a linear relationship of the concentration of the compound to its peak height (and area) in the chromatogram. A "calibration curve" is constructed by injecting a standard compound at known concentrations into the HPLC, measuring the peak height (or area) response, and plotting peak height (or area) versus amount injected. The concentration of compound in the sample is determined by measuring the peak height of the component in the chromatogram of the sample and comparing that value to the calibration curve to determine the amount of the compound present in the sample.

When using the chromatogram for quantitative and/or qualitative analysis there are three important concerns. The first is that the LC system must be operating properly. Second, the HPLC must have appropriate detectors to "see" the compounds. Lastly, there must be a separation with appropriate spacing of the peaks so that the peak height (or the peak area) can be conveniently and accurately measured. This third issue implies that there must be retention of the compounds of interest.

A PROPERLY OPERATING SYSTEM

In spite of what some may say, running an HPLC is knowledge intensive. An HPLC does not run itself, and a computer does not make it run better. The output of the chromatograph is only as good as the knowledge, control, and interaction of the operator. The operator must understand how the system works and how the transducers and other "readouts" are helpful in determining that the hardware is properly functioning. The operator needs to study the flow path of the system and identify the strengths and potential weaknesses of each junction, manifold, pump head, and valve. This information is available from the manufacturer of the equipment in the form of manuals, service notes, and so on.

Figure 3-2 shows a basic HPLC component system in more detail than shown in Figure 1-1. Your HPLC system may have more components than those shown in Figure 3-2, but it is sufficient to discuss the interplay and

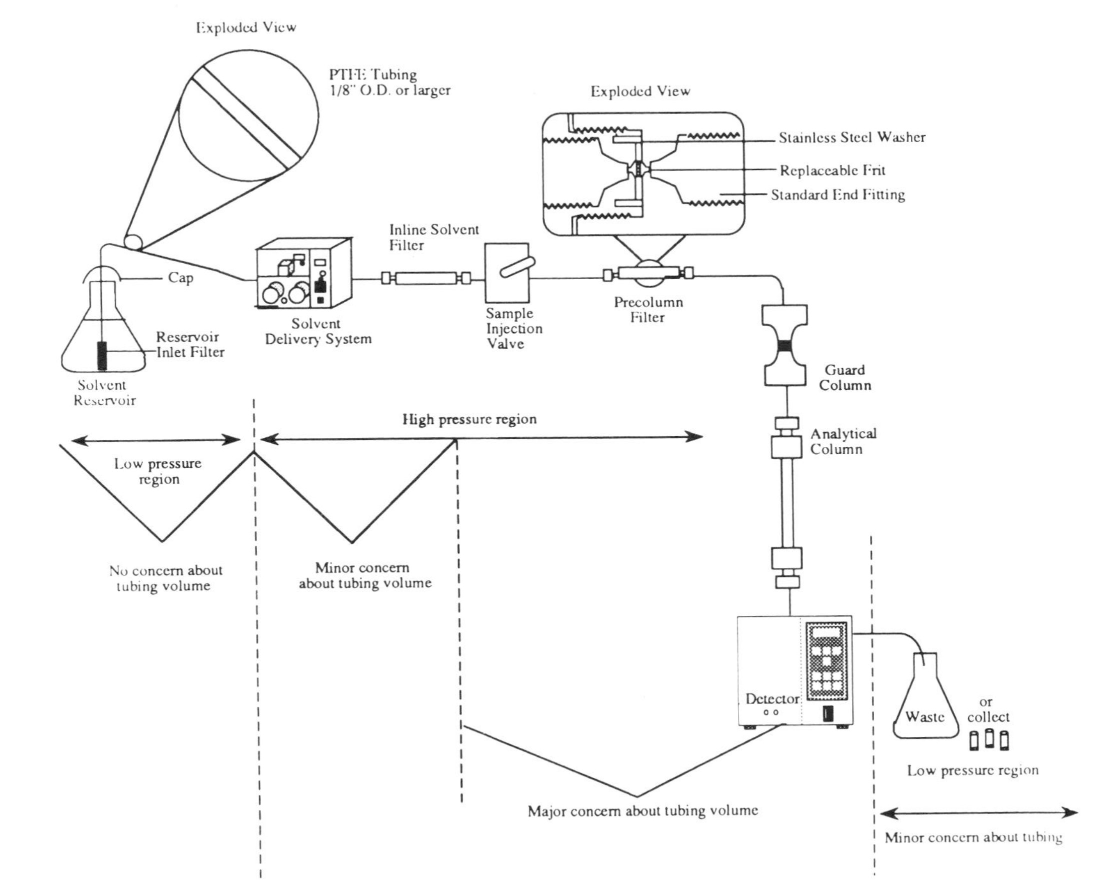

FIGURE 3-2. A typical isocratic HPLC. Plumbing considerations involve understanding the critical and noncritical bandspreading regions.

interdependence of the modules that comprise an "HPLC system." Note that an HPLC is a sophisticated plumbing device and understanding the transport of fluid in terms of mixing and smoothness of flow is important in constructing a properly operating system.

Reservoir

In the basic system the mobile phase is contained in a "reservoir." This reservoir may or may not be pressurized (to approximately 3–6 psi). The mobile phase should be prefiltered before being placed into this reservoir, yet a filter is usually placed in the reservoir to be a "sinker" to hold the inlet at the bottom of the reservoir and to remove any accidental contamination of coarse particulate matter (20–30 μm) before it reaches the solvent delivery system (the pump). Since this is a "low-pressure line," Teflon® tubing ($\frac{1}{8}$ in. OD or larger) is usually used to ensure that sufficient volume of mobile phase reaches the solvent delivery system. Addition considerations for mobile phase preparation are given in Chapter 6.

Solvent Delivery System

The solvent delivery system is necessary to provide a constant (precise), pulse-free flow of mobile phase through the HPLC. Flow rate should be accurate because it influences the reproducibility of retention times and peak heights (or peak area) in the chromatogram. However, requirements for precision and accuracy of the flow rate may vary depending upon the application. Flow rate is usually expressed in milliliters per minute. High-pressure capability is needed to overcome the resistance to fluid flow arising from the tightly packed particles (stationary phase) in the column. Therefore, the pumping system should be able to deliver a wide range of reproducible flow rates against back pressures in the system of up to 6000 psi (408 atm). The solvent delivery system should have a low internal volume to facilitate rapid mobile phase changeover with a minimum of mobile phase volume. Frequent mobile phase changes are often necessary to optimize a separation and the operator gains valuable operating time if a mobile phase change can be accomplished quickly. This low internal volume criteria is important also when a gradient HPLC is required. This is discussed in Chapter 7.

There are many types of solvent delivery systems to choose from and the choice of the solvent delivery system should be based upon the requirements of the application (see Fig. 1-6) and budgetary constraints. Some solvent delivery systems can even compensate for the fact that at high pressures many solvents are compressible and as such could have a volumetric change of as much as 6%. However, this feature might not be important for an individual application. Solvent delivery systems should be mechanically reliable, and should be made of materials that are not damaged by the mobile phases used. Leaks should be easily detectable and drainage should be such that the leakage will not damage other components. Since seals will wear, they should be easy to replace. Filters are often contained within a solvent delivery system to prevent small particles from lodging within other components. These filters should be kept clean during operation. The pressure output from the solvent delivery system is usually conveniently accessed and monitoring it is a good way to determine if leaks or clogged filters are present. This technique is discussed in the diagnostics section of this chapter. A gradient HPLC has additional operational considerations in the solvent delivery system(s) which are discussed in Chapter 7.

The mobile phase exits the solvent delivery system under high pressure (300–6000 psi) and should pass through an "in-line filter" (approximately 2 μm) so that any debris resulting from seal wear is removed prior to entering the injector. If the debris is small enough and is not removed, it could pass through the injector and deposit on the inlet of the column. An in-line filter will catch and remove these particles before they can cause a plugging problem later. However, the pressure will rise if the inlet filter becomes clogged. (Refer to the discussion on diagnostic monitoring later in this chapter.)

Injector

The function of the injector is to place the sample into the high-pressure flow in as narrow a volume as possible so that the sample enters the column as a homogeneous, low-volume "plug". To minimize spreading of the injected volume during transport to the column, the shortest possible length of tubing should be used from the injector to the column. Remember, the "shortest length" is not the operative term, and the phrase "shortest possible length" implies that up to a certain length of tubing no deleterious effect on band-spreading of the sample volume will occur.

There are two main types of injectors—the fixed loop and the variable volume loop. The fixed loop has the advantage of a high precision of injection volume but the loop must be overfilled by several times its volume, that is, a 10-μL loop requires 30–50 μL to ensure that it is filled completely with sample. The fixed loop also must be "washed" with several times its volume of mobile phase before another sample is loaded into the loop. The fixed-loop injector, while having few moving parts, does have rotating surfaces which will "shear" whatever material is used to seal the individual ports at their junctions. This can contribute debris to the flowing mobile phase. Also, the act of "injecting" involves an interruption of flow that may result in a pressure surge to the column. If a new sample size is desired, another loop must be added. Fixed injectors are often used in QC or routine operation where the sample size is defined and will not change.

The variable volume injector, as its name implies, can accommodate volumes from 1 μL to 2 mL. This has an advantage in that for limited sample sizes and for large preparative work, the same injector may be used with no mechanical changing of loops. As with all injections, the variable volume injector requires a syringe to place all of the sample into the loop but, in this case, the precision of the injected sample is that of the operator using the syringe. One manufacturer of the variable loop design uses valves, not rotating flat surfaces, which minimizes the wear and particles resulting from "shearing." Also, this valve maintains continuous flow at all times so that no pressure increases surge into the column. Variable volume injectors are preferred in research where sample injection sizes vary from one application to another.

Choosing an injector is determined by the application and the preference of the individual. Each injector design has been successful for specific applications and maintenance/operational preferences. A discussion of how to use large injection volumes to your advantage is given in Chapter 6.

Precolumn Filter and Guard Column

Using a filter to remove particles from the sample prior to placing it into the injector is a good habit if the nature of the sample allows it. However, after the injector, it maybe useful to install a "precolumn filter" and/or a "guard column." The precolumn filter (approximately 2 μm) is designed to remove particles (debris) coming from the injector before they reach the guard column or analytical column. In the precolumn filter the frit (filter) is easily replaced without disturbing the guard column or column. The precolumn filter is designed as an efficient filter with no mixing that would cause spreading of the sample volume prior to entering the column. The guard column is, in essence, a disposable (or sacrificial) top of the main analytical column. The guard column is the "final filter," both mechanical and chemical. In addition to removing debris, it also can adsorb undesirable sample components that otherwise might irreversibly bind, "coat," and possibly change the stationary phase of the analytical column. Guard columns are used primarily to protect expensive analytical and preparative columns. Although economy alone is a persuasive argument for the use of guard columns, the need for a long stable life of the analytical column to obtain reliable and reproducible results is perhaps even more important. The guard column should not significantly change the efficiency of the analytical column. Guard columns are designed with relatively small total volumes and minimal dead volumes, so that they do not cause band broadening. The ratio of guard column to analytical column volume should be such that adding a guard column decreases system efficiency by no more than 5–10%. Generally, a column volume ratio of 1:15 to 1:25 is satisfactory. Although larger guard columns would have the capacity to trap impurities from more sample injections, excessive band broadening often results.

Guard columns may be filled either with pellicular material (see Fig. 3-13) of the same bonded phase as the analytical column or with the identical packing material as in the analytical column. Pellicular packings (35–40 μm) are much more dense than porous microparticles, and thus are easily packed dry using the tap-fill method. The capacity of pellicular packings is lower than that of microparticles and this can sometimes be an advantage when using them as guard packings because of their low cost and because in certain situations the very low surface area (10 m^2/gm) can have an insignificant adverse effect on bandspreading. Microparticulate packings (less than 10 μm) have the advantage of increased capacity and efficiency, but require special equipment for slurry packing. Often, they can be purchased in the form of prepacked disposable inserts for use in a holder module. In either case, the guard column should be easy to replace or repack.

When developing a method employing a guard column, it is important to keep in mind that guard columns have a finite capacity for impurities. The number of sample injections or, in some cases, a change in back pressure can indicate when a change of guard column is required (see the section on Diagnostics in this chapter). The primary consideration is that you know, or estimate, the capacity of the guard column for the compound(s) to be collected so that undesired species are not eluted onto the analytical column. Deterioration of peak shape or resolution usually means that a guard column change is overdue.

Guard columns and in-line filters alone will not ensure long life for the analytical column. They are designed to perform a specific function for a finite period of time. They, too, will be short-lived if shortcuts are taken in sample preparation and mobile phase quality control. Use of a guard column does not mean that other good analytical practices can be neglected. Shortcuts mean shorter life for *any* column.

General Considerations in System Configuration

The width of a peak observed in a chromatogram is the result of broadening (or bandspreading) that takes place both within and outside of the column. Bandspreading that occurs outside of the column, for example, in column-connecting tubes, is called extracolumn bandspreading and can reduce the apparent system efficiency that the column alone would be capable of exhibiting. (Column efficiency is defined and discussed later in this chapter.) Therefore, it is desirable to reduce extracolumn bandspreading by using connecting tubes with small diameters, such as 0.005 or 0.010 in. instead of 0.020 in. However, as tubing size decreases below 0.010 in., plugging becomes more probable. Bandspreading in connecting tubing can also be reduced by using short lengths. The challenge for the chromatographer is determining how short the length should be. This is determined by the application because the magnitude of loss in efficiency resulting from the bandspreading in connecting tubing is relative to the total bandspreading in the system—injector, column, and detector(s). The chromatograms in Figures 3-3 and 3-4 illustrate this point. Note in these figures the various efficiency losses due to increased tubing might be considered by some to be quite

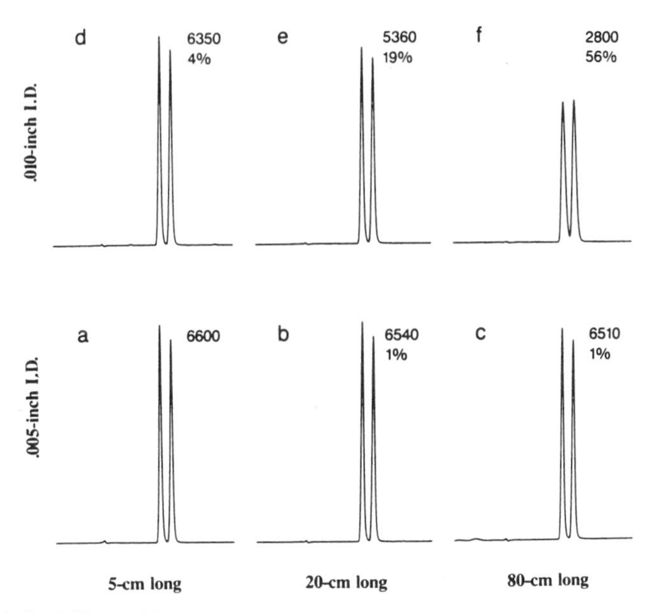

FIGURE 3-3. Effect of bandspreading upon performance. Column: C_{18} packing (5 μm), 4.6 mm ID × 100 mm; mobile phase of 80% acetonitrile in water; 1 mL/min flow rate; 10-μL sample. Loss of resolution is caused by substituting larger injector-to-column connecting tubes. The control (a) uses a 0.005 in. (0.13 mm) ID × 5 cm connecting tube. Chromatograms b and c show the effect of longer tubes. Chromatograms d–f show the effect of larger diameter and longer tubes. The loss of resolution is evidenced by increased peak width (observed indirectly by noting the decrease in the peak height) and increased height of the valley between the peaks. Next to each chromatogram is the plate number observed for the first peak and the percent reduction from the control due to the substituted tubes. (Reproduced with permission of Rheodyne.)

tolerable. Therefore, as stated earlier, while connecting tubing broadens peaks, tubing must be used; so use connecting tubing appropriately.

Bandspreading can also occur in a "cavity"—a place where an excessive amount of volume should not be. Examples of cavities are shown in Figure 3-5. By making proper tubing connections, these bandspreading contributions can be eliminated. Often the situation as shown in Figure 3-5*a* arises from mismatching fittings and ferrules from one manufacturer into the body of another manufacturer. Presently, there is no "standardization" of HPLC fittings, ferrules, and tubing and, as shown in Figure 3-6, the various manufacturers have different designs. Therefore, careful attention to detail is necessary to insure that everything fits properly. Labeling fittings, tubing, and bodies is helpful for keeping track of what is compatible.

Each component in an HPLC (Fig. 3-2) impacts several performance parameters, as shown in Table 3-1. The plus signs in the table denote impact—not necessarily a positive impact. A double plus sign means that the

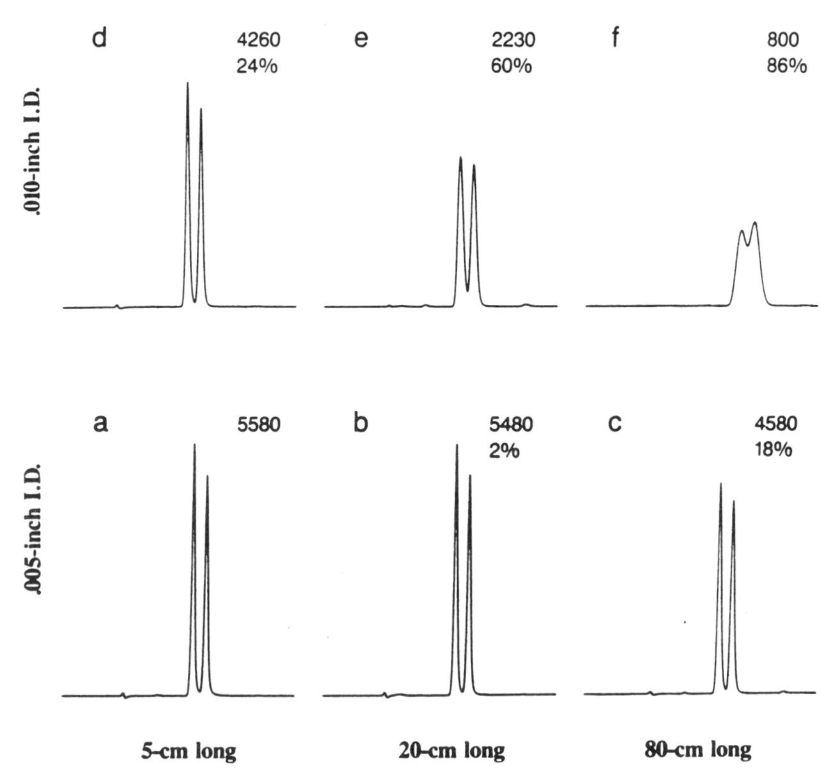

FIGURE 3-4. Effect of bandspreading upon performance. Same as Figure 3-3, except with a column and system having much lower dispersion: 2 mm ID × 100 mm column; 200 μL/min flow rate; and a 2 μL sample. (Reproduced with permission of Rheodyne.)

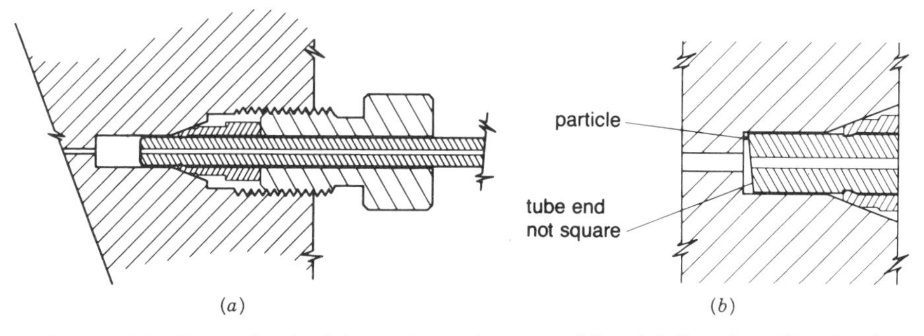

FIGURE 3-5. Example of mixing volume due to cavities. (*a*) Creation of cavity due to mismatched fitting. The gap results from inserting, into a deep port, a tube that was originally made up in a shallow port. The scale drawing shows a 1/16 in. OD × 0.0010 in. ID tube and a 0.090 in. gap, creating a 4.5-μL cavity. (*b*) Creation of cavity due to trapped particle and tube end which is not cut square. This represents a 1/16-in. tube, a bore of 0.010 in. ID, a port bore of 0.013 in., and a cavity volume of about 0.6 μL. (Reproduced with permission of Rheodyne.)

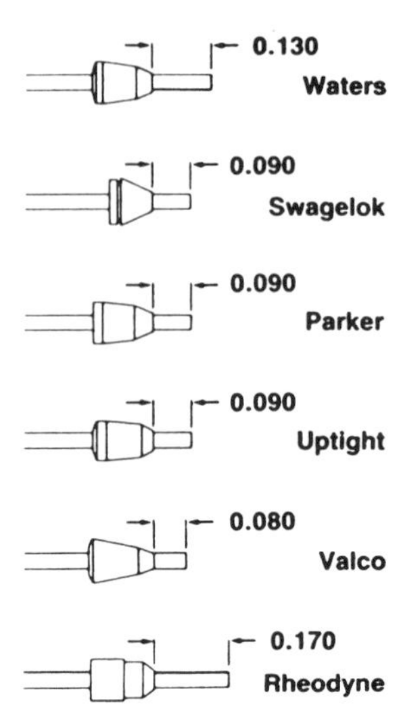

FIGURE 3-6. Tubing and ferrules. Examples of different lengths of tubing that extend past the ferrule after different manufacturers' fittings are assembled. Tubing extensions, shown in inches, are experimental values, not necessarily manufacturers' specifications. (Reprinted from reference 1 with permission.)

component has a major influence on that particular parameter. Another way of interpreting this table is that if the operator is having trouble controlling a parameter, several components could be at fault. For instance, if retention time reproducibility was not where the operator desired it, this problem could be caused by one or more of the following: mobile phase, solvent delivery system, gradient former, column, integrator, or temperature controller. Being able to decide the most probable culprit is an activity called "troubleshooting." Effective troubleshooting involves identifying the most probable causes of a problem and doing appropriate experimentation to isolate and confirm the cause. Table 3-1 should be helpful in this process.

Diagnostics

System pressure is the most informative diagnostic to monitor to ensure that the HPLC is working properly. The "real-time" pressure readout (often referred to as the "pressure trace") should be monitored at all times since anything that influences flow will be reflected in the system pressure. As valves are switched, shifts and changes in pressure occur. As flow rate changes, pressure will change. Short-term cyclic pressure fluctuations are indicative of flow pulsations. The amplitude of these pulsations is proportional to the accuracy of flow rate delivery and, hence, influences the accuracy of retention times and peak heights. An example of pressure trace and the various types of diagnostic information that can be obtained by looking

TABLE 3-1. Relationship Between Components in an HPLC and Its Influence on Performance Parameters[a]

	Efficiency	Separation Factor (α)	Specificity (Detector)	Quantitative Reproducibility	Retention Time Reproducibility	Sensitivity	Linearity	Dynamic Range	Throughput (Preparative)	Speed	Baseline Stability	Total
Mobile phase	+	++			++						+	6
Solvent delivery system	+			+	++	+			++	++	++	11
Gradient former		++		+	++					++	++	9
Injector	+			++			++		+			6
Column	++	++		+	++	+			++	++		12
Detector	+		++	++		++	++	++	+	+	+	14
Integrator	+			++	+	++	+	+		+	+	10
Temperature controller	+	+			+							3
Total	8	7	2	9	10	6	5	3	6	8	7	

[a] The + and ++ denote relative impact. Absence of a sign implies that there is no significant effect.

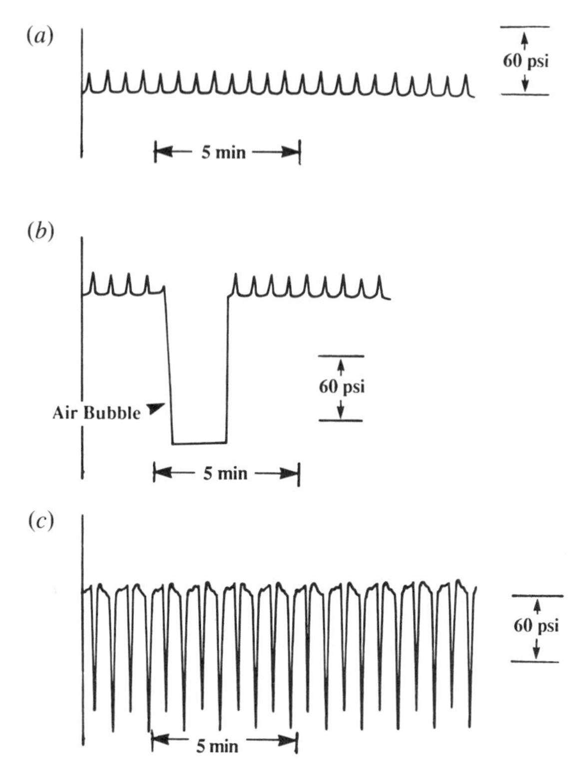

FIGURE 3-7. A pressure trace as a diagnostic aid. (*a*) A properly performing solvent delivery system. (*b*) A pressure dip probably due to a bubble being "pumped." (*c*) An improperly "pulsing" solvent delivery system which probably has bad check valves. Note that the relative pulsation for a "good" and "bad" solvent delivery system must be determined for each system. Each solvent delivery system will be somewhat different.

at the pressure trace are shown in Figure 3-7. Each LC system is different, so it is necessary to "know your pressure trace" and what it means in terms of instrument problems and diagnostic alerts.

As can be seen from Figure 3-7, a pressure readout/trace which is occurring in real time is very valuable to verify proper operation, or to spot problems that otherwise might not be identified by looking at the chromatogram. If a real-time pressure readout is not available the operator is at a big disadvantage. The emphasis should be on a "real time" readout which is not "smoothed" in any digital or electronic fashion. The pressure readout is indeed the chromatographer's most valuable diagnostic tool and a constant reading is indicative of good control of flow rate.

Precise flow control is essential for correct qualitative and quantitative analysis; and since the chromatogram is measured from time of injection to the time of the peak apex occurrence, the repeatability of this value needs to be determined to verify that the retention time of an unknown peak occurs at the same time as that of a known standard. This reproducibility information is crucial in quantitative analysis. Flow control is also important because the detectors in common use in LC are concentration dependent, and flow-rate variations will cause changes in both peak height and peak area measurements as well as in peak shape. Therefore, the flow rate should always be measured independently of the chromatogram as part of good laboratory practice. One way to do this measurement is to collect the flow output into a calibrated burette for a specified time. The volume collected divided by the time will result in the flow rate in units of milliliters per minute. At a flow of 1 mL/min, an appropriate time of recollection is 5 min. Another approach is to collect the flow for a specific time, weigh the amount collected, and convert to volume per unit time.

After verifying that the LC is delivering a constant flow, the repeatability of the injector must be determined since another source of error may arise in the sample injection. The precision of the peak area and/or peak height should be determined for the standards and should be about ±1%. Additionally, symmetrical peak shapes are desired and generally obtained in LC. If the peaks in the chromatogram are "fronted" or "tailed," it is an indication that a chemical or physical problem exists in the system and the operator should not use this type of chromatogram. Instead, the system should be adjusted so that the chromatogram resembles symmetrical peak shapes. Such adjustments may require that the operator examine the hardware and column in order to troubleshoot and identify the specific problem to be fixed. The operator may need to rely on instrument and/or column manufacturers for help in troubleshooting and correcting improperly operating systems.

"SEEING" THE COMPOUNDS

Having a "correct" chromatogram means using an appropriate detector which can "see" the compounds in the sample. The chromatogram must reflect the content of the sample for successful use of the HPLC. The two most common detectors in LC are the ultraviolet/visible spectrophotometer (UV/Vis) and the differential refractive index (RI) detector. The UV photometric detector at a fixed wavelength of 254 nm is commonly used because it is often cheaper than a full-capacity spectrophotometer, and is usually applicable since UV-absorbing compounds often have some absorbance at 254 nm even if they do not have a maximum absorbance there. When chromatography is to be carried out on non-UV absorbers, a differential RI detector is the popular choice. This detector measures all compounds in the eluent that differ in refractive index from the eluting solvent. Since all compounds have a refractive index, the RI detector is sometimes called the "universal detector" because it has the potential to detect all compounds. However, the RI detector is not always useful since the difference in refractive index between a sample and the mobile phase might be too small to sense.

Wide Scope vs. High Sensitivity

The choice of detectors in LC is often a trade-off between wide scope and high sensitivity. For instance, the refractometer is readily available and easy to operate; it can detect most compounds (wide scope), but it often has a lack of sensitivity for many compounds. A variable-wavelength UV detector offers a good choice for solutes that have some UV absorbance capability. Absorption at a specific wavelength results in a more selective and sensitive detection mode than a refractometer, but only for UV-absorbing compounds. In other specific cases the fluorescence or electrochemical detector can be used. These have a high sensitivity for individual compounds but are also limited in the number and type of compounds they can detect (narrow scope).

Unless a sample is reasonably well defined, there is a chance that solutes will not be detected, and a combination of detectors may be necessary. Dual detection almost always provides more than double the information obtained with either detector. For research purposes, the most valuable information to aid in interpreting separations data can be obtained from both the UV and RI detectors in series. The RI response correlates more closely with concentration while the UV response is selective and reflects both concentration and the molar absorptivity of the detected solute. A very large UV peak, therefore, may indicate an exceptionally good UV absorber or a high concentration of low or intermediate molar absorptivity material or both.

The valuable information about the nature of a sample obtained with dual detectors can be discussed in relationship to the chromatograms shown in Figure 3-8, from which several observations can be made. First, a UV peak (peak A) with no corresponding RI peak or a much smaller one is caused by either one or both of the following conditions:

1. A very strong UV absorber is present in minute quantities. Since the maximum sensitivity of the UV detector is approximately 15 times greater than that of the RI detector, a trace of a powerful UV absorber can cause a significant peak with the UV detector but be invisible or just barely visible with the RI detector.
2. An eluting compound has a refractive index similar to that of the mobile phase. Since the RI detector is only sensitive to the difference in refractive indices, compounds with refractive indices similar to that of the solvent do not show up as a peak.

The second observation is that peak B appears to be a good UV absorber and is not present in a very large amount since the RI response is small or it has an RI value similar to that of the eluent. Peak C has an RI peak without a

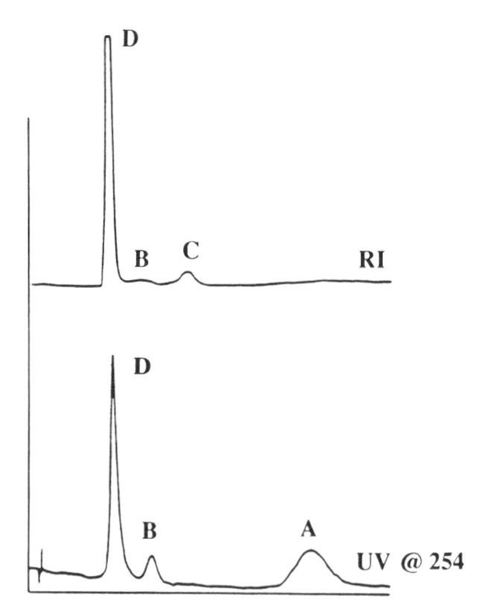

FIGURE 3-8. Comparison of response traces from a UV photometer (254 nm) and an RI detector.

corresponding UV peak which indicates that the eluting compound is not a UV absorber and must lack the functional group(s) that cause UV absorption at the wavelength used.

The third observation is that peak D has an RI peak much larger than the UV peak which is caused by:

1. A weak UV absorber, or
2. Two unresolved compounds. One is at a trace level and is a strong UV absorber and the other is a non-UV absorber present in much higher concentration.

The combination of UV and RI detection can be useful in providing information about unseparated pairs.

Detecting the compound may also be impeded by not correctly operating the detector. Often the operator's manual is helpful in ensuring that the detector is truly "seeing" the sample. For instance, does the detector have the proper light source intensity (is it on)? Is the detector set to the appropriate wavelength for detecting the compound(s)? Are you operating in the linear range so that the response is directly proportional to concentration? A linearity plot alerts the operator to concentrations at which a constant detector calibration factor can be used (2). Is the noise level of the baseline within the specifications? Only if the operating specifications are verified before work is begun can it be insured that the results will represent reality.

Time Constant

A common adjustment made to minimize baseline noise is to increase the time constant of the detector. This may be done with a switch on the detector, an input to the data system, or it may be preset in the detector, data system, or computer. In all cases the time constant will influence the response of the detector. To understand this impact it is necessary to remember the definition of the response time of the detector. Response time is the time required, after a stepwise change in the composition in the detector cell, for the output signal to reach a predetermined percentage (often 63%) of the new equilibrium value. If the response time of the detector is not referenced, it is probably measured at the 63% level. If it is not referenced to the 63% value, it will be stated at what percentage the time constant was measured. Thus, if the time constant is 1 sec, it will take 1 sec to attain 63% of the value of the new signal. When a time constant is used, the output signal that represents what is in the detector cell lags the true (equilibrium) signal. In certain situations this could result in a misrepresentation of reality.

Detector response should be faster than the rate of change of solute concentration in the flow cell to avoid loss of information and distortion of peaks. An example of the effect of response time upon peak distortion is shown in Figure 3-9. In this case the response time was changed by adjusting

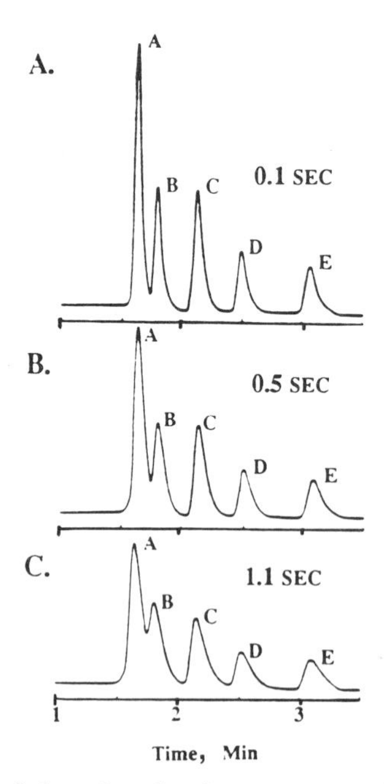

FIGURE 3-9. The effect of changing the time constant of the detector on the peak shape. Time constants of A = 0.1 sec, B = 0.5 sec, and C = 1.1 sec are denoted on the chromatogram.

the time constant of the detector. The time constant on a detector was intended to filter or "smooth out" background noise. From Figure 3-9 we see that although it is nice to have noise-free chromatograms, it is more important to have an accurate representation of peak profile. For this reason, older detectors that require large time constants for a low noise signal may not be suitable for use with some HPLC analyses which have fast separation times. All chromatograms should be run with no time constant (or lowest possible setting), and the time constant should be adjusted so that the peak response it not affected.

THE SEPARATION

Three things must occur in order to obtain an effective separation: (1) Sample components must be retained on the column; (2) The sample components must be segmented from one another; and (3) The sample components must have relatively narrow bandspreading (peaks). (For a more detailed discussion see pages 325–328.) Without these essential conditions resulting from physical and chemical interactions, the separation will not occur within a reasonable operating time. Therefore, attaining the separation depends heavily upon choosing an appropriate column and mobile phase. In describing the conditions that must occur on the column, chromatographers use specific terms which are part of the nomenclature or "jargon" of chromatography.

Capacity Factor

Once the appropriate detector is chosen and the hardware system is operating correctly, the next ingredient for a successful chromatogram is retention of the components in the mixture. The measure of the retention of a compound on a column is referred to as the capacity factor and is a measure of retardation of the compound in terms of the number of column void volumes it takes to elute the apex (center) of the peak. This measure is called k' (pronounced kay prime) and is simply the ratio of the elution volume of the component (V_l) to the void volume of the column (V_0), which is expressed as

$$k'_1 = \frac{V_1 - V_0}{V_0} \tag{3-1}$$

Obviously, to increase retention, the sample must be in an environment where it prefers the stationary phase. This is accomplished by changing the polarity of the mobile phase. But too much retention wastes analysis time. The optimum value of k' for the best separations is in the range of 2 to 10. Control of retention is discussed in this chapter and in Chapters 4 and 5.

Selectivity

Since the "name of the game" of LC is the separation of two of more components, the *relative capacity factors* of two components become a measure of the column's ability to discriminate between them. This ability to discriminate between the two components and selectively retard them is referred to as the "selectivity" of the column. The selectivity is expressed in terms of a "separation factor" (or alpha) which is defined as

$$\alpha = \frac{k'_2}{k'_1} = \frac{V_2 - V_0}{V_1 - V_0} \qquad (3\text{-}2)$$

If the separation factor is unity, the peaks coincide, and no separation has occurred. If the separation factor is 1.3, the column selectively retards one component 30% more than the other. The larger the α value, the easier the HPLC separation is to achieve; however, an α value of 1.1–1.4 is typically desired. As we show later in this chapter, because resolution is influenced by three factors, separations can be attained for α values smaller than 1.1. In some modes of modern HPLC, meaningful resolution can be achieved with α values as low as 1.05, which means that the column retards the second component only 5% more than it retards the first component.

Efficiency

Another key feature of attaining a separation is the width of the peaks at their baselines (see Fig. 3-1). Wider peaks indicate bandspreading and generally occur at longer residence times on the column (higher k' values); however, other column conditions can also influence this characteristic. The rate of bandspreading (also referred to as zone spreading) can be measured and is referred to as the efficiency of the column. When bandspreading is small, a column is said to have high efficiency. For fast separations, narrow baseline widths (minimum bandspreading) are certainly desirable, particularly when α is small.

Martin and Synge (3) introduced the important concept of *theoretical plates* into chromatography. Their concept was derived from partition theory and random statistics, and was related to similar ideas developed for extraction and fractional distillation. They supposed that the column could be divided into a number of sections called theoretical plates, and that solutes (dissolved compounds) could be expected to achieve equilibrium between the two phases (mobile and stationary) that exist within each plate. The chromatographic process, like an extraction process, can be visualized to occur when mobile phase (solvent) is transferred to the next plate, where a new equilibrium is established. Theoretical plate numbers of 1000 or more are common for HPLC columns, which means that 1000 separate equilibria must be established to obtain the same degree of separation by solvent extraction procedures. Because sample zones in HPLC are nearly Gaussian in shape, they could consequently be defined by the equation that describes the standard error function (see Chapter 6, Figs. 6-3 and 6-4). As a result, the measure of band width for a chromatographic peak is the same as the standard deviation of the distribution, σ. Therefore, the "height equivalent to a theoretical plate" (HETP) (also plate height H), and plate number N, were defined by Martin and Synge in terms of column length and band width as

$$H = \frac{\sigma^2}{L} \tag{3-3}$$

$$N = \frac{L^2}{\sigma^2} \tag{3-4}$$

where σ (band width) and L (length) must be expressed in the same units. Note that these equations do not depend upon a chromatogram for calculation, but allow column efficiency to be determined at any time, provided the solute zones can be seen on the column. The more common form of equation (3-4) takes into account that zone width W is approximately 4σ (see Fig. 3-1), so that the equation for theoretical plates becomes

$$N = \frac{L^2}{(W/4)^2} = 16\left(\frac{L}{W}\right)^2 \tag{3-5}$$

When N is calculated from a chromatogram, L and W must be converted to time-based units, terms cancel, and the theoretical plate equation is simply

$$N = 16\left(\frac{t_R}{W}\right)^2 \tag{3-6}$$

where W is now peak width at the baseline of the chromatogram in the same units as t_R. The width, W, is determined by drawing lines tangent to the front and back of the peak and determining the distance between the points where the two tangent lines intersect the baseline, as shown in Figure 3-10.

Plate height H is usually calculated from N according to the equation

$$H = \frac{L}{N} \tag{3-7}$$

and is useful when comparing the efficiencies of two columns of different length. Chapter 6 contains an additional discussion of plate height calculations and implications for the practicing chromatographer.

If molecules did not spread apart as they travel through the column, peaks would be narrow and resolution would be easy to accomplish. However,

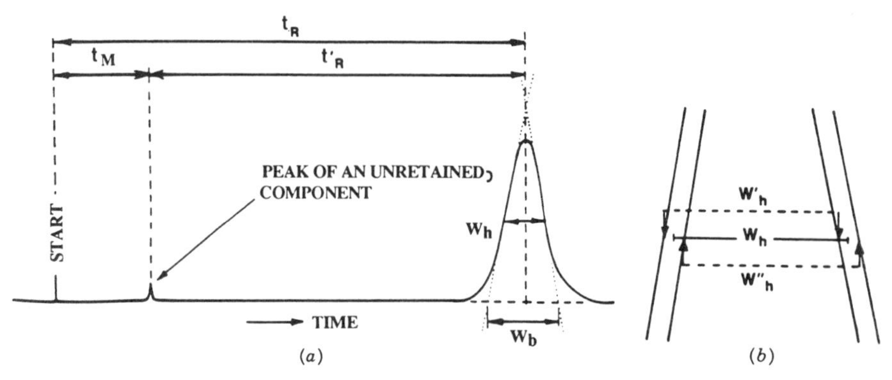

FIGURE 3-10. Determination of the column efficiency. (*A*) Use of the chromatogram. (*B*) The procedure used to measure the peak width, at W_h, to account for the thickness of the pen. Use the average of the two determinations, W_h' and W_h'', to determine W_h.

bandspreading cannot be eliminated, only minimized. As expressed by Martin and Synge, several major factors contribute to chromatographic bandspreading:

> The height equivalent to a theoretical plate depends upon the factors controlling diffusion and upon the rate of flow of the liquid. There is an optimum rate of flow in any given case, since diffusion from plate to plate becomes relatively more important the slower the flow of liquid and tends, of course, always to increase the HETP. Apart from this, the HETP is proportional to the rate of flow of liquid and to the square of the particle diameter. Thus the smallest HETP should be obtainable by using very small particles and a high pressure difference across the length of the column. The HETP depends also on the diffusibility of the solute in the solvent employed, and in the case of large molecules, such as proteins, this will result in serious decrease in efficiency as compared with solutes of molecular weights on the order of hundreds. (3)

Contributions to the theory of zone spreading were also made by Lapidus (4), Van Deemter (5), and Giddings (6).

The contributions to zone spreading can be expressed with a simple relationship often referred to simply as the Van Deemter equation:

$$H = A + \frac{B}{\mu} + C\mu \tag{3-8}$$

where H is the plate height and μ is the average mobile-phase linear velocity. Remember that the smallest possible H means maximum efficiency and minimum zone spreading. The *flow inequality term, A,* arises from flow velocity differences (flow splitting) within the packed bed. The A term is also referred to as the *eddy diffusion term* because the flow splitting around a particle is analogous to eddy streams around rocks in a brook. The A term can be very large for poorly packed columns and/or large particles. However, it can be minimized by using good packing techniques and selecting particles that have a small diameter and a narrow particle size distribution. The *longitudinal diffusion term, B,* arises because all solutes tend to travel from a region of high concentration to one of lower concentration. A nonchromatographic example of this is the dissolution of a sugar cube in a glass of water without any stirring. In chromatography, spreading by longitudinal diffusion can occur in both the mobile and stationary phases. In liquids, the contribution to H from this term is small except at very low flow velocities which are seldom encountered in practice. The *mass transfer or lateral diffusion term, C,* arises from slow diffusion of solute molecules into and out of the particles. The contribution to H from C can be minimized by using small-diameter, porous particles or particles having a thin porous shell (pellicular) and working at moderate flow velocities of about 1 cm/sec. When the B and C terms are minimized, several authors have shown that the theoretical limit of H is $2d_p$, where d_p is average particle diameter in the packed column. (Therefore, $2d_p$ is the minimum value of the A term.) The individual contributions to band broadening and the chromatographic results are shown in Figure 3-11.

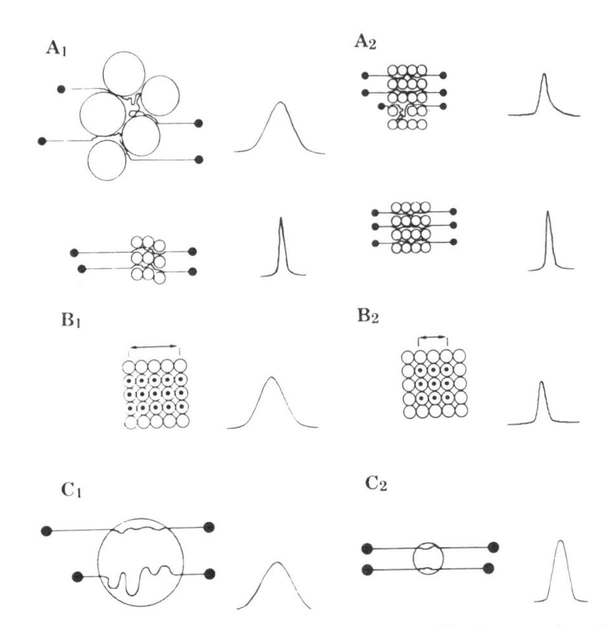

FIGURE 3-11. Schematic representation of the contributions to bandspreading and the chromatographic result. Contribution of flow inequality (A term): A_1 is the effect of particle size and A_2 is the effect of packing nonuniformity. Contribution of diffusion (B term): B_1 is at very slow flow and B_2 is at typical flow. Contribution of mass transfer (C term): C_1 is in a large particle and C_2 is in a small particle.

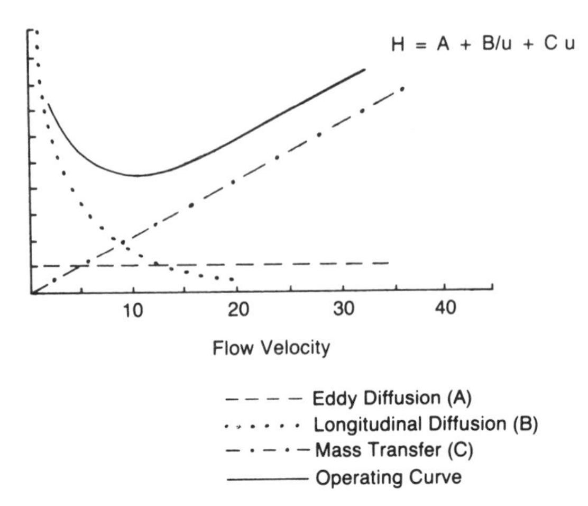

FIGURE 3-12. A van Deemter plot.

The net, or total, contribution to bandspreading is a summation of all of the terms and can best be described by looking at a plot of H versus the linear velocity μ, as shown in Figure 3-12. This plot is referred to as a *van Deemter plot* and the solid line in Figure 3-12 is a typical shape. The dotted lines are the individual contributions which are summed to give the operating curve. Van Deemter plots generally show a large or steep slope for large, porous particles and for pellicular particles; small or flat slopes are typical for small, porous particles. A flat van Deemter curve means that linear velocity can be increased to shorten analysis time without losing significant column efficiency and resolution. Thus, the practical use of LC columns may not be at the highest efficiency values (lowest point in the valley in Fig. 3-12). If an adequate separation is obtained using higher flow rates (linear velocities), plates may be "wasted" in order to save time. This is discussed further in Chapter 6.

When segmented by their efficiency contributions, there are three basic particle types of packing material shown in Figure 3-13. Each has its own advantages and disadvantages. Fully porous particles (Fig. 3-13, far right side) are usually in the range of 37 to 50 μ and are similar to materials used in TLC. They are relatively inefficient (large C term) and have very large surface areas, typically 300 m^2/g and ranging as high as 500 m^2/gm. Their primary application is preparative-scale separations, where the large surface areas make possible very large sample loads. In preparative work, the efficiencies are less important than in analytical work. Also, these particles are relatively inexpensive and have a low resistance to flow (low back pressure).

Pellicular particles (Fig. 3-13, center) are commonly available in the 37 to 50-μm size range. The name "pellicular" implies that these particles are composed of a solid core which is coated with a porous layer. These particles have an advantage in that mass transfer is much faster than for porous

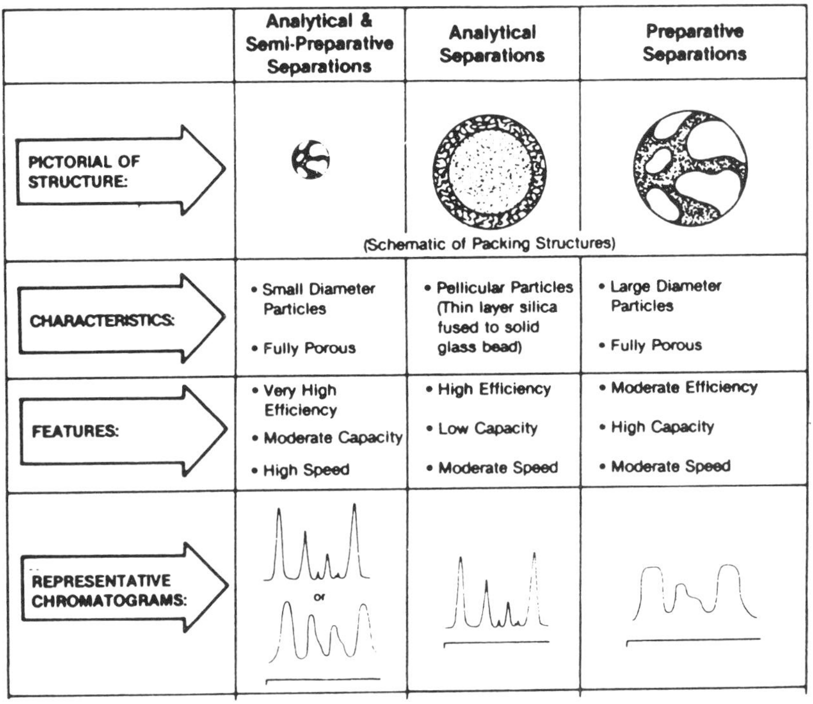

	Analytical & Semi-Preparative Separations	Analytical Separations	Preparative Separations
PICTORIAL OF STRUCTURE:	(Schematic of Packing Structures)		
CHARACTERISTICS:	• Small Diameter Particles • Fully Porous	• Pellicular Particles (Thin layer silica fused to solid glass bead)	• Large Diameter Particles • Fully Porous
FEATURES:	• Very High Efficiency • Moderate Capacity • High Speed	• High Efficiency • Low Capacity • Moderate Speed	• Moderate Efficiency • High Capacity • Moderate Speed
REPRESENTATIVE CHROMATOGRAMS:	or		

FIGURE 3-13. Available particle types for HPLC.

packing supports, so they can be useful in analytical separations. They have a disadvantage, however, in that their low surface areas (typically 10 m^2/gm) make them undesirable for preparative scale applications. Pellicular packings are generally used in small columns ("guard columns") which are placed before the analytical column. Guard columns are used in situations where small amounts of unwanted material might contaminate the analytical column; therefore, the small, pellicular column "guards" or protects the analytical column without causing unwanted bandspreading of the sample components (peaks). The guard column needs to be replaced periodically when it becomes contaminated or coated with unwanted material, which is not an issue since pellicular packings are less expensive than their small-particle ($\leq$10 μm) counterparts. In certain situations, long pellicular columns (2 mm ID $\times$ 60 cm) may be used for analytical work.

Porous, small particles (Fig. 3-13, far left side) are generally in the less than 10-μm size range (also available in smaller sizes) and are popular because very high column efficiencies are attained, owing primarily to the shorter distances the solute molecules must diffuse within the packed bed. As a result of the high efficiencies, most analytical work is done on these particles. Most often the researcher chooses to buy these columns prepacked and, as such, the columns cost more than the pellicular or large-particle ones. As particle size decreases, the columns are less rugged in routine operation. This does not imply that they cannot or should not be used. They can be used; however, increased attention must be given to proper usage to insure a satisfactory lifetime for porous, small-particle columns.

Particle size should be selected on the basis of the trade-offs which the individual can tolerate. The consideration of efficiency, however, should not be made at the expense of selectivity (phase). The ideal situation is to have the correct stationary phase in the proper particle size for the application at hand.

Resolution

As was discussed in Chapter 1 resolution, R, is a measure of the distance between two adjacent peaks in terms of the number of average peak widths than can fit between the band (zone) centers. Assuming symmetrical (Gaussian) peaks, when $R = 1$, peak separation is nearly complete with only about 2% overlap. This case was shown in Chapter 1, Figure 1-4. Resolution results from the physical and chemical interactions that occur as the sample travels through the column. It should, therefore, be no surprise that resolution may also be expressed in terms of the contribution of the individual column characteristics: separation factor (selectivity, α), efficiency (narrowness of peak, N), and capacity factor (residence time, k') of the first component. The equation that describes this interrelationship is

$$R = \frac{1}{4} (\alpha - 1) \sqrt{N} \left(\frac{k'}{k' + 1} \right) \tag{3-9}$$

It is worthwhile to examine this relationship (3-9) for a better understanding of the influence of significant variables on separation performance. Note that of the three factors governing resolution, only the selectivity, α, relates to more than one component. It is also the most complex factor of the three, and it generally cannot be predicted how changes in operating conditions influence this value. The α values need to be experimentally determined by "trial-and-error" type changes in the mobile phase composition. Column and mobile phase contributions to selectivity are discussed in greater length in Chapters 4 and 5.

The efficiency term improves resolution through a square-root factor; therefore, the most appropriate starting point is to use a column with a moderately high number of theoretical plates or N value. Since N generally varies linearly with column length, the most straightforward approach to improving resolution is to increase column length. However, because increasing N has a square-root effect upon increasing resolution, a point of diminishing returns is reached where the increase in analysis time and/or increase in back pressure does not justify the accompanying increase in plates. As we will learn in Chapter 6, a variation on this approach to improved efficiency is merely to reduce flow rate without changing column length or to "recycle" the sample components through the same column. Both result in an increase in time.

The third factor to change resolution is the k' of the first compound. Of course, when the k' of the first component is changed, so is the k' value for the second component, but the effect of both k' values is accounted for in the selectivity term which was just discussed. Control of k' is achieved by the mobile phase composition. A reduction in the solvent strength of the mobile phase will increase k'. This happens because the balance in attraction for the sample is shifted in favor of the stationary phase. If all k' values are too low, components elute too quickly; thus, the solvent is too strong. Conversely, high k' values reflect long elution times, and can be reduced by increasing the solvent strength of the mobile phase. In LC, varying the mobile phases is preferable to changing the adsorbents, especially when optimizing the separation of a new application. Increasing k' will generally improve resolution. However, there is a preferred range of k' beyond which the $(k'/k' + 1)$ term approaches a constant value. This optimum range for k' is 2–5, which means the components should elute between 3 and 6 column volumes.

Selectivity vs. Efficiency

By looking at the resolution equation (3-9), one can understand the interrelationship of selectivity and efficiency and the accompanying effect upon resolution. If our goal is to obtain a resolution value of 1 between two peaks (see Chapter 1, Figure 1-4) and we have a k' value for the first peak of 2, we can illustrate the role of selectivity and efficiency by inserting those values into equation 3-9, which then simplifies to

$$\alpha - 1 = \frac{6}{\sqrt{N}} \qquad (3\text{-}10)$$

We can solve equation 3-10 for various values of selectivity (α) and this is done in Table 3-2.

TABLE 3-2. Relationship of Selectivity and Efficiency[a]

α	Difficulty of Separation	N
1.6	easy	100
1.35	moderate	300
1.11	difficult	3,000
1.035	very difficult	30,000

[a] Resolution of 1, and $k_1' = 2$.

As can be seen in Table 3-2, a high selectivity requires only a very few plates to achieve a separation. In other words, the chemical interaction can be a more powerful contribution to improving resolution than the physical interactions. Certainly, modern HPLC is capable of generating 30,000 plates with several high-efficiency columns in series; however, a more effective approach would be to adjust the chemical interactions through a change in the mobile phase composition to move the selectivity value to a range ($\alpha \sim 1.11$) where only one column (3000–8000 plates) would be needed. Of course, compounds having a very large selectivity can be easily separated on a modern HPLC; however, other separation techniques could also be used. For instance, a typical low-pressure, open-column LC would achieve 100 plates, while a typical TLC separation would achieve only 300 plates. It is important to remember that the N term in the resolution equation depends mostly upon physical phenomena and has a predictable contribution to attaining resolution. The α and k' terms, on the other hand, are not easily predicted and depend upon chemical interactions in the chromatographic system. The α and k' terms can have a profound influence on the separation and, if sufficiently large, obviate the need for a high number of plates in a separation.

MOLECULAR DISTRIBUTION BETWEEN MOVING AND STATIONARY PHASES

Chromatographic selectivity is based heavily on the phenomenon that each component in a mixture ordinarily interacts with its environment differently from other components under the same conditions. In fact, the separation performed using a separatory funnel and two immiscible solvents employs the same principle that is the basis of liquid chromatography—the relative preference of compounds in one phase for a second phase. This implies that with the proper mobile phase conditions and column packing, various components in the sample will travel through the column differently than others, resulting in the desired separation.

Liquid chromatography is clearly based upon a solubility phenomenon. When a solution is in contact with another immiscible solvent or solid surface, virtually every component in solution will tend to distribute itself between the original phase and the new phase. If the mobile phase selected for the sample is a strong solvent for all of the components, there will be little driving force for the components to associate with the stationary phase. Consequently, to achieve a separation there must be a good balance between the attraction for the sample components by the mobile and stationary phases. As a "band" of sample solution is pumped through the column packed with very fine porous particles, the correspondingly narrow channels cause the solution to be divided into thin films. This assures a high-contact frequency between the moving solution and stationary surfaces inside the pores and provides ample opportunity for the mobile and stationary phases to reach equilibrium. As the portion of mobile phase rich in a specific component moves past the equilibrated packing surface, the next increment of mobile phase containing less of this specific solute will tend to extract a portion of the adsorbed component back into the moving phase. Since almost every solute component will enter into this "back-and-forth exchange" to different degrees and at differing speeds from all other sample components, separation is achieved as the sample components pass through the column. In a properly selected system, none of the components will remain on the packing when sufficient mobile phase has passed through the column after the sample.

Chemical interactions form the basis of the HPLC modes used for chromatographic separations. The whys and hows of these "chemistries" need to be comprehended in order to become a "chromatographer." Although no text can ever replace the actual experience of doing chromatography, an understanding of the general types of interactions that can occur in the column(s) will prepare the newcomer for separation development, which is discussed in Chapters 4 and 5. In the following pages we discuss terms that aid the chromatographer in defining the types of interactions which occur in the separation.

Normal and Reverse Phase

In normal-phase chromatography, the retention is governed by the interaction of the polar parts of the stationary phase and solute. For retention to occur in normal phase, the packing must be more polar than the mobile phase with respect to the sample. Therefore, the stationary phase is usually silica and typical mobile phases for normal phase chromatography are hexane, methylene chloride, chloroform, diethyl ether, and mixtures of these. In reverse phase the packing is nonpolar and the solvent is polar with respect to the sample. Retention is the result of the interaction of the nonpolar components of the solutes and the nonpolar stationary phase. Typical stationary phases are nonpolar hydrocarbons, waxy liquids, or bonded hydrocarbons (such as C_{18}, C_8, etc.) and the solvents are polar aqueous–organic mixtures such as methanol–water or acetonitrile–water. In the strictest interpretation, normal and reverse phase are terms which only relate to the polarity of the column and mobile phase with respect to the sample as shown in Table 3-3 and drawn schematically in Figure 3-14.

TABLE 3-3. Normal and Reverse Phase Modes

	Normal Phase	Reverse Phase
Packing polarity	High	Low
Solvent polarity	Low to medium	Medium to high
Sample elution order	Least polar first	Most polar first
Effect of increasing solvent polarity	Reduces elution time	Increases elution time

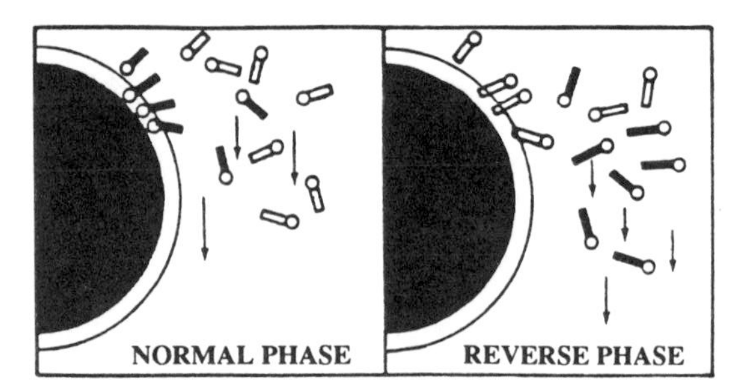

FIGURE 3-14. Comparison of normal phase to reverse phase. The circles depict polar groups and the rectangular ends depict nonpolar groups.

Liquid–Solid Chromatography (LSC)

The basis of separation in liquid–solid chromatography is shown schematically in Figure 3-15. Polar solids such as silica gel, alumina, porous glass beads, and some bonded phases (e.g., cyano group) may be used, but silica is the most popular since it is durable, readily available, and an economical adsorbent. The functional retention on bare (unmodified) silica gel packings involves interactions of the sample with the polar hydroxyl groups on the surface of the silica particle (Si–OH). Because these interactions are adsorptive, another common term used to describe this type of chromatography is adsorption chromatography. Adsorption chromatography and liquid–solid chromatography are synonymous. Since the silica surface is very polar, retention and, therefore, the separation is based upon differences in the polar functionalities of the various sample molecules adsorbing onto the

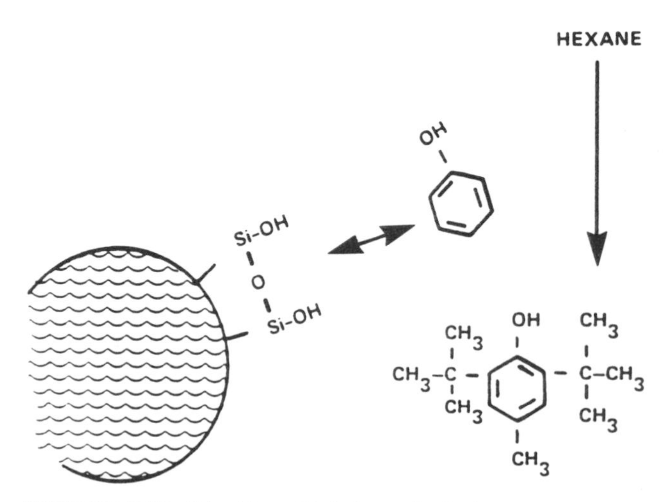

FIGURE 3-15. Liquid–solid (adsorption) chromatography.

packing surface. The most polar molecules are more strongly retained (adsorbed) to the silica surface. Liquid–solid chromatography is most useful in separating classes of molecules (e.g., ketones from aldehydes) and is not useful in separating members of an homologous series. Figure 3-15 depicts the relative adsorption of a phenol and a hindered phenol onto silica in a hexane mobile phase. The hindered phenol will elute first from the silica because this compound is less polar, and because the access to the molecule's hydroxyl group is sterically hindered compared to the phenol. Liquid–solid separations are based upon the polarity of molecules and the physical adsorption of functional groups onto the surface's active sites. If a molecule has a permanent dipole moment, it will be attracted by dipole–dipole interactions. If the sample molecule has any size to it, a dipole can be induced in it by another dipole (perhaps the packing). This is known as the polarizability of the molecule and this will also retard its movement and result in retention. The rigid, fixed adsorption sites on the packing cause the interaction of corresponding functional groups in a sample molecule to vary with the geometry of the molecule. When the packing sites are appropriately matched to the polar sites on the sample, that isomer is more strongly retained. This makes LSC appropriate for separating positional isomers. Generally, LSC is commonly used in the normal phase mode of chromatography (see Table 3-3).

Bonded-Phase Chromatography (BPC)

Phases made by covalently bonding a molecule onto a solid stationary phase are intended to prepare "liquid coatings" which will be permanent. Silica is a reactive substrate to which various functionalities can be attached or bonded. The functionalities most widely bonded to silica are the alkyl (C_{18} and C_8), aromatic phenyl, and cyano and amino groups. The most popular bonded phase is the octadecylsiloxy support commonly referred to simply as a C_{18} phase or an ODS phase. Bonding the silica with a "nonpolar" functional group alters the surface chemistry of the silica particle to an essentially "nonpolar" one. However, since the starting surface is silica, a totally "nonpolar" surface will never be attained. This, in itself, is not a problem but it does mean that bonded phases are unique entities and not easily classified as liquid–solid or liquid–liquid surfaces.

Bonded phases may be used in both normal and reverse phase chromatography. When normal phase chromatography is done on bonded phase packings, the packing is more polar than the mobile phase. Polar bonded phases such as the cyanopropyl and aminopropyl functionalities are popular for this use. These bonded phases are less subject to changing retention times of compounds because water is adsorbed from the mobile phase onto the stationary phase, a frequent concern when using bare silica packings for normal-phase separations.

The C_{18}, C_8, and phenyl bonded phases are most often used in the reverse phase mode where the more nonpolar sample components interact more with the "relatively" nonpolar column packing and thus elute later than polar sample components. As a result of their versatility in accomplishing so many separations, bonded phases have been responsible for the tremendous growth of analytical LC over recent years. Today, bonded phases are the most popular LC packing because most sample components are soluble in mixtures of aqueous–organic mobile phases and can be separated in the reverse phase mode. Furthermore, these bonded phase packings are faster to equilibrate, use less organic solvent, and have more versatile surface chemistries than bare silica. It has been estimated that 60–90% of all analytical LC separations are done on bonded phases in the reverse phase mode. In fact, the term "reverse phase" is often used synonymously with reverse phase BPC.

Ion-Pairing Chromatography (IPC)

Bonded phase packings are very versatile in another way, in that mobile phase additives can adsorb onto the nonpolar surface and impart additional selectivity to the interactions between the sample and the stationary phase. One example of this is "ion pairing" where an ionic organic molecule (a pairing ion) is included in the mobile phase. The nonpolar component of the mobile phase will be adsorbed to the nonpolar, bonded stationary phase. Since the mobile phase contains the organic ion, an equilibrium is set up in the column resulting in an ionic mobile phase and an ionic characteristic on the stationary phase. Separations can then occur on the basis of electrostatic (charge) and nonpolar interactions. In this fashion, a bonded nonpolar stationary phase can be used to separate ionic components and nonpolar compounds on the same column. Because the ion-pairing technique is quite attractive and easy to use, it has been used for many applications instead of ion exchange.

Ion Exchange Chromatography (IEC)

Ion exchange separations are based upon attractive ionic forces between molecules carrying charged groups of opposite charge to those charges on the stationary phase. Separations are made between a polar mobile liquid, usually water containing salts or small amounts of alcohols, and a stationary phase containing either acidic or basic fixed sites. The separation depends upon the ionic nature of the compound (e.g., pK_a), the polarizability of the molecule, the solvation shell of the molecule, and the relative attraction of the compound for the ion exchange surface. After the sorption of the charged sample, desorption is brought about by increasing the salt concentration (ionic strength) in the mobile phase or by changing the pH of the mobile phase. Both ionic strength and pH can minimize the charge (electrostatic) attraction between the sample and the stationary phase, and hence, make the mobile phase stronger.

As in all modes of liquid chromatography, IEC separations result when sample components move through the column with different speeds. At low ionic strengths, all components with an electrostatic attraction for the ion exchanger will be tightly held on top of the column. When the ionic strength of the mobile phase is increased by adding a salt, the salt ions compete with the adsorbed sample ions for the bonded charges on the column. As a result, some of the sample components will be partially desorbed and start moving through the column. If the salt concentration is higher, the resulting ionic strength causes a larger number of the sample components to be desorbed, and the speed of the movement down the column increases. The stronger the charge attraction of the sample to the column is, the higher the ionic strength needed to bring about desorption.

At a certain level of ionic strength, no sample components are held by the charges bonded to the column. In this case, all sample components will elute at the mobile phase volume (V_0). Somewhere in between total adsorption and total desorption one will find the optimal selectivity for a given pH value of the mobile phase. Thus, to optimize retention in IEC, a pH value is chosen that creates sufficient charge differences among the sample components. Then, an ionic strength is selected that competes with these charge differences so that the relative movement of each component through the column results in the desired selectivity.

While ion pairing appears to be the technique of choice for small molecules, ion exchange is more popular for larger ionic compounds such as peptides and proteins. Using ion exchange packings, proteins are separated on the basis of their net surface charge. Since a protein consists of amino acids, this surface charge of a protein is dependent upon pH and can be changed from a positive to a negative charge as pH is changed. Through the use of pH and ionic strength, protein separations can be obtained. Often, because proteins have a large net charge and are, therefore, highly electrostatically held to the column, gradient chromatography using ionic strength and/or pH changes is effective in eluting a mixture of proteins.

Gel Permeation Chromatography (GPC)

Gel permeation chromatography is a mechanical sorting of molecules based on the size of the molecules in solution. In true GPC there is no attraction of the sample for the packing. Therefore, the maximum amount of mobile phase required for complete sample elution will equal the volume of mobile phase in the column (V_0 in Fig. 3-1). In other words, in true GPC, all sample components elute within one column volume and before the column void volume (V_0 in Fig. 3-1). Size separation is achieved with a porous packing material which is compatible with the mobile phase. The smallest components in the sample migrate into the smallest pores of the packing while the molecular dimensions of the higher-molecular-weight components prevent them from penetrating as far into the pores. For example, in Figure 3-16, in the case of a high-molecular-weight polymer, the largest molecules may be

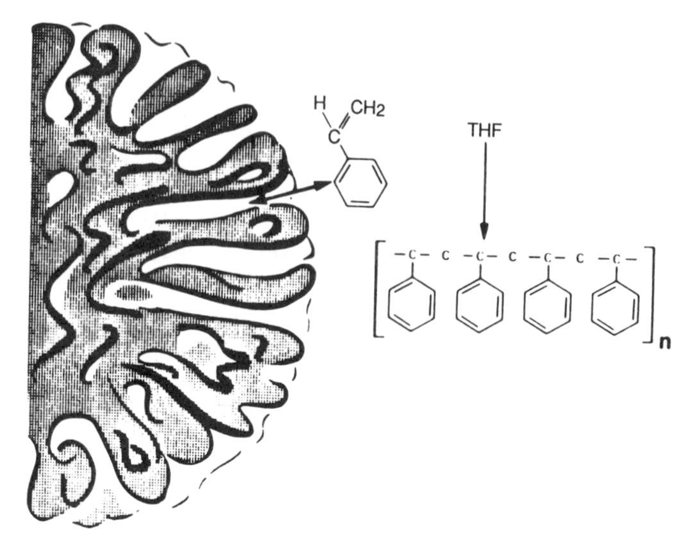

FIGURE 3-16. Gel permeation (size exclusion) chromatography.

excluded completely from the packing material and move only through the larger channels between the packing particles. For smaller molecules, the path through the column is longer and more tortuous since the small molecules enter the pores of the packing material and, as a result, become separated from the larger molecules in the process. This is best depicted by a hypothetical example of a separation of the components of chewing gum by GPC shown in Figure 3-17. The polymer is separated on the basis of size (big molecules elute first), the softeners, which are smaller molecular weight, elute next, followed by the smallest molecules, the flavorings.

The term *gel permeation chromatography* was originally used by polymer chemists to describe a separation of a polymer on the basis of size when using a porous stationary phase and an organic mobile phase. This separation was used to calculate the molecular weights of the polymer and to characterize the molecular weight distribution. When an aqueous mobile phase and aqueous compatible porous stationary phase (e.g., cross-linked dextran) was used to separate peptides and proteins, the biochemists used the term *gel filtration chromatography*. Later, attempts to unify terminology resulted in suggestions to use the term *size exclusion chromatography* (SEC) and *exclusion chromatography*. Unfortunately, a consensus has not been attained and we are left with a semantic issue in this mode of chromatography.

Summary

It should be clear that all separation processes require a type of interaction between the stationary phase, mobile phase, and sample. Each mode of LC

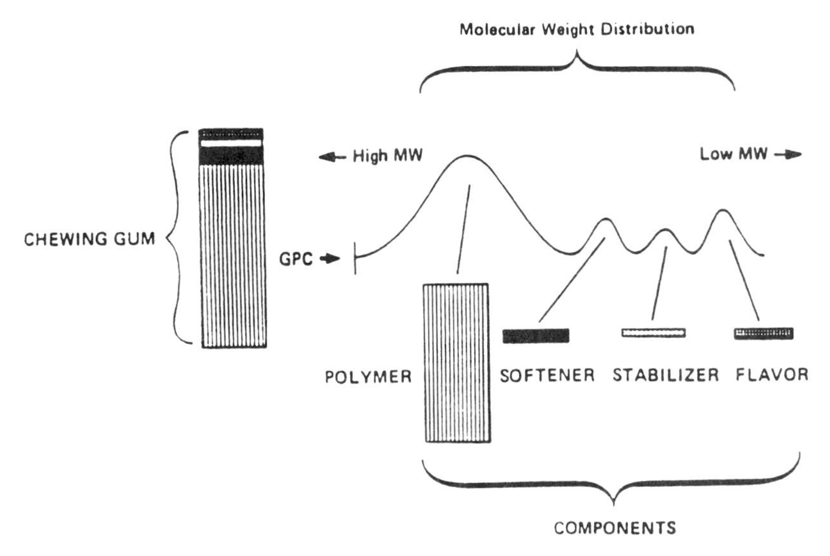

FIGURE 3-17. Gel permeation chromatogram of chewing gum.

differs in the type of interaction and occasionally several different modes may need to be used to solve a complex separation problem.

SUMMARY

The chromatogram embodies all of the knowledge that results from the HPLC. It is therefore important that the system is operating correctly before any type of analysis is attempted. While no one intentionally uses an HPLC that is not working properly, it is imperative to verify that it is functioning correctly. Additionally, it is important to choose a detector that will "see" the compound(s) of interest at the sensitivity level desired. Lastly, retention and separation must be attained.

The schematic in Figure 3-18 attempts to position the various mechanistic approaches possible for retention and, hence, separation of molecules with regard to the chemical properties of the molecules. A specific separation mode is applicable as a result of the physiochemical nature of the molecule (horizontal axis) and its molecular weight (vertical axis). As can be seen in this figure, there are numerous overlaps of the regions of separation modes, indicating that there are many possible ways of accomplishing a separation for a particular compound. There is no one right way to do a separation; many approaches are possible. The more one knows about the sample (e.g., polarity, solubility, molecular weight), the easier it is to select the retention mechanism. After choosing the retention mechanism, the chromatographer has three factors with which to experiment—the stationary phase, the mo-

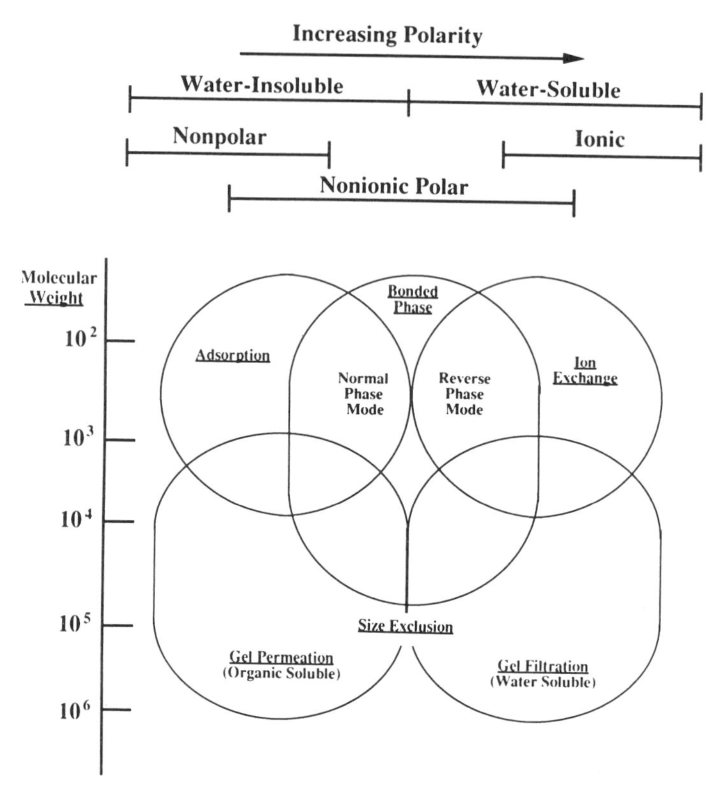

FIGURE 3-18. Areas of applications of HPLC separation mechanisms segmented on the basis of molecular weight (abscissa) and polarity (ordinate). (Adapted from reference 7 with permission.)

bile phase, and the column efficiency (narrowness of the peak). Often these factors are the most difficult to manipulate, requiring a knowledge base built around "column chemistries."

As implied by Figure 3-18, column selection (the mechanism and mode of HPLC) has many correct first choices. The correct "first choice" is a result of additional rules of thumb that are discussed in Chapters 4 and 5. Since no established "standardized" rules exist which apply to *every* separation problem, the new chromatographer will need to develop his/her own "guidelines." While at first this might appear disconcerting, it is really quite an opportunity. Experience gained by reading this text and doing the experiments in this book coupled with the large amount of available information on applications in the technical literature should transform the trial-and-error activity of methods development into an exciting problem-solving task which will eventually become a fulfilling and successful scientific venture.

REFERENCES

1. J. W. Dolan and P. Upchurch, *LC·GC*, **6,** 886 (1988).
2. C. A. Dorschel, J. L. Ekmanis, J. E. Oberholtzer, F. V. Warren, and B. A. Bidlingmeyer, *Anal. Chem.*, **61,** 951A (1989).
3. A. J. P. Martin and R. L. M. Synge, *Biochem. J.*, **35,** 91, 1358 (1941).
4. L. Lapidus and N. R. A. Amundson, *J. Phys. Chem.*, **59,** 416 (1955).
5. J. J. Van Deemter, F. J. Zuiderweg, and A. Klinkenberg, *Chem. Eng. Sci.*, **5,** 271 (1956).
6. J. C. Giddings, *J. Phys. Chem.*, **31,** 1462 (1959).
7. D. L. Saunders, *Chromatography: A Handbook of Chromatography and Electrophoretic Methods*, 3rd ed., E. Heffmann, editor, Van Nostrand Reinhold, Co., New York, 1975, Figure 5.1 on p. 81.

ACKNOWLEDGMENT

Figures 3-8, 3-9, 3-13 through 3-17 are courtesy of the Waters Chromatography Division of Millipore.

CHAPTER 4

APPROACHING THE PROBLEM

The general strategy
- Step 1. What is your job?
- Step 2. Know your unknown
- Step 3. Pick the packing
- Step 4. Achieve the separation

General tactics
- Easy as 1, 2, 3, 4
- An example
- Differences in size of the sample components
- Normal and reverse phase
- Differences in stationary phases
- Differences in the mobile phases
- Differences in ionic functionality of the sample components
- What happens after choosing the LC packing?

Summary
References
Acknowledgment

After reading the first three chapters, you know what LC is, what HPLC can do as a research tool, and how to interpret the chromatogram. However, you are probably wondering how to develop a separation so that you may begin solving your separation problems. The biggest hurdle confronting both novice and experienced liquid chromatographers is how to approach the problem. What is the first step to take with a sample that has not been analyzed previously? The material contained in the following pages discusses the basic approaches to the problem, reiterates a general set of guidelines for selecting the first system to try, and presents a systematic series of steps to follow that depend on the results of the first initial injections.

Remember, the "blessing" of LC is that there is a wide selection of solvents and packings which leads to an amazingly large number of correct combinations for accomplishing any given analysis. But this is also a "mixed blessing." With so many ways to do something, choosing the starting point may be frustrating. However, using a strategy based upon a few key concepts will enable you to select an appropriate operating mode. Then, by applying another sound strategy to the method development of the separation, a speedy resolution of your separation problems should result.

THE GENERAL STRATEGY

The first rule of thumb to remember is that there is no single approach to developing a satisfactory separation. Most problems can ultimately be solved by more than one mode of LC. Therefore, a general strategy for the analyst to follow is presented.

Step 1. What Is Your Job?

Define the goals of the separation. What degree of resolution needs to be achieved? How many compounds need to be resolved? It is crucial to have a clear statement of the problem to be solved. Is it necessary to obtain a quantitative separation of each compound present in a multicomponent mixture, or is there only one component of interest in the sample? Is it necessary to achieve a general profile of the sample for routine screening purposes, or must there be a specific quantitative separation for collection and subsequent analysis by other methods?

Step 2. Know Your Unknown

This statement may at first sound contradictory as the starting point for selecting a first mode of LC. However, the more information known about a sample, the better a basis there is for beginning. You must at least know the solubility characteristics of the sample, since it must be in solution to be injected. Knowledge of the best solvents for the sample is a consideration in selecting the first packing to try. Consider and list all of the available facts about the sample. If possible, write the molecular structure of all the compounds that are known or suspected to be present. If the molecular structure is not known, by using IR spectroscopy, if available, you might determine or verify the functional groups in the sample. Consider initial screening results with TLC. Ask yourself additional questions such as:

Can classes of materials be separated by size?

Are all the compounds very similar, differing only in isomeric structure?

Is the separation a one-time problem, or will it be adopted as a standard analytical method for routine work?

Depending on each of these requirements, a satisfactory separation may be achieved after two or three injections. This is especially so when you are only seeking a profile or looking for a single compound, or where the separation will not become an established quality-control method. Alternatively, more extensive work may be necessary for total resolution of all components or for defining a reproducible control method.

Last, but not least, search the literature for the separation of your compounds and similar compounds. Much of the world's scientific information is contained in more than 3000 electronic databases now available from "information utilities" or "database vendors." These systems provide on-line databases and the software necessary for search and retrieval. These on-line services permit you to use information even though the database is not in your possession. Databases now serve nearly every major field of science and technology as well as specialized areas such as robotics, oil spills, and genetic engineering. Most chemical libraries subscribe to the services of one or more major database vendors, for example, *Dialog, Nexis,* and *STN International* (Chemical Abstracts Service). Among the several hundred databases available through various vendors are: chemistry—*Chemical Abstracts* (1969–present), *Heilbron,* and CJACS (American Chemical Society journals in full-text format); medicine and biosciences—*Medline* (corresponding to index medicines) and *Biosis* (biological abstracts); agriculture and foods—*Foods Science and Technology Abstracts* and *Agricola;* technology—NTIS (the National Technical Information Service), *U.S. and World Patent Abstracts,* and *Science Citation Index.*

Databases can and should be searched using a variety of terms, singularly or in combination, such as: CAS Registry Number; author name; patent assignee; specific substance; words in title or abstract; range of years; document type (such as review article); and other articles citing a given article. The possibilities are almost endless and you can request the required amount of information which you desire for each of your answers for example, bibliographic reference or full abstract. The average search requires 10–15 minutes of on-line time. During those few minutes, the user may search through several million records equivalent to 20 years of *Chemical Abstracts*. Search results are generally mailed to libraries the evening of the day the search is done and the results generally arrive within 4 days. If necessary, search results can be printed on line; however, this adds some expense. Requests for searches should be as specific as possible by including (1) all known synonyms for compounds; (2) specifying if by HPLC, GPC, or other method; (3) noting the dates to be searched; and (4) requesting specific databases to search (if known). An example of one page from a search on "HPLC and lactulose" (a sugar) is shown in Table 4-1. Libraries can be a significant source of preliminary information, which will shorten the development time and eliminate the need to "reinvent the chromatographic wheel" for the required application.

Step 3. Pick the Packing

Since solubility is the key in developing a separation, it should be no surprise that in answering the second question (What do I know about my sample?), listing the solvents in which the sample is soluble is important in choosing the mode of LC most promising for your separation problem. Once the

TABLE 4-1. Sample Page from Literature Search on "HPLC and Lactulose"

STN INTERNATIONAL®

CA FILE SEARCH RESULTS - PAGE 3

L2 ANSWER 1 OF 5

AN CA105(15):130163c
TI Liquid-chromatographic method for estimating urinary sugars: applicability to studies of intestinal permeability
AU Delahunty, Thomas; Hollander, Daniel
CS Dep. Med., Univ. California
LO Irvine, CA 92717, USA
SO Clin. Chem. (Winston-Salem, N. C.), 32(8), 1542-4
SC 9-3 (Biochemical Methods)
SX 14
DT J
CO CLCHAU
IS 0009-9147
PY 1986
LA Eng
AN CA105(15):130163c
AB Sugars of exogenous origin excreted in the urine can be rapidly quantified by HPLC. A simple extn. with an ion-exchange resin is used to prep. the sample for anal. Aliquots (20 μL) are chromatographed on a cation-exchange column at 85°, with H_2O as the mobile phase. Sugars are detected with a refractive index detector. Lactulose, rhamnose, and mannitol all give discrete peaks and a linear response up to 5 g/L, with anal. recoveries from urine of 80, 62, and 80%, resp. Precision is good, the relative std. deviations for lactulose, rhamnose, and mannitol being 2.9, 4.0, and 5.6%, resp. The only endogenous compd. consistently present in the chromatograms is urea, which does not interfere. However, glucosuria, if present, could interfere with the lactulose estn. This method may be a simple, labor-saving means of quantifying urinary sugars in the clin. lab.

****

(Reproduced with permission of American Chemical Society.)

mobile phase and packing are chosen, make your first try. It may not work, but on the other hand, if there is some evidence of separation, consider yourself fortunate. If it does not work on the first try, you have a lot of company, including even experienced chromatographers!

Step 4. Achieve the Separation

Manipulate the variables using sound tactics to achieve more satisfactory results. A successful separation is achieved by establishing the proper balance between attraction of the mobile phase and packing for the sample. In a LC column, mobile and stationary phases both compete for the sample. If the sample is more like the mobile phase than like the packing in terms of polarity (or other measures of solvating strength) there will be little retention. When this occurs, it is necessary to make the eluent less like the sample, generally by changing the polarity of the mobile phase. An alternative is to change the packing to a type more similar to the sample; but, this approach is more costly and time consuming. Most separations are achieved by matching the polarity of the sample and packing and by using a mobile phase which has a markedly different polarity. Depending upon the results of the injection, changes can be made in instrument operating conditions, column length, solvent composition, packing type, and separation mode. Once there are some indications of a separation, similar (but less pronounced) changes are used to improve resolution.

GENERAL TACTICS

Easy as 1, 2, 3, 4

As you can see, developing a separation is as easy as following Steps 1–4. Step 1 and 2 are fairly straightforward; however, Steps 3 and 4 require some guidelines. Figure 4-1 is intended as a general guide for choosing which mode of LC is the first choice (Step 2) for methods development. By asking yourself questions associated with Figure 4-1, a straightforward selection of the appropriate packing is possible.

The first question in Figure 4-1 considers the molecular weight of the sample. Once the sample is considered in terms of its molecular weight, the next question concerns its solubility. If the component molecular weight range, number, and type of functional groups are known, Figure 4-1 is a basis for selecting the initial packing. For example, if the compounds are small molecules, differ in molecular weight by 10%, and are soluble in tetrahydrofuran, then small-molecule GPC may be used. Chapter 10 contains an experiment that illustrates this technique. Looking again at Figure 4-1, we see that low-molecular-weight compounds containing ionizable species are often best separated by ion exchange; however, another separation mode

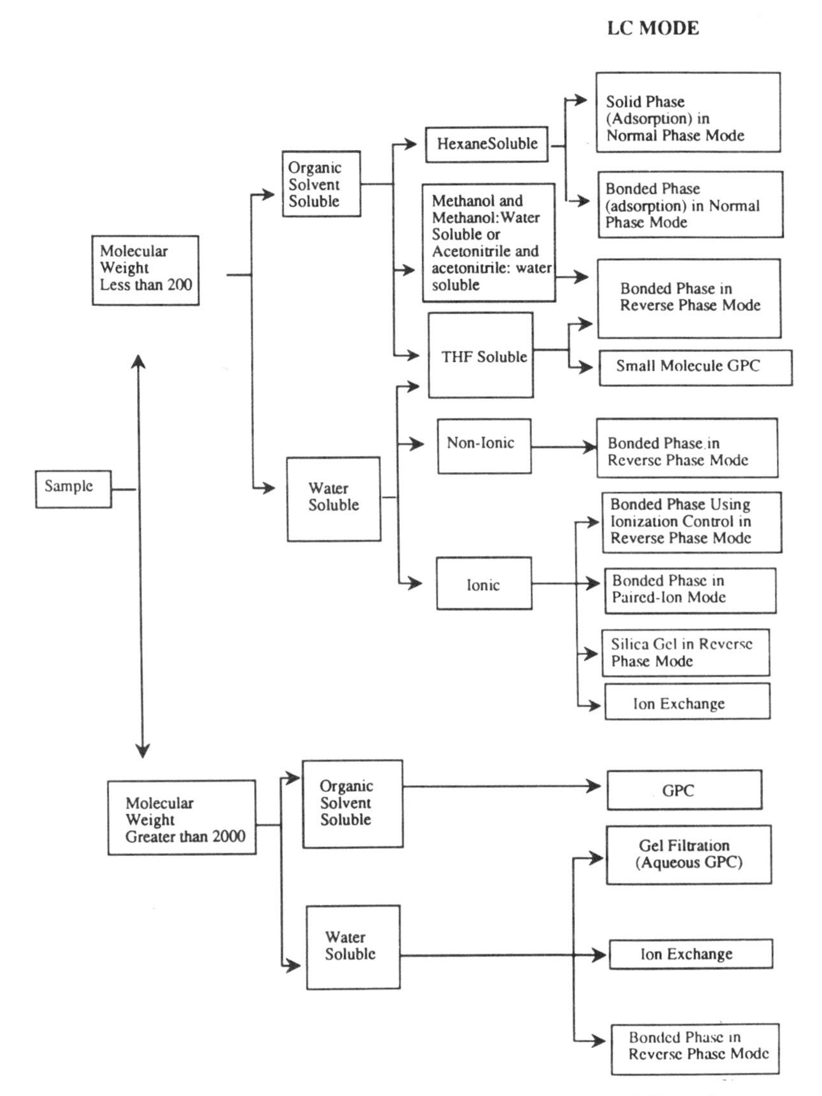

FIGURE 4-1. Decision tree guide for choosing an LC mode.

such as reverse-phase ion pairing may be equally effective.

If two paths are possible, knowledge about the sample can guide the movement on the "decision tree" in Figure 4-1. Consider whether the differences between sample components are primarily steric or if the components differ in polarity or solubility. If steric, an adsorption packing is probably the best starting point; otherwise a bonded-phase packing may be a better first choice. Your choice should be made based on convenience, personal preference, and experience, plus the solubility considerations given previously. Even if the first solvent–packing combination chosen is not effective, the guidelines presented in Figure 4-1 should lead you to a second or third choice which will result in a satisfactory solution to your problem.

Occasionally, choices are not always as clear cut as indicated in this guide, especially for compounds that are both organic soluble and nonionic, water-soluble compounds of molecular weight less than 2000. Specific applications require an understanding of the chemistry of the species being analyzed. This understanding comes from experience. As you become more experienced in using HPLC through such activities as running the experiments listed in Chapters 8–16, you will speedily arrive at appropriate methods for your application.

Successful separations can be carried out only by planning and careful experimentation, the details of which are discussed extensively in Chapter 5. In one sense there are *too* many ways to achieve a separation in LC. But while this makes the first choice of where to start difficult, the good news is that there are many ways to achieve success. By looking at Figure 4-1 it is obvious that the use of the reverse-phase mode in LC has broad applicability and is, in fact, the most used mode of LC. Reverse phase is used for 80–85% of the separation problems encountered by users of HPLC. For this reason, the majority of Chapter 5, on developing methods, is devoted to reverse-phase examples. Additionally, Chapter 11 is a useful experiment to experience the method development aspect of this mode. Chapter 9 is a useful experiment to experience method development in the normal phase mode.

An Example

As an example of using the decision tree, assume that the problem is to separate the polynuclear aromatics shown in Figure 4-2. Following the decision tree in Figure 4-1 results in two choices: reverse phase or normal phase. With 5- and 6-ring polynuclear aromatics, the first inclination might be to avoid using reverse phase. However in this example, it proved necessary to use reverse phase to provide adequate resolution for all of the sample components. The chromatogram in Figure 4-3 is an example of a successful reverse-phase separation. Because of the wide range of compounds present, this separation required an increase in solvent strength during the analysis. This gradient approach is discussed in Chapter 7. Also note that each of the compounds is detected at its absorbance maximum in order to have high sensitivity. Detecting at different wavelengths during a chromatogram is possible using a modern HPLC detector that is capable of switching detection wavelength during the run.

Two separations of polynuclear aromatic compounds are shown in Figure 4-4. Using the decision tree (Fig. 4-1) for the compounds in Figure 4-4*a* and *b* results in two choices: reverse phase and normal phase. Because it is easier

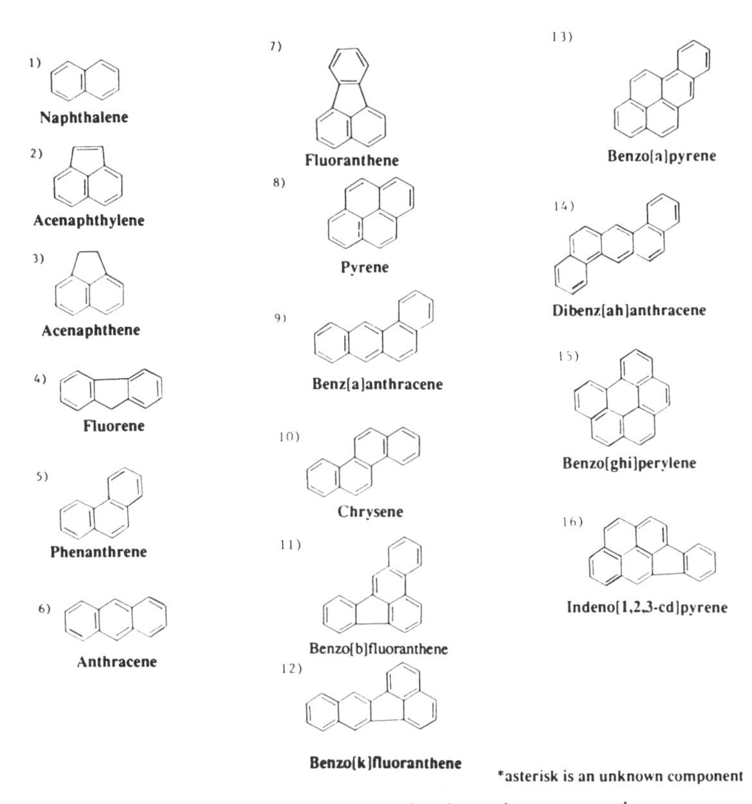

FIGURE 4-2. Structures of polynuclear aromatics.

to implement, reverse phase was chosen for the mixture in Figure 4-4*a*. This separation is much simpler than the one in Figure 4-3; however, note that a gradient was also necessary to accomplish the separation of the isomeric 4-ring compounds. A different separation problem, however, could lead to a normal-phase separation, as shown in Figure 4-4*b*. As implied from these examples, experience will be the teacher as to what mode of LC is the best choice. In reverse-phase chromatography, the sample and packing are very nonpolar and would be an attractive surface for the nonpolar polynuclear aromatic sample, while the solvent is relatively polar. This is in contrast to normal-phase liquid chromatography in which a polar packing such as silica is used with nonpolar solvents.

We have just seen how the decision tree can lead to two choices for a separation and the choice is based upon preference. In the examples given above, the separation of very nonpolar compounds was done in the reverse-phase mode. There are instances in which very polar compounds can be separated successfully by reverse phase, particularly when the major differences between the components are in the nonpolar portions of the molecules. Conversely, separation of some nonpolar compounds might best be

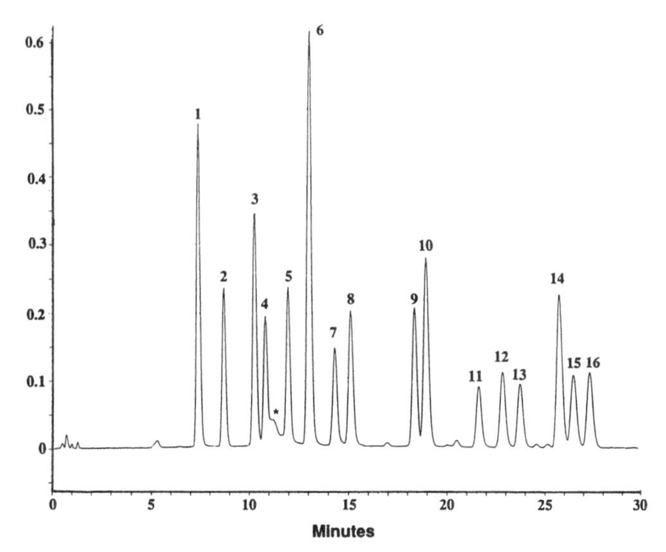

FIGURE 4-3. Separation of polynuclear aromatics. Numbered peaks are listed by compound name and structure given in Figure 4-2. The wavelength of detection is the absorbance maximum of each compound: (1) 219 nm; (2) 228 nm; (3) 225 nm; (4) 210 nm; (5) 251 nm; (6) 251 nm; (7) 232 nm; (8) 238 nm; (9) 287 nm; (10) 267 nm; (11) 258 nm; (12) 240 nm; (13) 295 nm; (14) 296 nm; (15) 210 nm; and (16) 251 nm. Detection wavelengths were switched automatically to the absorbance maximum of each compound.

achieved on polar packings in the normal phase if the differences between the nonpolar compounds are in their polar groups. Obviously, separations of many compounds can be made by either normal- or reverse-phase systems.

Differences in Size of the Sample Components

As was shown in Figure 4-1, when dealing with a mixture, a useful starting point is initial classification according to molecular size of the various components. If an unknown mixture has a variety of molecular weights, it is useful to first use GPC to segment the mixture on the basis of molecular weight (size). Gel permeation chromatography is a very predictable mode, mechanically sorting the compounds based on their dimensions in solution.

FIGURE 4-4. Two approaches to the separation of polynuclear aromatics. (*a*) Reverse-phase separation of isomeric 4-ring polynuclear aromatics using a gradient of 70/30 (v/v) to 100/0 (v/v) acetonitrile/water as shown beneath the chromatogram. Column: C_{18}; detection at 254 nm. (*b*) Normal-phase separation of aromatic hydrocarbons. Column: μPorasil (silica, 10 μm) 3.9 mm ID × 30 cm (2 columns); mobile phase: hexane; flow rate: 8 mL/min. (Fig. 4-4*b* reproduced from reference 1 with permission.)

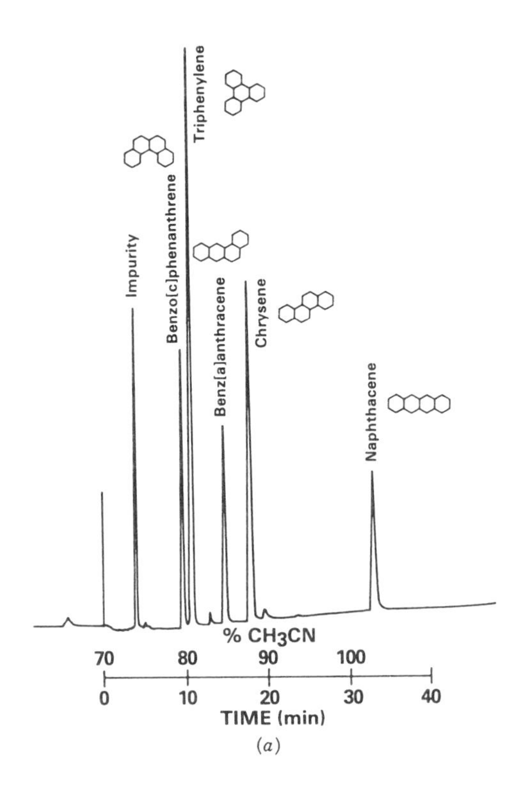

Triphenylene
Impurity
Benzo[c]phenanthrene
Benz[a]anthracene
Chrysene
Naphthacene
% CH3CN
70
80
90
100
0
10
20
30
40
TIME (min)

(a)

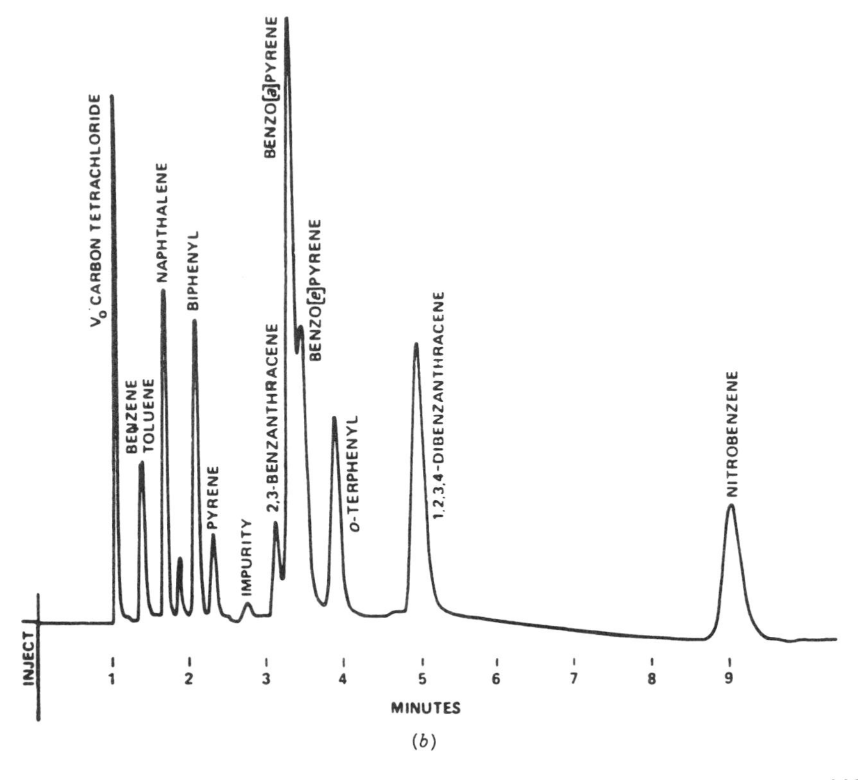

V0 CARBON TETRACHLORIDE
BENZENE
TOLUENE
NAPHTHALENE
BIPHENYL
PYRENE
IMPURITY
2,3-BENZANTHRACENE
BENZO[a]PYRENE
BENZO[e]PYRENE
o-TERPHENYL
1,2,3,4-DIBENZANTHRACENE
NITROBENZENE
INJECT
1
2
3
4
5
6
7
8
9
MINUTES

(b)

Each fraction (size) obtained can be subsequently analyzed by another LC mode (if necessary) with less difficulty than might be presented by the original starting mixture.

When the probable molecular weights (or sizes) of the sample components are known selecting the particular type of column porosity is usually straightforward (Step 3). Column packings are chosen by referring to the manufacturer's literature, an example of which is shown in Figure 4-5, to match the molecular weight range of the components. For species in the 100–1000 molecular weight (MW) range, differences on the order of 40–50 MW units should exist between components for discrete resolution. Column lengths are determined by the magnitude of the differences. Longer column lengths of a given exclusion-limit packing are required as the size differences between components diminish. The solvent for the sample is the final choice in GPC. The solvent must be of sufficient polarity to eliminate any attractions between the sample and the stationary phase. Tetrahydrofuran (THF) is a very popular GPC solvent because of its strong solvating power and its compatibility with the popular porous polymer stationary phases.

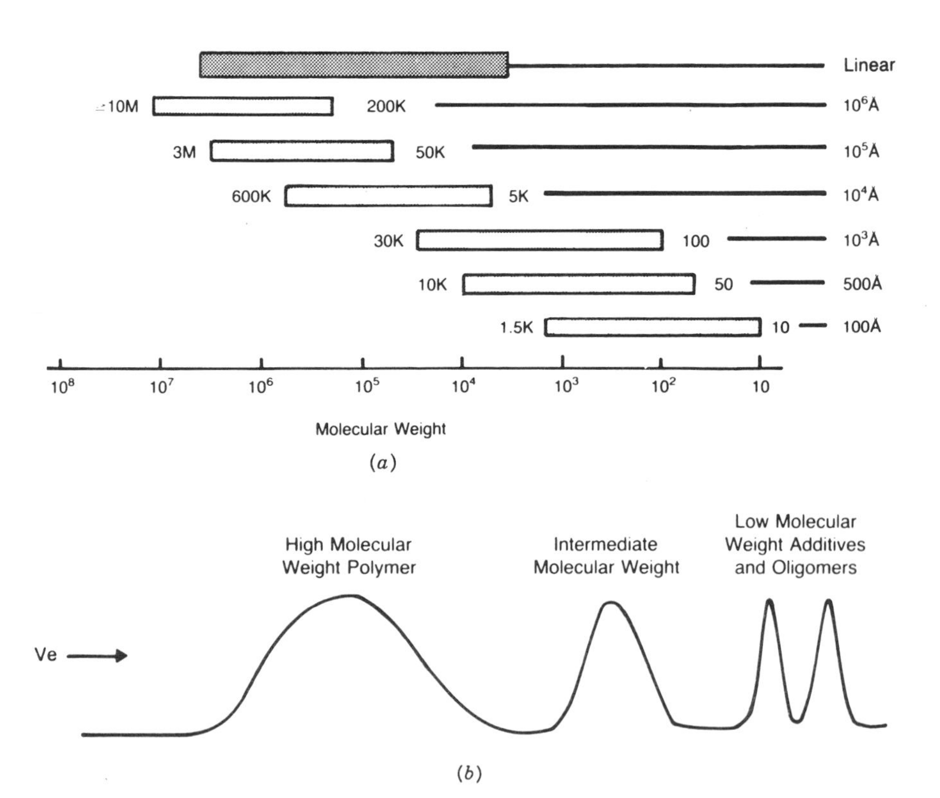

FIGURE 4-5. Effective molecular weight range and column type for size separations (*a*) and an example chromatogram of a gel permeation separation (*b*).

TABLE 4-2. Typical Stationary Phases for LC

Packing Type	Surface Functionality	Polarity
Solid phase (Adsorption)		
silica	—Si—O—Si—OH	High
alumina		
acid	—O—Al—Cl	High
basic	—O—Al—O—Na	High
neutral	—O—Al—O—Al	High
Bonded phase[a]		
cyano	—C≡N	Intermediate
diol	—OH and—C=O	Intermediate/low
ether	$—(CH_2—OCH_2)_n—CH_2OH$	Intermediate/low
	Aromatic—O—	Low
	Aromatic—N—	Low
octyl	$—(CH_2)_8H$	Low
phenyl	$—C_6H_5$	Very low
octadecyl	$—(CH_2)_{17}—CH_3$	Very low

[a] Common names are noted where appropriate.

Normal and Reverse Phase

Chapter 3 summarized the basic considerations governing all retention in normal and reverse phase (Table 3-2). However, sometimes it is difficult for newcomers to chromatography to have a feel for the interrelationship between the sample retention and solvent polarities when used in the normal and reverse phase. This relationship between elution order of the sample components is shown in Figure 4-6. The solvent polarities generally used in normal phase range from low to medium. Considering a two-component sample mixture (A and B) in the normal phase mode, upper portion of the diagram, the lower-polarity sample component (B) elutes first. Increasing the solvent polarity from low to medium with a given packing reduces the elution volume of the components. Conversely, in reverse phase (lower portion in Fig. 4-6), the polar sample component (A) elutes first, and, increasing the solvent polarity from medium to high increases the elution volume. As you develop separations, you will regularly use and observe these principles to solve your particular problems.

Differences in Stationary Phases

Once a mode of LC is chosen (Step 2), another uncertainty arises when deciding which specific packing should be used (Step 3). It is helpful to consider the available unbonded-phase (adsorption) and bonded-phase pack-

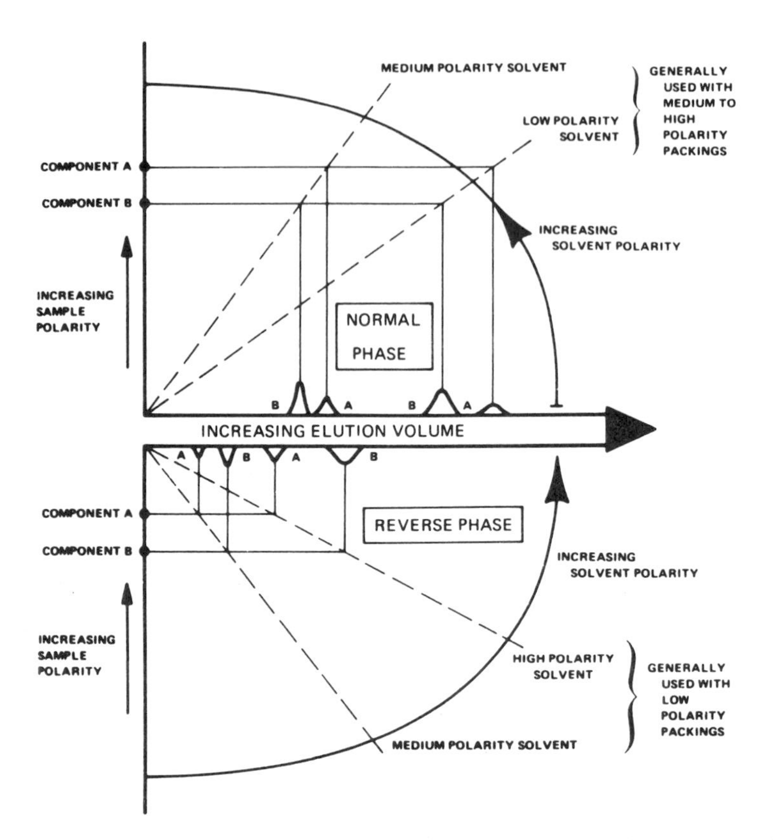

FIGURE 4-6. General retention relationship between sample and solvent as a function of polarity.

ings as a "continuum" of available packing surface polarities which can be used to achieve separations. Table 4-2 lists the common types of stationary phases from high to low polarity.

Examples of how these polarity differences are manifested in chromatography are shown in Figures 4-7 to 4-10. As shown in Figure 4-7 the longest alkyl chain packings are the most nonpolar and exhibit the longest retention for a sample whose separation is based upon nonpolar interactions. Of the three packings in Figure 4-7, the cyano (CN) bonded phase is the least hydrophobic (most polar) and exhibits the lowest retention times. The phenyl bonded phase is of intermediate nonpolar nature and the test compounds exhibit longer retention on this packing compared to the CN bonded phase. The C_{18} bonded phase is the most nonpolar (least polar) of these three phases and the longest retention for the sample molecules occur on this packing. If the C_{18} phase is assigned a retention value of 1, the phenyl phase is about two thirds and the CN phase has about one fourth of the retentiveness of the C_{18} phase. Also, notice that there are no reversals in retention

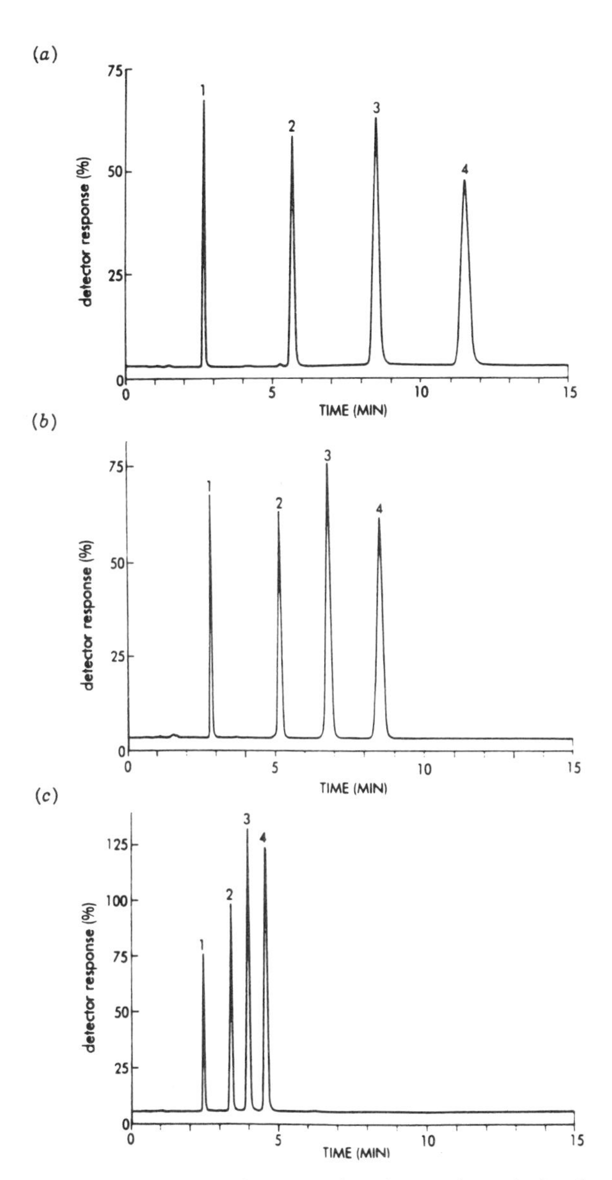

FIGURE 4-7. Comparison of relative retention due to the polarity (hydrophobicity) of C_{18}, phenyl, and cyano (CN) bonded phases. Column: Nova Pak family; (4μ) 3.9 mm ID × 15 cm. (*a*) C_{18} phase. (*b*) Phenyl phase. (*c*) CN phase. Mobile phase—acetonitrile: water, 35:65; flow rate: 2 mL/min; detection: 254 nm; sample: 1. benzyl alcohol, 2. acetophenone, 3. *p*-tolualdehyde, 4. anisole. Nova-Pak CN is the least hydrophobic and therefore the least retentive, followed by the more hydrophobic Nova-Pak Phenyl, and Nova-Pak C_{18} with the most hydrophobicity and the longest retention times.

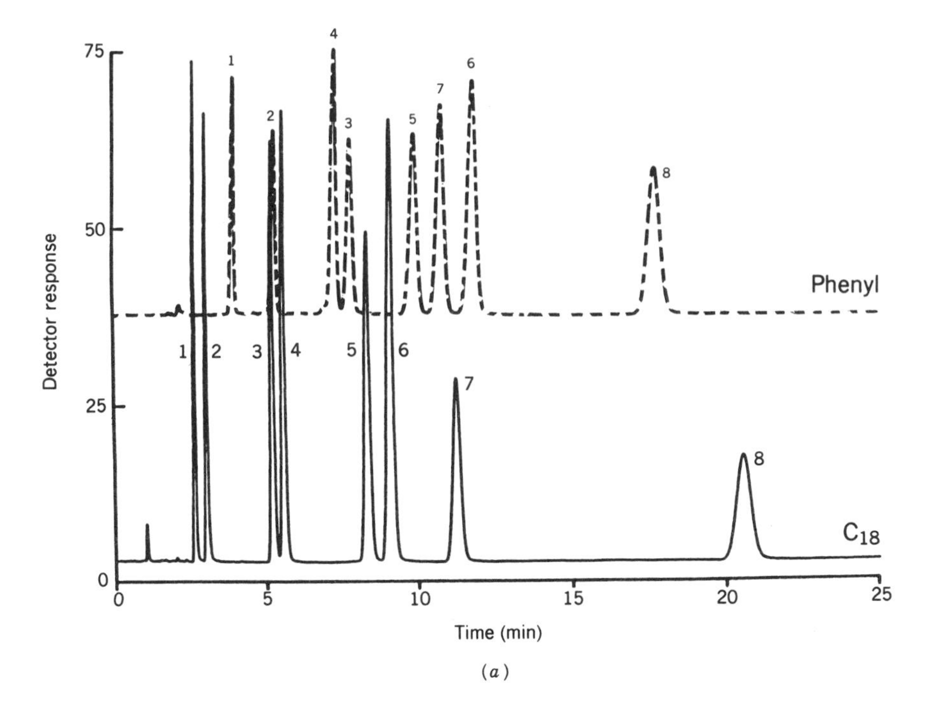
Detector response
75
50
25
0
Phenyl
C18
Time (min)
0
5
10
15
20
25
1
2
3
4
5
6
7
8

(a)

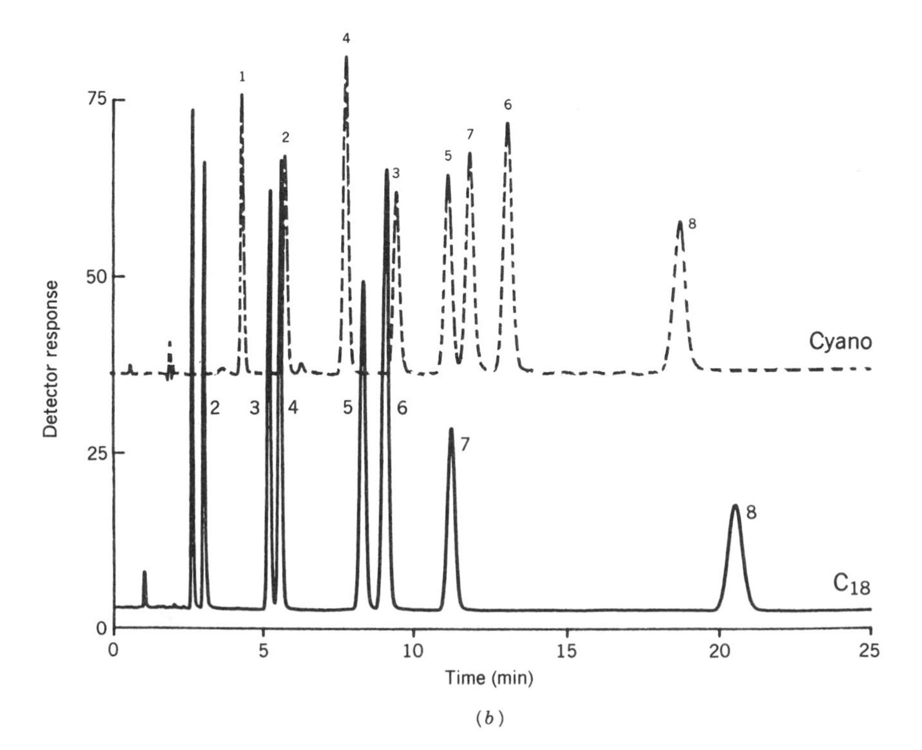
Detector response
75
50
25
0
Cyano
C18
Time (min)
0
5
10
15
20
25
1
2
3
4
5
6
7
8

(b)

FIGURE 4-8. Comparison of optimized separation on phenyl and cyano (CN) phases compared to a C_{18} phase. Columns: 8 mm ID × 10 cm Radial Pak cartridges containing Nova-Pak bonded phase. (*a*) Nova-Pak C_{18} using a mobile phase of acetonitrile: water (35:65), and Nova-Pak phenyl using a mobile phase of methanol: acetonitrile:water (33:8:59). (*b*) Nova-Pak C_{18} using a mobile phase the same as in *a* and Nova-Pak CN using a mobile phase of tetrahydrofuran:water (10:90). Flow rate: 2 mL/min. Detection: 254 nm. Sample: 1. benzyl alcohol; 2. 2-phenoxyethanol; 3. anisaldehyde; 4. acetophenone; 5. *p*-tolualdehyde; 6. *p*-methylacetophenone, 7. anisole; 8. phenetole.

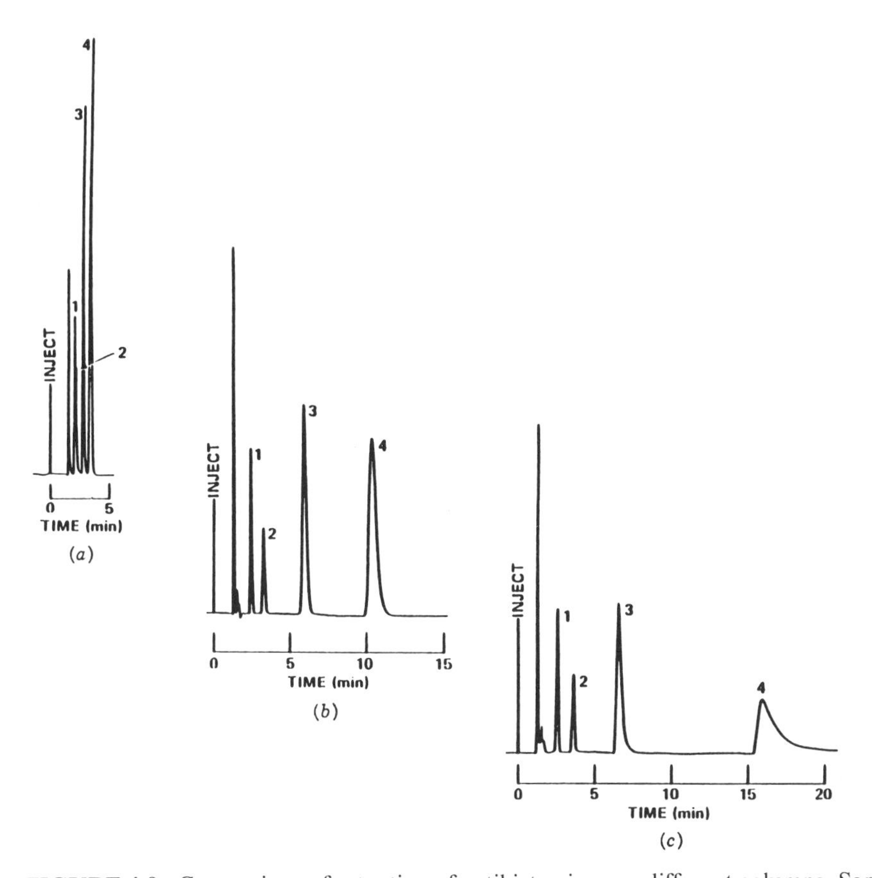

FIGURE 4-9. Comparison of retention of antihistamines on different columns. Sample components: 1. phenylephrine; 2. phenylpropanolamine; 3. naphazoline; 4. chlorpheniramine. Columns: 4 mm ID × 30 cm μBondapak phase. (*a*) Cyano phase. (*b*) Phenyl phase. (*c*) C_{18} phase. Mobile phase: methanol/water containing 0.005 M heptane sulfonic acid (PIC-B-7). Flow rate: 2 mL/min in *a* and 3 mL/min in *b* and *c*. Detector: UV at 254 nm, 0.1 AUFS.

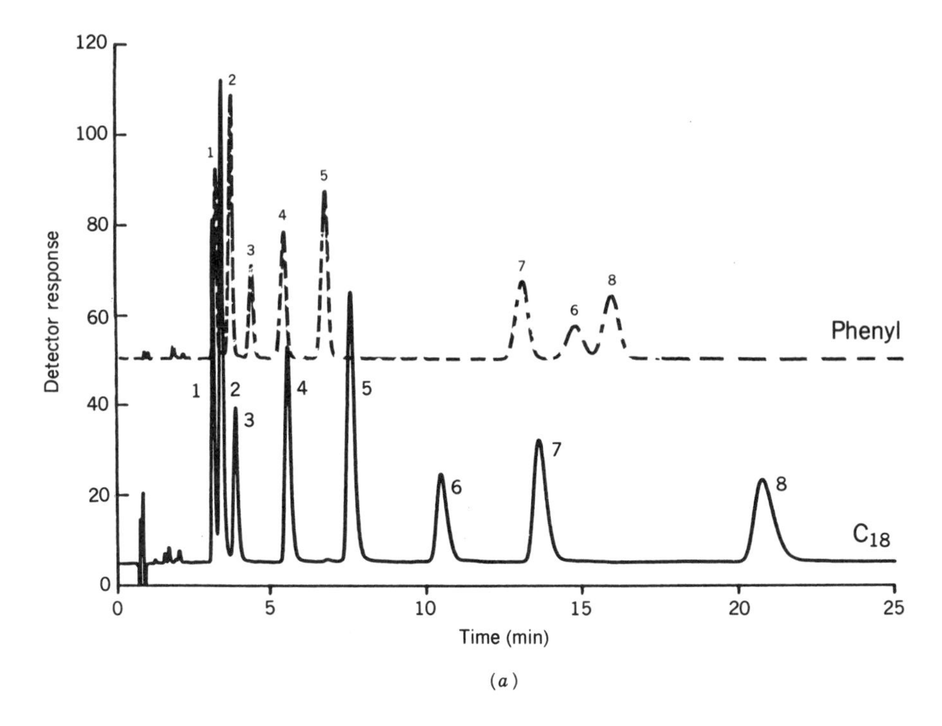
120
100
80
60
40
20
0
Detector response
1
2
3
4
5
7
6
8
Phenyl
1
2
3
4
5
6
7
8
C18
0
5
10
15
20
25
Time (min)

(*a*)

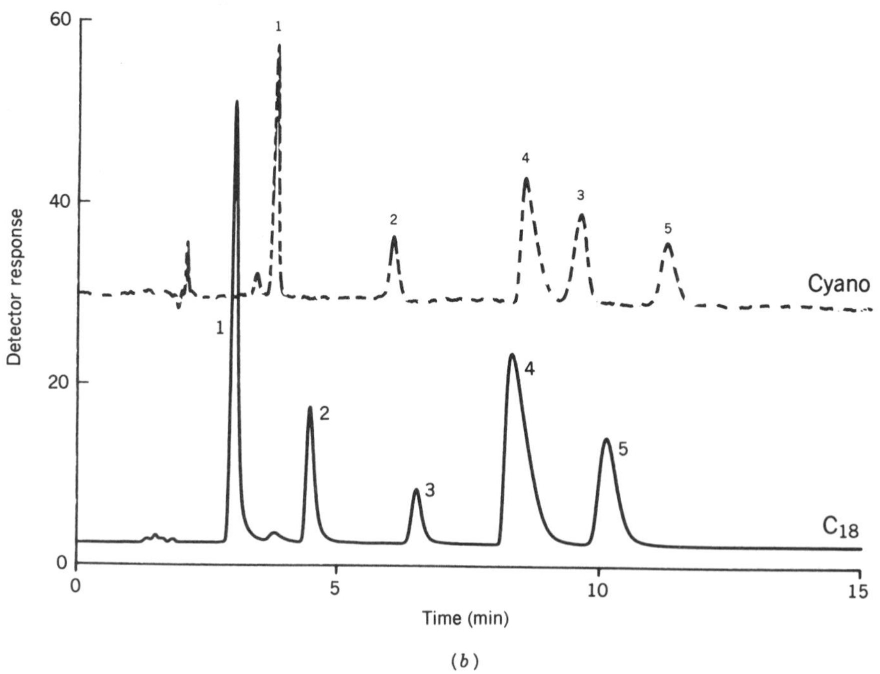
60
40
20
0
Detector response
1
2
4
3
5
Cyano
1
2
3
4
5
C18
0
5
10
15
Time (min)

(*b*)

FIGURE 4-10. Comparison of retention on different phases. Columns: 8 mm × 10 cm Radial Pak containing Nova-Pak bonded phases. (*a*) The mobile phase for the C_{18} column is acetonitrile:water (17.5:82.5) with 5 mM tetrabutylamine (PIC A); the mobile phase for the phenyl column is acetonitrile:water (17.5:82.5) with 5 mM tetrabutylamine (PIC A). Flow rate: 4 mL/min. Detection: 254 nm. Sample—cephalosporin antibiotics: 1. cephradine; 2. cefotaxime; 3. cefazolin; 4. cefuroxime; 5. cefoxitin; 6. cefoperazone; 7. cefamondole; 8. cephalothin. (*b*) The mobile phase for the C_{18} column is tetrahydrofuran:methanol:water (10:27.5:62.5) with 5 mM octanesulfonic acid (PIC B8); the mobile phase for the CN column is tetrahydrofuran:methanol:water (5:22:73) with 5 mM octanesulfonic acid (PIC B8). Flow rate: 2 mL/min. Detection: 254 nm. Sample—Beta-adrenergic blockers: 1. atenolol; 2. nadolol; 3. pindolol; 4. metoprolol; 5. timolol.

order. Most retention-order reversals are due to polar interactions with the stationary phase (e.g., electrostatic) in addition to the nonpolar interactions. When this occurs, the separation is due to a "mixed mechanism."

While it is tempting to assume from this example that the C_{18} is the "best" phase, it is incorrect. Generally a separation can be optimized on any stationary phase. Figure 4-8 illustrates this point by showing an eight-component mixture adequately separated on the CN, phenyl, and C_{18} phases using different mobile phases. Note that all three packing materials are able to separate the eight compounds with baseline resolution. Therefore, a "less retentive" packing does not mean a "less valuable" packing. Less retentive packings require weaker mobile phases to elute all of the compounds. In reverse-phase separations, this generally implies using a mobile phase with a higher water content. The role of mobile phases in developing a separation is discussed in Chapter 5.

A comparison of retention for common antihistamines on a C_{18}, phenyl, and CN column using a mobile phase containing a paired-ion additive is shown in Figure 4-9. As was observed in Figures 4-7 and 4-8, the compounds have the longest retention on the C_{18} column followed by the phenyl and the CN even when the mobile phase contains an additive.

Sometimes the relative retentions do not strictly follow the trend exhibited in Figures 4-7, 4-8, and 4-9. This is the situation in Figure 4-10. Looking at the comparison of the CN and C_{18} phases (Figure 1-10*a*) shows a reversal of the location of peak 6 with respect to peaks 7 and 8 for the cephalosporin antibiotics. In other words, there is a mixed mechanism causing retention. The example in Figure 4-10*b* also shows a reversal; however, since the mobile phases on each column are slightly different with regard to the organic component and content, it is not clear that this retention time reversal is due to a column- or mobile-phase-induced mixed mechanism.

Differences in the Mobile Phases

Just as the polarities of the sample and packing enter into the separation process, so does the polarity of the mobile phase. A successful separation is achieved when a proper balance is established between the retention of the sample components for both the packing and moving solvent. If the attraction of the sample is too strong toward the packing, excessive elution volumes will result. If the sample components are all very soluble in the eluent and only marginally attracted by the packing, no retention on the column will occur.

The chromatographer must become comfortable with a relative understanding of a solvent's ability to be an effective mobile phase. One such relative ranking of a solvent's "strength" is its relative polarity. The relative polarities of commonly used LC solvents are referred to as the eluotropic series (shown on the left side of Fig. 4-11). While polarity is a useful basis for ranking solvents, other characteristics such as hydrogen bonding properties and dispersive (London) forces also contribute to solvating strength. All of these factors are taken into account in the "solubility parameter" (2,3). A discussion of the solubility parameter is beyond the scope of this text, but may be found in most recent physical chemistry texts and general reviews. Solvents in the eluotropic series with comparable polarity rankings may have substantially different solubility parameters.

Because the general guideline in chromatography is to balance the mobile phase strength (polarity) with the stationary phase strength (polarity) with regard to the sample's polarity, packings are most often used with dissimilar type solvents. In Figure 4-11, the packings are positioned next to the most commonly used solvents. Recalling that solubility parameter differences often exist within a given solvent polarity range, it is often expedient to screen a series of appropriate solvents with a single packing during the early stages of work on a separation to see if selectivity differences exist in the LC separation.

Difference in Ionic Functionality of the Sample Components

The use of ion exchange depends upon specific knowledge of the ionization characteristics of the functional groups in the sample. Ion exchange packings are strong or weak and either cation or anion exchangers. The terms "strong" and "weak" ion exchangers refer to the charge characteristics of the stationary phase as shown in Figure 4-12. Strong exchangers stay ionized over a wide pH range because a strong acid or base is the surface functionality. For example, a strong anion exchanger might contain a quaternary ammonium site that is fully ionized up to a high pH. A strong cation exchanger might contain a sulfonic acid site that is fully ionized above a pH of 1. Weak exchangers might be a carboxylic acid (cation exchanger) or an amino group (anion exchanger) ionized only over a restrictive range of pH. A summary of

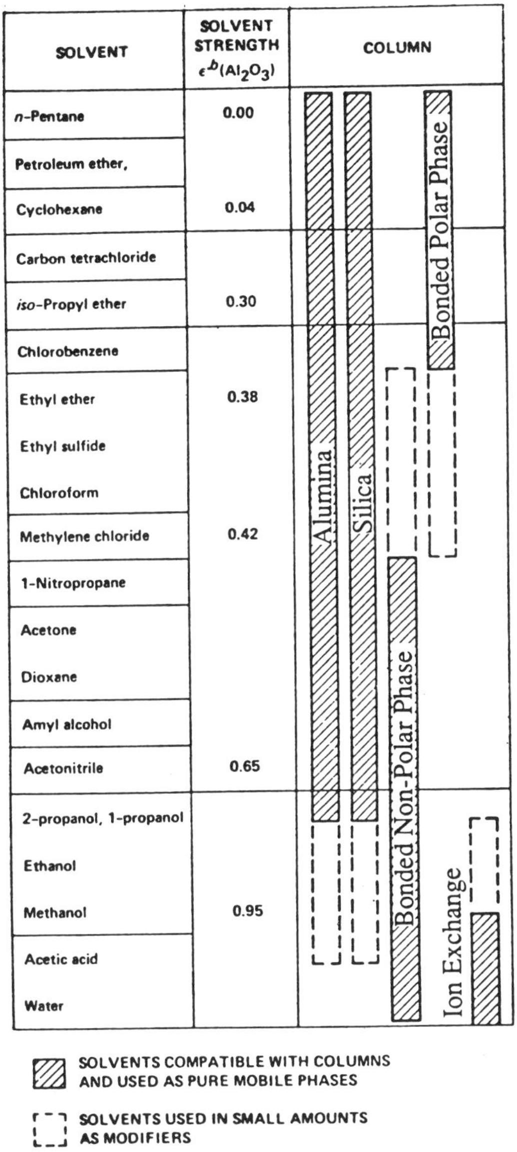

SOLVENT	SOLVENT STRENGTH $\epsilon^b(Al_2O_3)$
n-Pentane	0.00
Petroleum ether,	
Cyclohexane	0.04
Carbon tetrachloride	
iso-Propyl ether	0.30
Chlorobenzene	
Ethyl ether	0.38
Ethyl sulfide	
Chloroform	
Methylene chloride	0.42
1-Nitropropane	
Acetone	
Dioxane	
Amyl alcohol	
Acetonitrile	0.65
2-propanol, 1-propanol	
Ethanol	
Methanol	0.95
Acetic acid	
Water	

COLUMN

Alumina

Silica

Bonded Polar Phase

Bonded Non-Polar Phase

Ion Exchange

SOLVENTS COMPATIBLE WITH COLUMNS AND USED AS PURE MOBILE PHASES

SOLVENTS USED IN SMALL AMOUNTS AS MODIFIERS

FIGURE 4-11. Eluotropic series and column packing usage.

the guidelines for matching the various ion exchange packings is given in Table 4-3.

Ion exchange packings are available as porous beads or in pellicular form. The pellicular packings have very low exchange capacity, typically 60 microequivalents per gram of packing versus 4 milliequivalents with the porous types. A pellicular packing dictates the use of very dilute sample and buffer

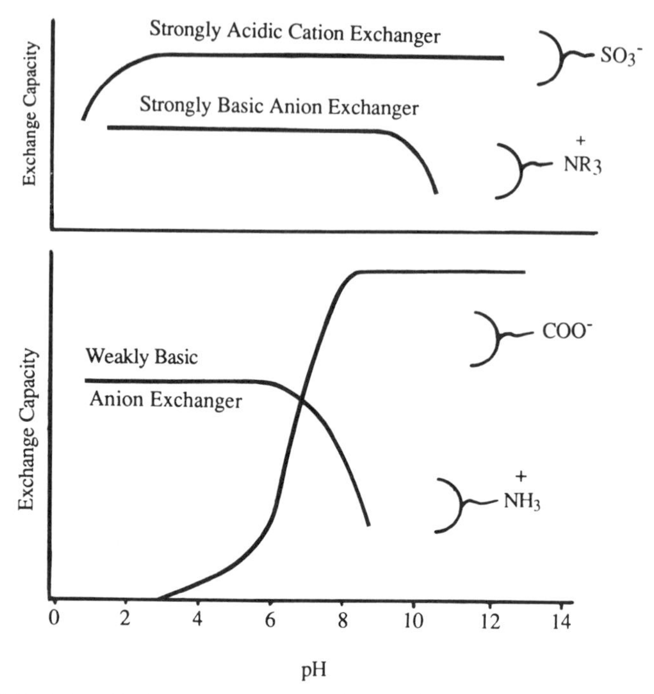

FIGURE 4-12. Exchange capacity of ion exchangers as a function of pH.

TABLE 4-3. Matching the Packing and Sample Type for Ion Exchange Separations

Type	Typical Packing Functional Group	Samples	Typical pH Range
Anion			
Strong	Quarternary ammonium salt	Free anions from acid or base salts	1–10
Weak	Amine salt	Strong and moderately strong acids, weak bases	4–8
	Diethylamino ethyl (DEAE)	Proteins	1–13
Cation			
Strong	Sodium sulfonic acid	Free cations from acid or base salts	1–10
Weak	Carboxylic acid	Strong and moderately strong bases, weak acids	4–8
	Carboxymethyl (CM)	Proteins	1–13

concentrations. In all ion exchange separations, the pH and the buffer composition and concentration can each have profound effects on the separation. A knowledge of the influence of these factors on the ionization of the sample and on the ionization of the packing is required. Temperature can also have a very significant effect on the performance of the ion exchange packings.

What Happens After Choosing the LC Packing?

After choosing the mode of LC (Step 2) and picking the packing (Step 3), the chromatograms most likely to occur after the "first try" are shown in Figures 4-13 and 4-14. The next actions (Step 4) to take in each instance to achieve satisfactory resolution are also listed with each chromatogram in the recommended order. While some of these action items may not be fully understood by the reader at this time, reading Chapters 5 and 6 should clarify the detailed meaning of each suggestion. Chapter 5 details the method development process. As you begin to develop the desired degree of separation, the changes will become more subtle.

There is one note of caution—while you are probing for a workable system with little knowledge of possible sample-packing interactions, some sample components may remain on the column. Consequently, if there is any doubt that compounds have eluted from each exploratory injection, a very strong mobile phase (solvent) for the sample should be flushed through the column to clean it thoroughly. (However, when using a solid-phase, adsorption column in the normal-phase mode, care should be taken not to "deactivate" the column by flushing a very strong solvent which will "irreversibly" adsorb to the column, e.g., water or methanol.) The next mobile phase used should be allowed to equilibrate with the column packing before proceeding with the next injection.

"Fine tuning" of the separation can also take the form of increasing the column efficiency by changing the flow rate. An example is shown in Figure 4-15. By decreasing the flow rate, the efficiency is more than doubled. This variable of slowing flow rate to increase plates should be used only as a last resort since the increase in efficiency is gained at the price of an increase in analysis time. Flow rate effects are discussed in Chapters 5 and 6 as well as in the experiment in Chapter 8. In the example in Figure 4-15, a twofold increase in plates comes at the expense of a fourfold increase in analysis time.

SUMMARY

After defining the separation problem, the next step is to consider the characteristics of the sample (structure and functional groups, molecular weight, and solubility), and to search the available literature. Using the knowledge at

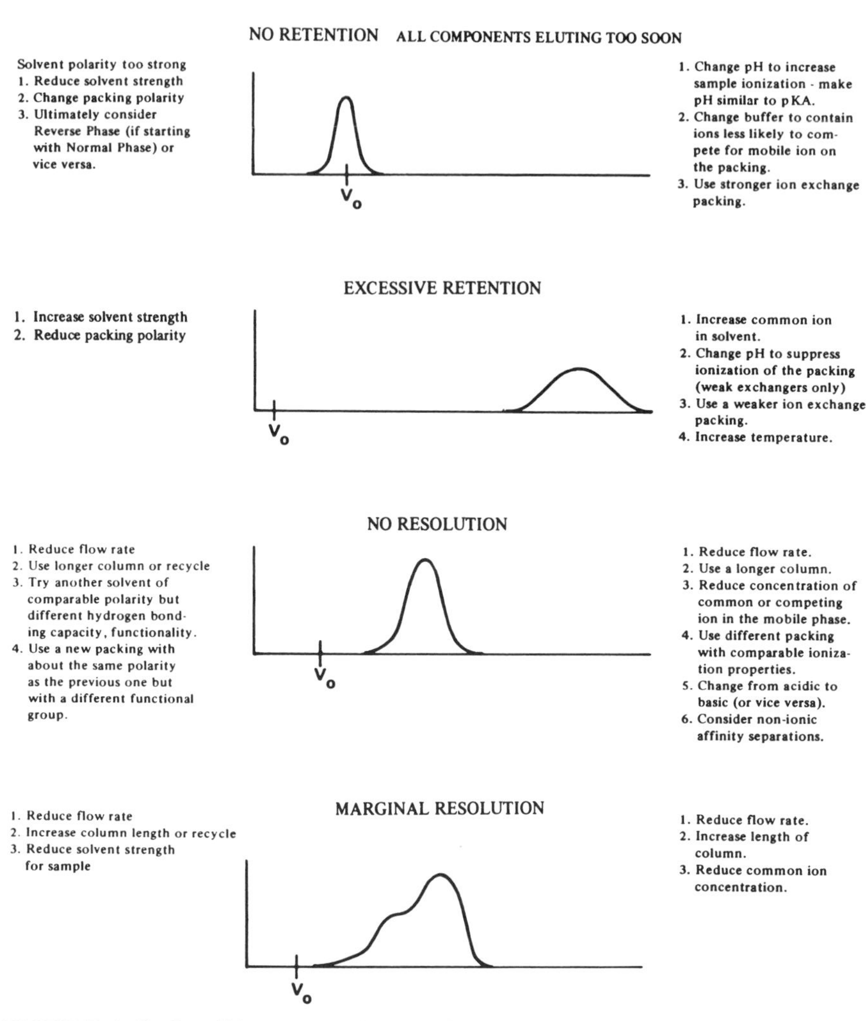

NON-IONIC MODES
ION EXCHANGE
NO RETENTION ALL COMPONENTS ELUTING TOO SOON
Solvent polarity too strong
1. Reduce solvent strength
2. Change packing polarity
3. Ultimately consider Reverse Phase (if starting with Normal Phase) or vice versa.
1. Change pH to increase sample ionization - make pH similar to pKA.
2. Change buffer to contain ions less likely to compete for mobile ion on the packing.
3. Use stronger ion exchange packing.
EXCESSIVE RETENTION
1. Increase solvent strength
2. Reduce packing polarity
1. Increase common ion in solvent.
2. Change pH to suppress ionization of the packing (weak exchangers only)
3. Use a weaker ion exchange packing.
4. Increase temperature.
NO RESOLUTION
1. Reduce flow rate
2. Use longer column or recycle
3. Try another solvent of comparable polarity but different hydrogen bonding capacity, functionality.
4. Use a new packing with about the same polarity as the previous one but with a different functional group.
1. Reduce flow rate.
2. Use a longer column.
3. Reduce concentration of common or competing ion in the mobile phase.
4. Use different packing with comparable ionization properties.
5. Change from acidic to basic (or vice versa).
6. Consider non-ionic affinity separations.
MARGINAL RESOLUTION
1. Reduce flow rate
2. Increase column length or recycle
3. Reduce solvent strength for sample
1. Reduce flow rate.
2. Increase length of column.
3. Reduce common ion concentration.
Vo
Vo
Vo
Vo

FIGURE 4-13. Possible results after the first injection in the retention mode and suggested action items.

hand and Figure 4-1, choose the "best" LC mode based upon exploiting the size, ionic functionality, or polarity differences that exist in the sample components.

If you know nothing about the sample except its solubility, refer to Figure 4-1 and attempt to choose a mode complementary to the polarity of the solvent. For instance, if the sample dissolves in heptane and chloroform, choose a normal-phase mode. If the sample is soluble in methanol or water/

NO COMPONENTS RETAINED

Packing exclusion limit too low, with all sample components excluded and eluting at the solvent front.
Use a higher exclusion limit packing.

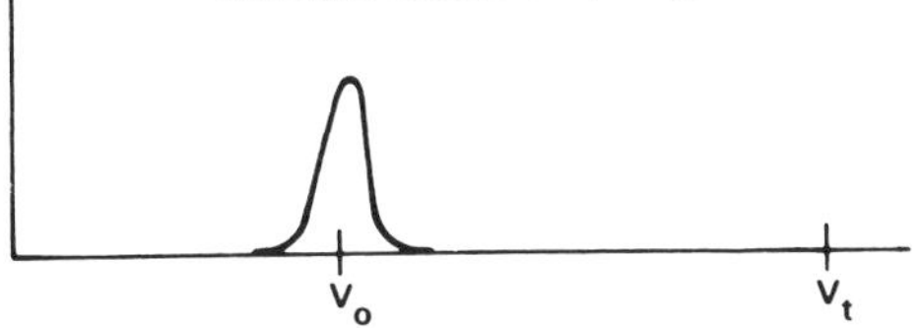

NO COMPONENTS EXCLUDED

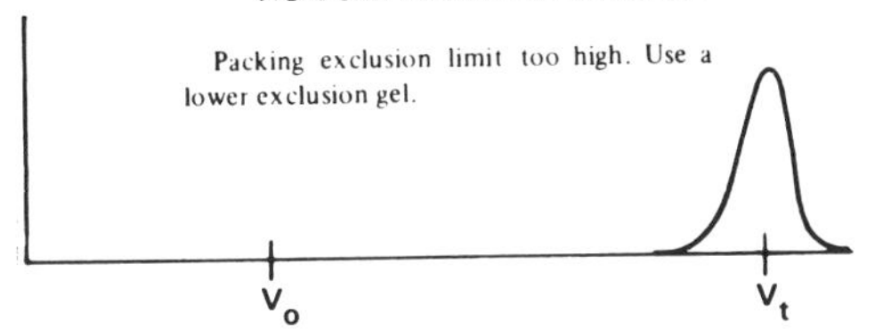

SOME ELUTION IN MORE THAN ONE COLUMN VOLUME

Some components adsorbing on the packing. Use a stronger solvent system to insure exclusion as the only operative mode.

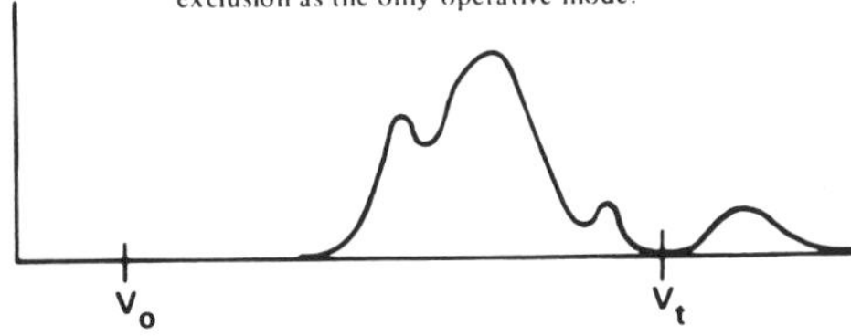

RESOLUTION MAY NOT BE ADEQUATE

Use more columns with the exclusion limit corresponding to the retention zone in this chromatogram. Consider recycle.

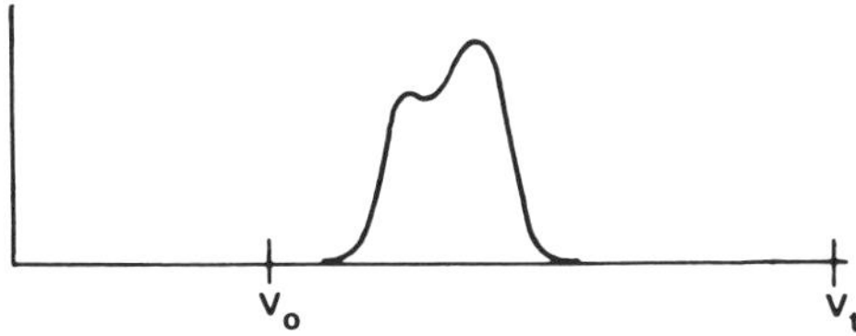

V_o - Interstitial Volume
V_t - Total Column Void Volume - All components will elute within V_t in true GPC.
V_t-V_o - Packing pore volume

FIGURE 4-14. Possible results after the first injection in the gel permeation mode and suggested action items.

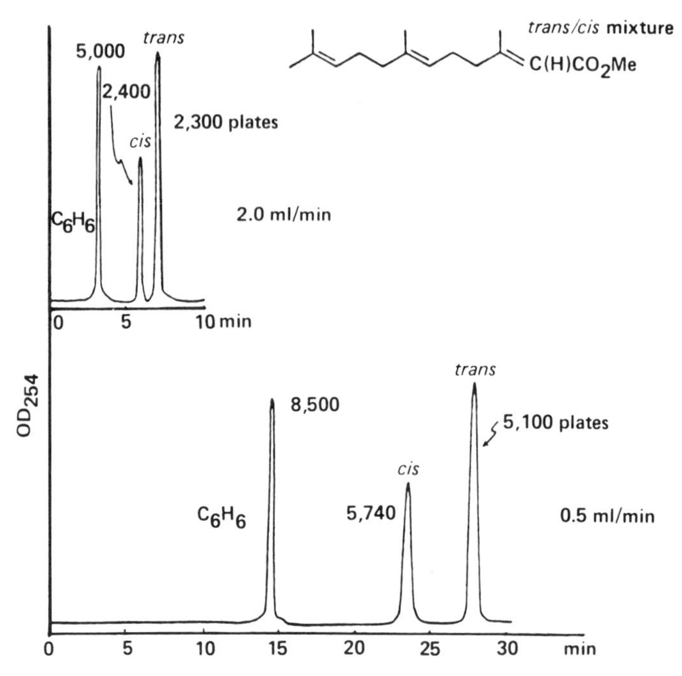

FIGURE 4-15. Effect of flow rate upon resolution.

methanol, choose a reverse-phase mode. Selecting the specific packing can be facilitated by referring to Tables 4-2 and 4-3 and Figures 4-5 and 4-11. Choosing the initial mobile phase for your sample is the final activity. Figure 4-11 is helpful in choosing solvents most frequently used with certain packings. These solvents are blended to make the initial mobile phase. After the initial column/mobile phase choice is made, make the first injection and develop the separation. Begin with a mobile phase that is a strong solvent for your sample. This tends to bring all your components off the column rapidly and give you a basis for increasing your elution volumes by subsequent adjustment of solvent strength.

According to the fundamental resolution equation (Chapter 3, equation 3-9) improving resolution results from the impact of the physiochemical changes on k', α, and/or N. When developing a separation, changing certain parameters will impact k', α, or N as shown in Table 4-4. Some of the items listed in Table 4-4 are obvious contributions, others are not so obvious. Remember that changing only one parameter may make dramatic changes upon resolution (e.g., mobile phase), whereas other changes (e.g., temperature) may have only a small effect upon resolution. As you read about the examples of developing a separation in Chapters 5 and 6, referring back to Table 4-4 will help you to determine how a certain parameter change influences resolution and specifically which term in the resolution equation is being affected.

TABLE 4-4. Parameters that can be Changed to Influence the Terms in the Resolution Equation

	For Bonded Phase, Liquid–Solid, and Liquid–Liquid
k'	Strength of mobile phase (polarity, hydrogen bonding, etc.) Strength of packing (surface area, amount of stationary phase and functionality) Temperature
α	Chemistry of mobile phase (polarity, hydrogen bonding, etc.) Chemistry of packing (functionality) Chemistry of sample (e.g., derivatization)
N	Flow rate (linear velocity) Column length Average particle size of the packing Particle size distribution How well the column is packed Pellicular vs. porous particles (mass transfer term) Volume of injection Mass of injection Viscosity of injected solution Viscosity of mobile phase (mass transfer term) Temperature (as it controls solvent viscosity) Diameter of column (ease of packing) Shape of particle (spherical or irregular and ease of packing) Extra column effects (dead volume) Residence on column (longer k' gives lower N) Mixed mechanisms (ionization, tautomerism, competitive adsorption)
	For Gel Permeation
k'	Pore size of packing
α	Chemistry of sample (hydrogen bonding, if molecule is extended or coiled, and making derivatives)
N	Same as parameters given above for N
	For Ion Exchange
k'	Ionic strength of mobile phase Capacity of packing (number of equivalents per gram, porous or pellicular) pH (effect on packing) Temperature
α	Chemistry of mobile phase (type of buffer and/or counter ion) Chemistry of packing (strong or weak exchanger and pH) Chemistry of sample (complexing agents and derivatives) pH (effect on ionization of sample and of packing) Addition of organics to mobile phase Temperature
N	Same as parameters given above for N

REFERENCES

1. R. V. Vivilecchia, R. L. Cotter, R. J. Limpert, N. Z. Thimot, and J. N. Little, *J. Chromatogr.,* **99,** 407 (1974).
2. L. R. Snyder, Chapter 2 in *Techniques in Chemistry,* 2nd ed., Vol. III, Part I, A. Weissberger and E. S. Perry, Eds., Wiley-Interscience, New York, 1978.
3. L. R. Snyder and J. J. Kirkland, *Introduction to Modern Liquid Chromatography,* Wiley-Interscience, New York, 1979, p. 259.

ACKNOWLEDGMENT

Figures 4-3, 4-5 through 4-10, 4-13 through 4-15 are courtesy of the Waters Chromatography Division of Millipore.

CHAPTER 5

DEVELOPING THE SEPARATION

Reverse phase
- Finding an appropriate mobile phase
- Developing a mobile phase to improve selectivity
- Using nonaqueous mobile phases
- Developing a separation
- Summary

Tactics for separating ionic compounds
- Reverse phase for ionic compounds
- Ion suppression chromatography
- Paired-ion chromatography
- General approach to the separation of ionic compounds
- Models for ion-pair retention
- Using a paired-ion reagent of the same charge as the sample
- Summary of use of paired-ion chromatography
- Separations on bare (unbonded) silica gel
- Ion exchange chromatography
- Ion exchange separation of proteins

Small molecule gel permeation chromatography
- Calibration curves
- Solvent effects
- Column efficiency
- Applications
- Preparative capability
- Summary

Normal phase
- Stationary phases

Mobile phases
Finding an appropriate mobile phase
References
Acknowledgment

When developing a separation, the most influential force is the chemical interaction of the mobile phase, sample, and stationary phase. This is a three-way interaction: (1) mobile phase–sample; (2) mobile phase–stationary phase; and (3) sample–stationary phase. The ability to alter the selectivity of a column by changing the composition of the mobile phase is a major reason why LC is such a versatile and, hence, powerful separation tool. When doing retention chromatography, the mobile phase is "interactive" since it is involved in the three-way chemical interactions. Only in the GPC (size separation) mode is the mobile phase noninteractive. With experience, scientists can easily develop an intuitive knowledge of those chemical interactions that are mainly dependent upon mobile phase and sample properties. At the point when an individual understands the basis for the mechanisms of separation, he/she becomes a chromatographer.

Some may tell you that learning how to develop an appropriate chromatogram is a frustrating, trial-and-error process which can waste time, solvents, and money. This, however, is not true if the scientist understands the strategy and tactics of "method development." Developing a separation involves an appreciation of a logical strategy which is applicable to all modes of separation. This chapter presents guidelines and examples for such a strategy.

Over 80% of the separations that are published use bonded-phase columns in the reverse-phase mode, which indicates that most chromatographers find the reverse-phase mode the most useful for solving their separation problems. Therefore, this chapter emphasizes examples of such separations. However, remember that while the major focus in this chapter is upon the use of bonded phases in the reverse-phase mode, this separation strategy is applicable to all modes.

REVERSE PHASE

"Which bonded-phase column should I use for reverse-phase chromatography?" This is an appropriate question to ask when developing a separation. However, the answer is not very sophisticated. Simply stated, the C_{18} bonded phase is the most appropriate because of its wide versatility for retention of most compounds. In addition, since it is the oldest column type available, it has a long history of documented application usage. Furthermore, it is a rugged and long-lived (6–12 months) column. Some may find it surprising that one single reverse-phase column can be applied to approximately 80% of the reverse-phase separation problems. However, remember that in HPLC the mobile phase is highly influential in accomplishing the separation. This is unlike GC, where the mobile phase (gas) is inert and the column supplies the only interaction with the sample.

As a result of its wide versatility, the C_{18} column has acquired a reputation as the "work horse" of reverse-phase HPLC. Other phases give fine adjustments to retention and selectivity control, as was discussed in Chapter 4. For instance, if retention is too long on the C_{18} column, substitution of a C_8 column results in decreased retention in the same mobile phase. Or, if a separation requires too much organic component in the mobile phase (for your liking), substitution of a C_8 (or C_4) column results in a comparable separation using a lower organic component in the mobile phase. As a rule of thumb, the first choice for a general purpose, bonded-phase column used in the reverse-phase mode is the C_{18} (octadecyl or ODS) phase. As your experience in chromatography increases, additional phases are useful.

Finding an Appropriate Mobile Phase

When using a bonded-phase column in the reverse-phase mode, the search for the best mobile-phase composition for an adequate separation is straightforward and follows a generalized approach. First, identify a strong solvent (e.g., methanol, acetonitrile, or tetrahydrofuran) in which the sample is readily soluble. The second step is to choose a "poor" or "weak" solvent in which the sample has limited solubility. Most often this will be water. Once a "good" and "poor" solvent has been identified, an injection of the sample onto the column should be made using the strong solvent as the mobile phase. In this situation all of the components should elute as one peak at V_0. If there is retention when using a 100% "strong" solvent as the mobile phase, a *stronger* solvent needs to be found. For instance, if retention occurs in 100% methanol, try 100% acetonitrile as the "strong" solvent.

After it has been determined that there is no retention of the sample components in the 100% strong solvent, the mobile phase should be made "weaker" by using a blend of the strong solvent with the poor solvent. The sample is then reinjected into this new mobile phase and the retention behavior is noted. Sequentially weaker mobile phases should be prepared and evaluated until retention of the sample components occurs. Ratios of strong and poor solvent should be done in logical steps (e.g., 100%, 80%, 60%, 40%, 20%) so that retention of the peaks can eventually be "fine tuned" to the appropriate mobile phase, containing an exact blend of the good and poor solvents to achieve the desired resolution. An example of this strategy is shown in Figure 5-1 for a known mixture. Very little retention and separation occurred until 40% acetonitrile in water was used. At 30% acetonitrile, additional retention of the sample components is seen and at 20% acetonitrile the retention is more than is needed. Therefore, the eluent is "fine

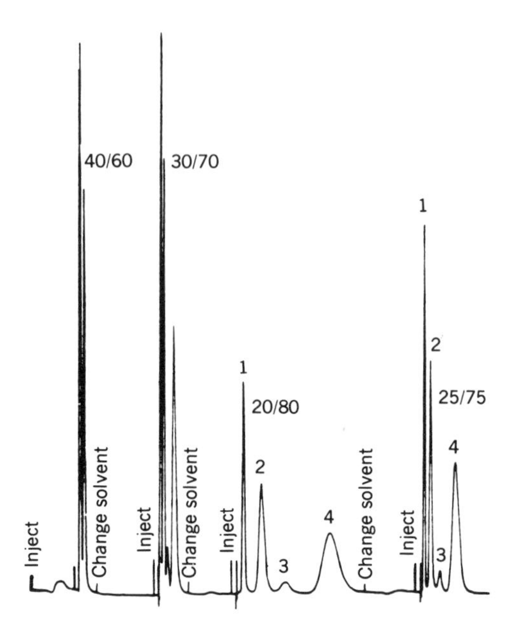

FIGURE 5-1. Determining an appropriate mobile phase for known standards. Column: Bondapak C_{18}/Corasil (pellicular); 2.3 mm ID × 61 cm. Mobile phase: acetonitrile/water (40/60, 30/70, 20/80, and 25/75 denoted on chromatogram). Sample components: (1) hydrocortisone; (2) hydrocortisone acetate; (3) estrone; and (4) methyl testosterone. Detector: UV at 254 nm.

tuned'' back to 25% acetonitrile/75% water, which results in a good separation.

This same approach can be used to develop a separation of an unknown mixture, as shown in Figure 5-2. In this example several of the components of a ''protein shampoo'' were separated by using the general approach of blending a poor solvent with a strong solvent to experimentally determine the appropriate mobile phase for the desired separation. The mobile phase, consisting of 60% methanol and 40% water, was too strong and the solvents were mixed to make weaker mobile phases until the desired separation was obtained. In this example, the mobile phase consisting of 40% methanol and 60% water was the best. In both examples (Figs. 5-1 and 5-2), the process of developing the separation was straightforward with no chromatographic surprises.

You will note in Figure 5-2 that the flow rate was run at 3 mL/min. It is a good policy in methods development to run at a rapid flow rate (within the pressure limits of the HPLC system) to minimize the time involved in developing a method. If we use the example of Figure 5-2 again, each mobile phase would involve approximately 20 min preparation time to blend the solvents and degas the mixture. The column would require approximately 10 min in equilibration before a run. Each chromatogram should be run three

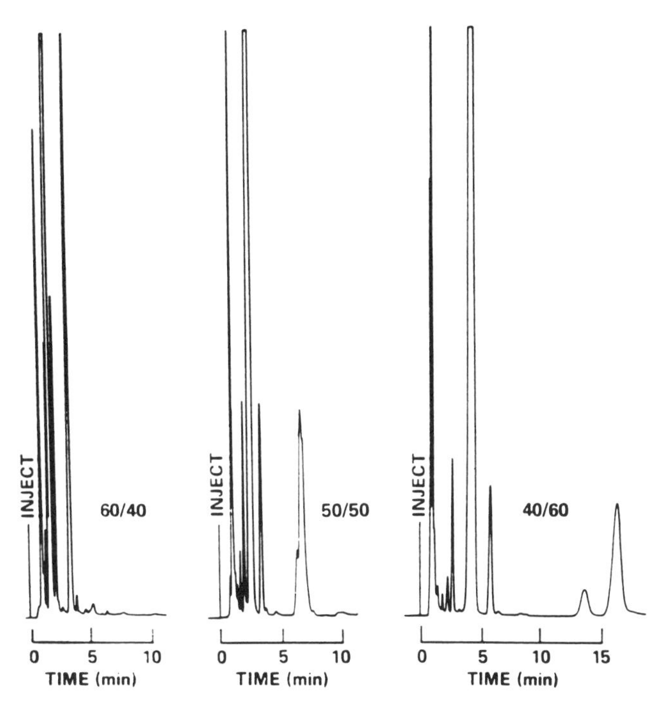

FIGURE 5-2. Determining an appropriate mobile phase for a sample of "protein shampoo." Column: μBondapak C_{18} (10 μm), 3.9 mm ID × 30 cm. Flow rate: 3 mL/min. Mobile phase: methanol/water (60/40, 50/50, and 40/60 denoted on chromatogram). Detector: UV at 254 nm. This is another example of k' optimization to obtain the desired resolution. In this case the solvent of 40/60 of methanol/water was required to obtain the resolution of the last two components.

times to ensure the reproducibility of retention times, which reflects an equilibrated column. Therefore, the method development would take approximately 4 hr, which is one reason why HPLC is one of the most cost-effective separations tools.

Developing a Mobile Phase to Improve Selectivity

When developing a separation, the first tactic is to search for the best binary solvent combination. But sometimes the first choice for the strong solvent does not work out as well as it did in the problems addressed in Figures 5-1 and 5-2. In certain situations blending two solvents may not result in a mobile phase in which a particular separation is possible. It is, therefore, necessary to try a new binary mixture. An example of this is the separation of the steroids shown in Table 5-1, where both water/methanol and water/acetonitrile combinations are used. Examining Table 5-1 shows that the hydrocortisone acetate and dexamethasone are particularly troublesome.

TABLE 5-1. Behavior of Retention (k' Values) in Various Mobile Phases

	Methanol/Water				Acetonitrile/Water				
	60/40 V/V	50/50 V/V	40/60 V/V	30/70 V/V	40/60 V/V	30/70 V/V	25/75 V/V	20/80 V/V	15/85 V/V
Hydrocortisone (CH_2OH, $C{=}O$, CH_3, OH, HO, CH_3, O)		0.61	1.6	5.9		0.6	1.1	2.2	8.4
Hydrocortisone acetate (H, O, $H{-}C{-}O{-}C{-}CH_3$, $C{=}O$, CH_3, HO, CH_3, O)		1.1	3.4			1.2	2.8	7.2	
Prednisone (CH_2OH, $C{=}O$, CH_3, OH, O, CH_3, O)		0.45	1.3				1.0	2.4	8.4
Prednisolone (CH_2OH, $C{=}O$, CH_3, OH, HO, CH_3, O)		0.57	1.6			0.46	0.91	2.3	8.8

Dexamethasone (structure: CH_2OH, C=O, CH_3, OH, HO, CH_3, F, CH_3, O)		1.1	3.0		0.88	2.0	5.0	19.4
Triamcinolone acetonide (structure: CH_2OH, CO, $C(CH_3)_2$, CH_3, O, O, HO, CH_3, F, O)		1.2	4.1		1.7	3.9	9.6	
Methyl testosterone (structure: OH, CH_3, CH_3, CH_3, O)	2.0	7.8	32.3	1.7	6.2	16.5		
Fluoxymesterone (structure: OH, CH_3, CH_3, HO, CH_3, F, O)				0.45	0.75	3.2	7.4	

[a] Chromatographic conditions: column—Bondapak C_{18}/Corasil, 2 mm ID × 61 cm; flow rate—0.8 mL/min; $k' = (V_e - V_0)/V_0$.

There is no separation ($\alpha = 1.0$) in 50% methanol and only a slightly better α value of 1.13 in 40% methanol. When it becomes apparent that the water/methanol binary solvent mixtures will not resolve the desired compounds in a reasonable time, it is appropriate to change to a new, different organic solvent/water mixture and determine if an adequate selectivity of the sample components is possible.

If methanol is not an appropriate solvent, generally the next choice is acetonitrile. In Table 5-1, various concentrations of acetonitrile in water result in improved selectivity for the key pair mentioned above. The best mobile phase in this table is the 25/75 mixture of acetonitrile/water in which the selectivity for the key pair is $\alpha = 2.8/2.0 = 1.40$. The retention for the methyltestosterone is long (16.5), but perhaps not too long considering the favorable selectivity in this eluent. Thus, by changing to a different organic solvent, a mobile phase could be found which gave enhanced selectivity.

The data in Table 5-1 also demonstrate several other behaviors typical of reverse-phase chromatography. First, the elution order of the sample components is from the most polar, eluting first, to the most nonpolar, eluting last. Second, as the eluent is made more polar (more water), the sample compounds are less soluble in the mobile phase, and hence more soluble in the stationary phase. Therefore, retention is increased. The third behavior is that for every compound and stationary phase combination, a variety of solvent mixtures can be used as mobile phases to elute a compound with approximately the same k' value. Mobile phases that have identical eluent strengths are referred to as *isoeluotropic* or *equieluotropic* for that compound. The general trend is for similar retention to exist in isoeluotropic mobile phases. However, isoeluotropic mobile phases sometimes have a different selectivity (α) for a pair of compounds due to secondary interactions between the solvent and analyte.

To emphasize the benefit of enhancing selectivity through the use of isoeluotropic mobile phases, refer again to Table 5-1. A quick inspection of this table reveals that a mobile phase of a 50/50 mixture of methanol/water gives approximately the same k' values for the compounds and a similar range of k' values as is found in a 30/70 mixture of acetonitrile/water. Specifically, for hydrocortisone these two eluents are isoeluotropic. Note that in the 50/50 methanol/water mobile phase, hydrocortisone acetate and dexamethasone have the same k' values and, therefore, $\alpha = 1.0$. In the isoeluotropic acetonitrile/water (30/70) eluent, the α value for those two compounds is $1.2/0.88 = 1.36$. In the acetonitrile/water mobile phase a separation is possible, while in the methanol/water mobile phase no separation occurs.

Occasionally, a binary mixture will not enable the separation to be attained. In this case a ternary mixture may be helpful. There is much in the literature about using more than one organic solvent in the mobile phase for attaining selectivity. Not only do chromatographers share insights about how and why solvents contribute to a separation, but there are also available software routines for computer-assisted development of mobile phases using

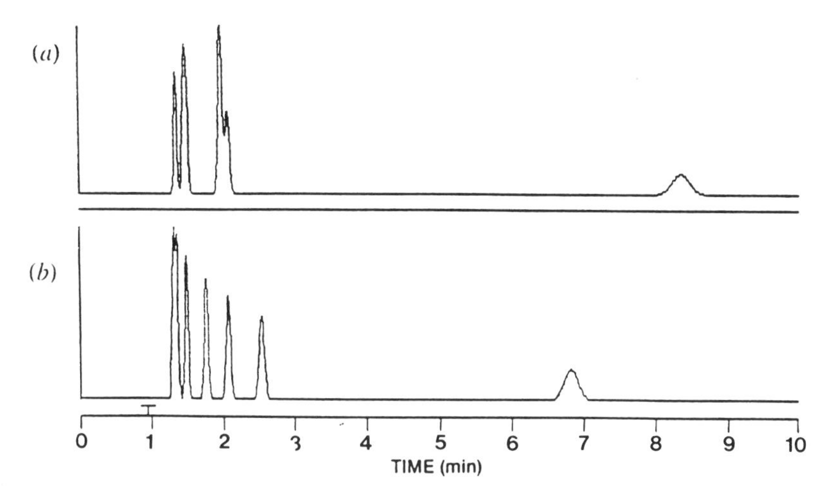

FIGURE 5-3. Separation of seven steroids on Bondapak C_{18}/Corasil II (pellicular) 2.3 mm ID × 61 cm. (*a*) Mobile phase: 50/50 methanol/water. (*b*) Mobile phase: 30/70 acetonitrile/water.

up to a quaternary solvent blend (e.g., methanol/acetonitrile/tetrahydrofuran/water). All of these routines are based upon the type of phenomena exhibited in Table 5-1. Because retention (log k') changes approximate linearly with solvent strength, one can plot log k' versus solvent composition and determine the best mixture of the two binary solvents (methanol/water and acetonitrile/water) which will result in a tertiary solvent (methanol/acetonitrile/water) as the eluent. A detailed description of this approach is beyond the scope of this book, but can be found elsewhere (1). However, if one looks at Table 5-1, the two mobile-phase combinations with good retention times are 50/50 methanol/water and 30/70 acetonitrile/water. The two chromatograms are shown in Figure 5-3. By plotting log k' versus solvent composition (Fig. 5-4) the best mobile phase is 14/27/59 acetonitrile/methanol/water and the chromatogram is shown in Figure 5-5.

Mobile phases that use three solvents are not a common occurrence and are needed only for very difficult problems. Also, developing three (or four) solvent mobile phases often require the use of computer programs. Therefore, the individual should evolve into using complex mobile phases as the experience with HPLC increases.

Using Nonaqueous Mobile Phases

Since reversed-phase separations are based upon differential solubility of the sample in the mobile and stationary phases, it is not essential to use water if solubility limitations exist. Often a blend of two organic solvents which are the strong and weak solvents can and should be used rather than one organic

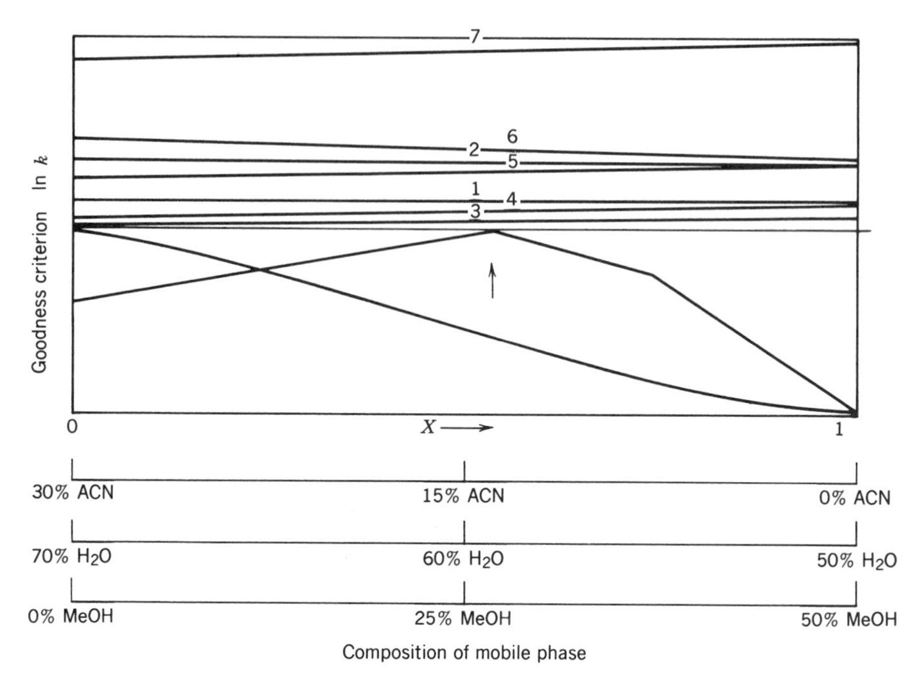

FIGURE 5-4. Computer plot of the log k' versus mobile-phase composition (top) and separation "goodness" criterion (bottom) versus mobile-phase composition for the seven steroids in Table 5-1. Optimum (highest value on "criterion" scale) denoted by arrow is predicted to be 14% acetonitrile, 59% water, and 27% methanol.

solvent (strong solvent) with water (poor solvent). For instance, if the sample is completely insoluble in water, the poor solvent must be something other than water. In the case of the separation of triglycerides (fats), water is definitely a poor solvent but it is "too poor" because the triglycerides are not even partially soluble. Other poor solvents are methanol and acetonitrile. Of these two, acetonitrile exhibits slight solvating power for the triglycerides. A good solvent for triglycerides is tetrahydrofuran (THF), which completely dissolves the sample. Thus, for the reverse-phase separation of triglycerides, a good solvent is THF and a poor solvent is acetonitrile. Remember, developing a reverse-phase separation does not require the use of an aqueous eluent, although some people disagree.

FIGURE 5-5. Optimized separation of the seven steroids in Table 5-1. Mobile phase: 14/27/59 acetonitrile/methanol/water.

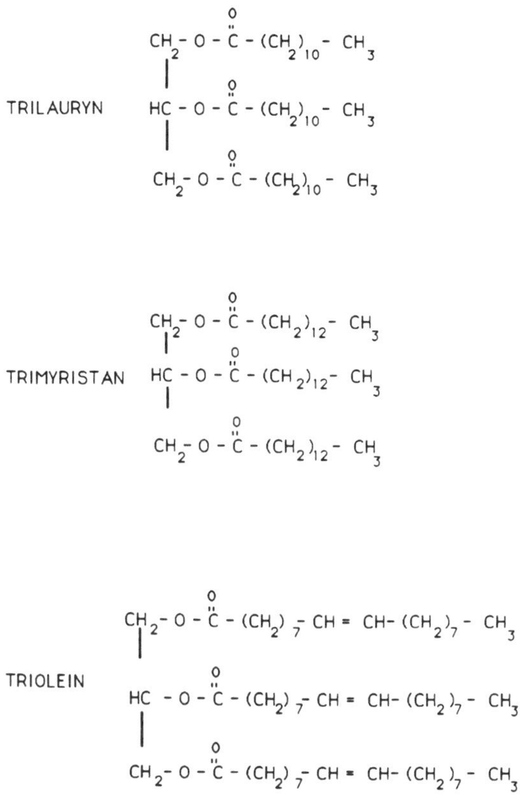

TRILAURYN

$CH_2-O-\overset{O}{\overset{\|}{C}}-(CH_2)_{10}-CH_3$

$HC-O-\overset{O}{\overset{\|}{C}}-(CH_2)_{10}-CH_3$

$CH_2-O-\overset{O}{\overset{\|}{C}}-(CH_2)_{10}-CH_3$

TRIMYRISTAN

$CH_2-O-\overset{O}{\overset{\|}{C}}-(CH_2)_{12}-CH_3$

$HC-O-\overset{O}{\overset{\|}{C}}-(CH_2)_{12}-CH_3$

$CH_2-O-\overset{O}{\overset{\|}{C}}-(CH_2)_{12}-CH_3$

TRIOLEIN

$CH_2-O-\overset{O}{\overset{\|}{C}}-(CH_2)_7-CH=CH-(CH_2)_7-CH_3$

$HC-O-\overset{O}{\overset{\|}{C}}-(CH_2)_7-CH=CH-(CH_2)_7-CH_3$

$CH_2-O-\overset{O}{\overset{\|}{C}}-(CH_2)_7-CH=CH-(CH_2)_7-CH_3$

FIGURE 5-6. Structure of triglycerides.

An example of a nonaqueous reverse-phase method development is the separation of the triglycerides trilaurin (TL), trimyristin (TM), and triolein (TO). The structures are shown in Figure 5-6. Remember that the first step in developing the separation is to use 100% of the strong solvent as the mobile phase. In 100% THF, all of the sample components elute in the void volume, V_0. By preparing and using mobile phases of sequentially poorer solvent power, retention of the three triglycerides begins to appear (Fig. 5-7). In a mobile-phase composition of 20% THF and 80% acetonitrile, an excellent separation is obtained as shown in Figure 5-7.

This type of reverse-phase chromatography is useful in comparing elution profiles of various oils. As seen in Figure 5-8*a* and *b*, the chromatographic profiles that represent the triglyceride content of olive oil and safflower oil are quite different. While the separation in Figure 5-8 is quite adequate for profiling oils for "identification," the "peaks" shown are not always single entities and often contain a mixture of two or more triglycerides.

In these examples, the solvents used contained no water and some have

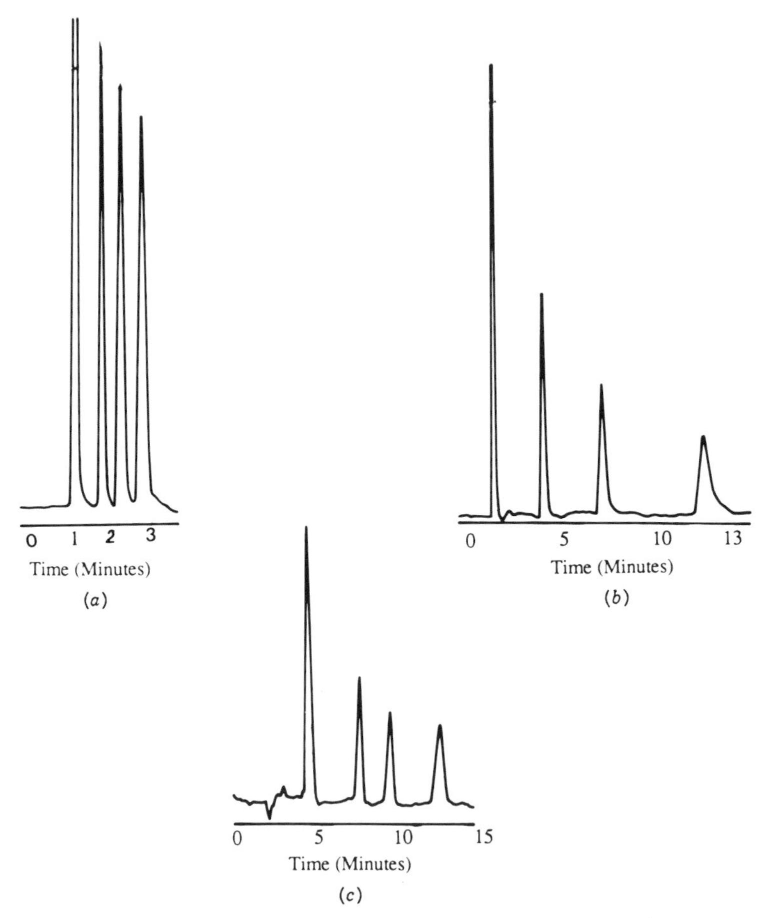

FIGURE 5-7. Separation of triglycerides. (*a*) Separation on a reverse-phase C_{18} column using a mobile phase of 40% THF in acetonitrile. (*b*) Separation on a reverse-phase C_{18} column using 20% THF in acetonitrile. (*c*) Separation on a porous polymer packing using 20% THF in acetonitrile.

generalized this approach under the jargon term *nonaqueous reverse phase* (NARP), implying that it is a special case of reverse phase. This, however, is false since the separation is developed in the same manner as are all reverse phase separations.

Developing a Separation

When developing a separation, a strategy of mobile-phase selection is involved. However, when the molecule has a polar and a nonpolar part, the tactics that support the method development strategy differ slightly from the previous discussions. To illustrate the method development approach to be

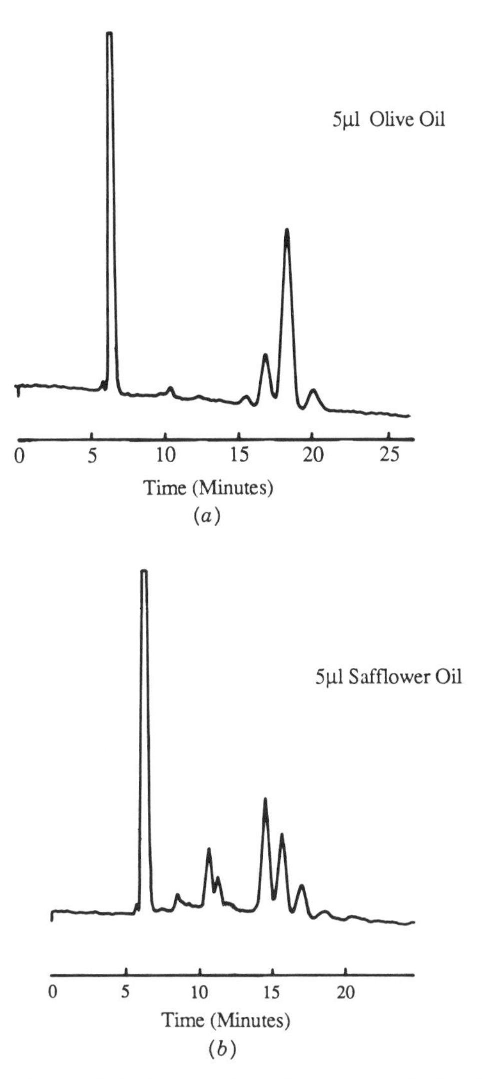

FIGURE 5-8. Comparison of elution profiles of commercial oils. Column: μBondapak C_{18} (10 μm), 3.9 mm ID × 30 cm. Mobile phase: tetrahydrofuran: acetonitrile (20 : 80). Detector: refractive index at 4X (Model R401). (*a*) Olive oil. (*b*) Safflower oil.

taken with this type of sample, three conjugated estrogens in a commercial pharmaceutical tablet will be separated from one another. The structures of the estrogens are shown in Figure 5-9.

In the earlier example, the focus was on how to develop a separation and it was assumed that the problem had been defined. It is instructive in this example to "get inside the chromatographer's head" and approach the problem from the beginning. (If the reader has forgotten the strategy of methods

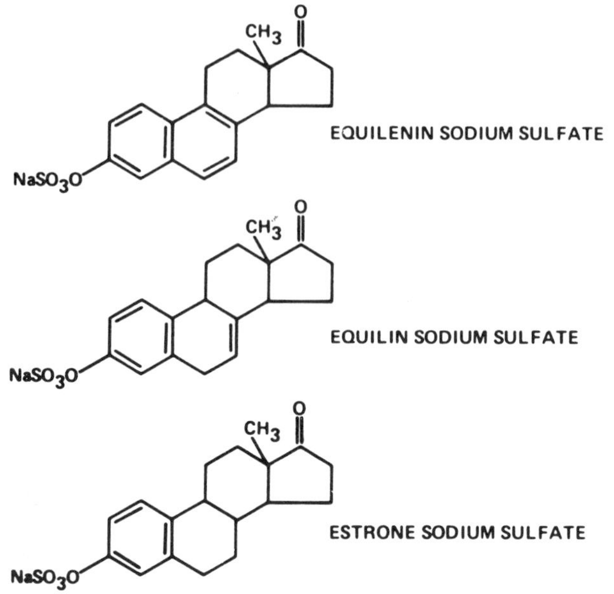

EQUILENIN SODIUM SULFATE

EQUILIN SODIUM SULFATE

ESTRONE SODIUM SULFATE

FIGURE 5-9. Structure of conjugated estrogens.

development, the appropriate sections of Chapter 4 should be reviewed before continuing.) Remember that, first and foremost, methods development philosophy is based upon a logical approach. The first step after defining the problem is to list the facts that are known about the problem/compounds. In this example, while all of the compounds are similar in structure, they differ in the degree of saturation in the second ring, which implies differences in the nonpolar part of the molecule. Also, the compounds all are sodium salts and have a minus charge (-1) when dissolved in aqueous solutions.

With the facts about the compound listed, it is time to choose the "best" approach to the separation problem. Using an ion exchange column would probably not be ideal since all of the compounds have the same charge and, unfortunately, the pK_a's of the molecules are not known. Therefore, because there are differences in the nonpolar part of the molecule, one should choose a stationary phase that will discriminate between these nonpolar differences. (If the pK_a's differed significantly, ion exchange could be another approach.) A nonpolar bonded phase such as a C_{18} is an appropriate choice. In many cases, the availability of any nonpolar column (C_8, phenyl, etc.) will decide what column is selected for a separation; however, as mentioned earlier in this chapter, since the C_{18} column gives good retention with a wide range of mobile phase strengths, it is the best general-purpose nonpolar column for reverse phase.

In this example, after a C_{18} column is chosen, the good and poor solvents must be selected. Since a nonpolar column is being used, water is chosen as the "poor" and methanol as the "good" solvent. However, acetonitrile or tetrahydrofuran might also have been chosen as the strong solvent. Sometimes personal preference may drive this decision; however, solvent strength follows the order methanol < acetonitrile < tetrahydrofuran. Therefore, choosing methanol first gives the additional flexibility of using one of the two stronger solvents if methanol does not work. Also, most people consider methanol a more friendly solvent with regard to safety, toxicity, disposal, and so on. But you have to be comfortable with your choice, so consider all aspects.

Since the starting point is a 10-μm C_{18} bonded-phase column and a mobile phase of water/methanol, sample preparation consists of dissolving three standards in a mixture of 50/50 water/methanol. This is an arbitrary choice since the final mobile phase is unknown. The sample solution is dissolved in the same solvent as the standards, mixed for a short period of time, and filtered to remove insoluble material in the tablet.

The next step is to begin chromatography with 100% methanol as a mobile phase using a reasonably rapid flow rate. Methanol is verified as the "strong" solvent by observing that after the injection of the sample, all components elute at the void volume; hence, there was no separation. As an eluent, methanol is too strong a solvent since the sample prefers being soluble in the mobile phase and is not attracted to the stationary phase. The eluent is then made "weaker" by adding a known percentage of water and the sample is injected. This process continues until retention begins to be apparent. In this example retention begins to be visible at a mobile phase composition of 50/50 methanol/water, as shown in Figure 5-10.

Inspection of the chromatogram in Figure 5-10 verifies that there is no significant retention and everything comes out at k' values of less than 1. The target is to place k' between 2 and 6; therefore, more retention is needed. Figure 5-11*a* shows the resulting chromatogram at 65% water and 35% methanol, and it is not much better. There are some slight differences in that more retention is occurring and the material is moving away from the solvent front; but k' is still less than 1. It is important to know whether the peaks that are moving away from the solvent front are the estrogens we are attempting to separate or if they are unknown "stuff" from the tablet. Therefore, the next step is to inject the standard of the estrogens and determine the retention for them. This is shown in Figure 5-11*b* and the standards have some retention, but more is still needed.

The amount of methanol is decreased to 20% and the chromatogram is shown in Figure 5-12*a*. The retention values of the standards are now in the k' range of 3 to 4, which is the optimum range for k' (2–6). While longer retention might be desirable, it is evident that the peaks for our standard compounds are not symmetrical. Since the samples are sodium salts, as the concentration of water is increased, the sample components will have increased ionic interactions with the stationary phase. These ionic interactions

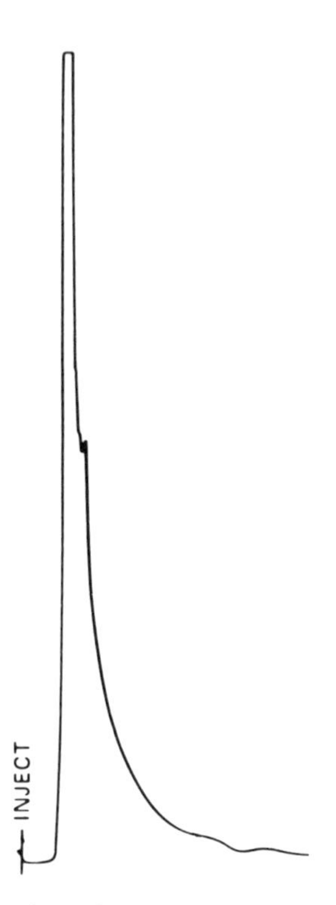

FIGURE 5-10. Separation of estrogens (tablet) in 50/50 MeOH/H_2O.

may give rise to a mixed mode of separation (nonpolar and ionic attraction), and in this example, this hypothesis is supported by the observation that the back of this peak is skewed (tailed).

Before continuing, it is helpful to determine what the sample looks like in this eluent. Figure 5-12*b* shows the chromatogram of the sample and, as was anticipated, the estrogens are being retained. The k' values are indeed in the optimum range and the estrogens are separating from the other materials in the tablet. So, part of the separation development problem is solved. However, as seen in Figure 5-12*b*, tailing is also in evidence. A higher concentration of water moved these peaks to longer retention times; however, more water also accentuated the tailing, which is undesirable. Therefore, because a good separation of the estrogens is still desired, the next approach is to change selectivity (the α term) and/or eliminate the tailing problem.

The change in α requires a change in the chemistry of the solvents, and there are a couple of approaches. A change from methanol to solvents like acetonitrile or tetrahydrofuran is possible. Or, focus can be placed on removing the ionic contribution to retention, which is believed to be the major cause of tailing. The water is the first place to start to change selectivity and, if that does not work, attention can then be turned to the organic portion of the eluent. Since these samples are sodium salts, the undesirable surface

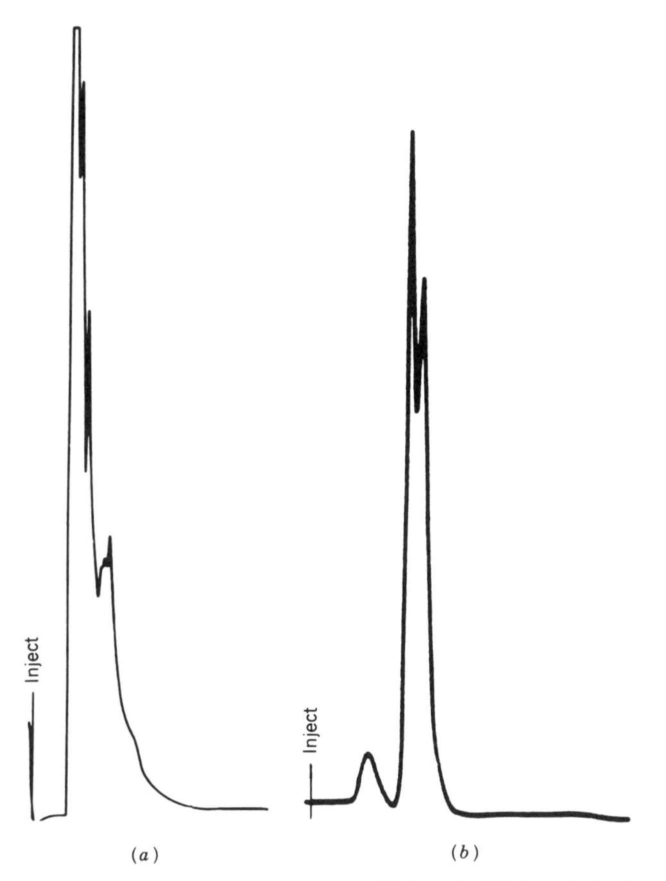

FIGURE 5-11. Separation of estrogens in 35/65 MeOH/H_2O. (*a*) Tablet. (*b*) Standards.

interaction is probably an electrostatic attraction with residual, unbonded surface hydroxyl groups. (A broader discussion of bonded-phase chemistry is given in Chapter 6.) Therefore, adding a phosphate salt to the water will effectively compete for the electrostatic sites, mask the sample's ionic interactions, and reduce any other mixed mechanisms that might be causing the tailing.

When a salt is added as a modifier: (1) it will drastically change k' because adding a salt to water increases the polarity of the water; and (2) the salt will also change the chemistry so that there will be a different α. Thus, by adding a salt to the aqueous portion of the mobile phase, we have an *entirely new* mobile phase and it is necessary to redevelop the separation using the same rationale as was used in Figures 5-10 through 5-12. Two new, important considerations must now be kept in mind. First, as the new eluent is used to develop the separation, the concentration of the phosphate buffer must be

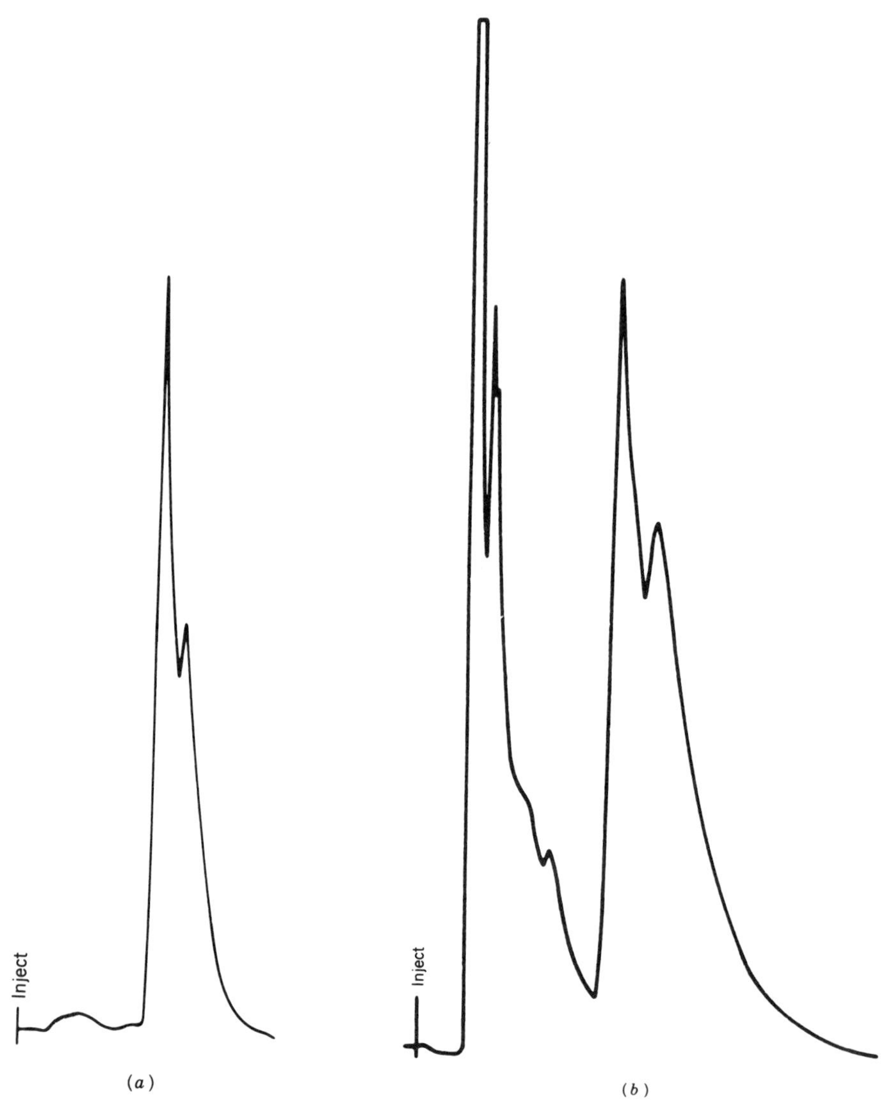

FIGURE 5-12. Separaton of estrogens in 20/80 MeOH/H_2O. (*a*) Standards. (*b*) Tablet.

kept constant in each mobile phase used. This means that fresh isocratic eluent must be manually mixed rather than blending acetonitrile with an aqueous salt solution using two pumps or a porportioning mixer. The second concern is to use a mobile phase in which the phosphate buffer is soluble. Because aqueous–organic mixtures are to be used and buffers are not often soluble in organic solvents, solubility of the aqueous buffer solution with the organic solvent must be evaluated before each mobile phase is used.

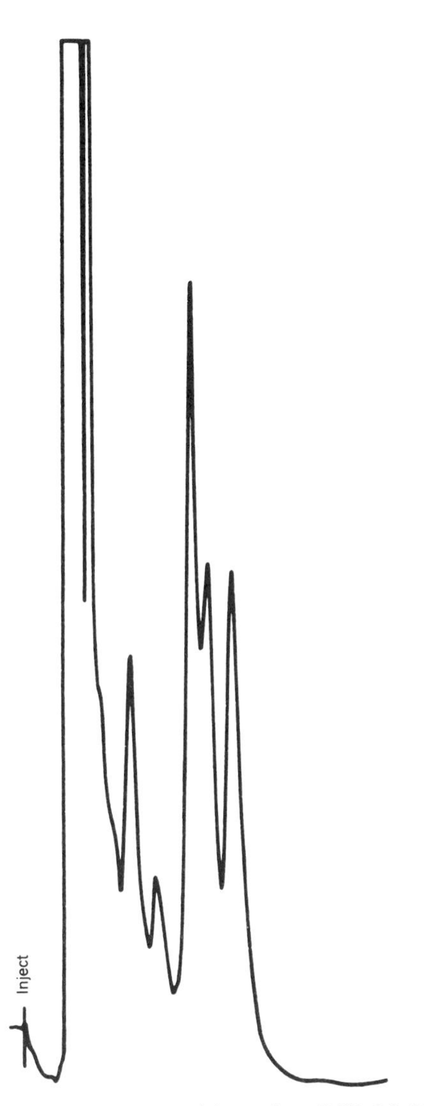

FIGURE 5-13. Separation of estrogen tablet using 50/50 MeOH/water (v/v). The buffer salt was 0.001 M KH_2PO_4.

In this example several high-organic aqueous mixtures with 0.001 M phosphate buffers were used as mobile phases. The retention of the steroid compounds was determined in various mobile phases (those used in Figures 5-10 to 5-12). The best separation is accomplished in a 50/50 (v/v) methanol/water mixture with 0.001 M phosphate buffer, and the chromatogram of the tablet is shown in Figure 5-13. The addition of the phosphate succeeds in eliminating the tailing problem and the k' values are in the proper range with better selectivity. Additional work may be focused on selectivity by using more water or, instead of methanol, choosing acetonitrile or THF.

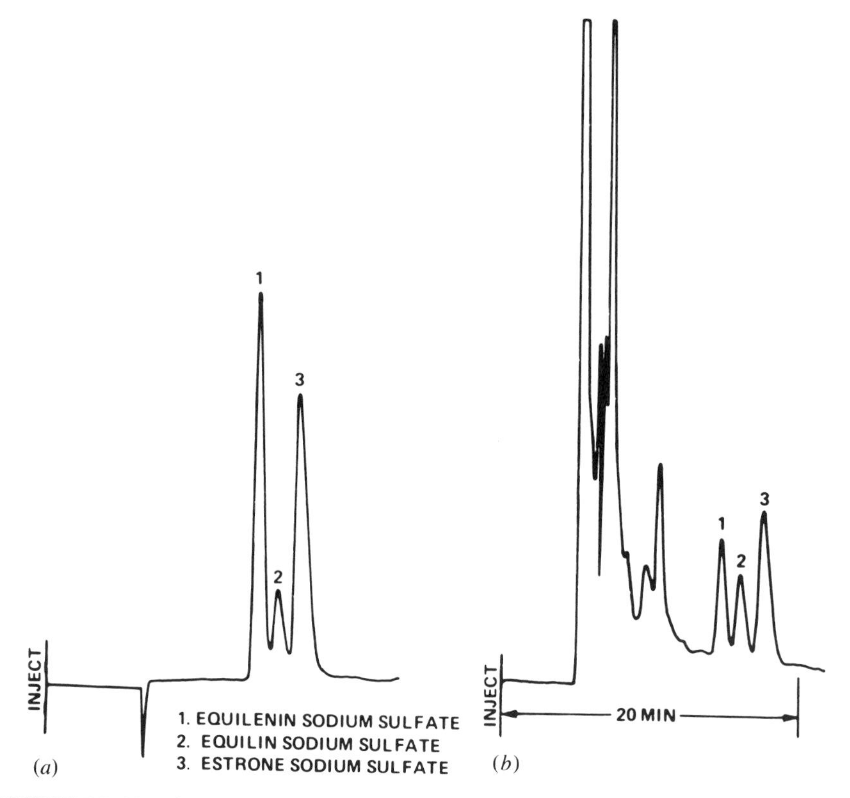

FIGURE 5-14. Final separation of estrogens. (*a*) Standards. (*b*) Tablet. The mobile phase is the same as Figure 5-13.

Another approach is to now give attention to the N term, the total number of plates. If the flow rate is decreased from 2 mL/min to 1 mL/min and two lengths of column are used, the result will dramatically increase N (this assumes that a second column is available, as in our example). If not, slowing the flow to 0.5 mL/min results in a higher plate count value (this is discussed in Chapter 6). The separation of the estrogen standards, with two columns in a series (a total length of 60 cm) using a flow rate of 1 mL/min with a solvent system of water/methanol (50/50) containing 0.001 M phosphate buffer is shown in Figure 5-14*a*. The final separation of the tablet components is shown in Figure 5-14*b*.

Summary

The problem was defined and a column was chosen that would exploit and accentuate the differences between the molecules. Next, "strong" and "weak" solvents were chosen which would be compatible with the column and would also result in solubility of the samples. The high-efficiency column was used at a fast flow rate of 2 mL/min. The standards (and it is very important to have standards) were dissolved in a solution of 50/50 methanol/water. The tablet was dissolved and filtered. The sample was injected and the chromatogram that resulted guided the judgment about what to do next. The initial decision was to adjust k' into the proper range for optimum resolution in a reasonable time by adjusting the ratio of water/methanol. When it was found that sufficient selectivity could not be attained and that there was a tailing problem, a change was dictated in the polar component (water) of the eluent. A phosphate buffer (0.001 M) was used because it would tend to suppress the ionic interactions of the sample with the stationary phase. The separation was redeveloped keeping the phosphate concentration constant at 0.001 M. After the k' values were reasonable, a decision was made that instead of working further on the selectivity, the total number of plates would be increased, so the flow rate was decreased and the column length was doubled.

In real life, one might choose to work further on the method to reduce solvent or column costs as well as analysis time. However, the important aspect of this example is to change one variable at a time and to be logical in your approach. In this example the initial focus was to obtain retention by adjusting k'. Next, the activity focused on the selectivity. Finally, N was increased. A more complete discussion of maximizing the use of plates is given in Chapter 6.

TACTICS FOR SEPARATING IONIC COMPOUNDS

The example just examined was illustrative of one type of ionic sample which a chromatographer may need to separate. However, a chromatographer will encounter a wide range of ionic compounds ranging from strong acids through strong bases, and it is indeed fortunate that a reverse-phase column can be used for the separation of this wide variety of ionic compounds. The strategy for approaching the separation is the same as before, but the tactics are modified depending upon the particular problem at hand.

Reverse Phase for Ionic Compounds

Before discussing the tactics to handle newer sample types, it is helpful to consolidate our thinking about the use of reverse-phase columns. The wide scope of solute retention which can be achieved using reverse-phase columns is shown in Figure 5-15. Compounds in the middle of the chart are nonionic or "neutral" at the pH being used and can be separated in organic/aqueous mobile phases. For solutes in the intermediate pH ranges (3–8), adjustment of the pH of the eluent using small concentrations of a buffer salt

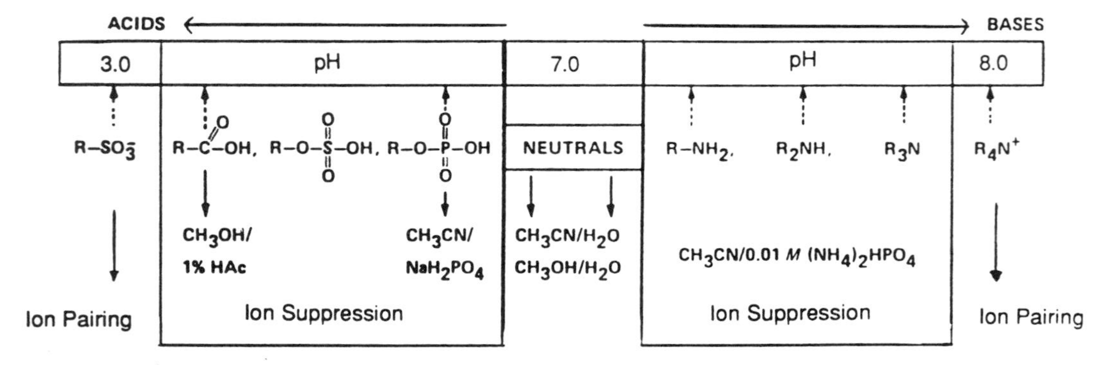

FIGURE 5-15. The scope of reverse-phase columns for the retention of neutral and charged molecules.

in the water to eliminate mixed-mode interactions and to suppress the ionization of the sample is needed so that the reverse-phase column can be made to retain ionic compounds. The tactics associated with this approach were discussed earlier in this chapter. This technique is termed "ion suppression chromatography." By adjusting the mobile phase to be in the pH range of 2 to 5, weakly acidic samples can be retained. Similarly, weak bases can be retained by ion suppression if pH control is used. An example of controlling retention through the use of a buffered mobile phase was the final separation shown in Figure 5-14.

An alternative to using ion suppression (with pH control) is to adjust the eluent pH such that solutes are fully ionized whereupon separations can then be carried out using an eluent that contains a suitable modifier. For example, if the pH is 6–8, the weak acids may be retained with the addition of an appropriate modifier to the mobile phase. This technique is called "paired-ion chromatography" or "ion-pair chromatography." Weak bases may be retained by ion pairing if the pH is below 7. Strong acids and strong bases can be retained throughout the usable pH range by ion-pairing techniques.

Ion Suppression Chromatography

As was discussed earlier in this chapter for resolving the mixture of estrogens, the separation of samples that ionize in aqueous solution can be optimized on a reverse-phase column by adjusting the ionic content of the aqueous eluent usually using a buffer in the mobile phase. Additionally, these ions (salts) in the mobile phase minimize the electrostatic interactions of the ionic sample with the stationary phase. The experimental result of using this technique is that the samples are eluted as sharper zones than they would have been in a similar mobile phase without the addition of a salt modifier. This technique is believed to control chromatographic retention through the suppression of the ionization of an ionic sample. Ion suppression for bonded-phase columns is most useful in the range of pH from 3 to 8.

The concept of ion suppression is shown graphically in Figure 5-16. The chromatographic retention of the sample is a sigmoidal function of the hydrogen ion concentration and parallels a titration curve for the acid or base. This behavior is related to the pH of the water portion of the eluent which is believed to control the ionization of the sample. A weak acid is retained longer at pH values less than its pK_a value (suppressed ionization) and elutes faster at pH values higher than its pK_a. A basic sample will elute faster at pH values below its pK_a value and will be retained longer at high pH values where its ionization is suppressed. The reason for the earlier elution of the ionic form of the sample as compared to the "neutral" form is that the ionic form is more soluble in the aqueous eluent and the "neutral" (unionized) form is more soluble in the stationary phase. At two pH units above or below the inflection point, the ionization and, hence, the chromatographic retention reach their limiting values. In practice, small changes in

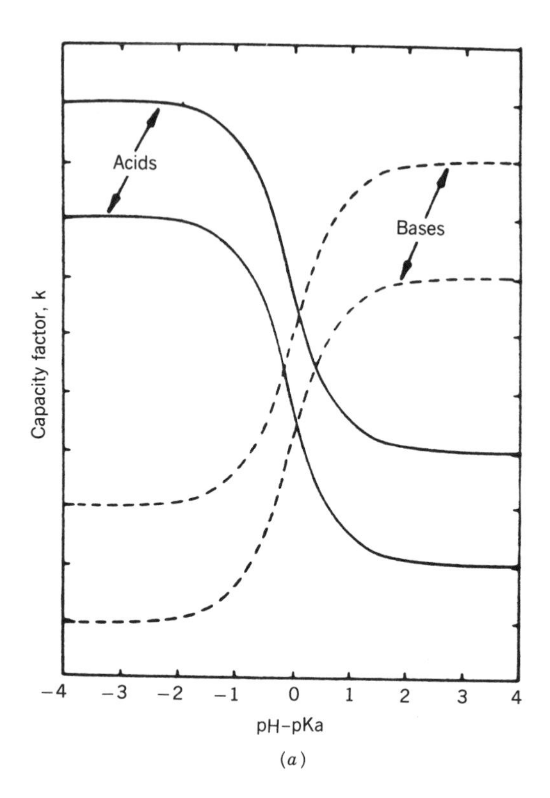
Acids
Bases
Capacity factor, k
−4
−3
−2
−1
0
1
2
3
4
pH−pKa

(a)

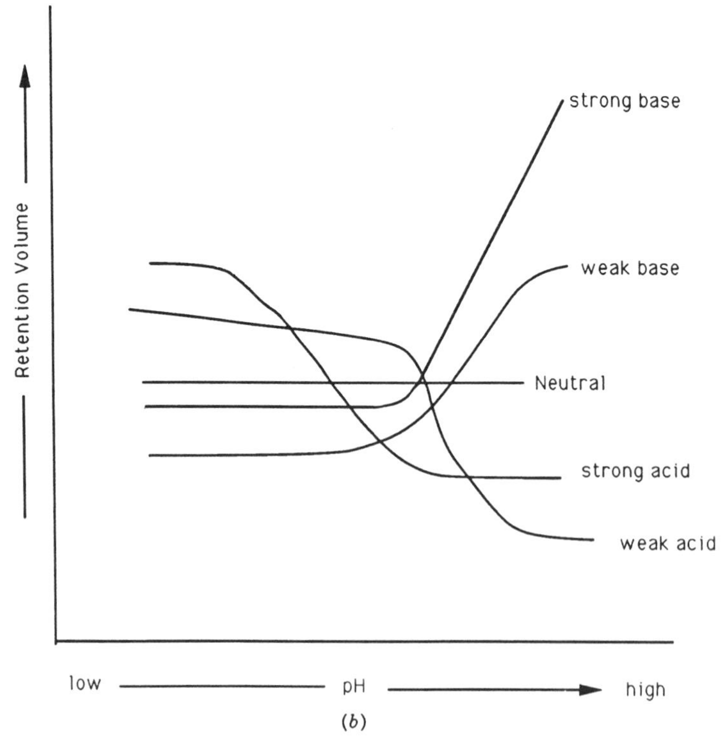
Retention Volume
strong base
weak base
Neutral
strong acid
weak acid
low
pH
high

(b)

FIGURE 5-16. Effect of pH upon chromatographic behavior of ionic compounds. (*a*) Relative capacity factor for acids and bases. (Reprinted from C. Horvath and W. Mellander, *J. Chromatogr. Sci.*, **15**, 393 (1977) with permission.) (*b*) Retention volume behavior as a function of pH.

the pH value near the pK_a value of the sample can result in large changes in the chromatographic retention of the sample; however, at the limiting value, small pH variations have essentially no chromatographic impact.

Figure 5-16*b* shows the typical behavior of retention versus pH for a strong and weak acid, for a neutral molecule and for a strong and a weak base (e.g., a molecule containing an amine group). Variation in pH does not change the retention of molecules that do not ionize. Strong acids show a significant decrease in retention volumes when the pH is increased. If the pK_a of the acid is low enough (e.g., below 2), there may not be a plateau region in the plot in Figure 5-16*b*. A weak acid behaves similarly to a strong acid but may show a slower decrease in retention volume until the pH exceeds the pK_a and the retention volume decreases. For strongly basic molecules, the retention volume will increase with increasing pH. The retention of weak bases will increase as the pH increases until the pH is one unit past the pK value. Above this point, the retention will be stable as pH continues to rise. Developing a separation when using ion suppression follows the standard strategy which has been described earlier. By changing the mobile-phase strength, the multicomponent mixture (Figure 5-17) of active ingredients in an over-the-counter headache medicine is easily resolved (Fig.

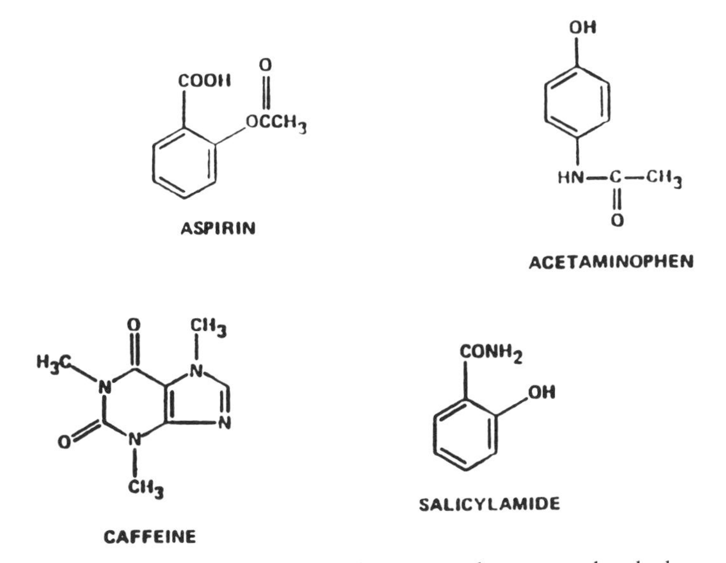

FIGURE 5-17. Structures of ingredients in an over-the-counter headache medicine.

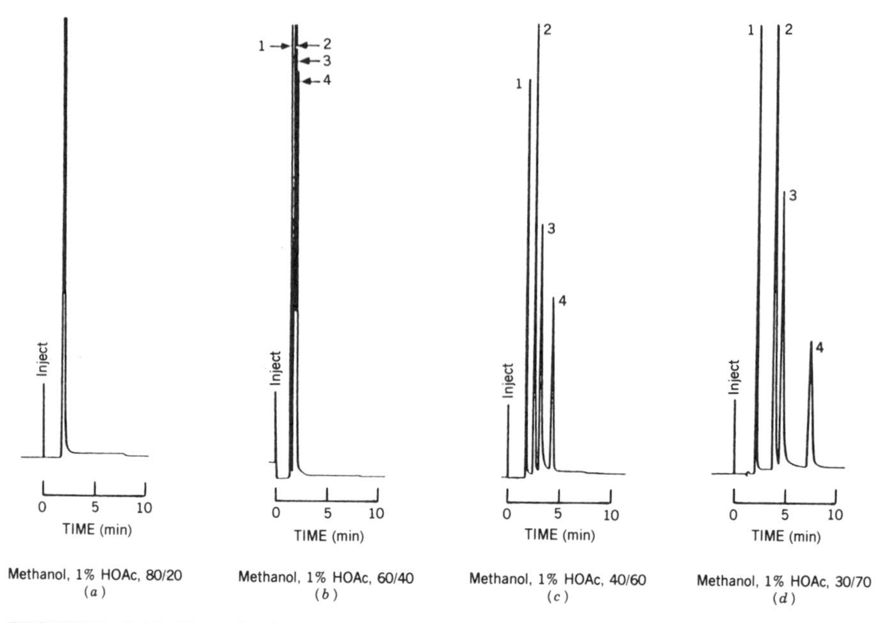

FIGURE 5-18. Developing a reverse-phase separation using ion suppression. Column: μBondapak C_{18} (10μ) 3.9 mm × 30 cm. Flow rate: 2.5 mL/min. Compounds: (1) acetaminophine; (2) caffeine; (3) salicylamide; and (4) aspirin (structures shown in Fig. 5-17). Detector: UV at 254 nm and 0.05 AUFS. Mobile phase is methanol combined with various amounts of water containing 1% acetic acid. (*a*) 80/20 (v/v). (*b*) 60/40 (v/v). (*c*) 40/60 (v/v). (*d*) 30/70 (v/v).

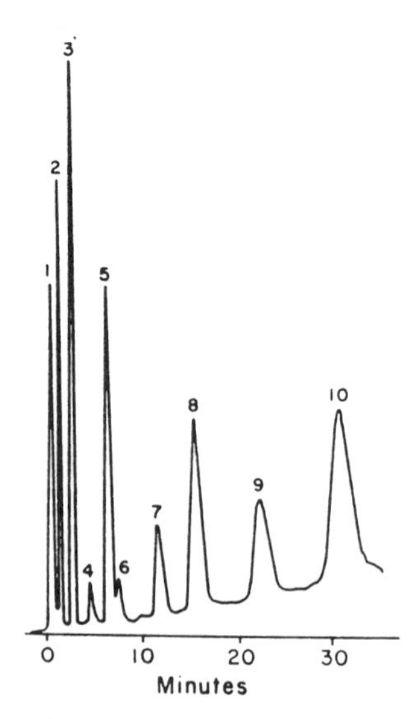

FIGURE 5-19. Separation of phenolic acids using the ion suppression technique. Mobile phase: 5% acetic acid in water. (Chromatogram reprinted from L. W. Wulf and C. W. Nagel, *J. Chromatography*, **116**, 271 (1976) with permission.)

5-18). Initially the mobile phase was too strong and elutes all of the compounds at V_0. The mobile phase strength is decreased stepwise until all four components are resolved. It is important to reemphasize that when using ion suppression, the additive used to adjust the pH must remain at a constant concentration during the method development process. Also, it should be mentioned that while acetic acid adjusts the pH to an approximate value of 3 and is often used in ion suppression, it is not a buffer. Therefore, use of a sodium acetate buffer (0.001 to 0.01 M) may be more appropriate for some bonded phases.

Ionic organic compounds are often water soluble while the unionized forms will be less water soluble. Because of this, ion suppression is commonly used to enhance the separation of organic acids. A typical example of the use of ion suppression is for the separation of phenolic acids shown in Figure 5-19. Without the addition of acetic acid to the eluent the acids would have eluted as broadly tailed peaks with little retention.

Paired-Ion Chromatography

Before paired-ion chromatography is discussed, it is illustrative to consider another solubility-based separation which relies on a similar approach. This separation is ion-pair extraction. Ion-pair extraction uses two immiscible liquids (aqueous and organic) often in a separatory funnel. An ionized compound (A^+_{aq}) that is water soluble can be made to favor solution into the organic phase during an extraction by using a suitable counter ion (B^-_{aq}) to form a "neutral" ion pair. Since the ion pair behaves as though it is a nonionic, neutral species, it will prefer to reside in the organic liquid layer, and the entire process can be described by the equation:

$$A^+_{aq} + B^-_{aq} = AB_{org}$$

Work in chromatography showed that when using bonded-phase columns, the addition of these ion-pair type reagents in the eluent substantially altered the retention of the ionic compounds while not significantly influencing retention of nonionic compounds. Because of the similarity of reagents that were used in classical ion-pair extractions, the technique of adding ionic alkyl reagents to an LC system has been termed "paired-ion" chromatography or "ion-pair" chromatography.

In addition to the usual variables of reverse-phase LC such as type of stationary phase and polarity of the eluent, the nature of the alkyl group (R group) on the pairing reagent is an important variable in ion-pairing situations. For this discussion, the compounds whose structures are shown in Figure 5-20 will be used as samples to illustrate the effect of the lipophilic nature (length of the R group) of the pairing reagent upon the retention of the samples. In the mobile phase which is used, these compounds are basic except for maleic acid and phenacetin. Therefore, as the length of the R

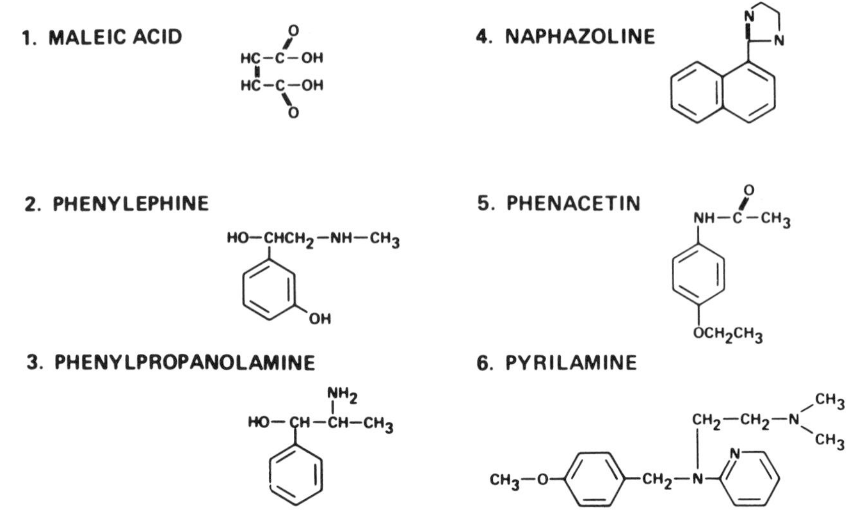

FIGURE 5-20. Structures of various antihistamines and decongestants. These compounds are referenced in Figures 5-21 to 5-24.

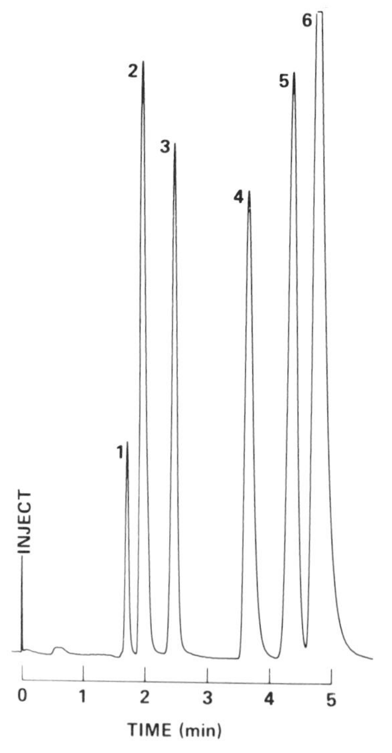

FIGURE 5-21. Chromatogram showing the separation of antihistamines and decongestants (refer to Figure 5-20) using pentanesulfonate. Mobile phase: methanol/water (50:50) with 5 mM pentanesulfonate and 1% acetic acid. UV detector: 254 nm. Flow rate: 2 mL/min. (Reprinted from reference 2 with permission.)

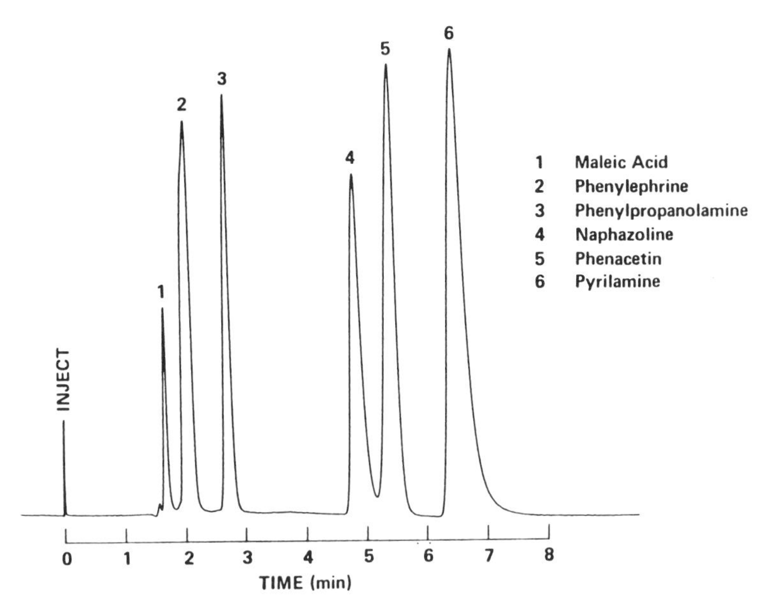

FIGURE 5-22. Chromatogram showing the separation of antihistamines and decongestants (refer to Figure 5-20) using pentanesulfonate. Mobile phase: methanol/water (45:55) with 5 mM pentanesulfonate and 1% acetic acid. UV detector: 254 nm. Flow rate: 2 mL/min. (Reprinted from reference 2 with permission.)

group (alkyl chain) is changed on the pairing reagent, the retention of maleic acid and phenacetin will not be affected. The other components are protonated in the eluent and the retention of these compounds will change as a function of the lipophilicity of the pairing reagent.

In Figure 5-21, the elution order for these compounds using a methanol/water mobile phase and pentanesulfonate as the ion-pairing reagent is shown. Increasing the resolution of these compounds can be accomplished by simply changing to a mobile phase with higher water content. This effect is shown in Figure 5-22 and is the usual reverse-phase behavior.

In addition to the solvent strength, one can also vary the lipophilic nature of the pairing reagent. In Figure 5-23 it can be seen that when using hexanesulfonate (increasing the length of the pairing reagent by one methylene group) all of the peaks except maleic acid and phenacetin are retained more than when using pentanesulfonate as the pairing agent (Figure 5-21). The same trend is evidenced in Figure 5-24, which shows the effect of using heptanesulfonate. The results of these chromatograms imply that as the chain length of the pairing reagent increases, the retention of a sample which is "paired" is also increased. In other words, the more lipophilic (nonpolar) the pairing reagent is, the greater the retention of the sample. A discussion of the mechanism of the separation can be found in Chapter 5.

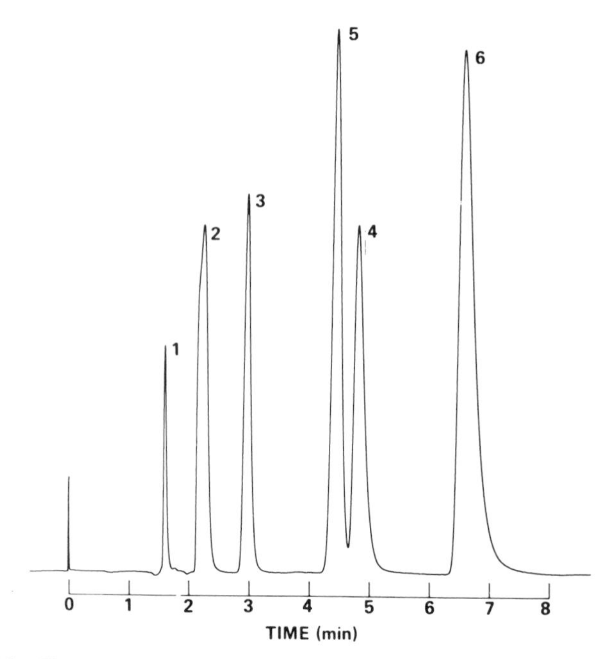

FIGURE 5-23. Chromatogram showing the separation of antihistamines and decongestants (refer to Figure 5-20) using hexanesulfonate. Mobile phase: methanol/water (50:50) with 5 mM hexanesulfonate and 1% acetic acid. UV detector: 254 nm. Flow rate: 2 mL/min. (Reprinted from reference 2 with permission.)

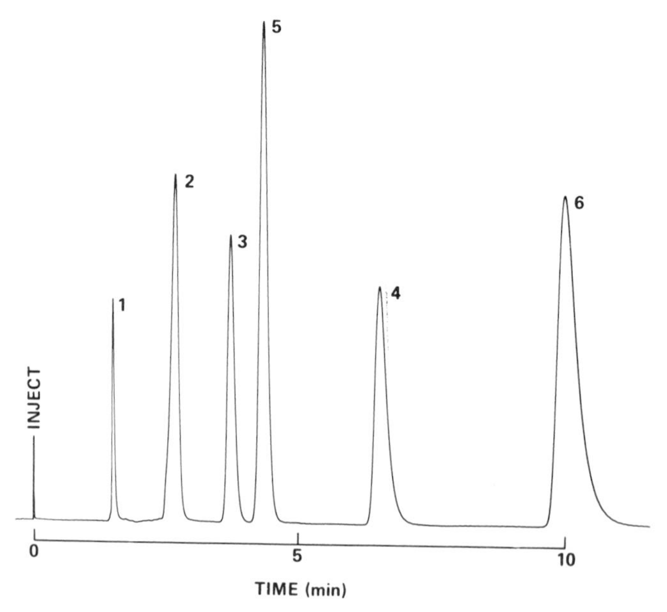

FIGURE 5-24. Chromatogram showing the separation of antihistamines and decongestants (refer to Figure 5-20) using heptanesulfonate. Mobile phase: methanol/water (50:50) with 5 mM heptanesulfonate and 1% acetic acid. UV detector: 254 nm. Flow rate: 2 mL/min. (Reprinted from reference 2 with permission.)

General Approach to the Separation of Ionic Compounds

The general guideline in developing a separation is to use as few variables as possible. Therefore, when dealing with an ionic or ionizable sample, one should try a typical reverse-phase mobile phase (organic solvent/water) first. This approach was discussed earlier for the estrogen tablet separation. If this doesn't work and/or if retention is too short or peak shapes are tailed, the researcher should try ion suppression. Often experienced individuals go directly to the use of ion suppression since their experience dictates that a typical reverse-phase mobile phase will probably *not* be successful. Therefore, these individuals "play the odds" in the hope of saving time in methods development. If ion suppression does not work and/or if retention is again too short or peak shapes are still undesirable, the researcher should try paired-ion chromatography in which control of the nature of the counter ion in addition to the control of ionization improves the manipulation of the corresponding retention of the sample. An example of an ion suppression and paired-ion reverse-phase separation is shown in Figure 5-25 for aspirin (no pairing possible) and phenylpropanolamine (ion pairing is possible).

Models for Ion-Pair Retention

There are essentially three popular hypotheses. The first model states that ion pairs form in the reverse-phase eluent and that these ion pairs travel through the column as neutral species. The second model states that paired-ion reagents operate by first adsorbing onto the bonded-alkyl stationary phase and then acting as adsorbed ion-exchange sites. These two models, the ion-pair model and the dynamic ion-exchange model, propose extreme situations. The third view, which is broader in scope than the previous two concepts, accommodates both extreme views without combining the two models. This proposal, the ion interaction model or "paired-ion model," emphasizes the importance of both adsorptive and electrostatic attractions in governing retention in reverse-phase, ion-pair systems. A complete discussion of this subject can be found elsewhere (2).

There is considerable debate concerning the exact model that describes the paired-ion phenomenon and it will continue, no doubt, for some time. It is important, however, to emphasize that theory guides experimentation; therefore, the importance of having a model is to understand the factors that control chromatographic retention, and thus to aid in the speedy and logical development of separations. Therefore, any of the three models discussed above can be useful in guiding your experimentation in methods development.

Using a Paired-Ion Reagent of the Same Charge as the Sample

Another means of controlling retention is through the addition of an ion-pair reagent which has the same charge as the sample. In this case, the capacity

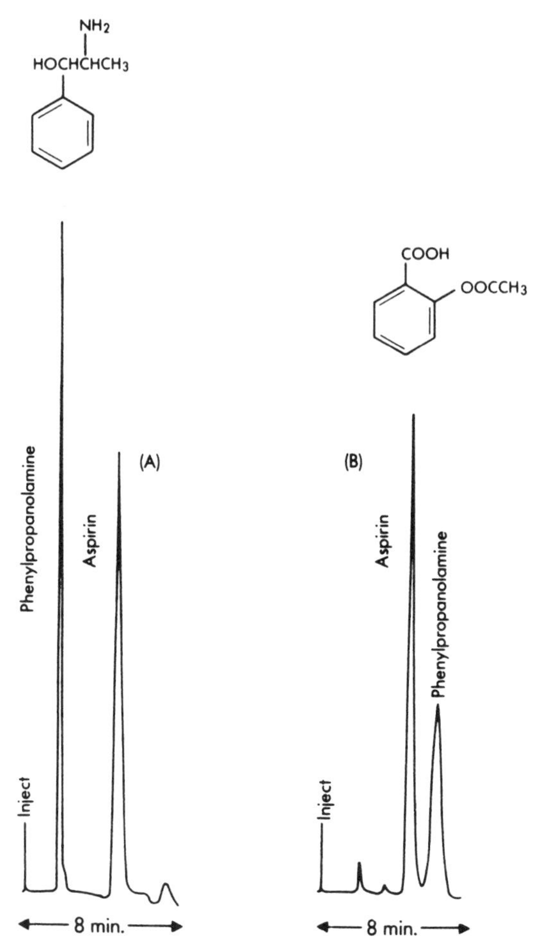

FIGURE 5-25. Separation of phenylpropanolamine and aspirin with and without the use of a paired ion. (*a*) Mobile phase contains no ion-pair reagent. (*b*) Mobile phase contains ion-pair reagent (5 mM heptanesulfonate).

factor k' of a lipophilic ionic sample can be reduced with the addition of ion-pair reagents of the same charge as the sample. Also, the tailing of the peak will be reduced from that when no modifier is present. This is not the "typical ion-pairing phenomenon," but is explained by the broader-scoped paired-ion model (2,3). A dramatic example of using this approach is shown in Figure 5-26.

Figure 5-26*a* shows the separation of local anesthetics at a pH of 3. The samples are positively charged at this pH. Figure 5-26*b* shows the effect of adding an ion-pair reagent, dibutylamine, which is protonated and is the same charge as the sample ions. In this solvent, the dibutylamine is adsorbed on the surface of the C_{18} column, and as a result, the nonpolar surface

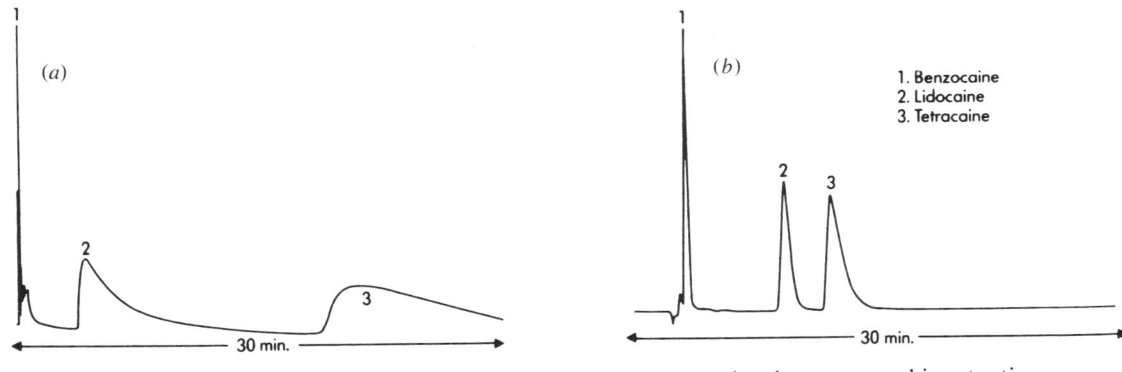

FIGURE 5-26. The effect of the charge of the ion pair upon the chromatographic retention of similarly charged ions. Column: Resolve C18 (10μ), 5 mm × 10 cm. Flow rate: 5 ml/min. Detection: UV at 254 nm. Mobile phase: acetonitrile/water (90 : 10). (*a*) The mobile phase contains no ion-pair reagent. (*b*) The mobile phase contains an ion-pair reagent (5 mM dibutylamine) of the same charge as the sample.

becomes positively charged. This reduces the available surface area and the retention volumes of the anesthetics are reduced and peak shape is improved. As a result of the repulsive ionic interactions involved in this separation, this behavior is sometimes referred to as the "competing base" effect, since another basic compound is added to the mobile phase and inhibits adsorption of the basic analytes onto the surface.

Summary of Use of Paired-Ion Chromatography

The general guideline is to use as few variables as possible when developing a mobile phase. However, it is also important to realize that there are several ways of accomplishing a separation. For example, let us examine a hypothetical acid, RCOOH (see Table 5-2). If the acid is weak acid, the ion-suppression technique can be used. Simply adding an acidic buffer leads to sharper chromatographic elution zones that might show slight peak tailing in the worst case. When ion suppression is not effective, the sample may be ionized and the ion-pairing technique used (refer to Table 5-2) to attain more retention.

For strongly acidic samples (those that are completely ionized at pH 2), it is necessary to take advantage of the strongly ionic nature of the sample. By adding a buffer to control pH and to ensure that the sample maintains ionic character, and by adding a suitable compound (in this case a quaternary ammonium salt) as a paired-ion reagent, the retention of the acid will increase.

Analogous considerations hold for the separation of bases, as shown in Table 5-2. Generally, for the separation of ionic compounds, the pH is kept in the range of 2 to 8, to optimize the useful life of the column/system.

TABLE 5-2. Techniques for Retaining Ionic Compounds Using Reverse-Phase Liquid Chromatography

Ionic Compound	Form at pH 2–5	Form at pH 7–8	Method of Analysis
Strong Acid, HA_s ($pK_a < 2$)	A_s (Ionic)	A_s (Ionic)	Paired-ion
Weak Acid, HA_w ($pK_a > 2$)	$H{:}A_w$ (Nonionic)	A_w (Ionic)	Ion suppression (pH 2–5) or paired-ion (pH 7–8)
Weak Base, B_w ($pK_a < 8$)	$H{:}B_w$ (Ionic)	B_w (Nonionic)	Paired-ion (pH 2–5) or ion suppression (pH 7–8)
Strong Base, B_s ($pK_a > 8$)	$H{:}B_s$ (Ionic)	$H{:}B_s$ (Ionic)	Paired ion

Separations of Bare (Unbonded) Silica Gel

The use of bare (unbonded) silica gel with aqueous eluents for the separation of basic lipophilic amines (e.g, pharmaceutical drugs) may be more appropriate than using the popular technique of reverse-phase LC on bonded-phase packings (3). This is quite significant since the "conventional wisdom" is that silica should not be used in aqueous eluents. Thus, instead of trying to eliminate the contribution to retention from surface silanol groups in bonded-phase packings (which many manufacturers claim they do—see Chapter 6 for further clarification), it is possible to exploit silanol groups in a positive manner by using bare silica gel as the adsorbent using aqueous eluents. Many separations of lipophilic amines which are very difficult on bonded reversed-phase packings are easily accomplished with good peak symmetry on silica. Retention of organic amines on silica using reverse-phase mobile phases appears to be dependent upon electrostatic and adsorptive forces and separations can be adjusted in the same manner as if the chromatographer was using a bonded-phase column. In other words, a silica column behaves as a "very weak" bonded-phase column in the reverse-phase mode for ionizable, basic organic compounds. In fact, when separating organic amines, silica packings behave quite nicely and predictably in the reverse-phase mode.

To illustrate the concept, Figure 5-27 shows the separation of four organic amines. These have been traditionally separated on bonded-phase columns using an inorganic salt in the eluent in order to reduce the tailing that would occur if the salt were not present. This has been discussed in prior sections in this chapter. Figure 5-27*a* shows a mixture of compounds on a C_{18} column with dibasic ammonium phosphate in the eluent. Tetracaine (peak 4), which is the most basic amine, is retained the longest with poor peak symmetry (tailing).

To understand the role of the surface silanols and their contribution to retention, these compounds were separated on a column that was purposely coated to half of the C_{18} level of the original C_{18} (Fig. 5-27*b*) (coated column had approximately 50% silanol content). If peak tailing is due solely to the silanol interaction, the peak symmetry should be worse than on the fully coated column. But there was no decrease in peak symmetry. In fact, retention of the tetracaine (peak 4) increased slightly while the peak symmetry improved. Thus, it appears that silanols do contribute to retention but the amount of silanols is not the main cause of peak asymmetry. The logical extension of this approach is to separate the compounds on unbonded silica gel, as shown in Figure 5-27*c*. In this situation, peak symmetry is quite good and retention is decreased.

The chromatograms in Figure 5-27 indicate that the surface silanols by themselves are not deleterious to the retention of organic amines when using organic, aqueous eluents containing an inorganic salt. In fact, silica gel itself appears to be the preferred adsorbent for the separation of these bases. The

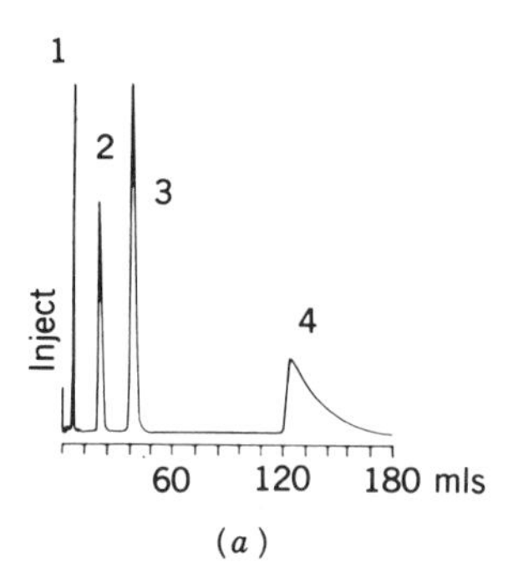

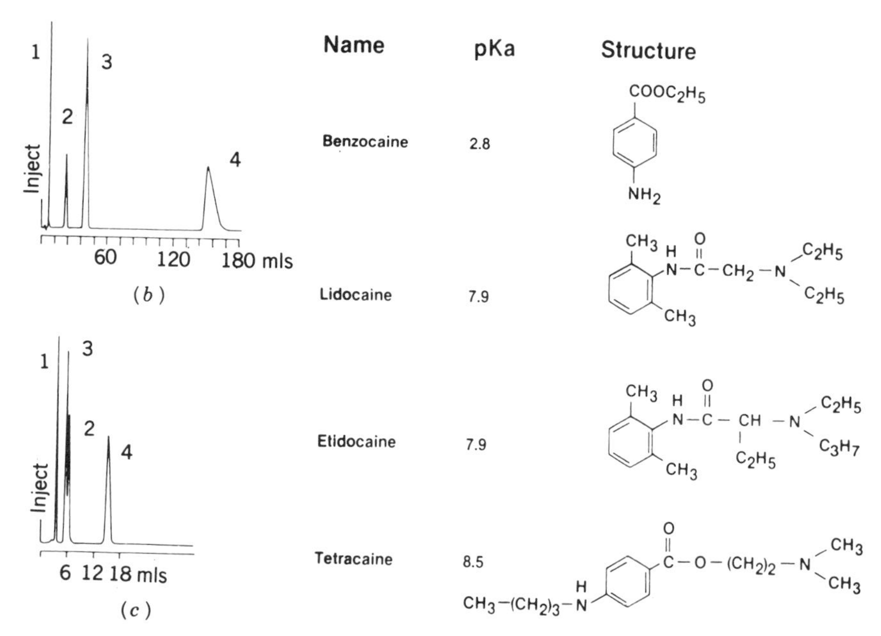

FIGURE 5-27. Effect of column surface upon retention of organic amines: (*a*) separation on a fully coated C_{18} column. (*b*) Separation on a column with 50% coating level of C_{18}. (*c*) Separation on a column of silica. The mobile phase is acetonitrile/water (60:40) with 4 mM dibasic ammonium phosphate (pH 7.8). The sample components are (1) benzocaine, (2) lidocaine, (3) etidocaine, and (4) tetracaine. (Reprinted from reference 4 with permission.)

key to good peak symmetry is not the presence or absence of residual silanols but more probably the accessibility of the surface silanol groups.

So to develop a separation on a bare silica gel column, the chromatographer first would know that the C_{18} approach had resulted in excessive retention and or tailing. Next the chromatographer would choose a pH at which the compounds are ionized. Often a pH of 7.8 is used since it is a basic pH within the "safe" range (i.e., no silica dissolution) and can be

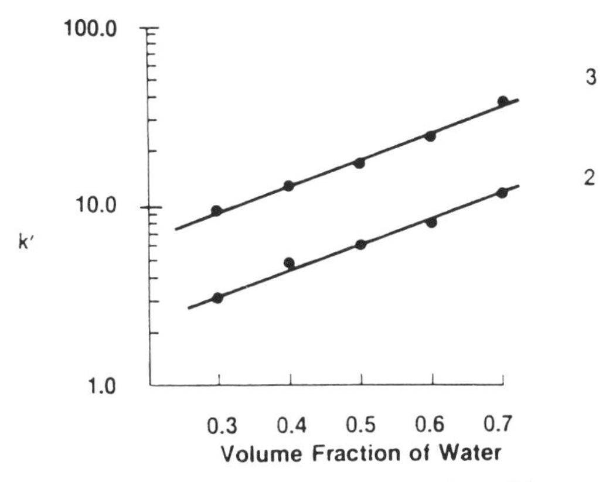

FIGURE 5-28. Effect of solvent strength upon retention with constant pH and ionic strength. The compounds are chlorpheniramine (2) and propranolol (3). The mobile phase is acetonitrile/water with 4 M dibasic ammonium phosphate at pH of 7.8. The column is silica. (Reprinted with reference 4 with permission.)

easily buffered. Then a suitable strong mobile phase is chosen. In this case, a good starting mobile phase should have as high an organic content as possible, but containing a buffer at a low level (5 mM) so that the salt in the mobile phase will not precipitate. Then the retention is adjusted by making the mobile phase weaker, as has been discussed for the C_{18} column. For the drugs chlorpheniramine and propranolol, a plot of retention versus aqueous composition is shown in Figure 5-28 and a typical chromatogram is shown in Figure 5-29. While Figure 5-29 shows retention that is far too long, it helps to illustrate the lack of tailing when using silica in this mode. Another separation developed using silica in the reverse-phase mode is shown in Figure 5-30. Present indications suggest that the lifetime of a silica column used in an aqueous mobile phase meets or may exceed that of a bonded-phase column (4,5).

Ion-Exchange Chromatography

Paired-ion chromatography is quite effective for small molecules with ionic and nonpolar constituents because of the dual forces of electrostatic and lipophilic attraction which can be controlled. Ion-exchange chromatography, on the other hand, effectively causes separations through *only* control of electrostatic attraction and is effective where the main differentiation in the sample molecules is the charged nature of each. Competition for the ionic sites on the stationary phase varies with the size of the ion and the number of charges (the charge/mass ratio). Large ions and multiple charged ions will bind more strongly than small, monocharged ions. The strategy for developing an ion-exchange separation is the same as in reverse phase, namely, by adjusting the mobile phase strength to compete with the station-

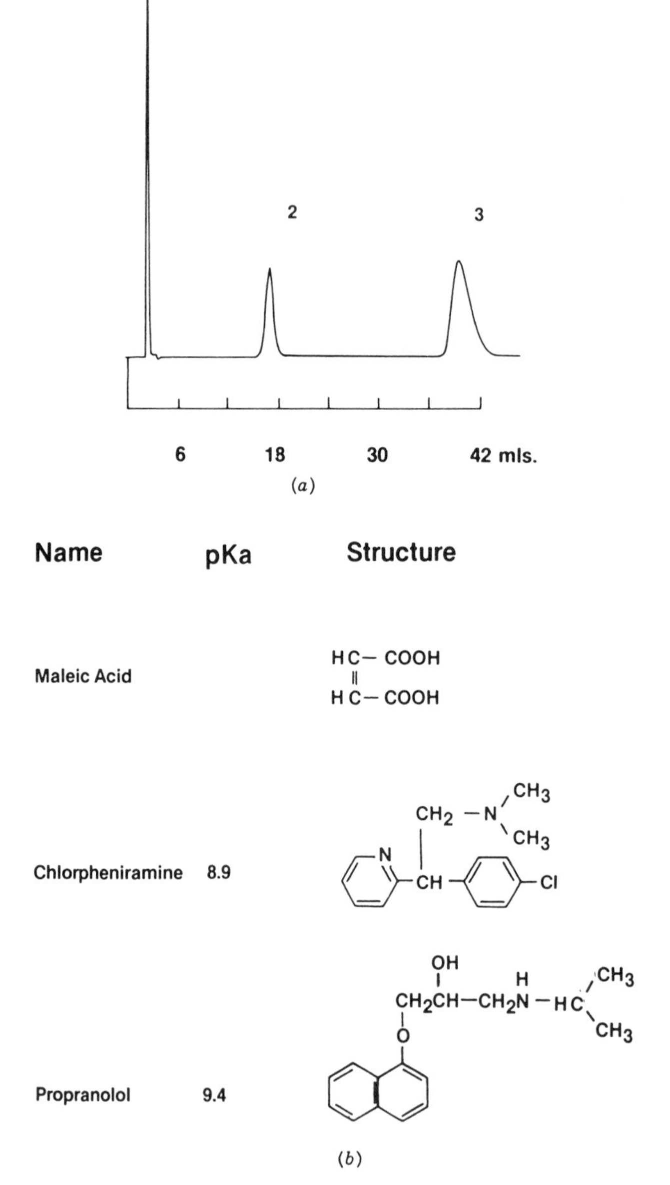

FIGURE 5-29. Retention of organic amines on silica. Separation of (1) maleic acid, (2) chlorpheniramine, and (3) propranolol. The mobile phase is acetonitrile/water (60:40) with 4 mM dibasic ammonium phosphate (pH 7.8). (Maleic acid is a V_0 marker). (Reprinted from reference 4 with permission).

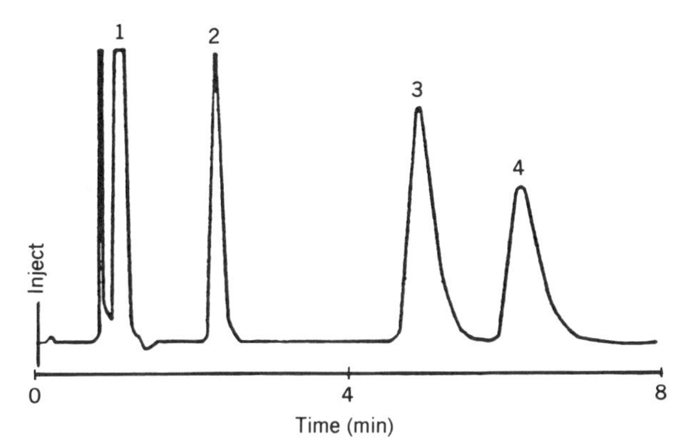

FIGURE 5-30. Separation of a 1:10 solution of cough syrup. Column: Radial Pak Silica (5 μm), 8 mm ID × 10 cm. Mobile phase: 75% methanol, 25% H_2O, 0.01 M $(NH_4)_2HPO_4$, pH = 7.8. Flow rate: 4 mL/min; 50 μL injection; detection at 254 nm; 0.05 AUFS. Key: (1) acetaminophen; (2) phenylpropanolamine; (3) chlorphenir-amine; (4) dextromethorphan. (Reprinted from Reference 5 with permission.)

ary phase attraction for the sample, but the tactics for implementing the strategy are different. Specifically, in ion exchange, by altering the pH or by increasing the salt content of the mobile phase, retention of the various analytes can be adjusted.

The pH effects the degree of ionization of both the stationary phase and the sample. There are two general forms of exchangers, strong and weak (see Fig. 4-8) and the pH affects each type of stationary phase differently. The *strong* exchangers, cation or anion, will be disassociated in a pH range from pH 1 to 13. The weak exchangers are a different story. For example, the NH_2 (amine) packing is a weak anion exchanger and has a pK_b of 6.5. Therefore, if this packing is used, the pH of the buffer should be below 5 for the surface to be ionic. A similar thought process would be used for a cation exchanger. When a weak exchanger is used, a pH change can "turn off" the ionization of the surface and, hence, influence the retention of ionized samples.

In addition to the pH effects on the degree of ionization of the stationary phase, pH affects the ionization of the sample. Samples that are strong acids form anions at low pH. Weak acids, however, require a higher pH to form anions. Thus, a strong anion exchanger retains a strong and a weak acid differently, based upon pH. As a general rule for anions, a starting pH should be 1.5 units greater than the highest pK_a value of the sample for the sample to be fully ionized. The pH can then be adjusted for appropriate retention. For cation separations, the trend is opposite that of anions and the initial pH should be 1.5 units lower than the lowest pK_b. For instance, if three analytes have different pH versus ionization profiles, the retention of the compounds at the various pH values and fixed ionic strengths will be as

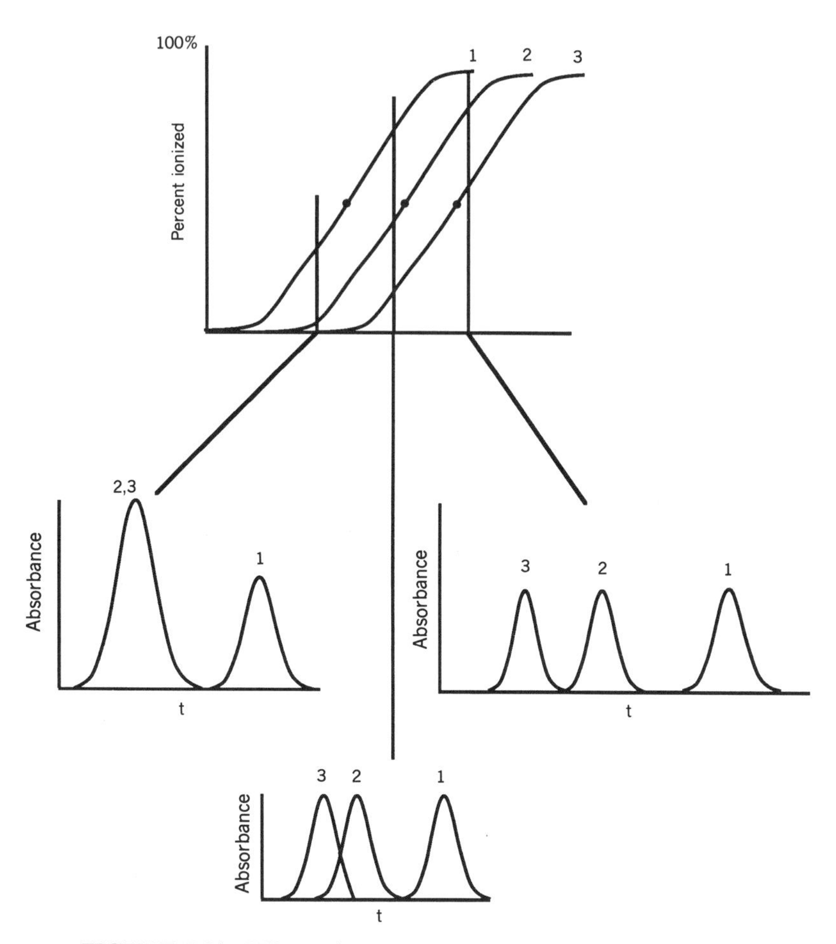

FIGURE 5-31. Effect of pH upon retention of ionic compounds.

shown in Figure 5-31. Of course, this is the behavior that would occur when the ion exchanger is a "strong" exchanger; that is, one whose surface charge is unchanged over the pH range used in Figure 5-31. The retention volume of an acid is decreased as the percent of ionization is decreased. At the pK of acid 3, it is at the midpoint of its percent of ionization. At this same pH value, acid 2 is 75% ionized and acid 1 is 100% ionized. Therefore, the order of elution is acid 1 < acid 2 < acid 3. If there are no other mechanisms at work, acids should elute from a strong anion exchanger in descending order of their pK values.

This behavior for retention in the ion-exchange mode can be contrasted to that which occurs in the ion-suppression, reverse-phase mode discussed earlier in this chapter (see Fig. 5-16). In the ion-suppression reverse-phase mode for strong acids, retention volume decreases when the pH is increased and for weak acids there is a constant retention volume with increasing pH until the pH is adjusted to be higher than the pK_a of the acid where the retention volume will decrease.

Once an appropriate ion-exchange packing and mobile-phase pH are chosen, an appropriate ionic strength needs to be found to enable a convenient elution time of the sample. As competing ions are increased in concentration, the eluting power of the mobile phase is increased. Buffer salts are often used to insure that a proper pH is maintained. Usually 20 mM is an appropriate starting concentration that can be adjusted higher or lower to attain the desired retention. The affect of increasing ionic strength can be explained on the basis of competition between sample ions and buffer ions for the fixed charged sites on the stationary phase. The higher the ionic strength (concentration of buffer) is, the greater the competition for the stationary phase. Therefore, sample ions are easily displaced and retention is decreased. Common buffers used in HPLC are phosphate, borate, nitrate, perchlorate, sulfate, acetate, and citrate depending upon the pH that is desired. (Chapter 6 has additional discussion on buffers.)

An example of two ion exchange separations are shown in Figure 5-32. The first separation is of phthalic acid, terephthalic acid and isophthalic acid using a 10-mM boric acid mobile phase adjusted to pH 9.2 using sodium hydroxide. The other separation in Figure 5-32 is of caffeine, *n*-acetyl-*p*-aminophenyl, aspirin, and salicylamide. In order for all of these compounds to be eluted in a reasonable time, the 10-mM boric acid solution also contains 20 mM sodium nitrate for ionic strength adjustment.

In summary, at least three sets of competing equilibria take place in an ion exchange system, so the ability to accurately predict what is happening is very difficult. First, remember that for weak ion exchangers, the pK_a and pK_b of the exchanger must be considered. Next, remember that whether the sample is adsorbed strongly or weakly on this column depends upon the nature of the carrier solvent. Lastly, remember that the most important variables for controlling retention are pH, counter ion concentration, temperature, and the type of counter ion used. The influence of these variables is generally:

1. *The pH of the Carrier Solvent.* The pH may be raised or lowered to control the degree of ionization of the sample. Ionized samples may be retained strongly while unionized samples cannot be retained by ionic interactions. The pH may also control the degree of ionization of the ionic sites on the column packing material. If the column packing is fully ionized, the sample may be strongly retained. If the column packing is unionized, there will be no ionic attraction.

2. *Counter Ion Concentration in the Carrier Solvent.* As a general rule, as the concentration of counter ion increases, the ionic adsorption of the sample decreases.

3. *Temperature.* As a general rule, as the temperature increases, the ionic sorption of the sample decreases. However, since ion-exchange resins

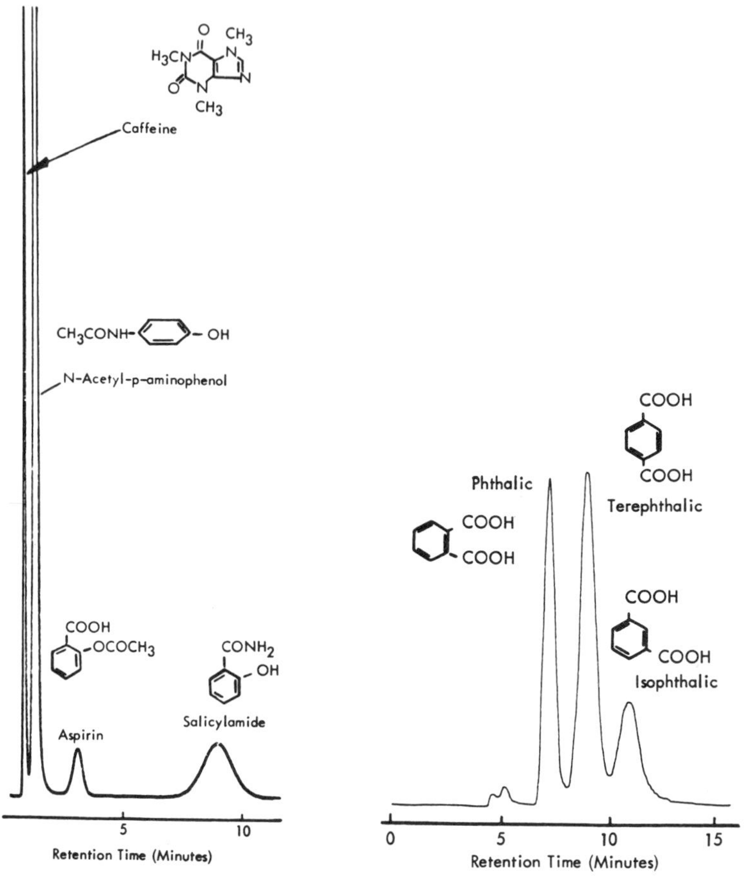

FIGURE 5-32. Ion-exchange separations on a strong anion-exchange packing (a methacrylate polymer containing quaternary ammonium groups coated on a pellicular packing). (Reprinted from DuPont Instruments Data Sheet, catalog number 820505.) See text for mobile phase.

contain reactive functional sites, experience has shown that if they are subjected to temperatures over 100°C for any period of time, these sites can be destroyed. Some ion-exchange materials may not even have this temperature stability. Care must be taken and the manufacturer's information must be followed.

4. *Type of Counter Ion Used.* Some types of counter ions are very strong adsorbers and therefore they compete very successfully for the ionic sites on column packing material. These strong counter ions elute samples more quickly than weaker counter ions.

The majority of the chromatographic separations using ion exchangers are carried out in aqueous solutions because of the superior ionizing properties of water. However, if an ion-exchange separation must be used, mixed solvents such as water/methanol should not be overlooked for additional retention control. Remember that the choice of buffer salt must be compatible with the equipment and soluble in the mobile phase being used.

Ion-exchange separations may become even more complex than they seem at first glance because of secondary interactions. For instance, in addition to ion exchange, adsorption to the packing matrix can be either reverse or normal phase in nature. Ion pairing of the sample could also occur in solution, which will tie up the ion and interfere with retention. Size exclusion effects can be important with large molecules like proteins. In this case size exclusion might prevent or reduce retention because the molecule cannot reach the active sites. Lastly, the ionic charges on the packing can be "poisoned" or "blocked" with strong interactions from other ionic "stuff" in the sample. When this happens, the column slowly loses retention and eventually is useless. Because of the ionic nature of the poisoning, extreme means may be needed to rejuvenate the packing, for example, unpacking the column and heating a polymer-based cation ion exchanger in acid.

Ion-Exchange Separation of Proteins

One particular application usage of ion exchange which is becoming very popular is the preparative separation of proteins, because the proteins can be recovered in their native forms (biologically active) after the separation. While the reverse-phase HPLC is the method of choice for separating protein fragments derived from enzymatic cleavage of a protein and for the analysis of proteins (refer to Chapter 2), the reverse-phase procedure denatures the proteins (no biological activity) because of the high organic content of the mobile phases. Therefore, for separations where biological activity is important, IEC should be used. Ion exchangers used for protein applications are often different from those used for small molecules. The packing materials are usually polymers or coated surfaces that are very compatible with water and proteins. The two most commonly used exchangers (anion and cation) are DEAE types where the packing surface contains diethyl aminoethyl groups and CM types where the stationary phase contains carboxymethyl groups. The functional group structures and their dependence on pH are shown in Figure 5-33. Both of these groups are capable of reversible ionic interactions with proteins.

A protein does not have a single charge; it has many charges on its surface that govern its retention on an ion exchanger. Unlike most small molecules which have an ionized and unionized form, a protein is made up of many amino acids, and as a result can often be both weakly acidic or weakly basic, depending upon the pH. Thus, the *net* charge on a protein is highly pH dependent and in going from low to high pH values, the net charge of the protein shifts from a positive to a negative value. As a rule of thumb, an ion exchanger can be selected on the basis of the pI of the protein. At a pH value

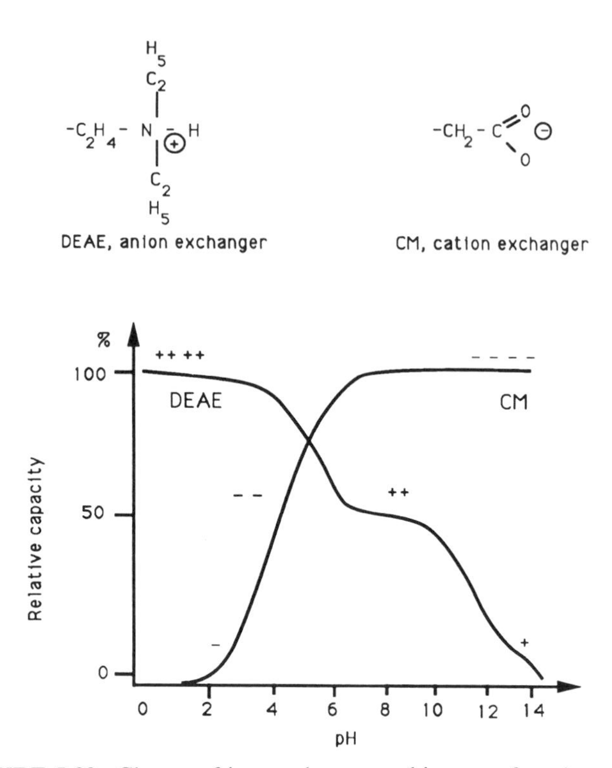

FIGURE 5-33. Charge of ion-exchange packing as a function of pH.

below the isoelectric point of the protein, the net charge is positive and it will bind to a cation exchanger (CM). Above the isoelectric point, the protein will desorb from a cation exchanger or it will sorb to an anion exchanger (DEAE). For instance, if the pI = 8.0, a protein can be retained on a cation ion exchanger using a mobile phase with a pH less than 7. If the pI = 7, a cation exchanger can be used below a pH of 6, or an anion exchanger can be used at pH above 8. If the pI = 5.5, an anion exchanger can be used with the pH above 6.5. The net charge versus pH profile is often unique for a given protein and, therefore, will contribute to selectivity.

In Figure 5-34 the pH–net charge curves for three hypothetical proteins are shown. There are many ways to accomplish the separation and hypothetical chromatograms on a CM and a DEAE ion exchanger are shown. At the lowest pH value, all three proteins are positively charged and sorb only to the CM ion exchanger. They are eluted in the order of net charge. At a slightly higher pH value, compound No. 1 is below its isoelectric point and is now negatively charged, while the other two retain positive charges. Separation of all three is possible on a CM column, while only one is separable on a DEAE column. At the next pH value, the only positively charged protein is No. 3, which sorbs to the CM ion exchanger. Because of their negative net

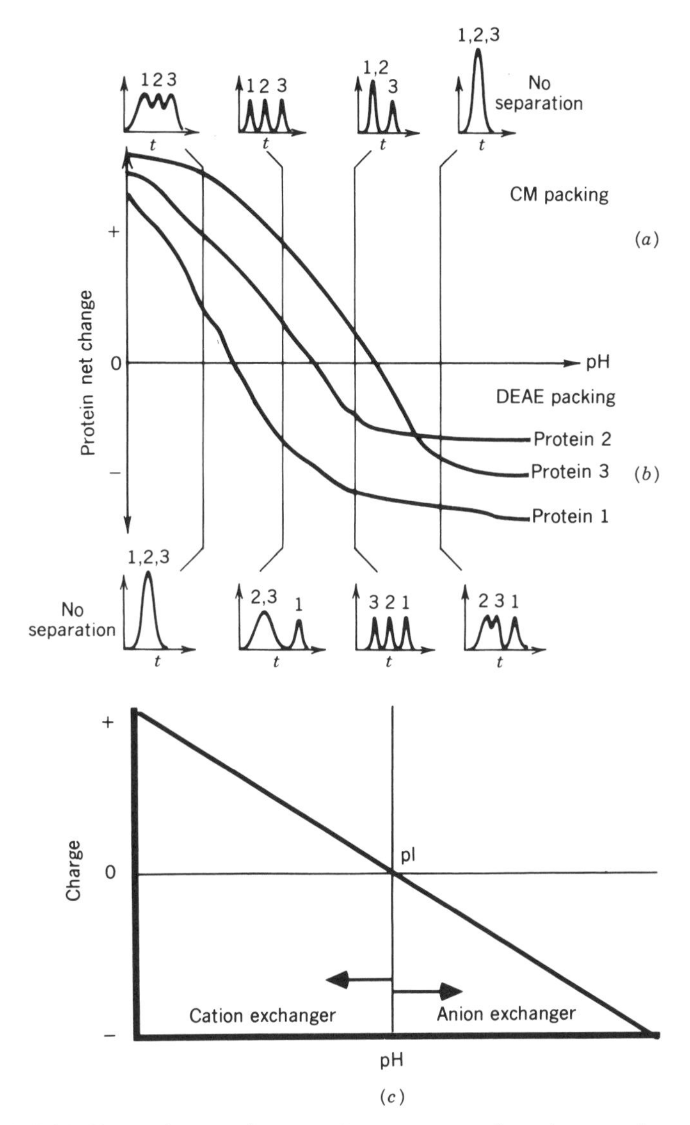

FIGURE 5-34. Dependence of separation on type of stationary phase (positive charged DEAE or negative charged CM) and pH. (*a*) Retention on CM packing. (*b*) Retention on DEAE packing. (*c*) Relationship of protein charge to the mode of ion exchange to be used. (*d*) Representation cation-exchange separation of proteins at two pH values. Mobile phase: 0–300 mM NaCl in a 20 mM sodium phosphate buffer. Flow rate: 1.5 mL/min.

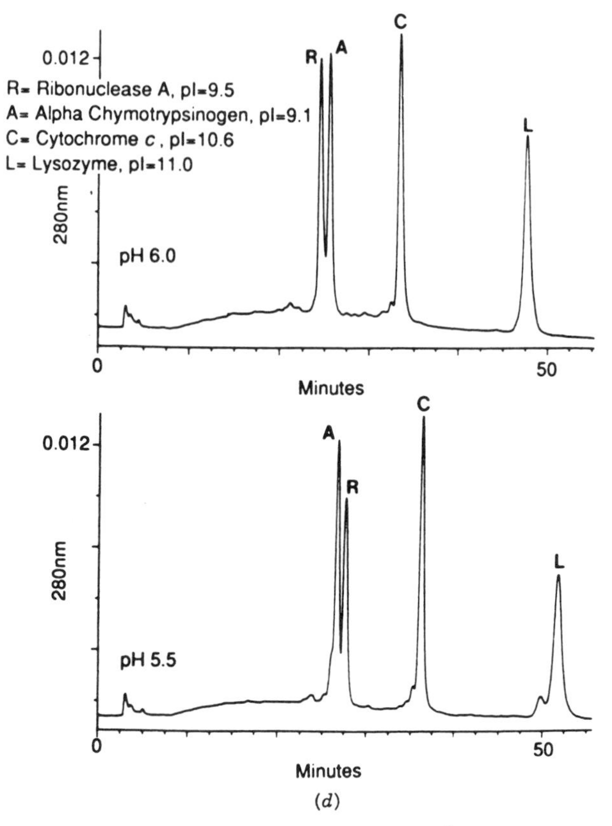

(*d*)

FIGURE 5-34. (*Continued*)

charges, both proteins No. 1 and 2 sorb to the DEAE ion exchanger and separation of all three is possible on a DEAE column. At the highest pH (alkaline), all three proteins can be adsorbed to the DEAE ion exchanger. Thus, by varying the pH of the mobile phase, one can greatly influence the selectivity in ion-exchange chromatography of proteins. The relationship between the pI of the proteins, the pH of the mobile phase, and the net charge on the protein dictates the ability of the ion exchanger to retain the protein (Fig. 5-34*c*).

While isocratic mobile phases can be used for ion-exchange separations of small molecules, protein separations are usually not accomplished with an isocratic system. Because of the high number of multiple ionic charges on a protein, desorbing the proteins often requires gradually increasing the ionic strength of the mobile phase, gradually changing the pH of the mobile phase, or both. These gradual changes as a function of time are referred to as gradient elution (refer to Chapter 7 for a discussion of this topic). An example of a gradient separation and the influence of pH upon the separation is shown in Figure 5-34*d*.

SMALL MOLECULE GEL PERMEATION CHROMATOGRAPHY

While most people think of GPC as belonging solely to "polymer people," this is not true. In fact, every chromatographic laboratory should take advantage of GPC's ability to separate small molecules (< 2000 MW) because small-molecule GPC (SMGPC) is a fast, easy-to-use method. All that is required for sample preparation is, literally, "dilute, filter, and shoot."

Historically, the potential for separating discrete small molecules has long been recognized (6–12), but since analyses took several hours, the use of SMGPC never was embraced in the laboratory. Today, however, technology has resulted in high-efficiency columns that can provide almost a 10-fold increase in efficiency (plates) over similar (older) columns of larger particle sizes. This means that small-molecule separations that once required several columns and several hours can now be performed on a single, high-performance column in a matter of minutes.

Today, SMGPC is very attractive as a simple and effective technique for the analysis of a variety of samples (13–15). The high-efficiency columns that are possible using modern GPC are evident from the separation of a mixture of *n*-alkanes shown in Figure 5-35. This is contrasted to the "early days" of GPC shown in Figure 5-36, when the same separation was possible but with a much longer analysis time. The separation in Figure 5-36 took $2\frac{2}{3}$ days, in contrast to the 20-min separation in Figure 5-35.

As was discussed in Chapter 3, the smaller the molecule, the greater the number of pores it can enter in the packing material. Larger molecules are excluded from some or all of the pores. The more pores a sample component enters, the longer it takes to elute from the column. This selective permeation into the pores is based on molecular size and results in the separation of the various sizes of sample components in a highly predictable manner.

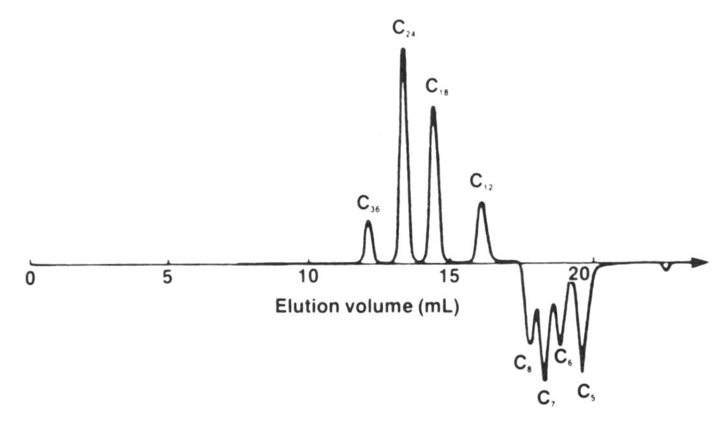

FIGURE 5-35. Separation of *n*-alkanes on two GPC columns. Column: 100 Å μ-styragel; 7.8 mm ID × 30 cm. Mobile phase: THF. Flow rate: 0.5 mL/min. Detection: refractive index. (Reprinted from reference 9 with permission.)

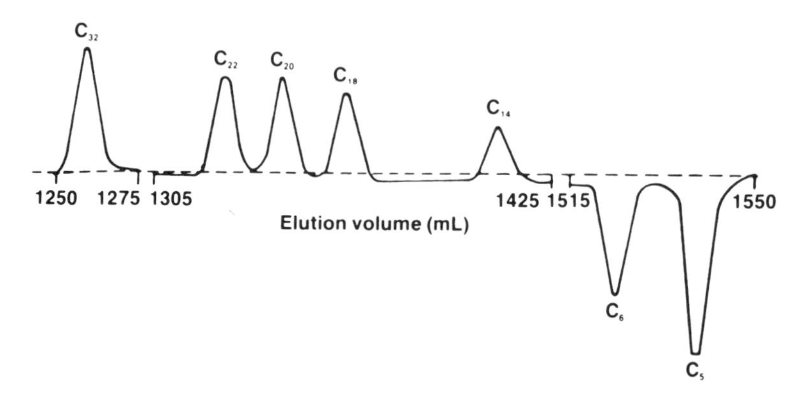

FIGURE 5-36. High-resolution GPC separation of hydrocarbons under analytical operating conditions. (Reprinted from reference 8 with permission.)

This predictability provides a major advantage over other LC modes because the maximum amount of solvent required to elute all sample components is equivalent to one column volume, and the elution volume of any molecule can be predicted for a given column if a calibration curve is available. Clearly, the most crucial ingredient for success in using SMGPC is selecting the proper type of column with the appropriate calibration curve.

Calibration Curves

In deciding when to employ SMGPC, one must consider the size and shape differences of the molecules to be separated—not just their molecular-weight differences. Molecules with small differences in molecular weight may have vast differences in molecular size in solution, and many researchers have underestimated the separating power of this LC mode by incorrectly correlating molecular weight with molecular size.

It has been a rule of thumb that two small molecules (< 2000 MW) that have a 10% size difference should be separable by high-efficiency GPC if a proper calibration curve (suitable range of pore size) is available. For example, Figure 5-37 shows the type of calibration curve that can be obtained using GPC columns for separations of the small molecules typically encountered in the pharmaceutical industry. The samples were simply dissolved in a suitable solvent, THF, and injected. Note that in the calibration curve for low-molecular-weight pharmaceuticals, most of the compounds do fall on the curve. The exceptions, however, prove the importance of thinking in terms of molecular size rather than molecular weight. This calibration curve also demonstrates an important, yet nontraditional (i.e., nonpolymer) use of SMGPC.

Two very obvious aberrations on the calibration curve, shown in Figure 5-37, are reserpine and vitamin A palmitate. Although reserpine has an

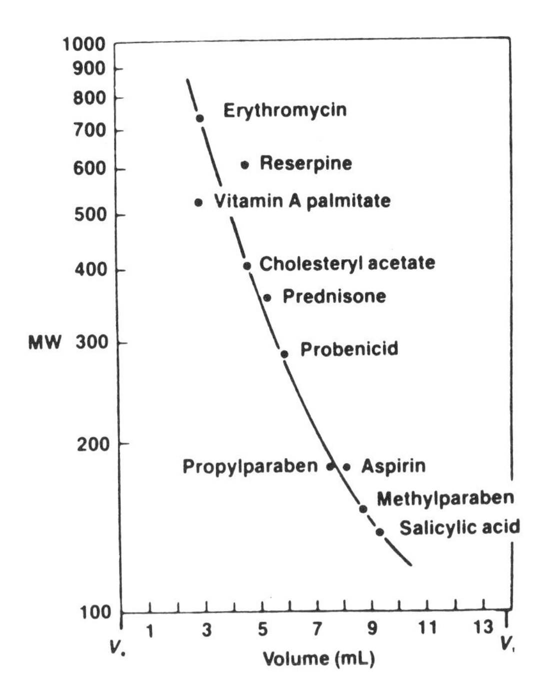

Vitamin A Palmitate

Reserpine

FIGURE 5-37. Calibration curve for some low-molecular-weight pharmaceutical compounds. Column: 100 Å Ultrastyragel; 7.8 mm ID × 60 cm (two columns). (Reprinted from reference 9 with permission.)

actual molecular weight of 608, its apparent molecular weight from the calibration curve is only 410. On the other hand, vitamin A palmitate, with an actual molecular weight of 543, appears to have a molecular weight of 760. A glance at the structure of the two compounds (see Fig. 5-37) indicates the reasons for these differences. Reserpine has an extremely compact structure, making this compound appear smaller in solution. In contrast, vitamin A palmitate is a very elongated, "rodlike" molecule, thus appearing larger than its molecular weight would indicate. So even though vitamin A palmitate is 12% smaller than reserpine in terms of molecular weight, in terms of size in solution (as seen on the GPC column), vitamin A palmitate actually appears 86% larger than reserpine. Therefore, the GPC separation of these compounds would be very easily accomplished.

The vital importance of size and shape as compared with absolute molecular weight is emphasized by another pair of "exceptions" on the calibration curve in Figure 5-37—aspirin and propylparaben. These two compounds have essentially identical molecular weights (180.2 and 180.1, respectively), but are easily separated by SMGPC. From the structures, it can be deduced that propylparaben acts as a larger molecule in solution than does aspirin. This is probably due to a solvent effect where the THF hydrogen-bonds to the polar groups in propylparaben. Thus, propylparaben should elute before aspirin, and as shown on the calibration curve, it does.

Small differences in the structure of the molecules result in changes in the elution volumes. Note the difference that shortening the side chain from propyl to methyl has on the elution volume of the paraben. Even such a minor difference is sufficient to permit the separation of these two compounds. The same effect can be observed if aspirin (acetylsalicylic acid) is hydrolyzed to salicylic acid. Reduction in molecular size increases the elution volume.

Solvent Effects

The mobile phase in GPC is deliberately chosen to be a strong solvent for the solute so that retention by mechanisms other than size exclusion (e.g., adsorption) will not occur. Therefore, the choice of solvent is not expected to greatly influence the chromatographic results. Solvent effects of two kinds do occur in practice. The first is related to the detector used. When detection is by differential refractometry (see Figs. 5-35 and 5-38), the separation of normal hydrocarbons by GPC might result in a chromatogram in which some peaks are positive and some are negative since the peak response depends upon the RI of the solute compared to the mobile phase. For example, in Figure 5-35 the lower-molecular-weight hydrocarbons have a RI less than that of THF and this explains why the peaks for species with carbon numbers less than 10 are negative and the higher hydrocarbons show a positive response. Obviously, this type of chromatogram (Fig. 5-35) complicates quantitation of the peaks, especially near the crossover from negative to

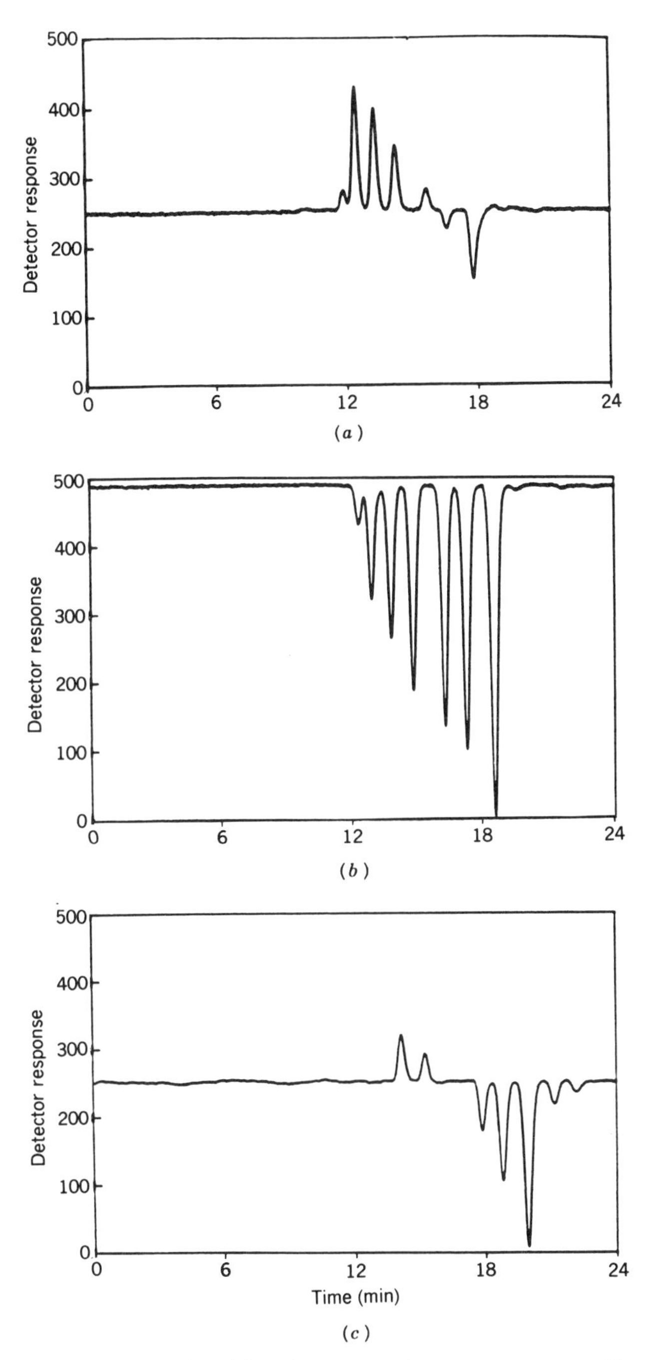

FIGURE 5-38. The effect of mobile phase upon detector response. The separation is a mixture of *n*-hydrocarbons on two columns 100 Å ultrastyragel, 7.8 mm × 30 cm) using an RI detector. Individual hydrocarbon peaks: C_{40} (limited solubility), C_{32}, C_{24}, C_{18}, C_{12}, C_7, and C_5. Mobile phase: (*a*) tetrahydrofuran; (*b*) toluene; and (*c*) methylene chloride. (The peak for C_{24} is missing since the RI for C_{24} is similar to that of methylene chloride.)

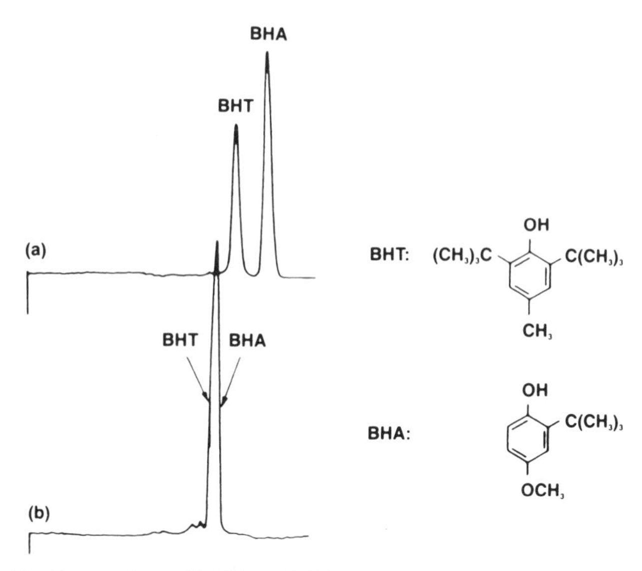

FIGURE 5-39. Separation of BHA and BHT on a SMGPC column using (*a*) chloroform and (*b*) THF as a mobile phase. Column: 100 Å ultrastyragel; 7.8 mm ID × 30 cm. (*a,b*) Flow rate: 1 mL/min; detection: UV, 254 nm. (Reprinted from reference 13 with permission.)

positive. However, if calibration cannot compensate for this behavior, this type of solvent effect can generally be avoided by choosing another mobile phase. In the case of *n*-hydrocarbons, the use of toluene (Fig. 5-38) produces a chromatogram in which all the peaks are negative. However, it should be noted that the difference in the refractive index of the hydrocarbons compared to toluene increases as one goes from C_5 to C_{36}, which explains the different responses for equal amounts of hydrocarbons injected.

The second type of solvent-related effect that commonly occurs is due to sample–mobile-phase interactions (such as hydrogen bonding) that can change the effective size of a molecule in solution. This was already discussed for the aspirin–propylparaben comparison. Because of the popularity of THF as a mobile phase for GPC, this sort of "differential solvation" must be kept in mind, particularly when polar solutes are analyzed. As demonstrated in Figure 5-39, knowledge of this effect can aid the chromatographer. Here BHA and BHT are fully resolved in $CHCl_3$ but coelute in THF. Both BHA and BHT are phenolic compounds, but the site on BHT is sterically hindered and apparently does not form a hydrogen bond with THF. The hydrogen-bonded BHA–THF complex that does form is apparently similar in size to BHT; hence the two solutes coelute in THF. In chloroform, BHA and BHT are resolved because of the differences in their molecular sizes.

Column Efficiency

Since there is no selectivity (α) involved in GPC, the ability to separate a pair of compounds depends upon the calibration curve and the efficiency of the column(s). One measure of the ability of the GPC column(s) to separate is the peak capacity of the column, which is defined as the number of resolvable peaks per chromatogram, n. The peak capacity (16) for a GPC column used in the analysis of small molecules is related to the number of theoretical plates (N) according to the equation

$$n = 1 + \frac{N^{1/2}}{4}(\Delta \ln V_E)$$

where $\Delta \ln V_E$ specifies the elution range of interest with regard to the calibration curve. For example, a 15,000-plate column could resolve 22 peaks to baseline; compare this with 14 peaks for a 6000-plate column or 10 peaks for a 3000-plate column. Because of the high number of plates per column, SMGPC may offer additional advantages over other separation techniques for particular applications.

Applications

Two representative examples of SMGPC separations are presented in Figures 5-40 and 5-41. The sample in Figure 5-40 was a rodent bait in which the active ingredient, warfarin, was to be determined. Quantitation of this component by SMGPC (15) was as reliable as the commonly used reverse-phase method (17), with the advantage of essentially eliminating sample cleanup because the extract was filtered and injected directly onto the GPC column.

In Figure 5-41 the active steroid (triamcinolone acetonide) and preservative (benzyl alcohol) are determined from a steroid cream. The higher-molecular-weight components of the cream base are well separated from the analytes. The ability to elute all the components of a cream or ointment in an SMGPC analysis gives this approach an important advantage over competing separation techniques that require sample preparation prior to the HPLC analysis.

Preparative Capability

Figure 5-42 shows a separation of an eight-carbon monocarboxylic acid from a nine-carbon dicarboxylic acid. As can be seen, injecting 10 mg of each acid (for a total of 20 mg) dissolved in 100 or 1000 μL (1 mL) gives essentially identical chromatograms, with the larger injection broadening the peaks only slightly. The implication of this work is that larger volumes, and therefore more sample, can be injected without influencing the purity of the separated peak.

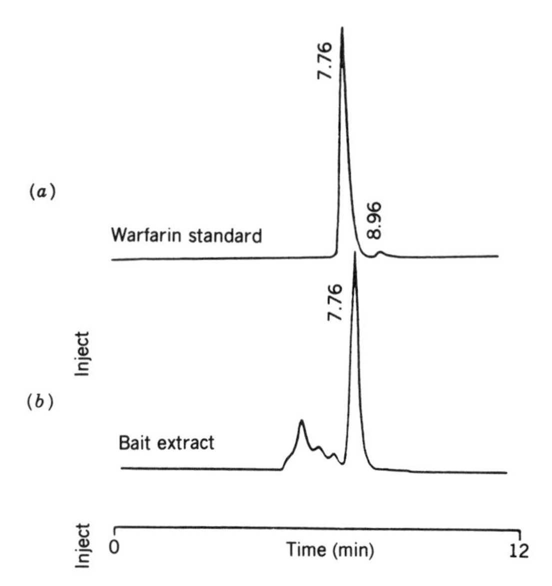

FIGURE 5-40. Small-molecule GPC determination of warfarin in grain bait. (*a*) Warfarin standard. (*b*) Bait extract. Column: 100 Å Ultrastyragel; 7.8 mm ID × 30 cm. Mobile phase: THF. Flow rate: 1 mL/min. Detection: UV, 280 nm, 0.2 AUFS. (Reprinted from reference 13 with permission.)

Once the maximum injection volume is decided, the maximum sample load should be determined. In this example, the amount of the mono- and dicarboxylic acids was increased from 10 to 100 mg (of each acid) per injection in 1 mL of mobile phase. At first glance, the overlap between peaks suggested that the electronics of the refractive index detector were saturated

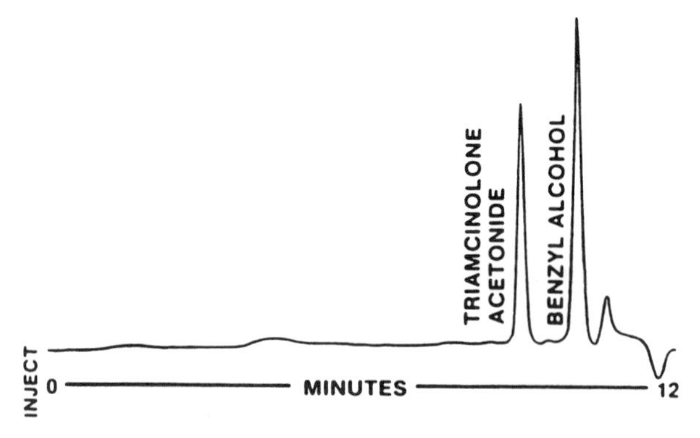

FIGURE 5-41. Determination of triamcinolone acetonide (peak A) and benzyl alcohol (peak B) from a steroid cream. Column: 500 Å Ultrastyragel; 7.8 mm ID × 30 cm. Mobile phase: THF. Flow rate: 1 mL/min. Detection: UV, 254 nm, 0.2 AUFS. (Reprinted from reference 13 with permission.)

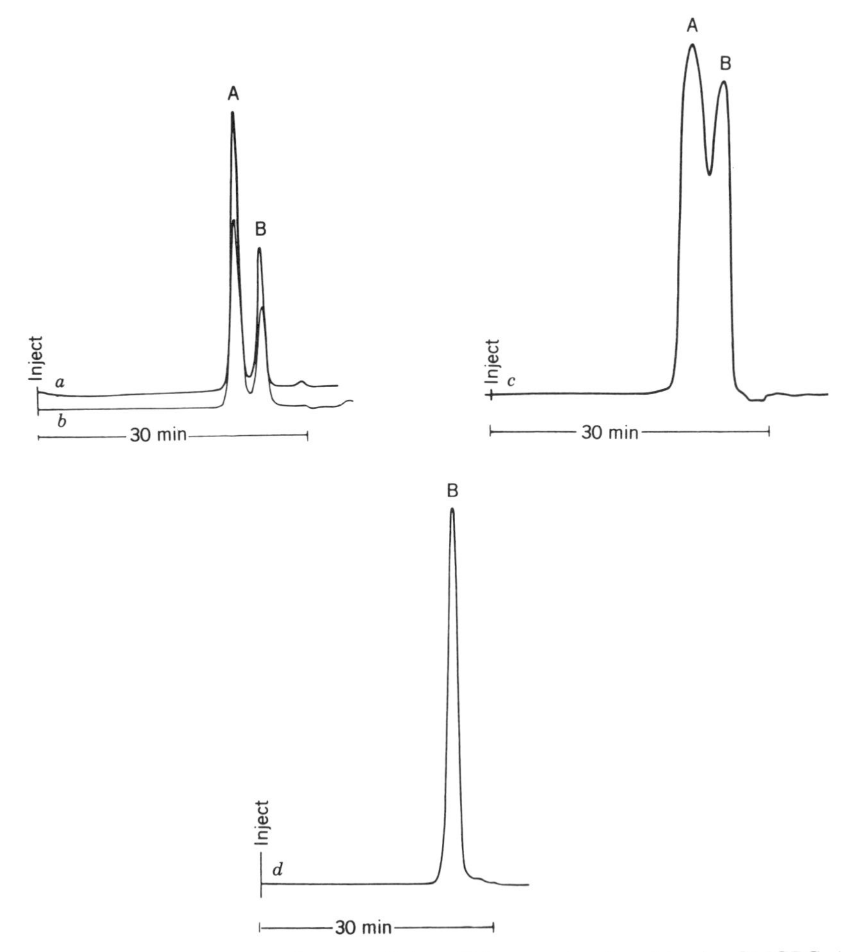

FIGURE 5-42. Separation of nonanedioic acid from octanoic acid using SMGPC. (*a*) A 100-μL injection containing 10 mg of each acid. (*b*) A 1-mL injection containing 10 mg of each acid. (*c*) A 1-mL injection containing 100 mg of each acid (note overload of RI detector). (*d*) Material from the second peak in (*c*) collected and reinjected. Column: 100 Å Ultrastyragel, 7.8 mm ID × 30 cm. (Reprinted from reference 9 with permission.)

and the response was nonlinear. To test this, the second peak was collected, diluted 10-fold (to put the detector back into the linear range), and reinjected. As can be seen from the single, sharp octanoic acid peak, there is no evidence of contamination from nonanedioic acid, indicating that high purity was preserved even when injecting 100 mg of each component. This separation also illustrates that meaningful preparative separations can be done easily in the GPC mode.

Another feature of SMGPC separations that has particular benefit in preparative work is the ability to "stack" or "piggy-back" injections. In GPC one column volume of solvent will elute all compounds, and peaks cannot catch up to or pass each other; therefore, the second injection can be made before the first sample has totally eluted from the column. This "stacking" of injections can significantly increase the throughput (grams of product per minute) for preparative work on such columns.

This technique also gives the chromatographer using GPC the ability to remove unwanted substances from the sample components. An example is the separation of pesticides from animal fat. Injection of the sample matrix of triglycerides (fats) is not compatible with the mobile phase and/or columns (both LC and GC) used to analyze pesticides. However, by using SMGPC to segment the high-molecular-weight fat component, the pesticide fraction can be collected, evaporated to dryness, and eventually analyzed by another analytical technique. This example of sequential analysis using SMGPC is the basis of an automated, commercial instrument used in pesticide residue laboratories.

Summary

Originally used chiefly as a method for characterizing polymers, GPC is also useful for separating small molecules from interfering matrices, such as those found in foods, pharmaceutical preparations, and natural products. In addition, GPC can easily detect the presence of high-molecular-weight components in a sample matrix, indicating the need for sample cleanup before the final analysis of the small molecules. Two questions should be asked: (1) "Are my sample components soluble in THF or another suitable GPC solvent?" and (2) "Is there a 10% difference in size between the sample components?" If so, we must choose the appropriate column to use. To this end, it is extremely useful to obtain literature from the column manufacturers which points out the name of the column to use for small molecules (< 2000 MW). Then it is a straightforward process to dissolve your sample, filter it, and inject it into the column. The separation of small molecules from high-molecular-weight substances present in the sample matrix is often the first step in "sequential analysis," in which the compounds isolated by GPC can then be further resolved by normal- or reverse-phase chromatography.

NORMAL PHASE

The C_{18} phases have emerged as the overwhelming first choice when developing new methods because, as noted in the first section of this chapter, developing a reverse-phase separation is straightforward. As a result, many separation successes are easily attained on a reverse-phase system. Unfortunately, this has led many to "forget" about using the normal-phase mode. In fact, because the theory of normal phase on silica gel predates the introduction of bonded phases (ca. 1972), many consider developing normal-phase HPLC separations as a "lost art."

Stationary Phases

A discussion of the stationary phases used in the normal-phase mode is found in Chapter 4 (see Table 4-1), but a few additional comments are in order. Silica gel was one of the earliest stationary phases used in chromatography. Its particle shape, surface properties, and pore structure were available in many variations, which made it very useful. Silica is wetted by nearly every potential mobile phase, is inert to most compounds, and has a high surface activity which can be modified easily with water or other agents. It further offers high sample capacity with high purity and excellent mechanical strength at a reasonable cost. Silica can be used to separate a wide variety of chemical compounds, and its chromatographic behavior, based upon extensive history, is generally predictable and reproducible.

When contemplating larger-scale *preparative* LC, underivatized silica gel, with its low cost, versatility, mechanical strength, good LC separation properties, and ease of scaleup and use, should be the first choice rather than the last. Additionally, the mobile phases used are volatile, which makes recovery of isolated compounds easy. Figure 5-43 shows an example of this approach for the separation of positional isomers using a methylene chloride–ethyl acetate mobile phase. Isolation of one of these isomers was part of a synthesis of polynuclear aromatic derivatives. Note that the resolution was large ($\alpha = 2.2$) in the analytical separation (Fig. 5-43*a*) since the goal was to eventually be able to do a preparative collection of material. The determination of the overload point was done on the analytical column (Fig. 5-43*b*). The final preparative separation (Fig. 5-43*c*) was accomplished on a preparative instrument resulting in 10 g being separated in 7 min with fractions 2, 4, and 5 being greater than 99.9% pure (18). Note that the loading per gram of packing is the same on both columns in Figure 5-43*b* and *c* and that the linear velocity on the analytical column is comparable to that on the preparative column. Those interested in preparative LC should consult reference 18.

Different polar functional groups bonded onto silica can also be used for normal phase. The most frequently used bonded phases are amino-, diol-, cyano-, and nitro-groups. These phases, when used in the normal mode, are similar to adsorption onto silica gel. The use of bonded-phase columns in the normal phase mode is shown in Figure 5-44. There is a different order of elution on the amino-bonded phase compared to the cyano phase; and different elution than on silica (ortho $<$ meta $<$ para), which is not shown. This is due to interactions with the functional groups on the packing rather than solely with the hydroxyl groups on the silica gel.

When used in the normal-phase mode, bonded-phase packings are stable to hydrolysis, have a shorter equilibration time to mobile phase changes than unbonded silica, and do not require water deactivation or humidified mobile

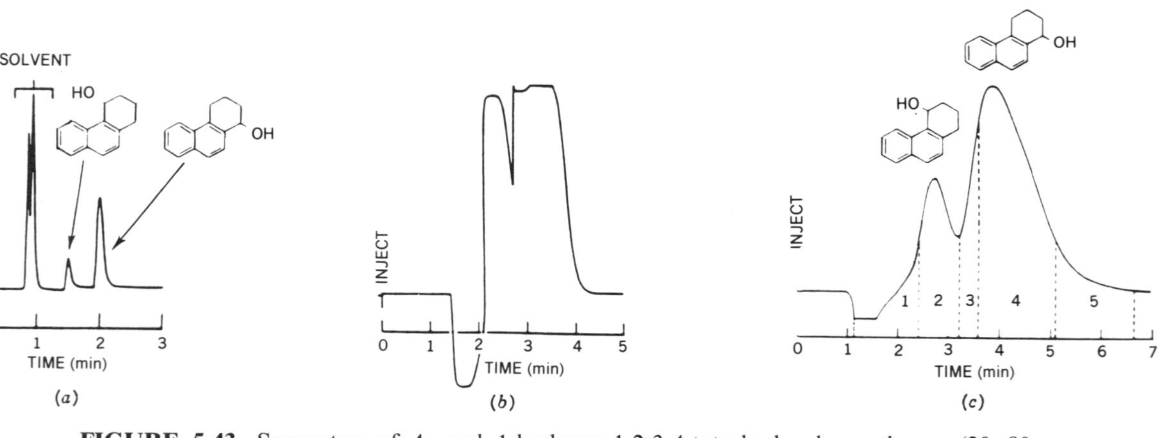

FIGURE 5-43. Separaton of 4- and 1-hydroxy-1,2,3,4-tetrahydrophenanthrene (20 : 80 synthetic mixture). (*a*) Analytical separation. Flow Rate: 4.0 mL/min. Column: μPorasil silica (10 μm), 3.9 mm ID × 30 cm. Mobile phase: methylene chloride/ethyl acetate (65 : 5). Detector: RI 16X. Sample size: 100 μg in 10 μL of mobile phase. (*b*) Small-scale preparative separation. Flow rate: 2.5 mL/min. Column and mobile phase: same as in *A*. Flow rate: 2.5 mL/min. Detector: RI at 128X. Sample injection: 54 mg in 1.75 mL of mobile phase. (*c*) Preparative separation. Flow rate: 300 mL/min. Column: 5.7 cm ID × 30 cm, PrepPak 500 Silica Cartridge, d_p = 55–105 μ. Mobile phase: same as in (*a*). Detector: RI, relative response 1. Sample: 10 g in 45 mL of mobile phase. (Adapted from reference 18 with permission.)

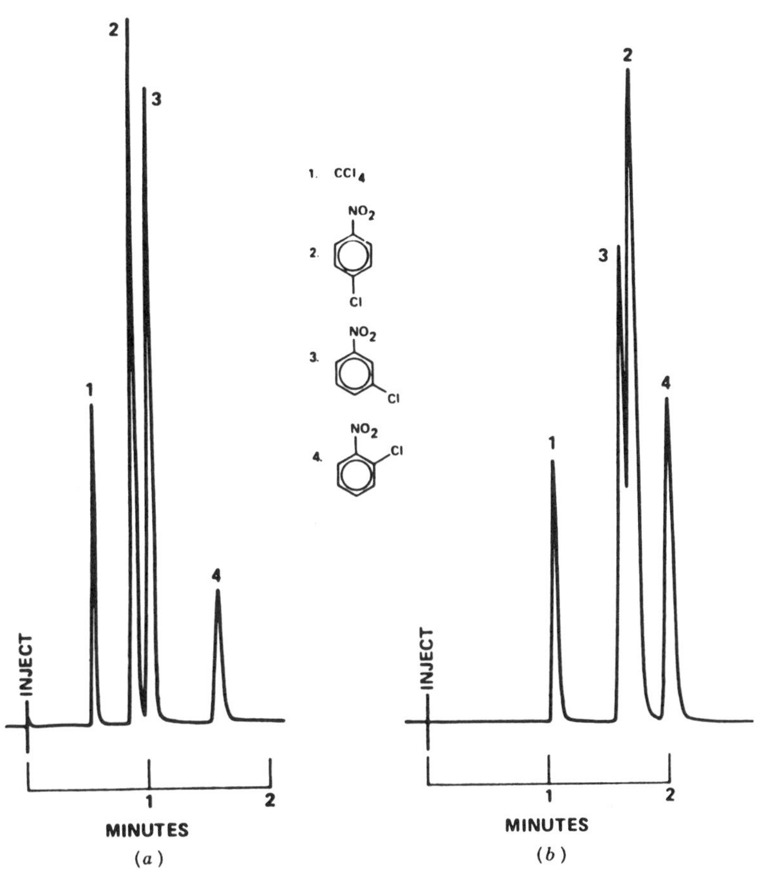

FIGURE 5-44. Comparison of the separation of nitrochlorobenzenes on two bonded phases. (*a*) μBondapak NH_2 (10 μm), 3.9 mm ID × 30 cm. (*b*) μBondapak CN (10 μm), 3.9 mm ID × 30 cm. Mobile phase: hexane. Flow rate: 6 mL/min. Detector: UV at 254 nm with 0.32 AUFS.

phases. Consequently, bonded phases can very rapidly be cleaned and methods development activities take a shorter time than on silica. Of course, bonded phases can also be used in the normal phase mode when doing preparative LC. However, the expense for the stationary phase is higher when compared to silica; but the use of volatile mobile phases is retained.

Mobile Phases

Mobile phases used for the normal-phase mode of LC are generally composed of only organic solvents. At first this makes the choice of a mobile phase quite foreboding since about 20 solvents come to mind as possible choices (see Fig. 4-7). How then can a strategy for developing a separation be implemented? The first answer is that a working theory of adsorption chromatography has been well developed, the details of which were summarized by Snyder in a monograph (19). The second answer is that most scientists using adsorption chromatography generally use only four or five key solvents as their main choices.

Before discussing a strategy of developing a separation, it is worth emphasizing a few points that were mentioned in Chapter 4. Solvent "strength" is a relative ranking of a solvent's ability to displace a given sample component from the stationary phase relative to other solvents. Contributing to the "strength" are the chemical characteristics of the solvent, which includes the nature of the functional groups, solvent polarity, dipole moment, hydrogen bonding ability, interactive and dispersive molecular forces, and other parameters which describe the physicochemical nature of a solvent. These parameters also contribute to the ability of a solvent to dissolve a particular compound. However, it is not strictly solubility, but the energy of the solvent–adsorbent interactions. For instance, alcohols are very strong eluents but are not good solvents for all solutes (e.g., saturated hydrocarbons). A solvent (mobile phase) displaces the solute (sample) from the surface because the solvent competes effectively for the adsorption sites.

A ranking of solvents or solvent mixtures in order of increasing eluent strength for a particular stationary phase is known as an *eluotropic series*. (In Chapter 4, Fig. 4-7 included an eluotropic series for alumina.) The list in Table 5-3 is the eluotropic series for various compounds on silica (20). Since silica is a polar material capable of hydrogen bonding and ion-exchange

TABLE 5-3. Eluotropic Series on Silica[a]

Solvent	k'
n-Hexane	0.00
2,2,4-Trimethylpentane (Isooctane)	0.0
Cyclopentane	0.05
1,1,2-Trichloro-1,2,2-trifluoroethane	0.1
Toluene	1.20
Dichloromethane (Methylene chloride)	1.30
Ethyl acetate	87.3
Methyl-*t*-butyl ether	90
Acetone	156
Tetrahydrofuran	160
Acetonitrile	170
Isopropanol (2-Propanol)	193
Ethanol (Ethyl alcohol)	377
Methanol (Methyl alcohol)	546
Water	1146
Acetic acid	8430

[a] From reference 20.

interaction with certain compounds, this eluotropic series reflects the increasing polarity and hydrophilicity of the solvents. Depending upon the method used to determine an eluotropic series, some change in the position of one solvent to another may occur (19). What would be ideal is a ranking of both the mobile phase and the solute for a stationary phase; but no eluotropic series exists which does so. Therefore, it is necessary to use the eluotropic series as guidelines for solvent strength.

Finding an Appropriate Mobile Phase

Any sample type that has been separated by normal-phase TLC is appropriate for a liquid–solid chromatographic (LSC) separation. An example of this is shown in Figure 5-45. In this separation, both the TLC plate and HPLC separations were done using the same mobile phase. Several good references on the early work in TLC (22,23) and on adsorption chromatography (24–26) should be consulted by those interested in a historical perspective of the use of normal-phase chromatography.

In general, the normal-phase mode is well suited for nonpolar samples and tends to differentiate molecules based upon the position of functional groups and the shape of the molecules. Retention is primarily determined by the attraction between the functional groups on the molecule and the hydroxyl (silanol) groups on the packing relative to the attraction of mobile-phase molecules to the stationary phase. Therefore, the relative adsorption of a compound increases as its polarity and number of functional groups increase. Thus, it should be no surprise that the contribution of an alkyl group to retention is very small. (This is contrasted to the reverse-phase mode where alkyl groups make a more significant contribution to retention.) As a result, normal phase is not used to separate the individual compounds in a homologous series or molecules that differ only in molecular weight and not in functional groups.

Just as we can rank the polarity of the solvent (the eluotropic series) with respect to the packing, the polarity of the *molecule* can be ranked based upon the presence (and/or position) of the major functional groups that it contains. Table 5-4 ranks some common functional groups in a qualitative fashion. As a rule of thumb, if a molecule has more than one functional group, the one with the largest polarity will determine the "polarity" of the molecule relative to other molecules in the sample mixture.

A typical application of using normal phase is shown in Figure 5-46 for the separation of aflatoxins (27). Aflatoxins are believed to be highly carcinogenic compounds and are produced by mold that grows on stored grain crops. Analysis of aflatoxins is important to confirm that the level of these compounds is below the maximum regulated by the United States Food and Drug Administration. The B group differs from the G group in chemical class while the differences within each group differ in degree of saturation. In this

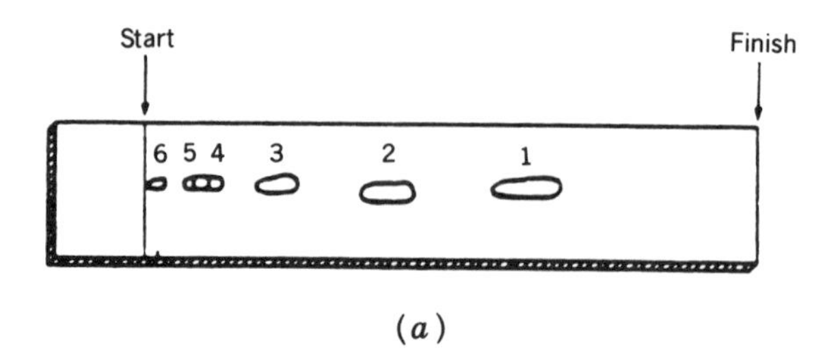

(*a*)

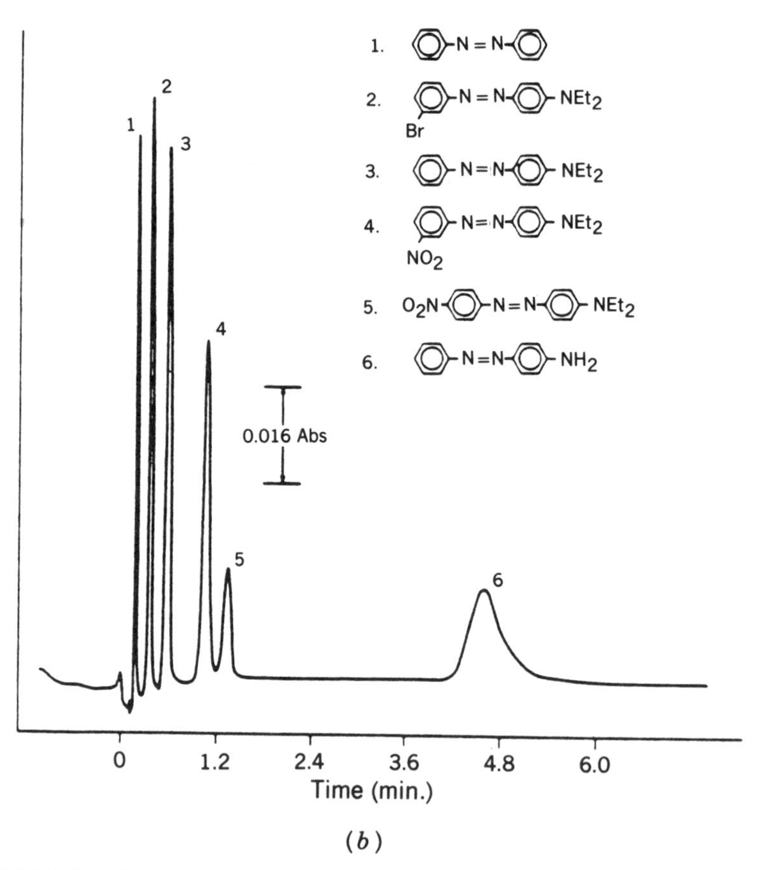

(*b*)

FIGURE 5-45. Example of using TLC as a guide for HPLC separation of Azo dyes. Analytes shown in the figure. (*a*) TLC plate: E Merck precoated TLC sheet, silica gel F-254. Solvent: 10% methylene chloride in hexane. Development time: 50 min. (*b*) Column: Micropak Si-10 (silica, 10 μm), 2.4 mm ID × 15 cm. Mobile phase: 10% methylene chloride in hexane. Flow rate: 2.2 mL/min. Detector: UV at 254 nm, 0.16 AUFS. (Reproduced from reference 21 with permission.)

TABLE 5-4. Approximate Order of the Polarity of Function Groups in Liquid–Solid Chromatography

Polarity	Group	Structure
Low	Methyl	CH_3
↓	Halide	F < Cl
	Ether	R—O—R
	Nitro	NO_2
	Ester	$—OC(=O)CH_3$
	Aldehyde	$—C(=O)H$
	Ketone	$—C(=O)—$
	Hydroxyl	—OH
High	Acid	$—C(=O)—OH$

example UV detection was used; however, fluorescence could be utilized for higher sensitivity.

As a result of the interactions involved between the molecule and the packing, LSC is used for the separation of molecules based upon chemical class (e.g., ketones from aldehydes) rather than for the separation of individual compounds within a class. A good example of the wide scope of compound-type selectivity in LSC is the separation of lipids into the following classes: hydrocarbons, cholesterol esters, triglycerides, diglycerides, monoglycerides, free steroids, and free fatty acids (28). Another popular use for LSC is the separation of isomers because the different geometry results in a relative adsorption difference onto the fixed adsorption sites on the packing surface. Often one of the isomers is more appropriately spacially matched to sitting on the surface than the other isomer. The order of elution is generally ortho < meta < para, as shown in Figure 5-47.

When developing normal-phase separations, the search for the best mobile-phase composition for an adequate separation can follow the generalized approach that was identified for reverse-phase separations. First, identify a strong solvent in which the sample is readily soluble. Second, choose a "poor" solvent in which the sample has less solubility. Once a "good" and "poor" solvent have been identified, an injection of the sample onto the column should be made, using the strong solvent as the mobile phase. In this situation, all of the components should elute as one peak at V_0. If there is retention when using a 100% "strong" solvent as the mobile phase, a *stronger* solvent needs to be found. After it has been determined that there is no retention of the sample components in the 100% strong solvent, the

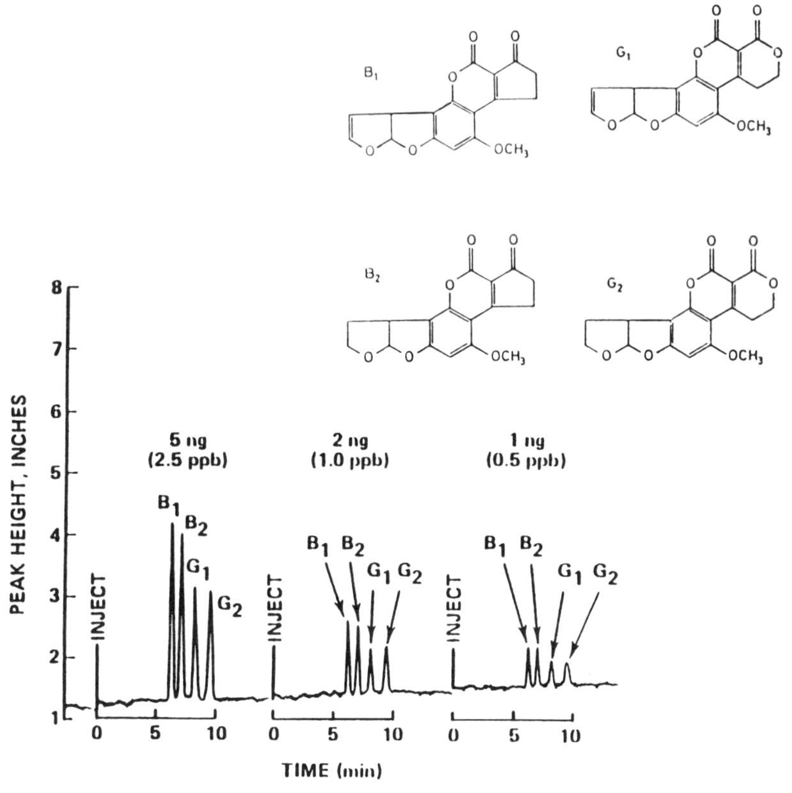

FIGURE 5-46. Separation of aflatoxins. This method was used to evaluate peanut butter and intended for use with other grain products. Column: μPorasil (silica, 10 μm), 3.9 mm ID × 30 cm. Mobile phase: $CHCl_3$ (H_2O Sat)/C_6H_{12}/CH_3CN, 25:7.5:1 (v/v).* Flow rate: 1.0 mL/min. Detector: UV : 365 nm, 0.005 AUFS. Structures are shown on top of figure. (Method described in reference 27.)

mobile phase should be made "weaker" by using a blend of strong solvent with the poor solvent. The sample is then reinjected into this new mobile phase and the retention behavior noted. Sequentially weaker mobile phases should be prepared and evaluated until retention of the sample components occurs. Ratios of strong and poor solvent should be done in logical steps so that, if necessary, retention of the peaks can eventually be "fine tuned" to the appropriate mobile phase containing the exact blend of the good and poor solvent for the desired resolution. An example of following this separa-

* Water saturated $CHCl_3$ is prepared by shaking 1000 mL $CHCl_3$ with approximately 100 mL distilled water four times for about 1 min each time, and equilibrating overnight before use. The water-saturated $CHCl_3$ is then mixed with cyclohexane and acetonitrile in ratios of 25:7.5:1 [v/v]. Either 1.5% ethanol or 2.0% 2-propanol is then added to the total volume to adjust retention time. Note: preparation of water-saturated solvents is discussed later in the text and a general-purpose procedure is suggested for preparing water-saturated mobile phases.)

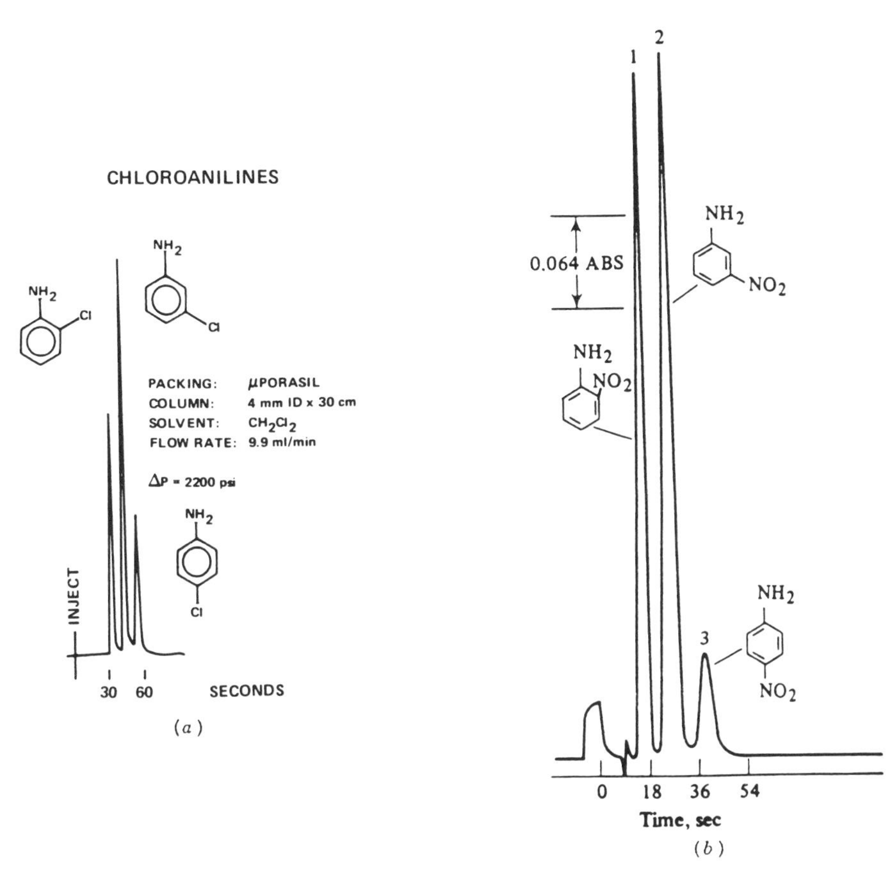

FIGURE 5-47. Separation of ortho, meta, and para isomers. (*a*) Column: μPorasil (silica, 10 μm), 3.9 mm ID × 30 cm. Mobile phase: methylene chloride. Flow rate: 9.9 mL/min. Detector: UV at 254 nm. (*b*) Column: Micropak AI-10, (alumina, 10 μm), 2.4 mm × 15 cm. Mobile phase: 40% methylene chloride in hexane. Flow rate: 1.58 mL/min. Detector: UV at 254 nm. (Chromatogram *B* reproduced from reference 21 with permission.)

tion development strategy is shown in Figure 5-48 for a sample of a synthesized steroid.

There might be a drawback to following this approach. Namely, in normal phase equilibration of the stationary phase takes more time than in reverse phase and it is, therefore, necessary to insure that the mobile phase has equilibrated before a new mobile phase is tested. This often requires waiting ten or more column volumes before injecting the sample and reinjecting the sample several times to insure that the retention time is stable.

Another approach to finding a mobile phase that is less empirical is to use the theoretical basis for understanding the energies of solute–solvent–stationary phase interactions and, thereby, achieve a given separation with less

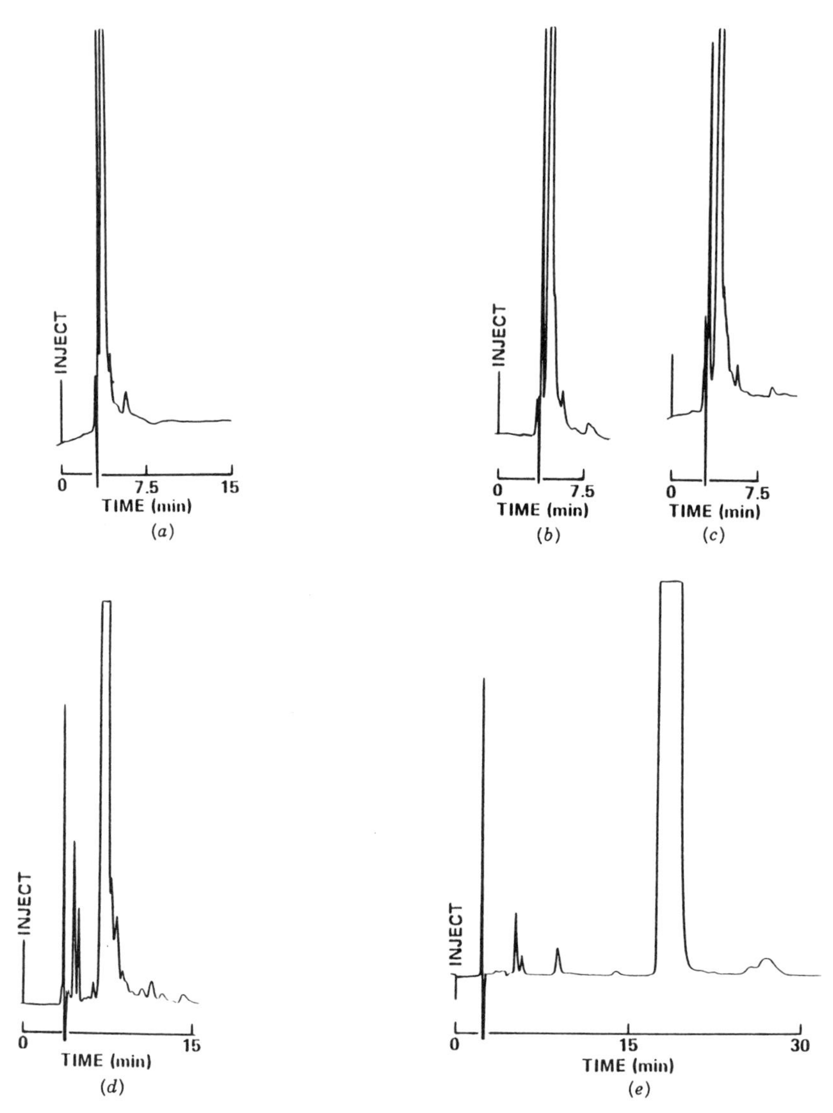

FIGURE 5-48. Developing a normal-phase separation. Column: μPorasil (silica, 10 μm), 3.9 mm ID × 30 cm. Flow rate: 1.0 mL/min. Detector: UV at 254 nm, 0.2 AUFS. Mobile-phase ratios of ethyl acetate/isooctane: (*a*) 100/0; (*b*) 90/10; (*c*) 75/25; (*d*) 50/50; (*e*) 20/80.

effort than the trial-and-error approach previously described. By using guidelines it is possible to choose a mobile phase which has a polarity approximately equal to or slightly weaker than the polarity of the sample components in order to have retention of the compounds. More specifically, by using Snyder's theory (19) it is possible to calculate an approximate solvent mixture which should be an appropriate mobile phase for a specific compound(s). In practice, the application of this theory requires the assembly of a number of sample adsorbents and solvent parameters, and a set of rigorous calculations. A simplification of Snyder's basic concepts was given by Saunders (29) and serves as a streamlined method of determining an approximate starting mobile phase.

As was discussed in earlier chapters, the ideal k' range for a separation is between 2 and 8. Snyder (19) defined a solvent strength parameter ε^0 as the adsorption energy per unit area of standard adsorbent and for a given sample and adsorbent log k' of the sample varies linearity with ε^0. Saunders defined a term E_3 as the solvent strength required to have a $k' = 3$ for a sample. Figure 5-49 shows values for a variety of compounds on a typical silica (300 m^2/g). When developing a separation, the first step is to refer to Figure 5-49 and the estimated E_3 values for the compounds of interest. Once E_3 is known, the individual can then refer to Figure 5-50 to obtain several alternative mobile phases for the initial separation attempt.

Consider naphthalene; for it to elute at $k' = 3$, a solvent strength (E_3) of 0.09 is needed using an active (dry) silica adsorbent (see Fig. 5-49). The bar in Figure 5-49 represents a range of solvent strength from 0.09 on dry active silica to -0.13 for a deactivated silica. Most practical systems will be partially deactivated and therefore the E_3 value would be midway in the range (bar) indicated in Figure 5-49. By grouping various parameters and comparing their relative effects on E, Saunders formulated the following approximate rules for estimating E_3 for many polyfunctional compounds on a fully activated column. The rules also apply to systems of equivalent deactivation. The listed rules are quoted from Saunders (29):

1. Saturated hydrocarbons and simple olefins are eluted with $k' < 3$ with all solvents. Fluorinated solvents are an exception (30).
2. Alkyl and halogen substituted aromatics have E_3 values within ± 0.05 unit of the parent compound.
3. Aliphatic compounds with 2 to 20 carbons will have E_3 values within ± 0.05 unit of the model compound ($R = C_6H_{13}$) used in Figure 5-49.
4. Difunctional compounds will have E_3 values larger than either of the monosubstituted compounds. For example, for benzaldehyde $E_3 = 0.23$ and for methoxybenzene $E_3 = 0.15$, but the E_3 value for methoxybenzaldehyde will be larger than 0.23 by an increment which depends upon the difference between values (ΔE_3) of the mono substituted compounds as follows:

ΔE_3	*Increment*
0–0.1	0.15
0.1–0.2	0.07
>0.2	0.00

In our example $\Delta E_3 = 0.23 - 0.15 = 0.08$; thus for methoxybenzaldehyde, the increment is 0.15 and $E_3 \approx 0.23 + 0.15 = 0.38$. Similarly, for diaminohexane $\Delta E_3 = 0.0$, $E_3 \approx 0.54 + 0.15 = 0.69$.

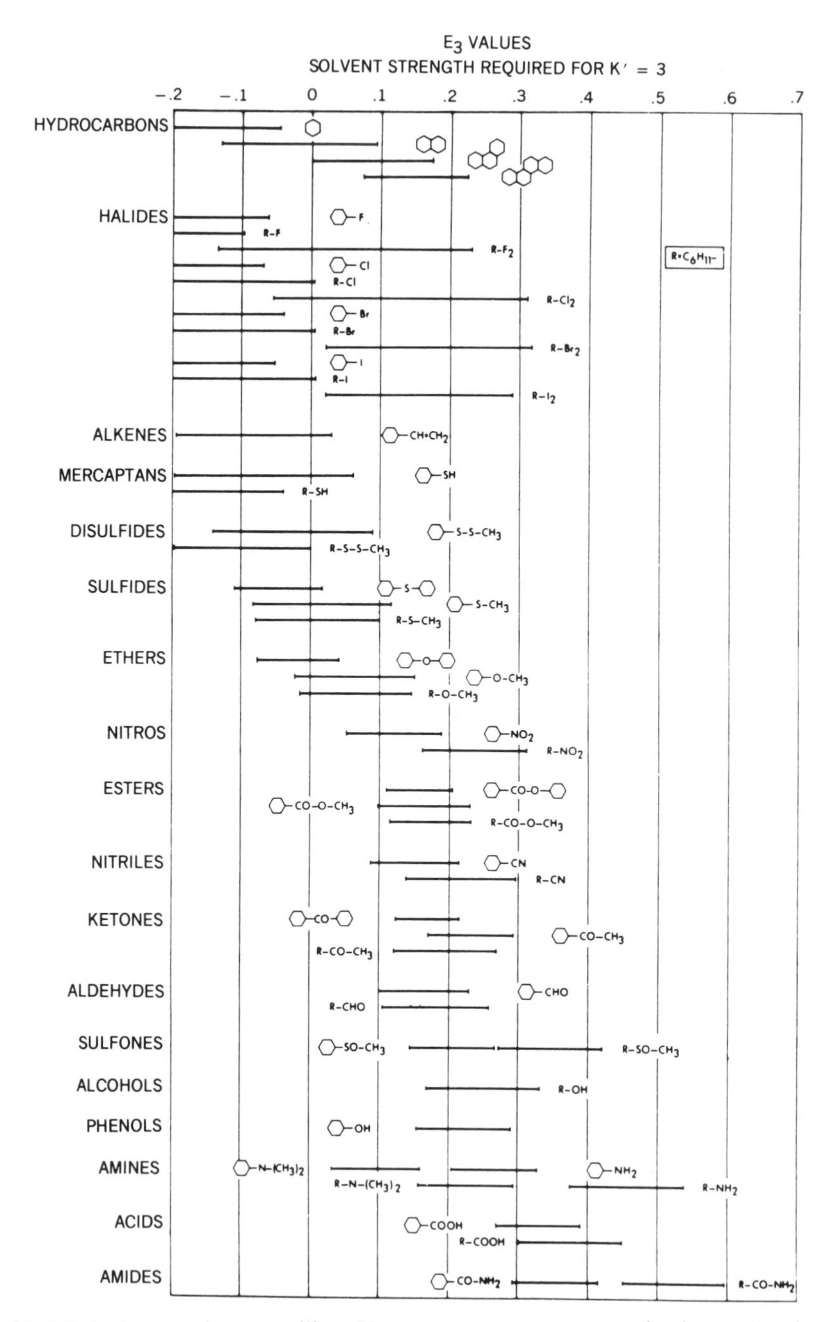

FIGURE 5-49. E_3 values on silica. Hexagons represent aromatic rings. (Reprinted from reference 29 with permission.)

5. Additional functional groups will add successively smaller increments to E_3, but the size of these increments cannot be easily predicted.
6. Addition of an aliphatic halogen or mercaptan moiety will increase E_3 by an increment 50 to 100% larger than the value indicated in 4 and 5 above. For example, for chlorohexyl ethyl ether $E_3 \approx 0.15 + 2\ (0.07) = 0.29$.
7. Compounds containing polynuclear aromatic moieties should be considered benzene compounds substituted with additional aromatic rings. For example, nitrophenanthrene should be considered a nitrobenzene ($E_3 = 0.19$) substituted with two additional rings, i.e., naphthalene ($E_3 = 0.09$). Therefore $\Delta E_3 \approx 0.1$ and the corresponding increment is 0.15, thus $E_3 = 0.34$.

As was mentioned, E_3 for a given compound will be largest when using dry solvents. The low end of the range bars plotted in Figure 5-49 is estimated to correspond to columns equilibrated with 75% water-saturated solvent, while the center of the scale corresponds to about 50% water saturation. Water saturation of 50% is commonly used for general work and is reported to have several other advantages (31). Water-saturated solvents may be prepared in a number of different ways. Saunders (29) recommends stirring the solvent mixture vigorously for 2–3 hours with 20 g/L of a mixture of equal quantities of water and silica gel. The silica serves to provide a high surface area to speed equilibration. However, mobile phases containing acetonitrile or methanol cannot be prepared with water saturated in this manner. These must be prepared using acetonitrile or methanol containing 3% (v/v) water, which should result in mobile phases with sufficient deactivation equivalent to about 50% saturation (29).

After the E_3 value for the compounds have been estimated using Saunders' rules and Figure 5-49, an appropriate mobile phase needs to be chosen by referring to Figure 5-50. In this figure, solvent strength is plotted across the top versus various binary mixtures of solvents in a manner similar to that shown by Neher (32,33) in TLC. The six solvents in Figure 5-50 are among the most useful for making mobile phases for adsorption chromatography (29) for several reasons: the whole range of solvent strengths is covered; viscosities are low, allowing low back pressure and high plate counts; UV cutoffs are low, permitting the use of UV detectors; and boiling points are low for rapid sample recovery.

In Figure 5-50, each horizontal line corresponds to a full range (0–100% v/v) of binary solvent mixtures. The first five lines represent mixtures of pentane with other solvents (line 1, pentane with isopropyl chloride; line 2, pentane with methylene chloride; line 3, pentane with ethyl ether; line 4, pentane with acetonitrile; and line 5, pentane with methanol). For the top line of this series (mixtures of pentane and isopropyl chloride) any solvent strength intermediate between pentane ($\varepsilon^0 = 0$) and isopropyl chloride ($\varepsilon^0 = 0.22$), can be estimated by reading from a vertical line dropped from the ε^0 scale. Thus, $\varepsilon^0 = 0.10$ is obtained by a 26% v/v isopropyl chloride in pentane; and $\varepsilon^0 = 0.20$ is obtained by 80% v/v isopropyl chloride in pentane. The

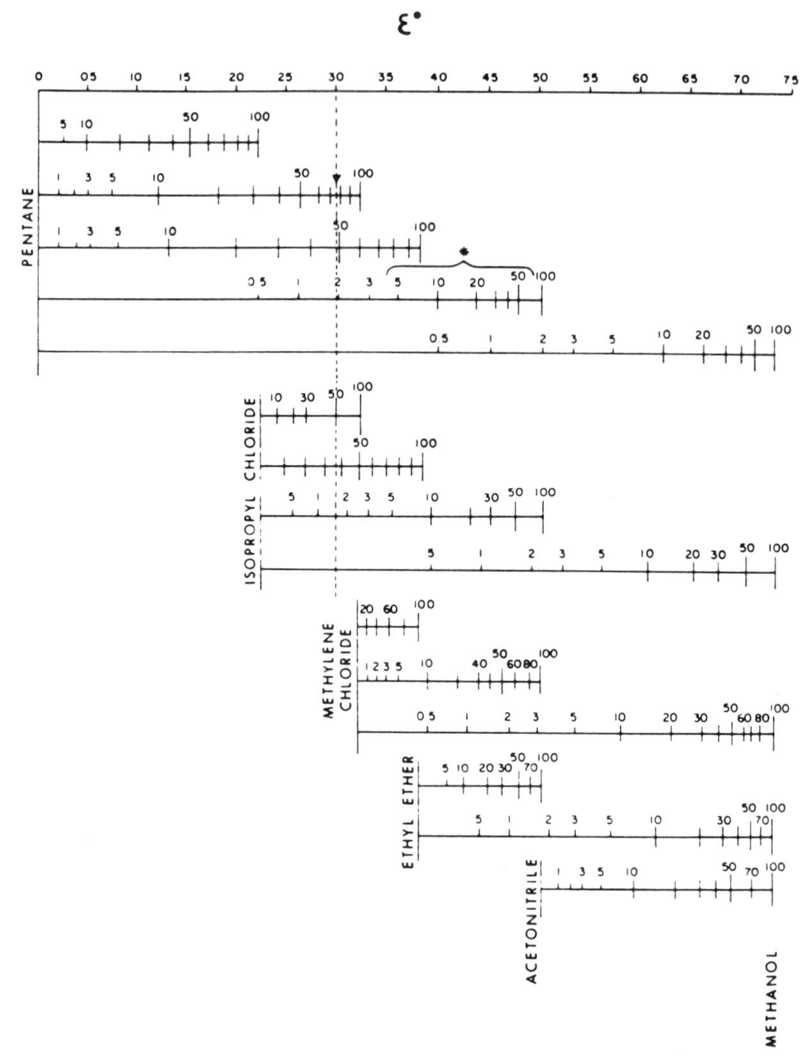

FIGURE 5-50. Mixed solvent strengths on silica. The * indicates that miscibility needs to be achieved by use of a small volume of an appropriate solvent such as chloroform. Use of chloroform is not recommended *per se*. An appropriate solvent must be in compliance with your organization's safety policy. (Reprinted from reference 29 with permission.)

other four lines in this series correspond to mixtures of pentane with methylene chloride ($\varepsilon^0 = 0.32$), ethyl ether ($\varepsilon^0 = 0.38$), acetonitrile ($\varepsilon^0 = 0.50$), and methanol ($\varepsilon^0 = 0.73$) (29). It should be pointed out that hexane can be substituted for pentane if the operator desires. This is important since some solvent delivery systems may have trouble pumping pentane. The second group of four lines in Figure 5-50, labeled isopropyl chloride, corresponds to binary mixtures of isopropyl chloride and the remaining four solvents (methylene chloride, ethyl ether, acetonitrile, and methanol).

For any desired solvent strength, Figure 5-50 indicates several possible binary mixtures. For example, the dashed line in Figure 5-50 intersects mixtures having $\varepsilon^0 = 0.30$. The arrow corresponds to 76% v/v methylene chloride in pentane. The following mixtures would also have $\varepsilon^0 = 0.30$: 49% v/v ethyl ether in pentane, 37% v/v ethyl ether in isopropyl chloride, and so on. The first choice for a sample with $E_3 = 0.30$ would be *any* of the mobile phases intersected by the $\varepsilon^0 = 0.30$ line.

By using Figure 5-50, the first mobile phase will probably be close to the optimum mobile phase, but not exact. Therefore, it is necessary to adjust retention by using more strong solvent to reduce retention and more poor solvent to increase retention. Additionally, Saunders found the following guidelines helpful in adjusting k' and selectivity. These guidelines are quoted directly from Saunders (29):

1. An increase in solvent strength ε^0 of 0.05 will decrease k' by a factor of 2–4 (19).
2. A decrease in ε^0 of 0.05 will increase k' by a factor 2–4 (19).
3. Selectivity will be greatest if the concentration of the stronger component of the solvent mixture is <5% (v/v) or >50% (v/v). For example, at $\varepsilon^0 = 0.30$, either 76% methylene chloride in pentane or 1.7% acetonitrile in isopropyl chloride would be expected to give maximum selectivity (34).
4. Substitution of ethyl ether or methanol for one of the other strong components can often improve selectivity by the formation of hydrogen bonds (34).
5. If a dry solvent results in a tailed peak or retention time which varies with concentration, a more highly deactivated solvent should be used. It should be remembered that this will shift E_3 to a lower value.

In addition to these rules of thumb, the following general considerations are helpful. These are listed in order of priority:

1. Begin with a single solvent, if possible, since it simplifies the separation system and permits economical solvent recovery.

2. If a single solvent will not accomplish the separation, try a mixture of two solvents of different types (e.g., on silica, try mixing a hydrocarbon and a halogenated hydrocarbon for fairly nonpolar solutes, or a hydrocarbon and an ether for moderately polar compounds). This simplifies mobile phase preparation and column equilibration.

3. If α is not satisfactory, try a binary mixture in which the two solvents are very different in individual eluent strength, with one solvent predominant (>80%) (e.g., on silica, try a small percentage of an alcohol in a hydrocarbon). Accurate preparation of mobile phases and column equilibration

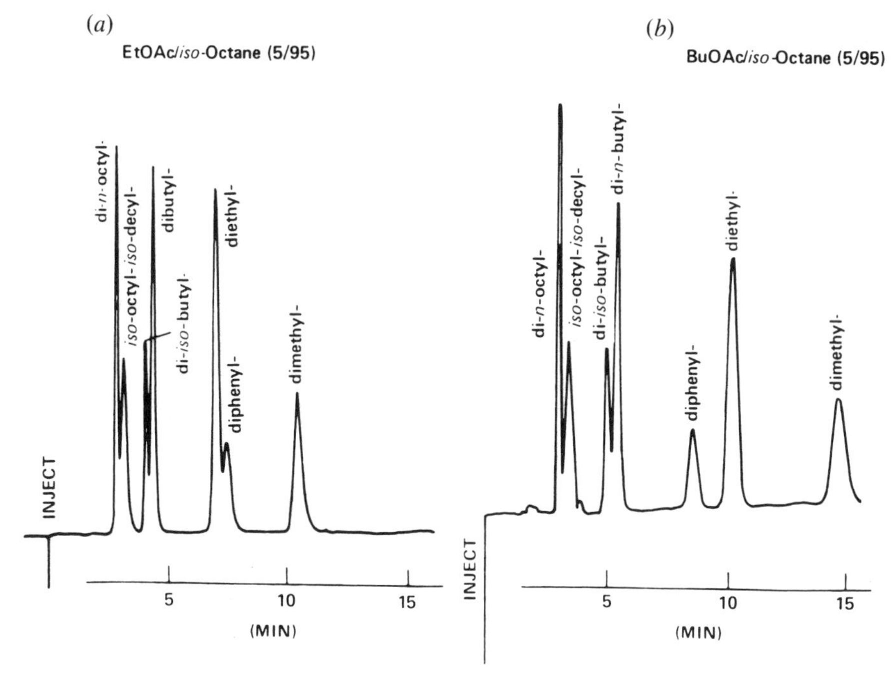

FIGURE 5-51. Effect of mobile phase polarity on retention of phthalate plasticizers. Column: μPorasil (silica, 10 μm), 3.9 mm ID × 30 cm; flow rate: 2 mL/min; detector UV at 254 nm. Mobile Phase: (*a*) is 5% ethyl acetate in isooctane, (*b*) is 5% butyl acetate in isooctane.

requires more careful attention as the percentage of the minor component decreases, especially below 5%.

4. If a binary mixture does not achieve sufficient separation, then explore tertiary systems, using knowledge of the nature of the sample and/or a method optimization scheme to save time in finding the best system, change the stationary phase, or switch to a different mode.

An example of the influence of mobile-phase polarity upon the retention and selectivity of sample molecules is shown in Figure 5-51. In normal phase, the most polar compound is retained the longest. This is reflected by the observation that the dimethyl phthalate is the most polar and is retained the longest. By changing from ethyl acetate to butyl acetate, the overall mobile phase is less polar and, hence, all of the compounds increase in retention. However, in addition, there is one change in selectivity between the diethyl and the diphenyl molecules.

In summary, normal phase is useful for oil-soluble compounds which differ in functional group type (class separation) and/or location of the functional groups (positional isomers) around an adsorption point on the mole-

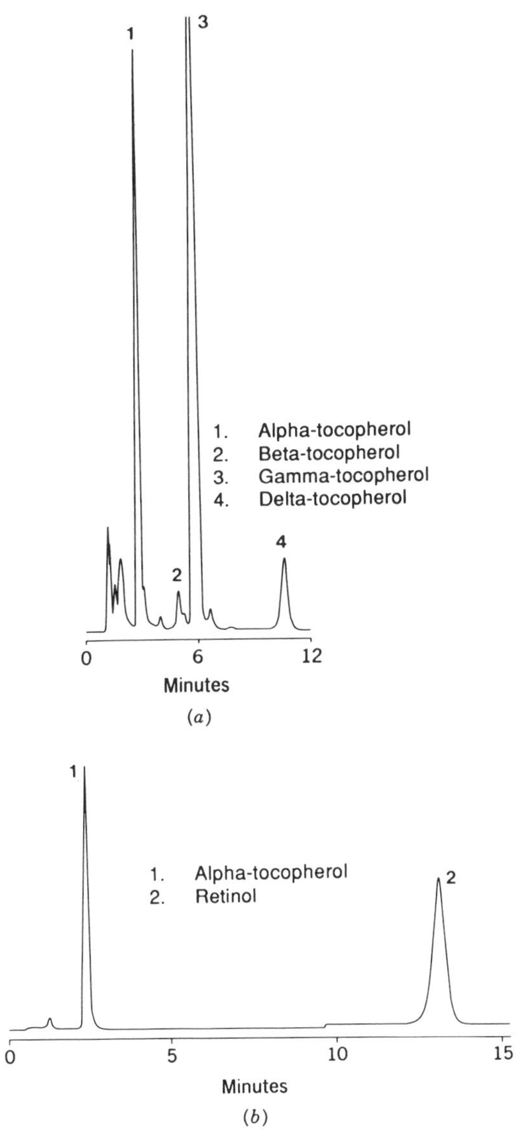

FIGURE 5-52. Examples of normal phase separations. (*a*) Corn-oil tocopherols. Sample: 10 μL of corn oil in 100 μL of mobile phase. Column: Nova Pak Silica (4 μm), 3.9 mm ID × 150 mm. Mobile phase: 0.3% isopropyl alcohol in isooctane. Flow rate: 1.0 mL/min. Detection: fluorescence 290 nm excitation and 335 nm emission. (*b*) Separation of vitamin E from vitamin A. Mobile phase: 0.5% isopropyl alcohol in isooctane. Other conditions are the same as those in *a* with the exception that retinol was detected with 365 nm excitation and 510 nm emission. (*c*) Structures of compounds.

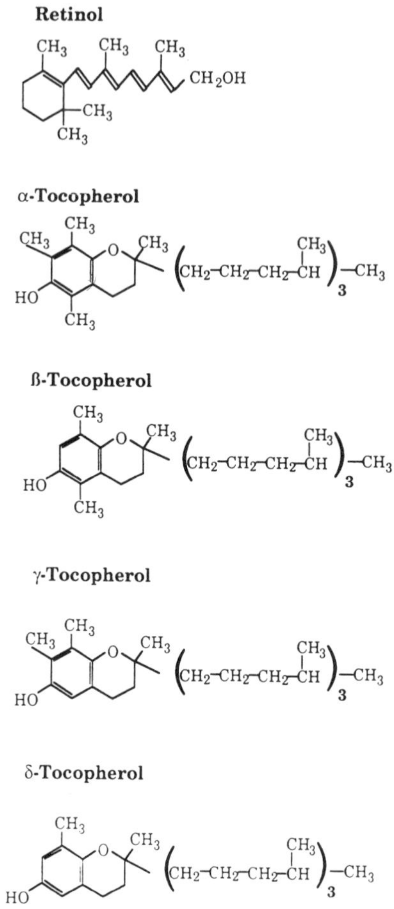

Retinol

α-Tocopherol

ß-Tocopherol

γ-Tocopherol

δ-Tocopherol

(c)

FIGURE 5-52. (*Continued*)

cule. The separation of oil-soluble vitamins are representative of an appropriate use of normal phase. Figure 5-52*a* shows the separation of positional isomers of various tocopherols (vitamin E) in corn oil while Figure 5-52*b* shows the separation of vitamins E and A.

REFERENCES

1. F. V. Warren, C. H. Phoebe, M. Webb, A. Weston, and B. A. Bidlingmeyer, *American Laboratory,* **22**(11), 17 (1990).
2. B. A. Bidlingmeyer, *J. Chromatogr. Sci.,* **18,** 525 (1980).
3. B. A. Bidlingmeyer, S. N. Deming, and B. Sachok, *J. Liq. Chromatogr.,* **5,** 389 (1982).

4. B. A. Bidlingmeyer, J. A. Korpi, and J. DelRios, *Anal. Chem.*, **54**, 442 (1982).
5. H. Richardson and B. A. Bidlingmeyer, *J. Pharm. Sci.*, **73**, 1480 (1984).
6. B. Cortis-Jones, *Nature*, **79**, 731 (1962).
7. J. Cazes and D. R. Gaskill, *Separation Sci.*, **2**, 421 (1967).
8. K. J. Bombaugh, W. A. Dark, and R. F. Levangie, *Separation Sci.*, **3**, 375 (1968).
9. B. A. Bidlingmeyer and F. V. Warren, *LC/GC*, **6**(9), 780 (1988).
10. R. V. Vivilecchia, B. G. Lightbody, N. Z. Thimot, and H. M. Quinn, *J. Chromatogr. Sci.*, **15**, 424 (1977).
11. A. Krishen, *J. Chromatogr. Sci.*, **15**, 434 (1977).
12. R. B. Walter and J. F. Johnson, *J. Liq. Chromatogr.*, **3**, 315 (1980).
13. F. V. Warren, Jr., B. A. Bidlingmeyer, H. Richardson, and J. L. Ekmanis, in *Size Exclusion Chromatography*, T. Provder, Ed., American Chemical Society, Washington, DC, 1984, p. 171.
14. J. Morawski, R. L. Cotter, and K. Ivie, paper presented at the Pittsburgh Conference on Analytical Chemistry and Applied Spectroscopy, Atlantic City, NJ, March 7–12, 1983, Abstract No. 890.
15. M. W. Andrews, J. Morawski, and A. T. Newhart, paper presented at the Pittsburgh Conference on Analytical Chemistry and Applied Spectroscopy, Atlantic City, NJ, March 7–12, 1983, Abstract No. 951.
16. W. W. Yau, J. J. Kirkland, and D. D. Bly, *Modern Size Exclusion Liquid Chromatography*, Wiley, New York, 1979, Chapters 2 and 4.
17. William Horwitz, Ed., *Official Methods of Analyzing of the AOAC*, 13th ed., AOAC, Washington, DC, 1980, method Nos. 141 and 142, p. 85.
18. B. A. Bidlingmeyer, Ed., *Preparative Liquid Chromatography*, Elsevier, Amsterdam, 1987.
19. L. R. Snyder, *Principles of Adsorption Chromatography*, Marcel Dekker, New York, 1968.
20. R. P. W. Scott, *Contemporary Liquid Chromatography*, Wiley-Interscience, New York, 1976, pp. 240–241.
21. R. E. Majors, *Anal. Chem.*, **45**, 757 (1973).
22. J. G. Kirchner, *Thin Layer Chromatography*, Wiley-Interscience, New York, 1967.
23. E. Stahl, *Thin Layer Chromatography. A Laboratory Handbook*, 2nd ed., Academic, New York, 1969.
24. H. G. Cassidy, *Technique of Organic Chemistry*, Vol. V, A. Weissberger, Ed., Interscience, New York, 1951.
25. V. R. Dietz, *Bibliography of Solid Adsorbents, 1943–53*, National Bureau of Standards Circular, Washington, DC, 1956, p. 566.
26. H. H. Strain, *Chromatographic Adsorption Analysis*, Wiley-Interscience, New York, 1942.
27. W. A. Pons, Jr., *J. Assoc. Off. Anal. Chem.*, **59**, 101 (1976).
28. L. J. Morris and B. W. Nichols, in *Chromatography*, 2nd ed., E. Heftman, Ed., Reinhold, New York, 1967, p. 466.

29. D. L. Saunders, *Anal. Chem.*, **46,** 470 (1974).
30. L. R. Snyder, *J. Chromatogr.*, **36,** 476 (1968).
31. J. J. Kirkland, *J. Chromatogr.*, **83,** 149, (1973).
32. R. Neher, in *Thin Layer Chromatography*, G. B. Marini-Bettolo, Ed., Elsevier, Amsterdam, 1964, p. 75.
33. R. Neher, *Steroid Chromatography*, 2nd ed., Elsevier, Amsterdam, 1964, p. 249.
34. L. R. Snyder, *J. Chromatogr.*, **63,** 15 (1971).

ACKNOWLEDGMENT

Figures 5-1, 5-2, 5-9 through 5-14, 5-17, 5-18, 5-25, 5-26, 5-34d, 5-44, 5-46, 5-47a, 5-48, 5-51, 5-52a, b and Table 5-1 are courtesy of the Waters Chromatography Division of Millipore.

CHAPTER 6

CONSIDERATIONS FOR PROPER OPERATION OF A LIQUID CHROMATOGRAPH

Differences in bonded-phase columns
- Differences in silica
- Bonding procedures
- The "best" column

Using the efficiency of the column to your advantage
- Calculation of efficiency
- Proper choice of flow rate
- Comparison of efficiencies of columns

Recycle chromatography
- Analytical recycle
- Preparative recycle

Introduction to statistics for chromatographers
- Accuracy
- Precision
- The error curve and the measurement
- Confidence limit
- Significant figures
- Correct values
- Precision of flow rates from solvent delivery systems
- Precision in LC analysis

Improved sensitivity for trace analysis

Recirculation of mobile phase
- UV and EC detection
- Baseline drift
- RI detection
- Conductivity detection
- Summary

Considerations when choosing solvents for mobile phases
Miscibility
How to use miscibility (M) numbers
Changing mobile phases
Use of buffers
Choice of buffer
Solvent compatibility
Removing particles and dissolved gases
Other solvent considerations
Solvent quality
Water quality
What is pure water?
Mobile phase preparation
Solid phase extraction
Conditioning
Loading the sample
Eluting the undesirable material
Isolating the analytes
Reconditioning
Application examples
Proper operation: Effect of flow rate
Proper operation: Loading during trace enrichment
Separation guidelines for solid phase extraction
Practical preparative liquid chromatography
Approaching the problem
Classifying the preparative problems
Peak shapes in preparative LC
Scaling-up to larger columns
Role of the detectors
Collection/recovery
References
Acknowledgment

DIFFERENCES IN BONDED-PHASE COLUMNS

The first item to remember when using modern LC columns is that *all columns of the same name can be different.* For a given compound a C_{18} (sometimes called ODS or octadecyl) column from one manufacturer will often behave differently compared to a C_{18} column from another manufacturer. If they behave similarly, thank your lucky stars, since this is the exception and not the rule. Even from the same manufacturer, a "Brand Name A" C_{18} column and a "Brand Name B" C_{18} column may behave differently for the same sample compounds.

An example of the types of differences that might exist between different manufacturers is shown in Figure 6-1. This separation is of theophylline,

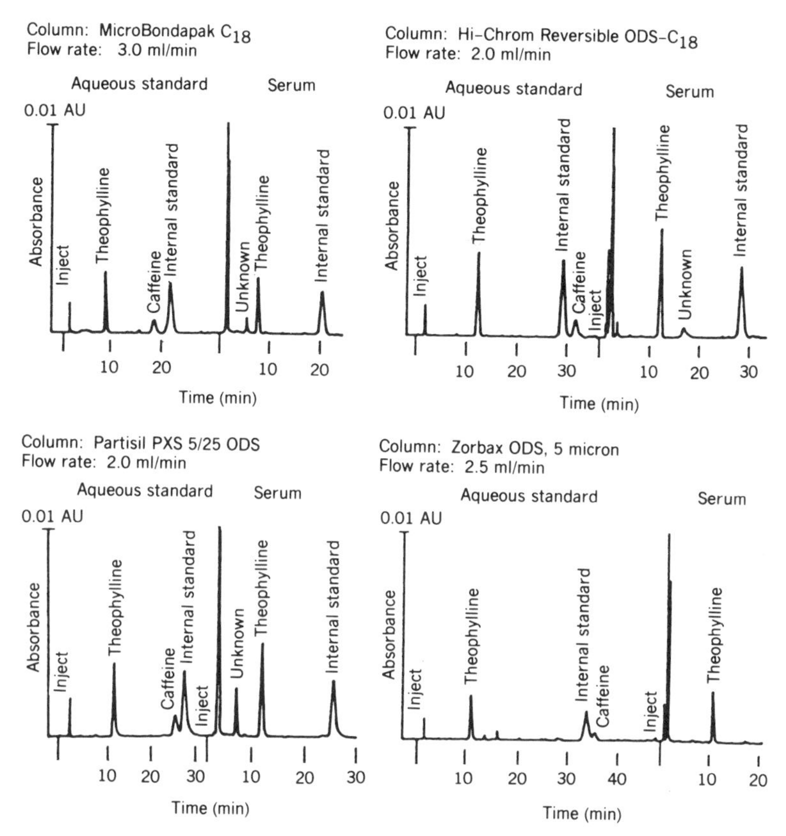

FIGURE 6-1. Chromatograms for separation of theophylline and caffeine. Mobile phase: 95/5/0.2 water/acetonitrile/glacial acetic acid. Detector: 280 nm, 0.02 AUFS. (Reprinted from reference 1 with permission.)

caffeine, and an internal standard (β-hydroxylpropyltheophylline). The importance of this separation was discussed in Chapter 2. The internal standard in this example was chosen because it is an uncommon drug and is not a known metabolite of any xanthine derivative. Since this study (1) did not adjust the flow rates to give the same linear velocity (see Fig. 6-5) for each column, it was necessary for the authors to tabulate the differences in relative resolution in order to "compare" columns. From Table 6-1 it is clear that none of these five columns have the same selectivity; however, the Partisil ODS column and the μBondapak C_{18} have the most similar relative resolution for theophylline and caffeine. The variations in selectivities for the compounds represented in Figure 6-1 and Table 6-1 may not be an issue for someone doing this work since all of the compounds in this mixture were separated on all of the columns. However, these differences in retention between columns might be considered significant if the assay is required in a

TABLE 6-1. Comparison of Columns[a]

Column (Manufacturer)	Resolution: Theophylline–Caffeine	Resolution: Caffeine–Internal Standard
μBondapak C_{18} (Waters)	8.9	1.9
Hi Chrom ODS-C_{18} (Regis Chemical Co.)	15.3	1.7
Partisil-5 ODS (Whatman, Inc.)	9.8	1.0
Zorbax ODS (E.I. duPont)	18.0	0.6

[a] Reprinted from reference 1 with permission.

fixed time. Also, differences in retentions on various columns may be more problematic if coelution occurs on one column and not the other.

Differences of the type observed in Table 6-1 exist between columns for several reasons: (1) the type of silica used; (2) the preparation of the silica before bonding; and (3) the type of chemical reaction used to bond the nonpolar alkyl (C_{18}) group to the silica surface.

Differences in Silica

Some chromatographers believe that silica is an "inert" backbone upon which the stationary phase is placed. For neutral, nonpolar compounds this is essentially true. However, for acidic and basic samples there is evidence that small differences in the starting silica gel will influence chromatographic peak shape and retention. While all silicas are composed of silicon dioxide and have SiOH groups on the exposed surface, depending upon how the silica is made, there will be surface heterogeneity of the types of silanols (Fig. 6-2*a*) and in the spacing of these various silanols on the surface. Also, the various types of silanols are believed to have different acidities. Obviously, if the silanol heterogeneity is different, the contribution of each silica to retention will be different. Even if the same silica is used by two manufacturers, if one "prepares" the silica before bonding (2,3), this can change the surface heterogeneity and make the prepared silica surface different from the original. To appreciate the differences in HPLC behavior, one must acknowledge that the silica is not "inert" (4).

Silica gel is a polymer of silicic acid and has the general formula $SiO_2 \times H_2O$. There are many ways to prepare silica gel. Let us consider only one preparation of silica gel to demonstrate how and where differences or variations in the silica may contribute to a different bonded (final) phase. A common way of making silica gel particles for LC is to use acidic hydrolysis of sodium silicate, followed by polycondensation dehydration of orthosilicic acid and grinding of the silica to make granular particles (e.g., 10 μm). If the reaction is not controlled (buffer, pH, temperature, etc.) the "hydrogel" will

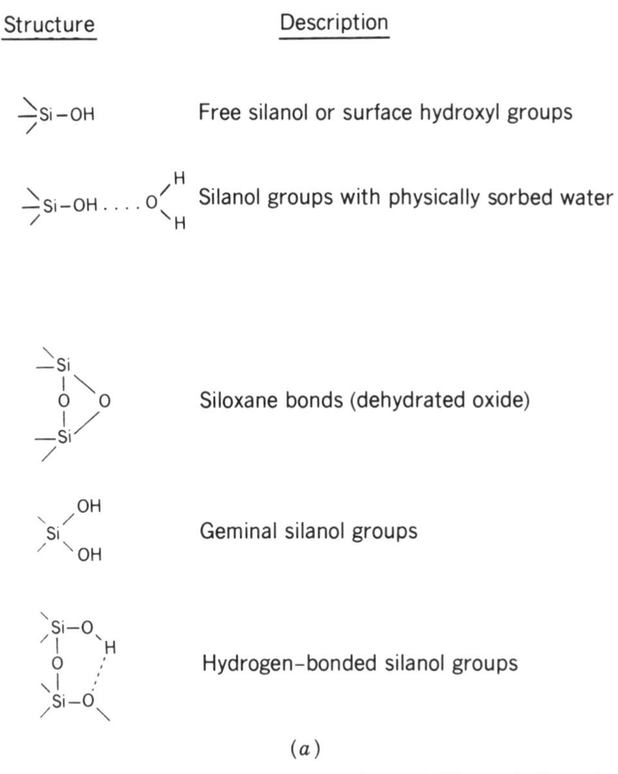

FIGURE 6-2. (*a*) Types of silanols on the surface of silica. (*b*) Reaction of various chlorosubstituted silanes with silica gel.

result in different surface areas and silanol populations. Often this silica is "prepared" prior to bonding, using any or all of the techniques mentioned in the literature (2,3) such as acid washing, heat treatment, and so on. Since this takes place before a bonding reaction, the result will be a different silica for each type of preparation. In the literature, each silica preparation is supported with "reasons" why the preparation is a preferred preparation. Therefore, it should be easy to understand that a broad range of silica substrates available in the marketplace results in a wide diversity of column performance.

Bonding Procedures

Once a silica is chosen, a reaction must be run to convert the silanols to surface-attached alkyl groups. This reaction may be accomplished as shown in Figure 6-2*b* with a mono-, di-, or trimethoxysilane. Each reaction results in a different nonpolar surface, but all are called a C_{18} (or ODS) column. When a chlorosilane is used, reaction with a monochlorosilane is preferred to produce a consistent reaction without polymerization. However, not all

Structures and Reaction	Description
$\geqslant Si{-}OH + Cl{-}Si(R_1)(R_2){-}R \rightarrow \geqslant Si{-}O{-}Si(R_1)(R_2){-}R + HCl$	Monochorosilane
$2 \geqslant Si{-}OH + Cl_2SiR_2 \xrightarrow[\text{reflex}]{\text{dry solvents}} (\geqslant Si{-}O)_2SiR_2$	Dichlorosilane
$2 \geqslant Si{-}OH + Cl_3SiR_2 \xrightarrow[\text{reflex}]{\text{dry solvents}} (\geqslant Si{-}O)_2Si(R)Cl$	Trichlorosilane in dry solvents
$2 \geqslant Si{-}OH + Cl_3SiR \rightarrow (\geqslant Si{-}O)_2Si(OH){-}R + (Cl_3SiR)_n \rightarrow (\geqslant Si{-}O)_2Si(R){-}O{-}[Si(R)(O)]$ polymer with HO–Si–R and –O–Si–R units	Trichlorosilane in solvents with protic impurities

(*b*)

FIGURE 6-2. (*Continued*)

manufacturers use monochlorosilanes, which explains some additional differences in C_{18} columns. Also, there are reports that this reaction can be done with a methoxysilane rather than a chlorosilane, in which case another type of final LC column may be produced.

If two manufacturers use the same silica, the same silica preparation, and the same monochlorosilane reaction, there could still be a difference, if one manufacturer does a second bonding reaction and one does not. Since the primary reaction with the original alkylsilane (e.g., monochlorooctadecyldimethyl silane) cannot cover 100% of the surface silanols owing to steric

TABLE 6-2. Possible Differentiation in Bonded-Phase Columns

Physical	Surface area
	Pore volume
	Pore diameter
	Pore size distribution
Chemical	Surface properties
	Number of silanols
	Types of silanols
	Acidity of surface
Bonding chemistry	Nature of silane C_1-C_x functional group
	Way of bonding
	Surface concentration of bonded groups
Chromatographic conditions	Composition of organic phase
	pH
	Modifier(s)
	Salts

constraints, a second reaction with a smaller alkyl silane (''capping'' reagent, e.g., monochlorotrimethylsilane) is used in an attempt to cover the remaining (free) silanol groups. This second bonding reaction is commonly referred to as ''end capping.''

Furthermore, two column packings both made with a C_{18} silane *can* be different because of variations in reaction conditions, residual water in the reaction solvent, the reaction solvent, the temperature, the extent of primary bonding, the extent of secondary bonding, and so forth. In addition to these items, Table 6-2 lists a summary of all of the physical, chemical, and chromatographic parameters that can contribute to making one bonded-phase column different from another. The large possible combination of these types of variations should convince you that ''not all C_{18} columns are alike.'' However, a manufacturer should control all of the variables closely enough to make a reproducible product from batch to batch. Understanding the differences in bonded-phase columns and characterizing the reverse-phase packings have been areas of active research (2–9) and continue to be areas where new knowledge is reported. While columns from a given manufacturer are not always ''identical,'' they are sufficiently reproducible that they can be used routinely to solve analytical problems.

The ''Best'' Column

There are so many types of C_{18} columns because it is not clear how to make ''the best'' bonded-phase column. Each manufacturer makes what they believe to be the ''best'' in terms of a manufacturable product. As a result, each chromatographer must develop a personal understanding of what columns are appropriate for the desired application need. This information is best gained from experience, but can be augmented by technical exchanges with colleagues and from information in the technical literature. One must make an initial column choice from an application perspective. Compare the compounds being separated on a vendor's column in the advertisements to the type of compounds you need to separate. What columns do your colleagues use? What compounds are being separated in the journals and the trade magazines that refer to your application? What type of column assistance will you need and what vendor can best supply it? All of these questions, and possibly more that you supply, need to be answered before you choose your first column. After working with the first column, your second and succeeding purchasing decisions should be easier since they will be based upon your own experience.

Some people feel that columns cost too much and, therefore, purchase the lowest-price column. This is one strategy for choosing the "best" column. However, when the nature of a column in a chromatographic system is considered, it makes sense that there should be little to no price sensitivity in the user. For instance, if an instrument sells for $20,000, a typical column sells for $300. Therefore, a column is only $300/$20,000, or 1.5% of the cost of an instrument installation. This is a minor part even if repeat buying is calculated. Second, keep in mind that without the column performance the entire instrument is worth precisely zero. Third, it is very hard to evaluate how a column will perform without prior experience with it—there is no easy believability to competitive claims that one C_{18} column will work as well as, or even the same as, another. In essence, a column is nothing more and nothing less than a user's attempt to solve an analytical problem. Users should not be sensitive to price, they should be concerned with time, effort, and repeatability. Therefore, most individuals should gladly pay for these benefits, because the "best" column does your separation.

USING THE EFFICIENCY OF THE COLUMN TO YOUR ADVANTAGE

As a result of high-efficiency columns, one of the benefits of modern HPLC is high-speed analyses. However, when high linear velocities are used for rapid method development or for fast analyses, theoretical plates are, in essence, tossed away as excess baggage. Generally, this is not a problem since there is more than enough efficiency to attain the separation owing to the good selectivity. Occasionally, however, a chromatographer may run into a difficult separation where it is desirable to use all of the efficiency that a column has to offer. In this situation, an understanding of the trade-offs of resolution and flow rate need to be appreciated in order to optimize the analysis. (One example was discussed in Chapter 5, Figs. 5-12 and 5-13.)

When a column is purchased, you should be aware of the manufacturer's claims about its efficiency. This is often a high value, for example, 15,000 plates per foot. How was this value obtained? Is this value sufficient? What does it mean? Since you may need to use these plates, you should be aware of the implications and proper use of the efficiency value of the column.

Calculation of Efficiency

The most common performance indicator of a column is a dimensionless, theoretical plate count number, N. This number is also referred to as an "efficiency value" for the column. There is a tendency to equate the column efficiency value with the "quality" of a column. However, it is important to remember that the column efficiency is only part of the "quality" of a column. The calculation of theoretical plates is commonly based on a Gaussian model for peak shape because the chromatographic peak is assumed to result from the spreading of a population of sample molecules resulting in a Gaussian distribution of sample concentrations in the mobile and stationary phases. The general formula for calculating column efficiency is

$$N = a(V_r/W)^2 \tag{6-1}$$

where N is the number of theoretical plates, V_r is the retention volume for the peak, W is the peak width in volume, and a is a constant that depends upon the height (from the baseline) at which peak width is measured. Figure 6-3 shows the relationship between the Gaussian peak profile, the height at which the peak width is measured, and the value of a. The plate count equations relate the narrowness of the peak (W) to its elution volume (V_r). It is also possible to use rigorous statistical moments to calculate N, but because a computer is needed, most chromatographers prefer the convenience of the manual methods shown in Figure 6-3.

Obviously, the choice of measurement–calculation method can influence the column efficiency value. For an ideal (Gaussian) peak, all the methods

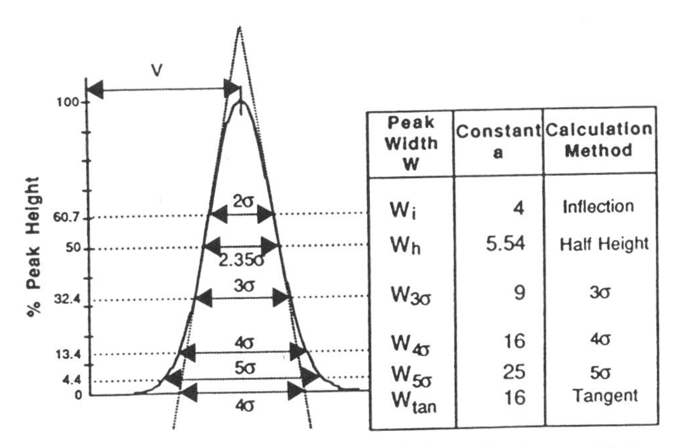

Peak Width W	Constant a	Calculation Method
W_i	4	Inflection
W_h	5.54	Half Height
$W_{3\sigma}$	9	3σ
$W_{4\sigma}$	16	4σ
$W_{5\sigma}$	25	5σ
W_{tan}	16	Tangent

FIGURE 6-3. Alternatives for calculation of column efficiency.

listed in Figure 6-3 will give the same result. However, when a peak has a "tailing" or "fronting" portion, as many do, then the choice of test procedure becomes very important. For example, the 5σ test method is a more stringent method for sensing tailing compared to the popular tangent and half peak height methods. A thorough discussion of the practical aspects of the measurement of plate height is given elsewhere (10). Regardless of the specific testing procedure used, several parameters influence the determination of column efficiency: the composition of the eluent, linear velocity, the solute used in the measurement of plate count, temperature, and the method chosen for measurement and calculation. It is therefore important that all efficiency values be accompanied by a statement of all the conditions under which the plate count value was obtained.

Proper Choice of Flow Rate

Chromatography is a segregation process and, as such, any mixing of sample molecules which occurs as they travel through a column is deleterious to achieving the separation. The plate count is a reflection of the mixing in a column and is dependent upon the velocity at which the chromatography takes place. As discussed in Chapter 3, this dependence is a result of three main physical phenomena which occur as molecules travel through a packed bed. These three band-broadening contributions which occur are: (1) eddy diffusion, which causes mixing as the mobile phase flows past the particles; (2) diffusion of the molecules in the liquid mobile phase, which is generally very low in liquids; and (3) mass transfer of the sample molecules from the liquid to the stationary phase, which is the time required for the molecules to reach equilibrium between the two phases.

For several real examples, the relationship of N to flow rate is represented in Figure 6-4. This plot, often referred to as a "van Deemter Plot" after the person who demonstrated this relationship in GC, implies that the flow rate of a column may be increased for faster analysis time with an accompanying loss of efficiency. However, because chromatographers often have more plates than are needed for the selectivity which they have, high flow rates are very popular because the time of analysis can be reduced while the separation is still suitable. When a particular separation requires more plates, lowering the flow rate toward the optimum value results in a tremendous gain in efficiency. For example in Figure 6-4*a*, at 2.5 mL/min the column showed 7300 plates. At a flow rate of 1.0 mL/min, one could expect an efficiency of over 10,000 plates per column (in the figure the actual value is 11,500 plates). At a linear velocity of 0.1 cm/sec one can expect optimal efficiency of 14,000 plates for this column. This is almost twice the number of the plates obtained at 2.5 mL/min, and this would improve the resolution by a factor of 1.414 (the square root of 2). However, this gain in increased efficiency performance is accompanied by an increase in the time of analysis by a factor of 4 (2.5 mL/min divided by 0.6 mL/min). Nevertheless, increas-

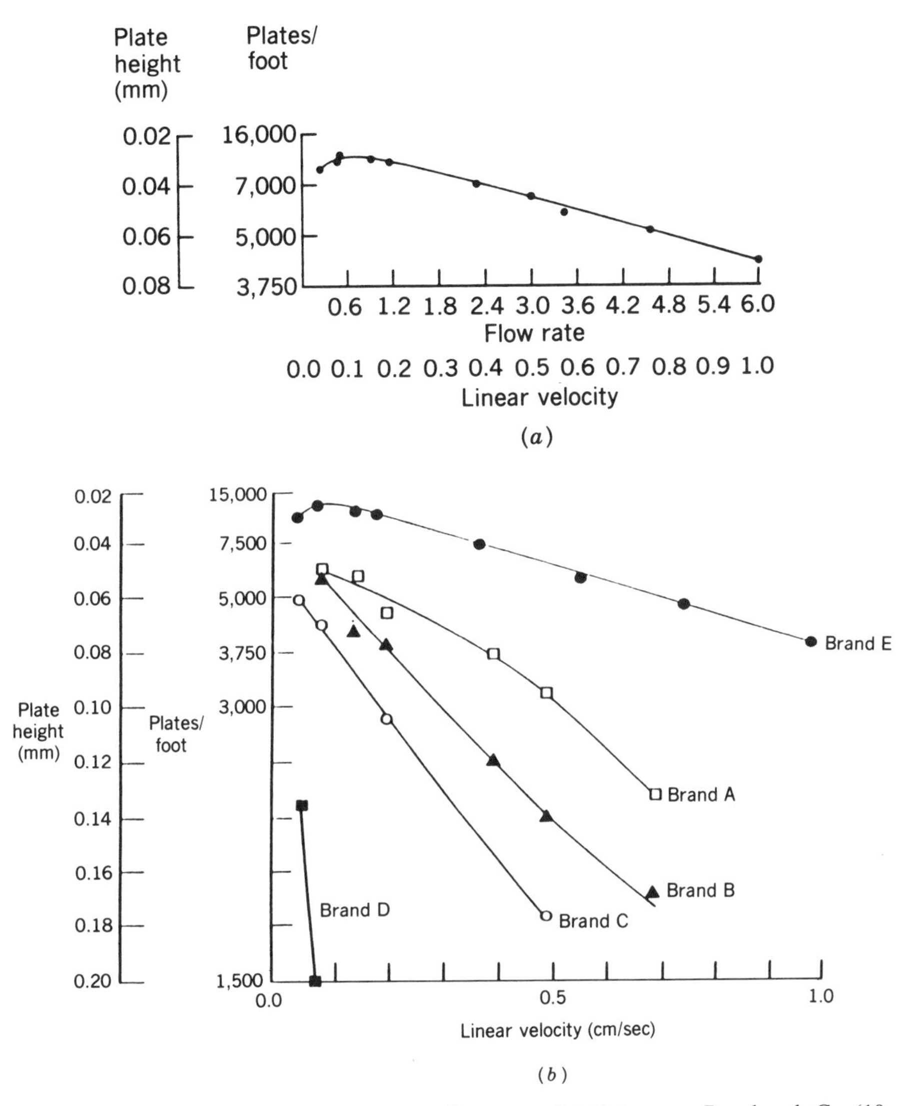

FIGURE 6-4. Column efficiency versus flow rate. (*a*) Column: μBondapak C_{18} (10 μm) 3.9 mm × 30 cm. Mobile phase: Acetonitrile: water, 60:40. Sample: Acenaphthene, $k' = 3.5$. (*b*) Comparison of reverse phase columns (10 μm). Conditions are the same as in *a*. (*c*) Comparison of normal phase columns: silica (10 μm). Mobile phase: hexane. Sample: nitrobenzene.

ing the analysis time by a factor of 4 is a small price *if* the chromatographer can now do the separation.

If, however, the flow is changed to a very slow flow rate, efficiency decreases again because of the contribution of diffusion to peak broadening (mixing). Obviously, as shown in Figures 6-4*b* and *c*, different columns have

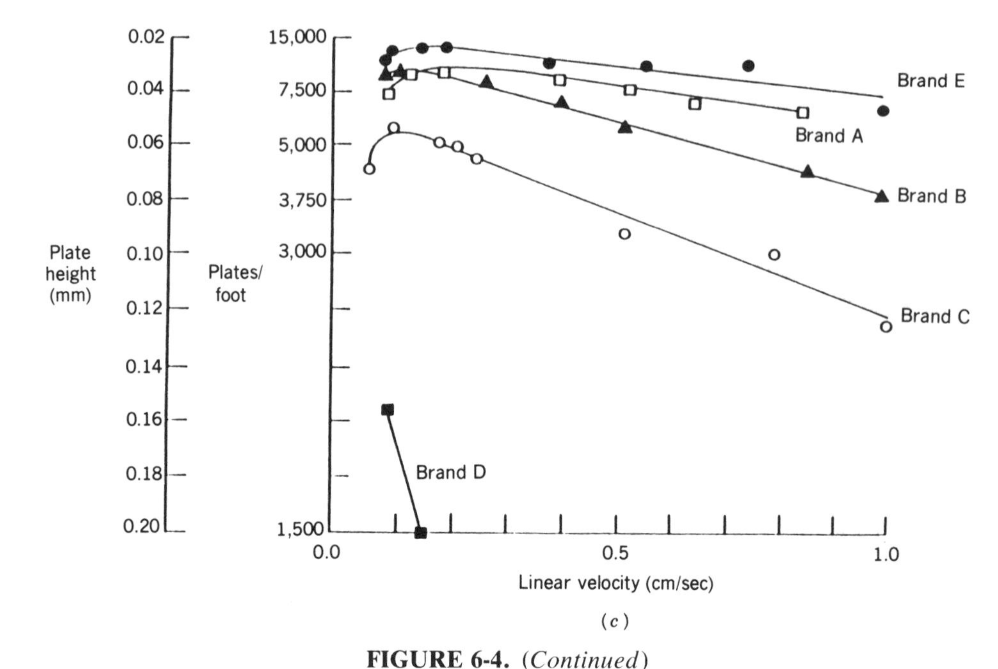

(*c*)

FIGURE 6-4. (*Continued*)

different velocity versus plate-height relationships depending upon the particle size of the stationary phase and how well the column was packed.

A word of caution is needed here. To develop a separation that is dependent upon a high number of theoretical plates is not ideal. A small decease in efficiency during use could make the column useless for that separation. Also, when replacing the column, remember that the next column may not have sufficient efficiency to be useful. Overall, the "proper choice" for flow rate is a trade-off decision which must be made in the context of the analysis requirements.

An example of increasing the efficiency by decreasing the flow rate is shown on a pellicular silica gel column in Figure 6-5*a*. (This figure was shown previously as Fig. 4-15.) This is the separation of a cis and trans isomer of synthetically prepared juvenile hormone mimics which are separated in the normal-phase mode. (Juvenile hormone mimics are used to stop or retard the maturation process of insects and, hopefully, control the insect population.) The presence of benzene is probably residue from the reaction solvent. At 2 mL/min the first peak corresponds to 5000 plates and the last peak corresponds to 2300 plates. By lowering the flow rate to 0.5 mL/min, there is a corresponding increase in efficiency in the separation. With the additional efficiency at the slower flow rate, it is possible to observe a small, additional shoulder (peak) under the first peak. However this was done at the expense of increased time (a factor of 4).

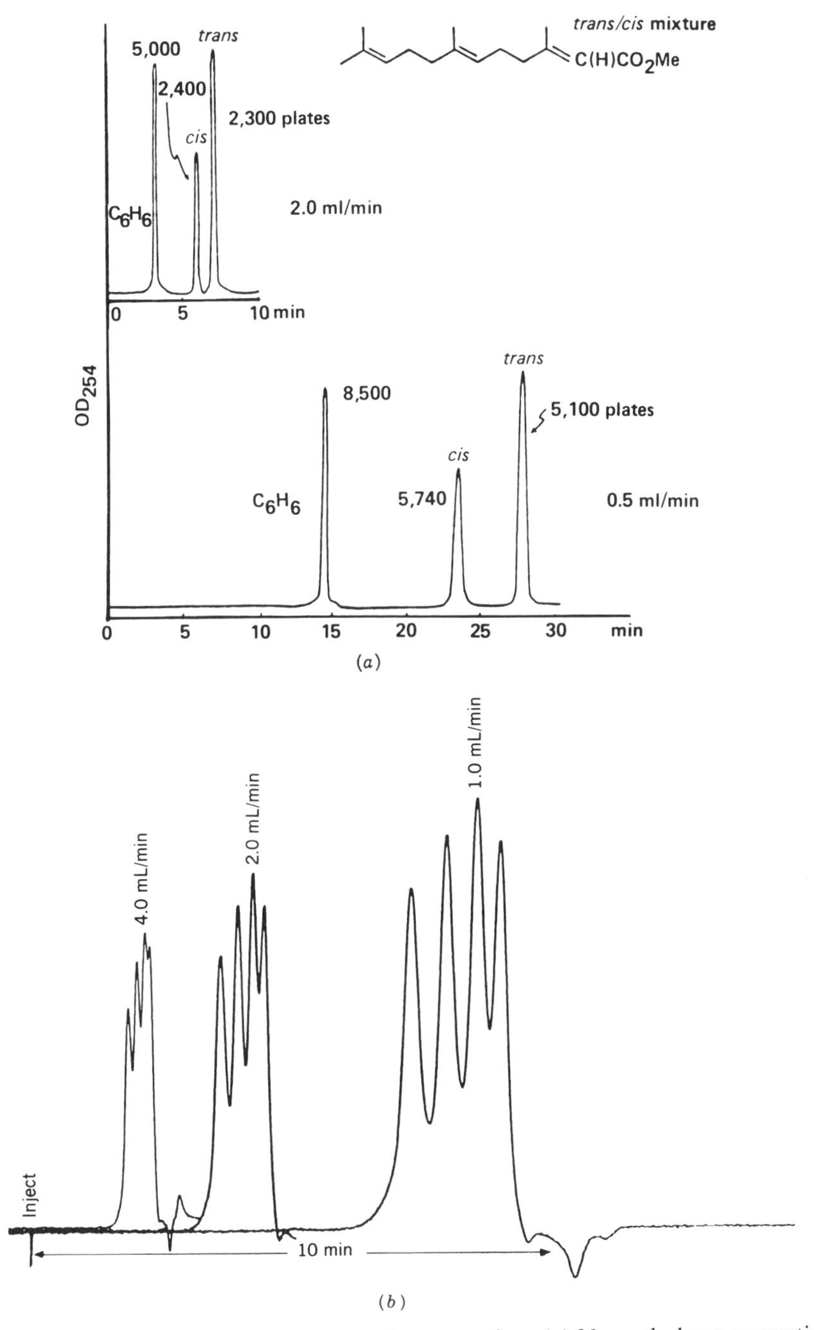

FIGURE 6-5. Effect of flow rate upon the separation. (*a*) Normal-phase separation of juvenile hormone mimics. Column: Corasil II (silica), 2 mm × 183 cm. Mobile phase: ethyl ether/hexane (1/199). Detector: UV Flow rate: 2.0 mL/min and 0.5 mL/min. (*b*) GPC separation of phthalates. Column: 100 Å μStyragel, 8 mm × 30 cm. Mobile phase: tetrahydrofuran. Detector: RI Sample: dioctylphthalates dibutylphthalate, diethylphthalate, and dimethylphthalate. Flow rate: shown on chromatogram.

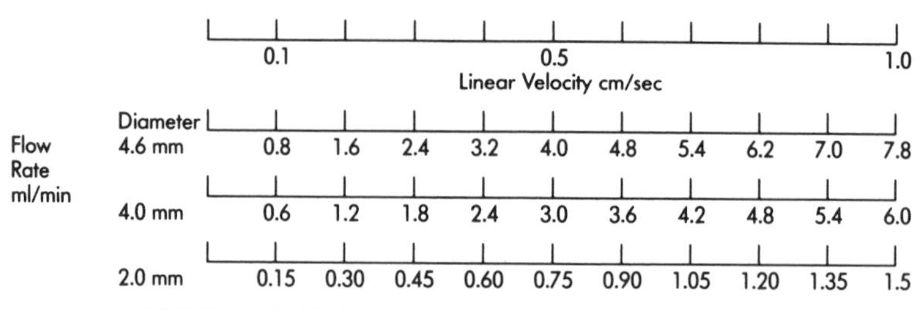

FIGURE 6-6. Guide to linear velocity versus flow rate.

An example of the effect of flow rate on increasing efficiency for GPC separations is shown in Figure 6-5*b*. There are approximately 6000 plates at 4 mL/min, 9300 plates at 2 mL/min, and 10,000 plates at 1 mL/min. This is quite a dramatic increase in plates—going from 4 mL/min to 1 mL/min—as can be seen by the increased resolution obtained in the separation of phthalates at the lower flow rate.

Comparison of Efficiencies of Columns

At the same flow rate, columns of varying IDs operate at different linear velocities; therefore, to compare the performance of several different columns, you need to operate each column at the same linear velocity. Figure 6-6 gives the relationship between linear velocity and flow rate for selected column diameters. This figure assumes the same packing density in each column. The linear velocity is marked on the top line. The corresponding flow rate for various diameter columns would be found on the lower line. For instance, a linear velocity of 0.5 cm/sec would occur at 4.0 mL/min on a 4.6-mm ID column, at 3.0 mL/min on a 4.0-mm ID column, and at 0.75 mL/min on a 2.0-mm ID column. Linear velocity can be calculated for each column by the equation

$$\mu = \frac{L}{t_m} \tag{6-2}$$

where μ is the linear velocity, L is the column length, and t_m is the time for the unretained peak. Therefore, if repeating a separation on a different diameter column, remember to use an "equivalent" flow rate. Otherwise, differences in the separation profile due to efficiency will be apparent.

RECYCLE CHROMATOGRAPHY

In addition to slowing the flow rate, another means of increasing efficiency of the separation is to add more columns. If this is inconvenient or cost prohibitive and more efficiency is desired, the chromatographer can use recycle chromatography. The main use of "recycle" is to improve the resolution of a separation by increasing the number of plates (efficiency) resulting from sequentially passing (i.e., recycling) the sample through the existing column or column set. Recycle can be used in both the analytical and the preparative mode.

Analytical Recycle

In the analytical mode, recycle has been used to check the purity of fractions that may be resolved by multiple passes through the column. In the preparative mode, recycle is used to increase the throughput (grams of purified material per time) of the separation. Recycle is used with isocratic mobile phase and not used in gradient work. The schematic of the recycle operation is shown in Figure 6-7*a* and it is accomplished by turning a valve, as shown in Figure 6-7*b*. Switching a valve to the "recycle" position is done as the peak emerges from the detector. As a result the sample is sent back to the pump in a closed loop, through the by-pass line, the injector, and the column and into the detector again.

When recycling "peaks," it is important to minimize the "dead volume," which can contribute to band spreading. Thus, the injector should always be in the "load" position to eliminate the sample loop, otherwise resolution will be lost because of mixing (band broadening) in this loop. During construction of many LC systems, 0.020-in. or 0.040-in. tubing is used from the detector to a recycle manifold. This is done in case a pressure-sensitive RI detector is used in this line. If a detector that can withstand high pressures, such as a UV detector, is used, a 0.009-in. tube can be substituted for better recycle results. But do not forget to replace this 0.009 in. tube with the original tubing when the RI is placed back into the system.

The benefits of using recycle in the analytical mode can be illustrated by referring to the chromatogram in Figure 6-8. The reacted mixture is separated into the parent compound (rubrene), its oxide, and the ozone complex. It was desired that both the ozone complex and the oxide be isolated as individual peaks for structure determination. Therefore, before each peak was collected and identified by an analytical technique such as IR, NMR, or mass spectroscopy, the peaks were recycled to attain high peak purity.

When the oxide is recycled as shown in Figure 6-9, no further separation of components is accomplished. This does not necessarily mean that only one compound exists under this peak, but it does mean that we have more confidence in the assumption that it is one compound. After eight passes through the column, the compound has experienced approximately 64,000 plates. This is a highly efficient separation and would be able to resolve compounds with a selectivity (α) value of 1.02. Also, a partial separation should be seen for smaller α values. So while it is not "proof positive" information that the compound is pure, an individual could collect this peak

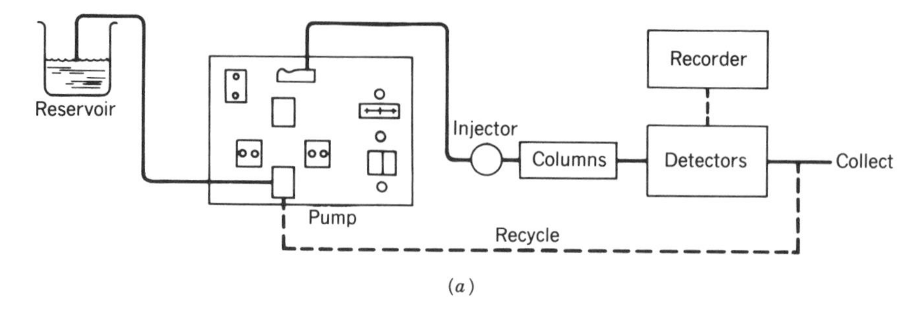

(*a*)

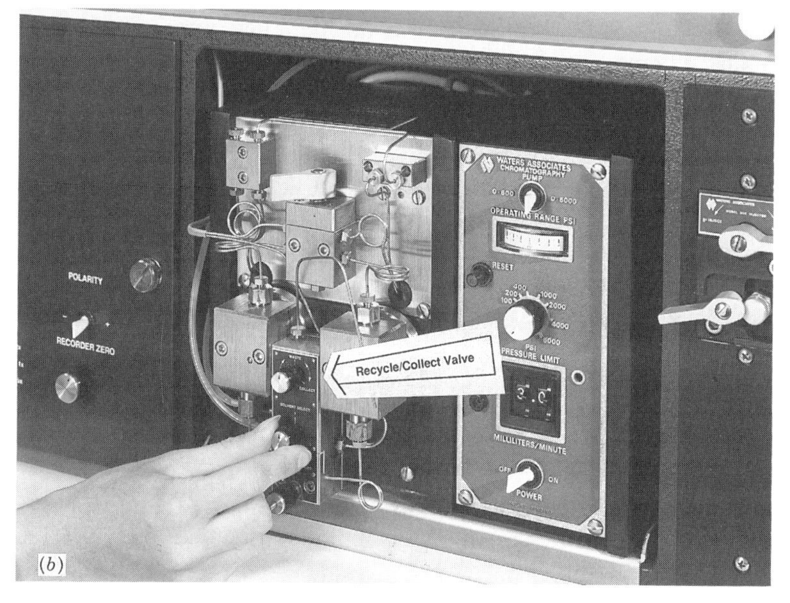

(*b*)

FIGURE 6-7. Recycle Chromatography. (*a*) Schematic. (*b*) Example of a recycle valve.

and move on to a structural determination method with a high degree of confidence that after eight recycles (64,000 plates) the peak is pure.

Upon recycling the ozone complex (Figure 6-10) it is observed that what at first pass seemed to be a single compound is eventually split into several peaks as a result of the influence of increasing N by approximately 8000 plates per pass of each recycle. After six recycles (48,000 plates), each of the three peaks were collected and found to be pure and the structure was determined.

The improvement in the resolution of a separation is a multiple of the square root of the number of plates which the sample encounters. Even if some band broadening occurs because the sample passes repeatedly through the pump, injector, or detector, it is still possible to attain a distinct separation, as demonstrated in the 14-pass recycle shown in Figure 6-11.

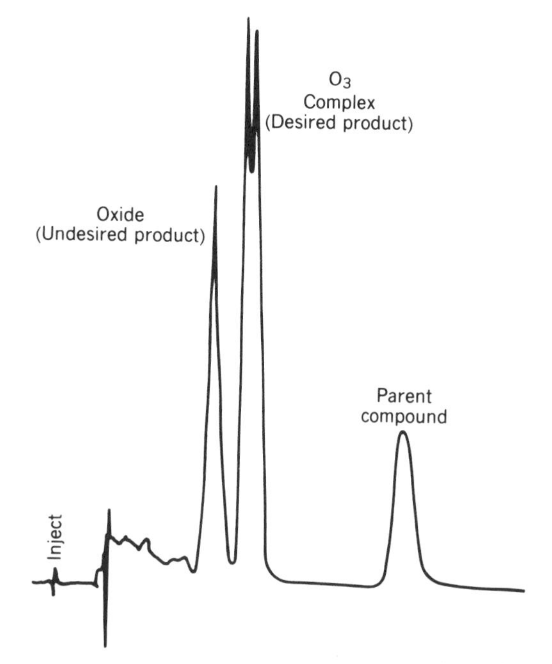

FIGURE 6-8. Initial separation of a reaction mixture.

This approach is quite useful where clean fractions are desired and used for further experimental work. When recycle is used, it is normal practice to increase detector sensitivity and decrease chart speed during appropriate cycles in order to accommodate for the effects of band broadening and to obtain a better visual judgment of the resulting chromatogram.

Preparative Recycle

In the preparative mode, baseline resolution is not necessary for recovery of sample components of high purity. In fact, for maximum throughput resolution of more than 0.7 obtained on the first pass through the column indicates

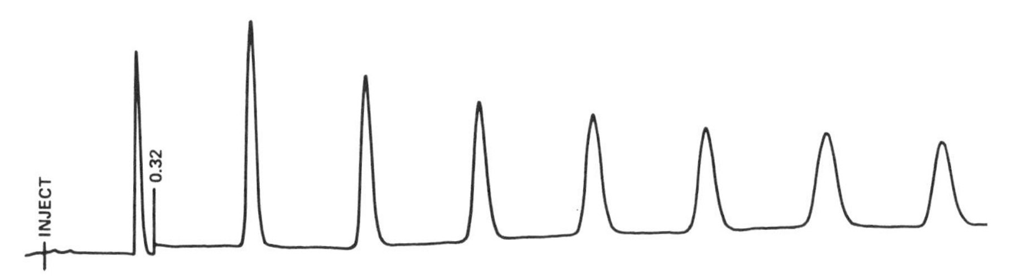

FIGURE 6-9. Recycle as a purity check of the oxide peak.

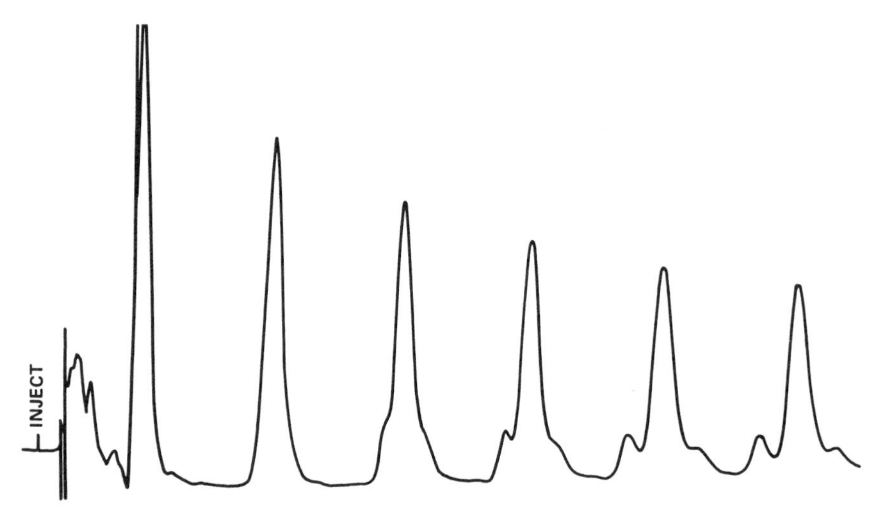

FIGURE 6-10. Recycle as a purity check of the ozone complex peak.

that the full loading capacity of the column is not being utilized. The overlapped compounds (R = 0.7) illustrated in Figure 6-12 show how pure material can be collected from partially resolved peaks on the first pass of a separation in the recycle mode. This technique is commonly referred to as "peak shaving." In peak shaving, sample material is collected separately from the leading and trailing peaks during each recycle pass. This results in better peak resolution on the next pass and maximized sample recovery (even for minor components) while minimizing solvent consumption. In the example in Figure 6-12, optimum collection points are indicated by solid vertical lines while the dashed lines show hypothetical pure peaks contained under the observed curve.

In summary, recycle results in increased throughput, reduced separation time, lower solvent consumption, and complete recovery of the compounds (if desired) in the preparation mode. Chapter 10 (Experiment 3) contains an experiment that illustrates the basic experimental techniques of accomplishing recycle chromatography.

INTRODUCTION TO STATISTICS FOR CHROMATOGRAPHERS

As a chromatographer, there will come a time when it will be necessary to gather data and present the summary of that data in order to demonstrate how good "your" chromatography is. The two things most people need to summarize for others interested in a chromatographic method are accuracy and precision of retention time, of peak height (or peak area), and of the total analysis.

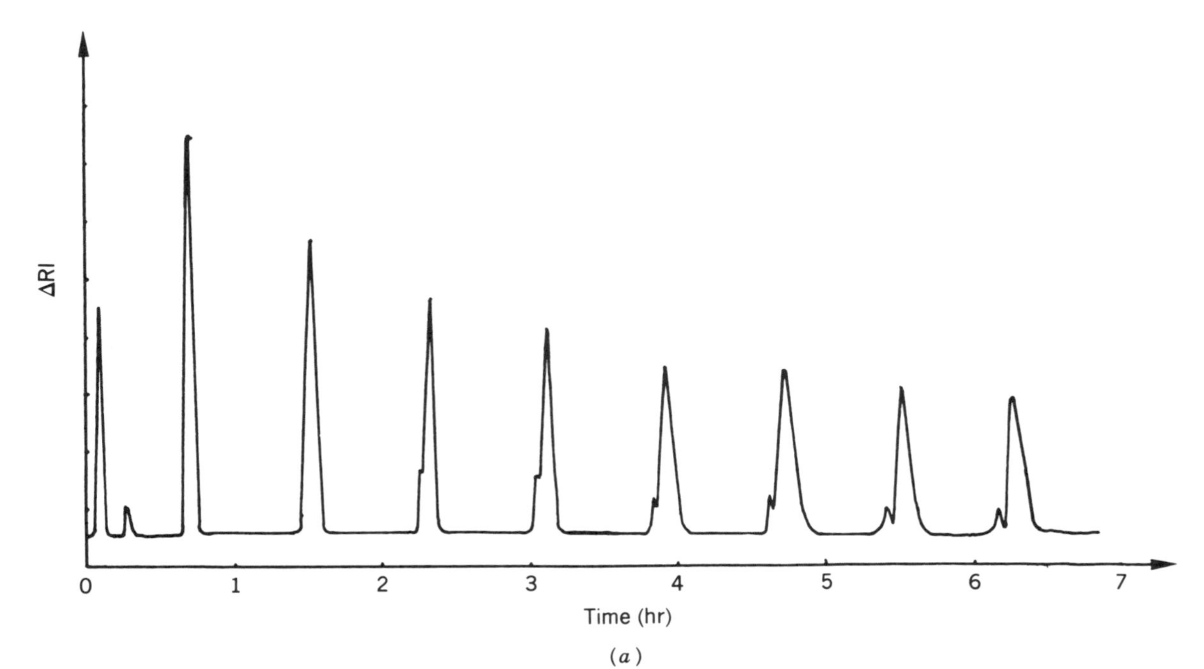

(a)

FIGURE 6-11. Demonstration of resolution improvement using recycle. (a) Separation of 26-hydroxycholesterol 3, 26-diacetate from human aorta. After seven recycles, resolution of the 25S-epimer (first peak) is achieved from the 25R-epimer (second peak). Column: μPorasil (silica, 10 μm), 3.9 mm ID × 60 cm (2 columns). Mobile phase: 2.5% (v/v) ethyl acetate in hexane. Flow rate: 1 mL/min. Detector: refractometer. (Reproduced from J. Redel, *J. Chromatogr.* **168,** 273 (1979) with permission.)

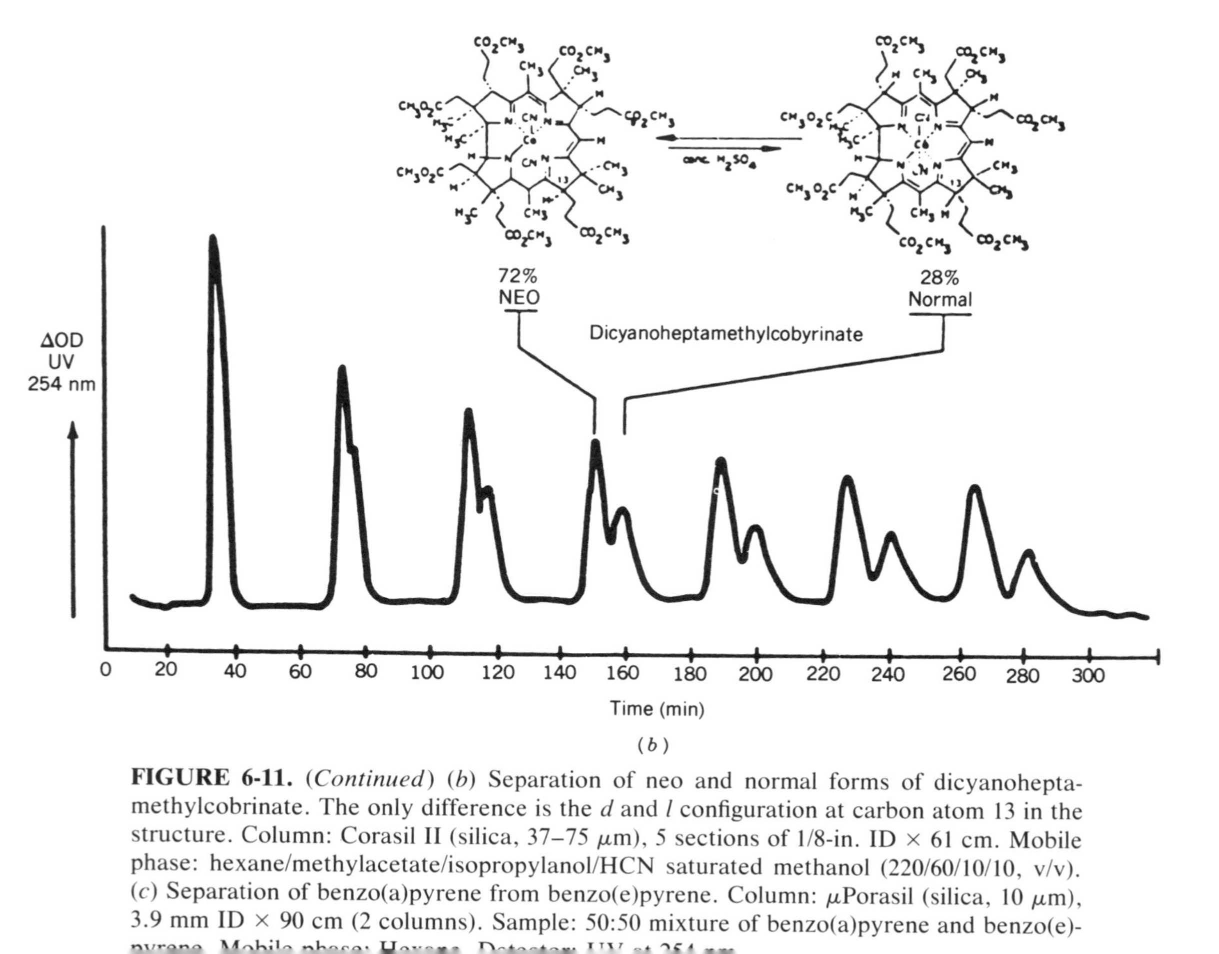

(*b*)

FIGURE 6-11. (*Continued*) (*b*) Separation of neo and normal forms of dicyanoheptamethylcobrinate. The only difference is the *d* and *l* configuration at carbon atom 13 in the structure. Column: Corasil II (silica, 37–75 μm), 5 sections of 1/8-in. ID × 61 cm. Mobile phase: hexane/methylacetate/isopropylanol/HCN saturated methanol (220/60/10/10, v/v). (*c*) Separation of benzo(a)pyrene from benzo(e)pyrene. Column: μPorasil (silica, 10 μm), 3.9 mm ID × 90 cm (2 columns). Sample: 50:50 mixture of benzo(a)pyrene and benzo(e)-pyrene. Mobile phase: Hexane. Detector: UV at 254 nm.

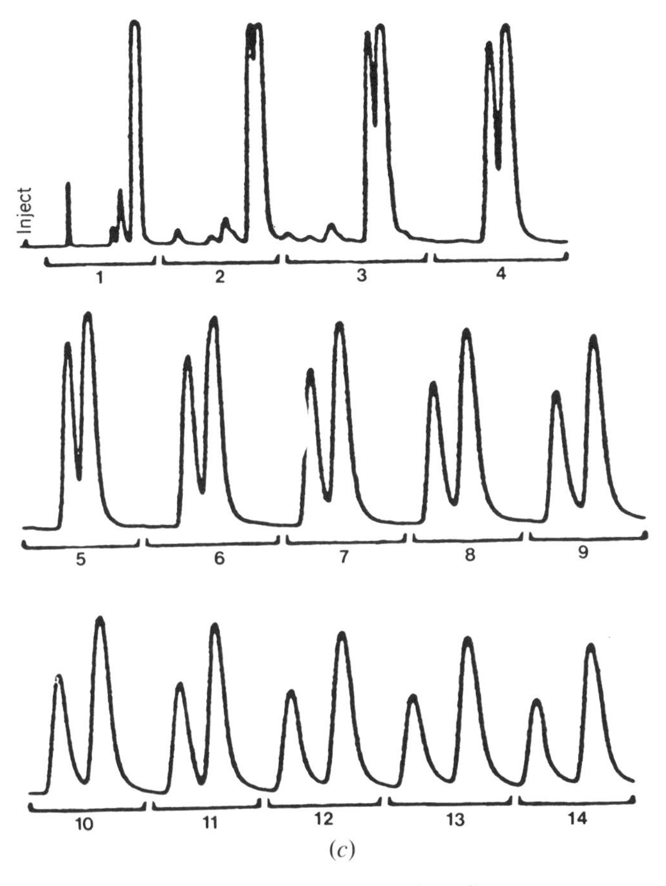

FIGURE 6-11. *(Continued)*

Accuracy

The term accuracy tells how close a single experimental result or a mean (average) of a group of results (replicates) is to the "correct" value. An accepted, true, or correct value must be available before accuracy can be discussed. Accuracy is usually expressed in terms of error:

$$\text{Absolute error} = \text{experimental value} - \text{correct value} \tag{6-3}$$

$$\text{Relative error} = \frac{\text{absolute error}}{\text{correct value}} \times 100 \text{ or } 1000 \tag{6-4}$$

When equation 6-2 uses 100, the relative error is in percent. When it uses 1000, relative error is in parts per thousand (ppt).

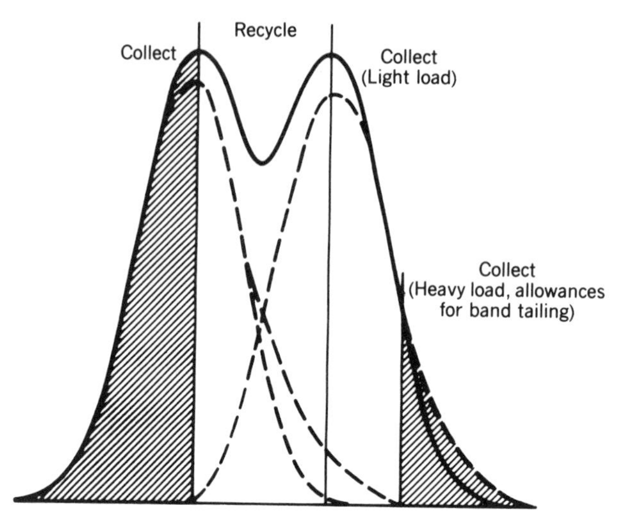

FIGURE 6-12. The peak-shaving technique using recycle. In the heavily loaded case, the first peak may have a much steeper front edge than illustrated. Also remember that since more mass can be removed from the leading peak during peak shaving, it is always desirable to have the component of interest elute first from the column (if possible).

Example Calculation

Correct answer = 12.30
Experimental values = 12.00, 12.10, 12.50 mg
Mean of experimental values: 12.00
12.10
12.50

Sum = 36.60 ÷ 3 = 12.20 (mean)
Absolute error = 12.20 − 12.30 = −0.10 mg

The negative (− sign) error indicates that the experimental value is lower than it should be. A positive (+) error indicates that the experimental value is higher than it should be.

$$\text{Relative error} = \frac{0.10}{12^*} \times 100 = 0.83\% \text{ (or 8.3 ppt)}$$

The idea of a "relative error" is to express the result in terms of the size of the thing that you are measuring. The 0.10-mg absolute error shown above is analytically much less serious when the value measured is 12 mg than if the same absolute error was made in measuring 1.2 mg. The relative error in this case would then be calculated as

* See section on Significant Figures.

$$\text{Relative error} = \frac{0.10}{1.2} \times 100 = 8.3\%$$

Although the absolute errors are the same, the second relative error is 10 times as large as the first.

Precision

Precision expresses the degree of scatter among a series of supposedly identical measurements made on the same thing, usually in the same way. If four measurements of the volume coming out of a solvent delivery system are made in a graduated cylinder and all give very similar values, the data are then precise. The data are accurate only if the mean of the four results agrees closely with the milliliter per minute value set on the pump. In real analytical situations where top-line equipment and reliable procedures are employed, good precision usually means good accuracy as well. If precision is good but the mean is not correct, a "determinate" (or constant) error is inherent in the analysis. This could result from a poorly calibrated instrument or an impurity in a reagent. Determinate errors can be found and eliminated or corrected.

Poor precision is caused by "indeterminate" (or random) errors. These are inherent in every analysis and are characterized by both plus and minus values (whereas determinate errors are unidirectional). Experienced analysts minimize their indeterminate errors so that they obtain precise results. Examples of this type of error would be measuring the injection volume in a syringe for an injection or measuring peak heights for calibration curve construction and sample interpolation from the curve. Indeterminate errors are unavoidable and come from estimating and carrying analyses to their maximum limits. For example, there would be no error in measuring volume in a 50-mL graduated cylinder to the nearest ± 1 mL, but there must be an error in estimating tenths of a milliliter by judging the meniscus position between the graduation lines. The analyst must estimate numbers to get enough significant figures in his final analysis answer.

It is because of these unavoidable, random, and indeterminate errors that replicates are run. The mean of a series of replicates should be more accurate than any single value because the indeterminate errors will tend to cancel (average themselves out) in the runs. It is statistically seldom worth running more than four replicates because the number of trials in the standard deviation, σ, and confidence interval calculations is a square-root term in the denominator. Figure 6-13 illustrates the term accuracy and precision.

The three most common ways of expressing precision, with examples, are as follows:

1. $$\text{Relative mean deviation} = \frac{\text{mean deviation}}{\text{mean}} \times 100 \text{ or } 1000 \qquad (6\text{-}5)$$

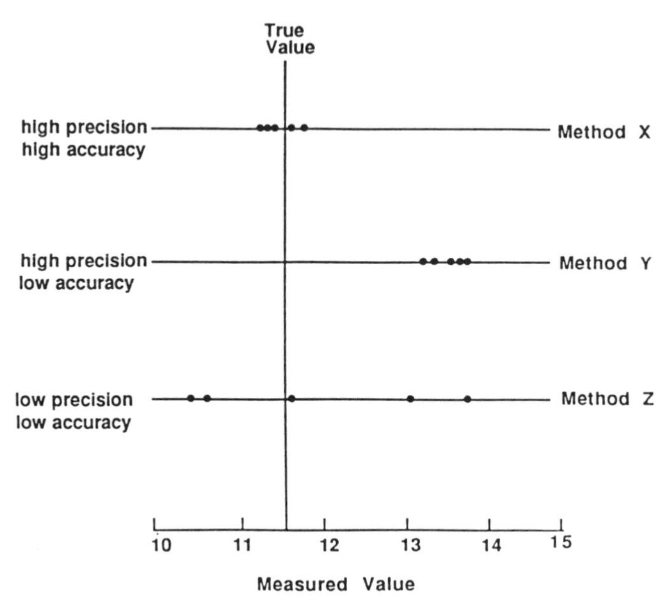

FIGURE 6-13. Accuracy and precision. (Developed from reference 11 with permission.)

2. Relative standard deviation $= \dfrac{\sigma}{\text{mean}}$

$$= \frac{\dfrac{(\textit{sum of deviations})^2}{\text{no. trials} - 1}}{\text{mean}} \times 100 \text{ or } 1000 \quad (6\text{-}6)$$

where σ is the symbol for standard deviation; regardless of the size of σ, $\pm 1\sigma$ will include 68.3% of the deviations for a series of replicates, $\pm 2\ \sigma$ will include 95.5%, and $\pm 3\ \sigma$ will include 99.7% of the deviations.

3. Variance $= \sigma^2$ (6-7)

Personal preference often dictates which of these measures of dispersion is utilized. Relative standard deviation is probably the most widely used. Variance finds its greatest application in calculations of the most probable uncertainty in the case of sums of differences, which is found by taking the square root of the sum of the individual variances. It is an advantage to use σ over σ^2 for expressing precision because the units of σ are the same as that of the quantity measured.

As an example, assume that the four replicates in an analysis give values of 20.10, 20.15, 20.30, and 20.40

	Value	Deviation*	(Deviation)^2	
	20.10	0.14	0.020	
		20.15	0.09	0.008
	20.30	0.06	0.004	
	20.40	0.16	0.026	
Sum =	80.95	0.45	0.058	
Mean =	80.95 ÷ 4 = 20.24			

The lower the values that measure precision, the less scatter there is in the values. Again, relative precision expresses the results in terms of the size of the measured quantity, which is much more meaningful than knowing only absolute precision.

The Error Curve and the Measurement

When an analysis is made, the objective is to find the true value (statisticians define this as μ). A single measurement is defined as X and if a large number of quantitative analyses are made, the measured values will not all be identical. The measured values, due to random error, are scattered around the true value, as shown in Figure 6-14. This figure is known as a "normal distribution curve," a Gaussian distribution curve, or an "error curve" and demonstrates how individual measurements X are scattered around the true value μ. From this curve it can be implied that, statistically speaking, the frequency of having random errors in HPLC quantitation are equally probable in the positive and negative direction and large errors occur less often than small errors. Also, since this curve represents the range of results that might occur, one can see that the breadth indicates the precision of the measurements. The breadth is determined by the standard deviation. A large standard deviation implies a less precise determination and a broader Gaussian distribution than a small standard deviation.

Because of random error, the true value μ and the standard deviation σ can only be known *exactly* if an infinite number of measurements is taken. Nevertheless, because of the symmetry of the error curve, positive and negative errors will cancel each other when the individual measurements are averaged. Therefore, the mean of the measurements, $\overline{X}$, is the best estimate of the true value μ. Obviously, the larger the number of measurements, the closer $\overline{X}$ will be to go μ. However, a point of diminishing returns occurs and the improvement in $\overline{X}$ results as the square root of the number of measurements which are taken (Fig. 6-14*b*). Similarly, the standard deviation which

* Deviation is the difference between the mean and each individual result. The mean deviation is the sum of the deviations divided by 4 = (0.45 ÷ 4 = 0.11). The relative mean deviation = 0.11/20 × 100 = 0.55% or 5.5 ppt. σ = standard deviation = 0.058/4-1 = 0.14. RSD = 0.14/20 × 100 = 0.70% or 7.0 ppt. Variance = σ^2 = $(0.14)^2$ = 0.020.

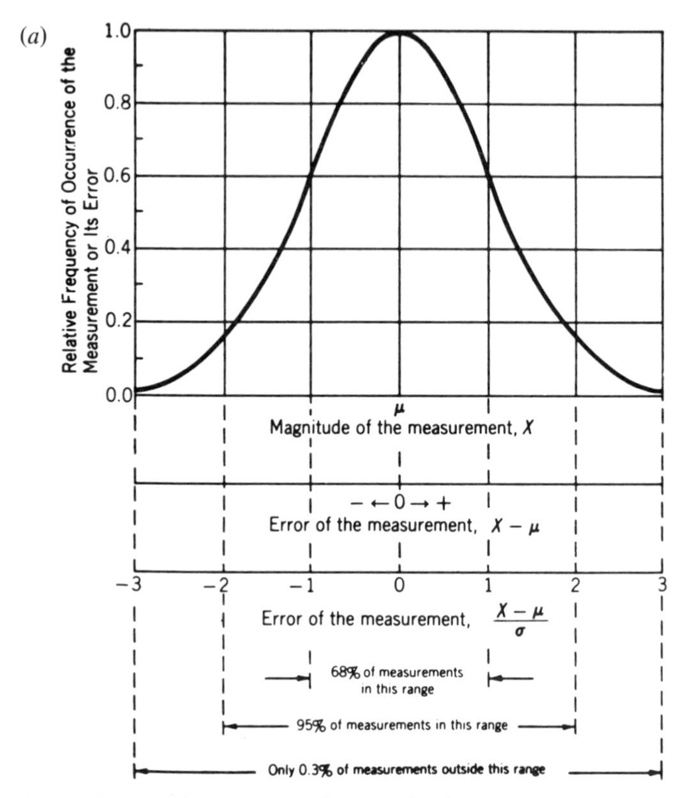

FIGURE 6-14. Theory of error. (*a*) The standard error curve. (*b*) Scattering of means and individual measurements about the true value. (*c*) Impact of reproducibility upon accuracy assuming only random errors. (Figure *A* and *B* reprinted from reference 11 with permission.)

is calculated is only the best estimate of the *true* standard deviation. Because of this, some prefer to use s for the calculated standard deviation and σ for the true standard deviation. What is important to realize is that the calculated value is an estimate of the true value.

By going from one to four measurements the reduction of error is improved by the square root, one would have to take 16 measurements to improve again by a factor of 2. Therefore, four determinations are considered to be the desired number since it is the best compromise between improving the measurement and the throughput of samples. However, each company or organization may have a specific protocol to follow when analyzing samples.

Confidence Limit

The term "confidence limit" or "confidence interval" is also widely used as a measure of precision (not accuracy!). Its calculation, which requires a

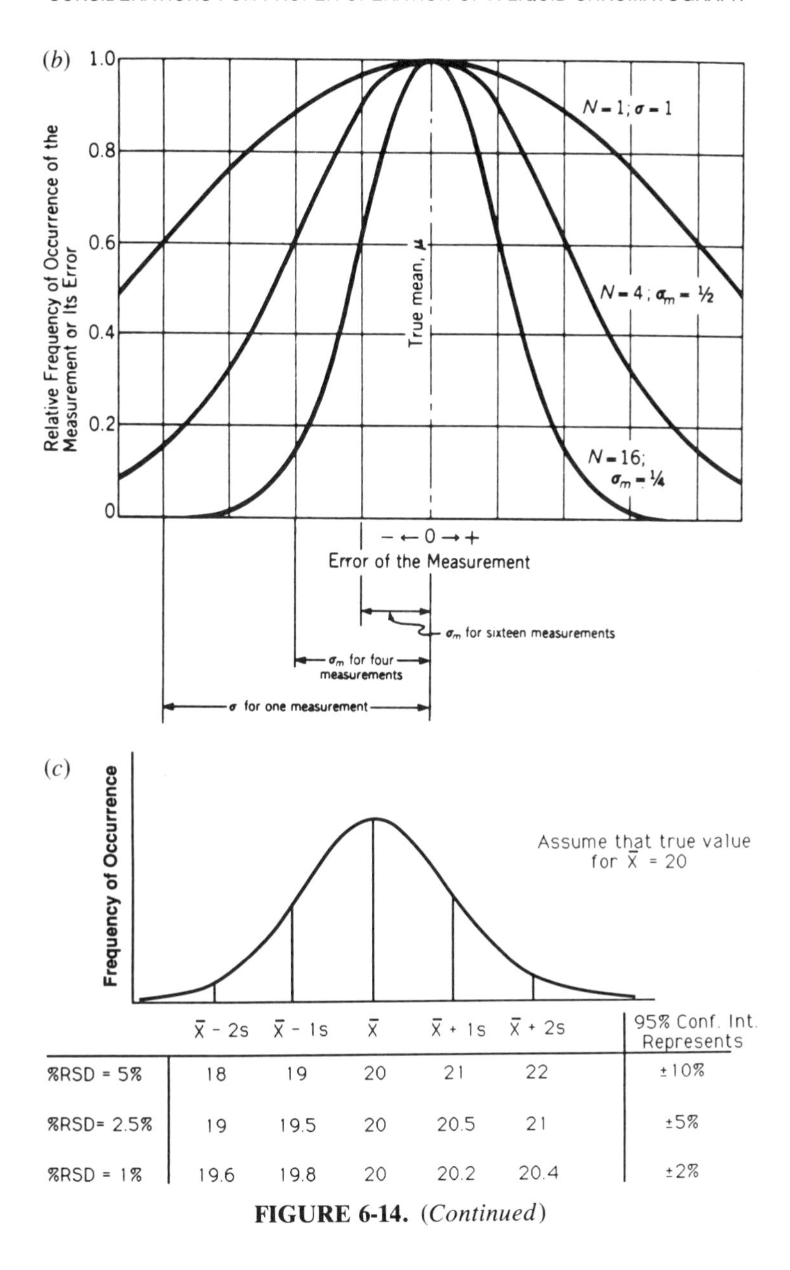

	$\bar{X}$ - 2s	$\bar{X}$ - 1s	$\bar{X}$	$\bar{X}$ + 1s	$\bar{X}$ + 2s	95% Conf. Int. Represents
%RSD = 5%	18	19	20	21	22	±10%
%RSD= 2.5%	19	19.5	20	20.5	21	±5%
%RSD = 1%	19.6	19.8	20	20.2	20.4	±2%

FIGURE 6-14. (*Continued*)

value for σ plus tabular constants, is rather complex so it will not be illustrated in detail. The definition is easiy confused and will be given so at least the term is understood when heard. The confidence interval is a range within which we can expect to find the true mean, with a given degree of probability. If a 95% confidence limit for an analysis is expressed as 1.80 ppm $\pm$ 0.20, this indicates that based on the determinations run, it is statistically possible to state the chances are 95 in 100 that the "true mean" will be in the range between 1.60 and 2.00 ppm. If the calculation is made at a lower percent (less odds for correctness desired), the range will be smaller. If we want the 99% interval, the range will be larger to include more conceivable values. We can also reduce the interval by running more trials, but more than four trials gains little compared to the effort expended. This is because, as stated earlier, $\sqrt{N}$ is the denominator of the equation for calculation of confidence limit. Therefore, the confidence interval is halved by employing the mean of four measurements ($N = 4$), but 16 measurements would be required to narrow the limit by another factor of 2 (see Fig. 6-14*b*).

Significant Figures

The scientist should record all experimental numbers that are certain plus one that is uncertain (estimated). For example, in reading a volume from a 50-mL graduated cylinder, where there are only lines every 1 mL, tenths are estimated so that no more than one decimal place can be given. A meniscus judged exactly one-half way between the 26- and 27-mL line would be 26.5 mL (not 26.50), or three significant figures. The 5 is estimated and uncertain. In smaller graduated pipets, where there are lines every 0.1 mL, the volume is recorded to the second decimal place, which is the estimated value.

When one encounters a recorded number, it is assumed that the last written digit is the uncertain one and that it is uncertain by ±1, unless more information is given. Hence, the number 145.1373 g means the measurement was made to the nearest ±0.0001 g (0.1 mg), the limit of a common laboratory analytical balance. This number has seven significant figures.

The value 17.20 mL has four significant figures and indicates measurement to the nearest ±0.01 mL (undoubtedly in a pipet or buret and not a graduated cylinder). If one expresses this as liters by dividing by 1000, the number becomes 0.01720. This also has four significant figures, since nothing can be gained or lost by a mathematical operation. This illustrates the rule that zeroes before the first digit are not significant, but only serve to locate the decimal place.

Zeroes after the first real digit should be significant, but people are very sloppy about this; 1.1 L of water (measured perhaps in a 2-L graduated cylinder) indicates measurement to ±0.1 L or the nearest 100 mL. Changing this mathematically to milliliters gives 1100 mL, but the precision of measurement has *not* increased. If it is written as 1100, this indicates uncertainty of only ±1 mL, which one could not obtain from a 2-L graduated cylinder. To express as milliliters but still indicate the correct number of significant figures, use exponential notation:

$$1.1 \text{ L (2 SF)} = 1.1 \times 10^3 \text{ mL (2 SF)}$$

Significant figures are often butchered during calculations. In addition or subtraction, the answer can have only as many *decimal places* as the number with the least places:

$$\begin{array}{rl} 92.70 \text{ g} & \\ + \underline{36.1 \text{ g}} & \text{(limiting number)} \\ 128.8 \text{ g} & \text{(1 decimal place, 4 SF)} \end{array}$$

In multiplication and division, the number with lease *significant figures* limits the answer:

$$\begin{array}{c} 0.0023 \text{ g} \times 1.9070 \text{ g} = 0.0044 \\ (2 \text{ SF}) \times (5 \text{ SF}) = (2 \text{ SF}) \end{array}$$

These are simplified rules which are not statistically exact but *usually* give the correct number of significant figures for the answer of a calculation.

If you refer to the section on Accuracy you will notice in the first example given on relative error calculation that the numerator was only 2 SF (0.10) after the subtraction step. Therefore, the denominator was rounded off to 2 SF (12) and the answer was expressed as 2 SF. The division by 3 to get the mean does not limit the SF to one because the 3 is part of a mathematical operation and not an experimental value. Some sources specify rounding off during a calculation while others say it should be done only at the end. In the relative error calculation, the answer is changed by 0.02% if rounding off is done at the end as shown below:

$$\frac{0.10}{12.30} \times 100 = 0.8130$$

or 0.81% after rounding off versus 0.83% previously shown.

Correct Values

Where do "correct" values come from for measuring accuracy of an analytical method? Philosophically, no one really knows the absolutely true value of a measurement, but the best we can do is to measure the sample by a widely accepted procedure and to compare the experimental value to this. In many cases, the accepted method will be published by an association or society which is involved in setting procedures. Two such organizations are the Association of Official Analytical Chemists (AOAC) and the American Society for Testing and Materials (ASTM). Another approach is to obtain already analyzed and certified samples from an agency, for example, the National Institute of Standards and Technology (NIST). Your experimental answer can then be compared to the provided results to test its accuracy.

In one's own laboratory, standard samples can be carefully prepared from pure chemicals to simulate actual samples, and the accuracy of your method calculated by comparing the answer you get to the amount put in the formulated sample. For very complex samples such as a crop or biological matrix, the best that can be done is "spiking" or fortification. A known amount of analyte is added to the sample, and the analysis is run to see how much is "recovered." In this case, percent recovery measures accuracy. Once the accuracy of a method is established, a good analyst should be able to use it on a routine basis, running 2–4 replicates on real samples, and be confident of *accurate* results.

Precision of Flow Rates from Solvent Delivery Systems

One of the concerns to chromatographers is how "good" a flow-rate precision can be obtained in LC to ensure good laboratory practice. A partial answer to this question can be found in a paper entitled "Precision of Contemporary Liquid Chromatographic Measurements" (12), where it was reported that with an optimized system, retention times could be measured with a standard deviation of 0.1% or less. This deviation was determined both on samples run on the same day and on different days. The article states:

> The small standard deviations realized for both retention measurements and retention volumes indicate that the pump is an extremely precise instrument.

But proper care must be taken, as reflected in the following paragraph:

> The . . . [pump] gave a flow of mobile phase over a period of about 12 hours with a standard deviation of 0.07%. This performance was considered amazingly good, but such precision can only be maintained *if the pump is operated with the necessary precautions*. The majority of chromatographers (and it must be said that until recently, the authors of this paper are to be included) treat precision LC pumps as just another piece of pumping hardware, whereas *they should be treated with the care and respect given an analytical balance.* [Italics are this author's]. It was found that to maintain the precision the following procedures need to be taken. The pump should never be allowed to run dry; otherwise abrasion between piston and cylinder produces small leaks. Any mobile phase used should be filtered through a 0.2 μm . . . filter, particularly if the solvent has been dried over activated silica gel or alumina. The usual method of filtering by a filter paper was found to be inadequate. The filter contained in the pump should be regularly changed. . . .

In another publication (13) it was demonstrated that with proper maintenance and operating care each of two HPLC solvent delivery systems was capable of delivering smooth, precise flow rates of as low as a few microliters per minute. While not all chromatographers will desire this high degree of precision, it is clear that if people observe good laboratory practice, promote cleanliness, and refer to the operator manual so that they take "proper care" of their pumps, excellent precision can be obtained for solvent delivery systems.

Precision in LC Analysis

Many LC users are curious about the precision they can expect from an analysis. This question is often answered with specifications concerning flow rate precision, injection volume precision, and so forth. However, the precision of the analysis is influenced by all of the variables in the HPLC system, including operator error. Often within a laboratory, the precision of the LC analysis is within ±1%. However, the larger question concerns the between-laboratory reproducibility of analyses.

To address this larger issue, ASTM Committee E-19 ran a cooperative program on the study of the quantitation of HPLC (14) which revealed that the technique is generally accurate and precise with relative standard deviations ranging from 6% to a worst case value of 16% for between-laboratory reproducibility of individual component analysis. Seventy-eight laboratories participated in the "round robin" test. A variety of reverse-phase columns was used (representing products from 12 manufacturers). Two solvent systems were used, either acetonitrile/water or methanol/water, in a number of ratios and in isocratic and gradient elution modes. Ultraviolet detectors were used exclusively. Two samples were each analyzed in triplicate. The first sample was an easily separated four-component mixture; the second was a more complex six-component mixture. Mean values of the analytical data submitted were consistent with the known concentrations of the components in each sample, indicating a highly satisfactory degree of overall accuracy. However, the spread of data, expressed as percent relative standard deviation, revealed analytical problems for some laboratories. Relative standard deviations for the whole data set ranged from 6 to 11% in the first sample. The problems were more serious in the more complex sample, with relative standard deviations ranging from 9 to 16%.

A statistical method that identified data that had significant deviation from the majority of the results ("outlier data") eliminated the results of five laboratories' analyses (6%) from the data set of the sample whose components were readily resolved. It eliminated 10 laboratories (13%) from the data set of the more complex and difficult-to-separate second sample. These laboratories need to be concerned about the quality of their chromatography. Unstable baselines, noisy detectors, poor resolution of components, and occasional carelessness were evident in most of their data returns. Removal of data from laboratories with large deviations from the norm reduced the relative standard deviations for the individual peaks in the first sample to a range of 3 to 5% and in the second sample to a range of 3 to 8%. This latter information was representative of the performance of about 90% of the participating laboratories.

It is important to realize that if there are other steps involved in an analysis in addition to the final quantitation using HPLC, the precision and accuracy of the HPLC analysis is *not* the issue. The issue is the precision and accuracy of the entire method. To give this proper perspective, the AOAC compiled the results of the analyses of 50 interlaboratory round-robin studies (15). Each study involved at least 20 laboratories with numerous types of analytical methods applied to various analytes. The conclusion was that the major factor leading to assay variability was identified as the concentration level of the analyte, not the variability of the final assay procedure. Precision of major components (concentrations greater than 10% of the weighed amount) were around the 1–2% region. When major components were at the 1% level, precision was only slightly worse than when they were at the 10% level. However, as the concentration of the analyte decreased, the interlaboratory coefficient of variation (percent RSD) followed an exponential increase with the value being 40% RSD at the parts per billion level. In conclusion, it is clear that at trace levels, validation of the entire analytical method is essential, not just the HPLC assay.

IMPROVED SENSITIVITY FOR TRACE ANALYSIS

Often a large sample injection (e.g., 200 μL) will experience only a little additional band broadening over that which would occur for a small injection (e.g., 10 μL). Thus, the large injection can result in a higher concentration for detection with no loss in resolution. (This assumes there is no mass overload, which is often a valid assumption in situations where insufficient sensitivity occurs due to very small concentrations of the sought-after substance. However, experimentally this assumption should be verified for each situation.) Also, improved chromatographic precision can result if there is smaller uncertainty of measurement with the larger volume injection.

A rule of thumb is that the injection volume can be as high as 30% of the volume of a peak that elutes from the column when using a small injection (e.g., 10 μL) and there should be no significant broadening of the peak with this larger injection. To understand this rule of thumb, one must consider the contributions to band broadening. At injection, the sample volume will be diluted to a volume that is mainly dependent upon the efficiency of the column. The final peak width, W_{peak}, observed in the detector is a result of the volume of the sample injected and of the spreading from the column, the detector, and the extra column effects.

$$W_{\text{peak}}^2 = W_{\text{inj}}^2 + W_{\text{col}}^2 + W_{\text{det}}^2 + W_{\text{ex}}^2 \qquad (6\text{-}8)$$

Dilution in the chromatographic system is defined simply by the efficiency of the column and can be calculated from the plate equation:

TABLE 6-3. Influence of Injection Volume Width (W_{inj}) on Peak Volume Width (W_{peak})

2500-Plate Column			8000-Plate Column		
W_{inj} (mL)	W_{col} (mL)	W_{peak} (mL)	W_{inj} (mL)	W_{col} (mL)	W_{peak} (mL)
0.01	0.80	0.80	0.01	0.45	0.45
0.1	0.80	0.81	0.1	0.45	0.46
0.2	0.80	0.82	0.2	0.45	0.49

$$N = 16(V_e/W_{peak})^2 \tag{6-9}$$

where V_e is elution volume, N is the plate number, and W_{peak} is the base width determined by the "tangent" method. Rearranging gives

$$W_{peak}^2 = 16V_e^2/N \tag{6-10}$$

Using equations 6-8, 6-9, and 6-10, we can calculate the effect of large volume injections. For example, assume that we have two columns, one column has a plate count of 25,000 and the compound elutes with a 10-mL retention volume, and the spreading is $W = 0.8$ mL. The second column has 8000 plates, and the column elutes with a 10-mL retention volume, and the spreading is $W = 0.45$ mL. For these two situations the effect of injection volume size can be shown in Table 6-3.

When the plate count of the column was 2500, an injection volume size of 200 μL is easily tolerated and the larger injection volume increased the peak height by 20 times over the 10-μL injection and increased the peak width only 2.5%. Thus, the increased sensitivity is a "free" benefit. For the higher-plate column, a 200-μL injection may still be satisfactory since it only broadens the peak by an additional 9%. In addition to doing a calculation, sample volumes can be injected to verify the calculated value. Some individuals may prefer to not do a calculation but to experimentally determine how much band broadening would be tolerated by injecting sequentially larger sample volumes. An example of this is shown in Figure 6-15*a* where an injection size of 500 μL might be used if there was enough separation between the carbaryl and other components in the sample.

The previous mathematical reasoning was for the "worst case situation" when large volume injections are made into a mobile phase of similar eluting strength. Other LC situations are more favorable for using large sample volumes. When injecting a sample in a solvent which is weaker than the mobile phase or when injecting in the initial mobile phase and using gradient chromatography to elute the peak of interest, large-volume injections are often crucial for accomplishing trace analysis. In these situations, the sample in the large volume is essentially held or "trace enriched" on the top of the column while the large sample volume of the weaker solvent, the solvent

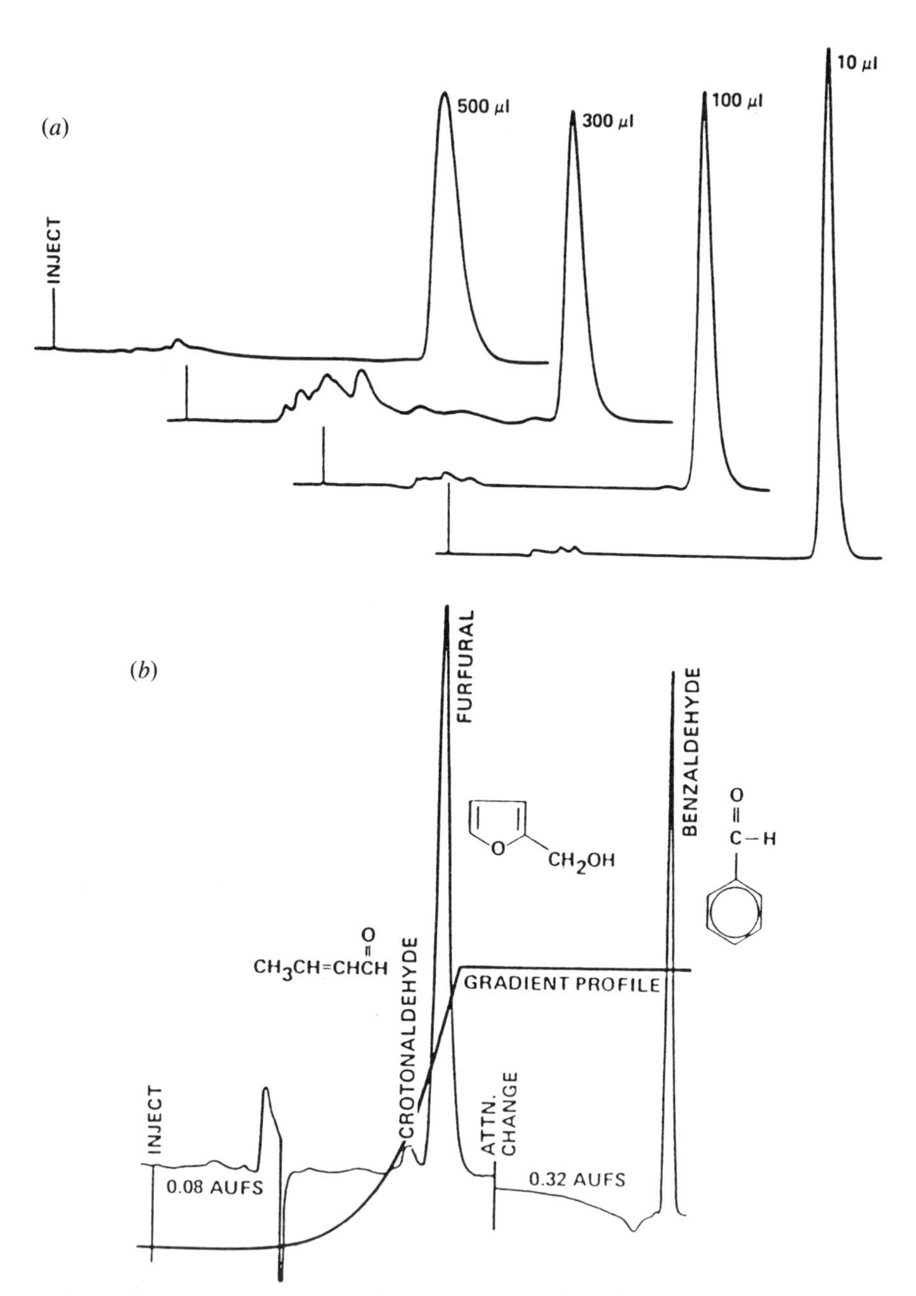

FIGURE 6-15. Trace analysis with large-volume injection. (*a*) Effect of sample volume on peak broadening in the isocratic mode. Column: μBondapak C_{18}, 3.9 mm × 30 cm. Mobile phase: 50% v/v water in acetonitrile. Sample: Carbaryl (1 μg per injection). Flow rate: 1.0 mL/min. Detector: UV at 254 nm. Trace analysis with large-volume injection. (*b*) Effect of large-volume aqueous injection in the gradient mode. Column: μBondapak C_{18}, 3.9 mm × 30 cm. Mobile phase: water and acetonitrile using a gradient of 5–50% acetonitrile with a curve shape as shown in the chromatogram for 6 min. Flow rate: 2.0 mL/min. Sample: 1 ppm of crotonaldehyde, furfural, and benzaldehyde, using a 500-μL injection.

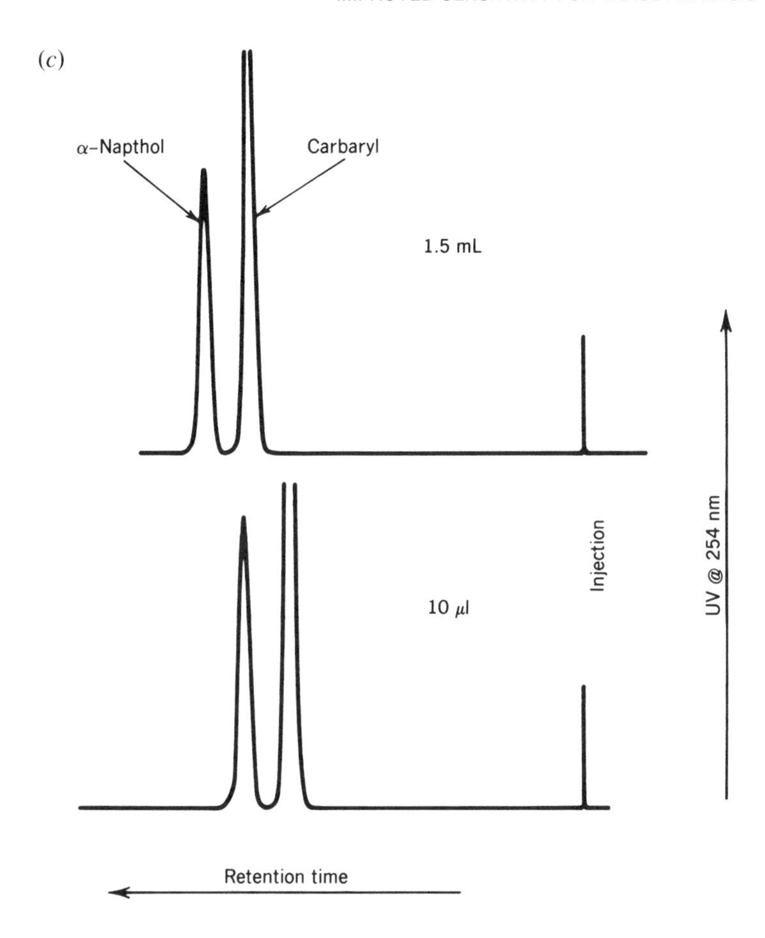

FIGURE 6-15. (*Continued*) (*c*) Effect of large-volume injection in isocratic mode. Comparison of injection of 10 μL of sample (bottom) with 1.5 mL of sample (top) containing the same concentration of sample of 1 ppb of carbaryl and α-napthol.

in which the sample was dissolved, travels down the column. The baseline of the chromatogram will exhibit a disturbance at V_0, but it will not affect the elution of the compounds of interest when gradient chromatography is used. An example of this is in Figure 6-15*b* for trace analysis where we are looking for one part per million each of benzaldehyde, furfural, and crotonaldehyde in water. The sample size was 500 μL. With the use of a gradient and trace enrichment the larger sample volume was easily injected and the organic materials were concentrated on the head of the C_{18} column and then eluted with a gradient using acetonitrile to obtain sharp peaks.

In an isocratic mode, the large sample volume will shift the retention time of the peak of interest by the volume of the sample injected. For example, if a strong mobile phase of 70% acetonitrile, 30% water is used to elute a pesticide of carbaryl where the retention time of a 10-μL injection of a standard is 6 min. An injection of a large volume of 1.5 mL of a water sample containing the pesticide would have a large baseline upset at V_0 and a retention time of the sample at 7.5 min. In this case, the peak width will be approximately the same as the original since there are essentially no extra contributions to band broadening by injecting a large volume of a weak solvent. This example is shown in Figure 6-15*c*.

Most people dismiss large-volume injections. However, experienced chromatographers realize when to use larger-volume injections to help enhance sensitivity and improve precision.

RECIRCULATION OF MOBILE PHASE

If solvent costs are a real issue in your laboratory, the focus should be on how to save solvents in chromatography rather than whether to buy a new system or column to solve the problem. The answer may be solvent recirculation. Most chromatographers are understandably cautious and do not recirculate mobile phases. However, there are a few references in the literature on recirculation that might change the "traditional" outlook.

UV and EC Detection

The practical use of solvent recirculation for a sensitive analysis, monitoring drugs in serum, was reported (16) and it was stated that the researchers "injected more than 400 samples over a seven-day period using cycled mobile phase with satisfactory reproducibility and sensitivity. The method worked well with both UV and electrochemical detection . . .", and could "save approximately 80% in solvent use." Some of the necessary precautions were taken, such as using a large volume of mobile phase (up to 4L) which is kept stirred and covered, together with regular calibration and baseline adjustment. This article should be read for further information on this approach to using solvent recirculation.

Baseline Drift

Baseline drift is an obvious objection which can be envisioned to occur when using recirculation. However, it has been reported (17) that the use of batch recirculation did not result in baseline drift when analyzing organic acids using UV detection. In this report 3.5 L of mobile phase was prepared and after about 3 L of column effluent was collected the mobile phase was recycled. It was possible to reuse the mobile phase about five times before fresh mobile phase was prepared. Samples included waste digester and artificial "gut liquids", foods (including soy sauce), and wines.

RI Detection

The use of solvent recirculation during the separation of sugars on amine-modified silica columns using RI detection has been reported (18). The gradual buildup of sugars slowly changed the RI of the mobile phase and thus affected the baseline drift and response of the differential RI detector. However, their data showed only a small change over 500 injections, and reasonable separations were achieved after 1000 injections of soft drink using only 1 L of solvent. The detector achieved its initial sensitivity upon being supplied with fresh solvent.

Conductivity Detection

Mobile phase reuse through recirculation is often used in the separation of inorganic ions using ionic mobile phases and conductivity detection (19). It was reported that it is possible to use recirculated mobile phases for 2–3 weeks in a lab analyzing ionic compounds of chlorides and nitrates using a 1-L reservoir. However, for less than parts per million work, this time is expected to be significantly less. The report encourages the use of mobile-phase recirculation to "simplify chromatographic operation by decreasing the time spent in the preparation of eluents and minimizing the volume of 'good' eluents discarded" (19).

Summary

Of course, chemical common sense should prevail when obvious problem solvents, such as unstabilized THF, are being used or when the sample may affect the solvent by reaction, changing buffer pH, and so on. Obviously, recirculation is impossible for gradients. Nevertheless, recirculation of mobile phase appears to be a practical means of reducing solvent costs to about one quarter of the cost of fresh solvent. Using the precautions advocated by the authors cited (16–19) and using the recirculation of eluent technique in a cautious and systematic fashion could result in significant savings. If you wish to consider recirculating mobile phases, consulting references 16–19 will clarify some of the considerations that should be taken into account.

CONSIDERATIONS WHEN CHOOSING SOLVENTS FOR MOBILE PHASES

In LC there is a relatively small number of stationary phase packing materials available. Therefore, it is necessary to vary solvent polarities to separate materials with vastly different k' values. Often this guidance comes from the tabulation of the physical properties of the solvents (Table 6-3). In addition to polarity of the final mobile phase, there are two additional major considerations when selecting solvent pairs for either isocratic or gradient elution chromatography. The solvents must be miscible with each other and "compatible" with the column packing material and the sample. Additionally, prior to inserting a new column–eluent combination in the LC system, the old mobile phase must be miscible with the new one. Otherwise the system must be flushed with an intermediate solvent that is miscible with both eluents so that the system is "cleared."

Miscibility

If the solvents are immiscible, the LC system will fail. If the pump will be delivering an eluent that is not soluble with the previous mobile phase or if the new mobile phase consists of two immiscible solvents, the net result is to have tiny "slugs" of different solvents traveling through the HPLC. Typical indications of this problem are: (1) erratic flow rate, (2) noisy baseline, and/or (3) baseline drift. To insure that these problems are not caused by a mismatch of solvents, refer to Table 6-4 for the miscibility numbers (M) and their use. The discussion on determining solvent miscibility using miscibility numbers is adapted from reference 20.

How to Use Miscibility (M) Numbers

The relationship between the M numbers is:

1. All pairs of solvents whose M numbers differ by 15 units or less are miscible in all proportions at 15°.
2. Each pair whose M number difference is 16 has a critical solution temperature between 25 and 75°C, preferably about 50°C.
3. A difference of 17 or more corresponds to immiscibility or to a critical solution temperature above 75°C.

By definition, the M number is a reliable means for predicting miscibility of any liquid with the standard solvent. When applied to other pairs of liquids selected from Table 6-3, it leads to a high percentage of valid predictions.

Wrong predictions may also be expected when either unusually strong or unusually weak hydrogen bonding or other types of interaction can occur between unlike molecules. The first effect, for example, may give rise to unpredicted miscibility between ethers or tertiary acids, amines, and hydroxylic solvents. Long-chain alcohols and carboxylic acids apparently lower their miscibility with low-numbered standard solvents in large part to hydrogen bonding. They are therefore likely to exhibit anomalous immiscibility when paired with appropriate solvents of low M number.

In a substantial number of cases, about 17% of those studied, immiscibility is encountered with solvents at both ends of the lipophilicity scale. This

TABLE 6-4. Physical Properties of Solvents

Polarity Index	Solvent	Viscosity CP (20°)	Boiling Point (°C, 760 Torr)	Miscibility Number (M)
−0.3	*N*-Decane	0.92	174.1	29
−0.4	Isooctane	0.50	99.2	29
0.0	*N*-hexane	0.31	68.7	29
0.0	Cyclohexane	0.98	80.7	28
1.7	Butylether	0.70	142.2	26
1.8	Triethylamine	0.38	89.5	26
2.2	Isopropyl ether	0.33	68.3	—
2.3	Toluene	0.59	101.6	23
2.4	*p*-xylene	0.70	138.0	24
3.0	Benzene	0.65	80.1	21
3.3	Benzyl ether	5.33	288.3	—
3.4	Methylene chloride	0.44	39.8	20
3.7	Ethylene chloride	0.79	83.5	20
3.9	Butyl alcohol	3.00	117.7	—
3.9	Butanol	3.01	177.7	15
4.2	Tetrahydrofuran	0.55	66.0	17
4.3	Ethyl acetate	0.47	77.1	19
4.3	1-Propanol	2.30	97.2	15
4.3	2-Propanol	2.35	117.7	15
4.3	Chloroform	0.57	61.2	19
4.4	Methyl acetate	0.45	56.3	15,17
4.5	Methylethyl ketone	0.43	80.0	17
4.5	Cyclohexanone	2.24	155.7	28
4.5	Nitrobenzene	2.03	210.8	14,20
4.6	Benzonitrile	1.22	191.1	15,19
4.8	Dioxane	1.54	101.3	17
5.2	Ethanol	1.20	78.3	14
5.3	Pyridine	0.94	115.3	16
5.3	Nitroethane	0.68	114.0	—
5.4	Acetone	0.32	56.3	15,17
5.5	Benzylalcohol	5.8	205.5	13
5.7	Methoxyethanol	1.72	124.6	13
6.2	Acetonitrole	0.37	81.6	11,17
6.2	Acetic acid	1.26	117.9	14
6.4	Dimethylformamide	0.90	53.0	12
6.5	Dimethylsulfoxide	2.24	189.0	9
6.6	Methanol	0.60	64.7	12
7.3	Formamide	3.76	210.5	3
9.0	Water	1.00	100.0	—

behavior is accounted for by assigning a dual M number. The first component is always less than 16 and defines the boundary of miscibility with a solvent with high lipophilicity. Converse statements apply to the second. The greater the differences between the two components, the more restricted the range of miscibility with other liquids. In the extreme case, the liquid is immiscible with all the standard solvents, and its M number is 0.32. Some fluorocarbons qualify for this exotic classification. Two liquids that both have dual numbers are usually miscible with each other.

For a solvent that is not listed, the procedure for estimating the M number is to determine the cut-off point for miscibility using a sequence of known solvents. A correction term of 15 is then either added or subtracted, whichever is appropriate. Solvents used in these tests should of course be "pure." After the miscibility number is determined, an experiment should be run to verify the solubility of the two solvents. For example, if a liquid has borderline solubility in dimethylsulfoxide (M = 9), the M number assigned is 24. On the other hand, if an unknown liquid was borderline soluble in *p*-xylene (M = 24) its M number is estimated to be 9. Reference 20 has a list of approximately 400 liquids and should be consulted by those who are interested.

Changing Mobile Phases

Prior to changing from one mobile phase to another, refer to Table 6-4 to determine the miscibility of the solvents to be used. The following considerations apply:

1. Changes involving two miscible solvents may be made directly. Changes involving two solvents that are not totally miscible, for example, from chloroform to water, require an intermediate solvent (such as methanol).
2. When changing from an aqueous buffered solvent to an organic solvent, first change to water.
3. Separate tubing and solvent filter assemblies should be used for each solvent. When a supply tube and filter have been used with one solvent, purge them of the previous solvent.
4. The entire instrument must be flushed without columns. Refer to the column manual from the manufacturer for the recommended procedure. Often using a 2-ft length of 0.040 in. stainless tubing to connect the injector outlet to the detector inlet is sufficient for flushing the LC system with new mobile phase. The system is flushed at a flow rate of 2.0 mL/min for approximately 4 min—2 min with the injector in the load position followed by 2 min in the inject position.

Use of Buffers

Some separations require addition of salts or buffers to modify retention. When converting a system from a buffered aqueous system to an organic system, a possible problem exists in that the buffer and organic solvent are miscible to a point, after which the salts precipitate from solution. It is important to keep this in mind when dealing with gradients that run from aqueous buffered solvent to organic solvents. Before such gradients are attempted, the upper percentage limit of organic solvent must be established. This can be done by taking a known volume of buffered solvent and adding incremental volumes of organic solvent until precipitation occurs. If you exceed this limit of organic solvent in a gradient run, you will start precipitating the salt in the system. This results in a build-up of pressure and eventual termination of flow through the columns.

It is important to keep in mind the operating limitations of all columns used for LC analysis. General precautions for solvent compatibility, flow limitations, and general care should be thoroughly read prior to any analysis. Failure to do so can result in misleading analytical information and often may terminate usefulness of the column(s). Careful attention will increase column life and allow you to return to a stored column with a knowledge of purging steps required prior to analytical use. Always tag columns with the last solvent passed through it. For lengthy storage, refer to the column maintenance booklet provided with each column.

Choice of Buffer

The need for pH control was discussed in the context of several examples in Chapter 5. Certainly, in addition to solvent strength, the choice of the buffer, when appropriate, is a key variable. Choice of the buffer needs to be made on the basis of the pH range desired and whether a volatile buffer is desirable. Table 6-5 gives a listing of several typical buffers useful in HPLC. This listing is segmented into general-purpose and volatile buffers and includes some specific preparation for LC use. As a general rule of thumb, the pH of the solution will be the mid value of the pH range. Adjustment can be made by using the corresponding acid or base.

After preparation, all buffers should be filtered through a 0.2-μm filter. The normal operating range of buffer concentration is 0.001–0.1 M; although, at this high concentration, precipitation might occur, especially with acetonitrile. With high buffer concentrations, it is preferable to use methanol as the organic phase. Always check the buffer:solvent ratio to be sure no precipitation occurs.

The use of volatile buffers is helpful in the life science area and in some preparative areas where recovery of the appropriate functions will be salt-free upon evaporation. Also, people doing mass spectrometry detection of LC fractions of basic compounds (e.g., drugs and antihistamines) can minimize tailing and have a mobile phase which can go directly into the mass spectrometer using volatile buffers. For instance, ammonium carbonate is convenient because it decomposes to NH_3 and CO_2 upon mild heating.

While some scientists approve of ammonium carbonate [$(NH_4)_2CO_3$], there is some controversy over the use of this buffer with reverse-phase

TABLE 6-5. Common Buffers Used in HPLC

Buffer	pH Range
General Purpose Buffers	
Potassium dihydrogen phosphate (monobasic)	1.1–3.1
Dipotassium hydrogen phosphate (dibasic)	6.2–8.2
Tripotassium phosphate (tribasic)	11.3–13.3
Sodium dihydrogen citrate	2.1–4.1
Disodium hydrogen citrate	3.7–5.7
Trisodium citrate	4.4–6.4
Sodium acetate	3.8–5.8
TRIS [tris(hydroxymethyl)aminomethane]	8.0–10.0
Ammonium chloride	8.3–10.3
Volatile Buffers	
Ammonium formate	3.0–5.0
Pyridinium formate	3.0–5.0
Ammonium acetate	3.8–5.8
Pyridinium acetate	4.0–6.0
Ammonium carbonate (used for reverse phase)	8.0 (adjusted)
Ammonium carbonate	5.5–7.5 and 9.3–11.3

Specific Solutions

Buffer	Molarity	Preparation (gL)	pH
$NH_4H_2PO_4$	0.01	1.15	~4.5
KH_2PO_4	0.01	1.36	~4.6
$NH_4C_2H_3O_2$	0.01	0.77	~6.0
$KC_2H_3O_2$	0.01	0.98	~7.1
$(NH_4)_2HPO_4$	0.01	1.32	~8.0

packings. Since a 0.1% solution of $(NH_4)_2CO_3$ has a pH of 8.3–8.5 depending upon the original pH of the water being used, the solution tends to dissolve the base silica. Therefore, in order to use ammonium carbonate successfully, it is recommended that the $(NH_4)_2CO_3$ solution be adjusted to pH 8.0 with dilute acetic acid. Ammonium acetate is another volatile buffer which may be used and a 0.01-M solution has a pH of ~6.0. If necessary, acetate buffers may be adjusted to pH 7.5–8.0 with dilute NH_4OH for the separation of bases. The rule of thumb for mobile phases is that the aqueous portion should be below a pH of 8.0. When the pH of the solution is above 8.0, the mobile phase has a tendency to dissolve the base silica which can result in short column lifetimes.

Solvent Compatibility

The solvent should always be compatible with the previous solvent in the HPLC system. If it is not, a "flush" solvent needs to be used that is compatible with the previous and future mobile phase. Also, the mobile phase must be compatible with the stationary phase. An example of noncompatibility using a solvent pair that is miscible would be a chloroform–methanol mixture using a polystyrenedivinylbenzene-type packing. The methanol causes shrinkage of the gel matrix resulting in a permanent loss of column efficiency. Another example of noncompatibility would be this solvent pair used on a silica gel column in the normal phase mode—this mixture would partially deactivate the column. In this instance, however, the column could be reactivated by oven drying at 105°C overnight. When in doubt, reference should be made to the discussion and charts that have appeared in earlier chapters concerning typical solvent–packing combinations (e.g., Fig.4-11).

Removing Particles and Dissolved Gases

Good practice in LC involves avoiding solvents or samples containing foreign particles. If the homogeneity of your sample is in doubt, a filter or a sample clarification kit should be used. Particles of approximately 0.45 μ or larger can be removed from either aqueous or organic samples with these kits. Additional information on these sample filtration kits is given in manufacturers' publications.

Once the mobile phase is mixed, it should be filtered and degassed. Even though the HPLC systems have filters routinely in line, the simple preventative maintenance practice of filtration can minimize clogging, filter changes, and associated expense. Also, routine filtering should minimize system troubleshooting. The filtration process is usually explained and documented in the equipment manufacturer's literature. The experimental set-up usually requires a vacuum pump to supply the suction to pull the mobile phase through the filter into the filtering flask. Filtration also has a secondary effect of degassing the mobile phase, which is often sufficient for many HPLC systems. However, if it is not, other degassing procedures should be used.

The main purpose of degassing is to remove dissolved air (gases) which, if not removed, may cause problems with proper operation of the HPLC system. If the dissolved gases are not removed before operation, the solvent may degas with time during operation and small bubbles will form in the connecting tubing between the reservoir and the solvent delivery system. Several small bubbles may combine to be a large bubble which can enter the pumping chamber and "air lock" a piston; or if the bubble becomes compressed and is pumped through the system, it may be sufficient to cause a change in retention times of the peaks. Also, the gas may reappear at the end of the system in the detector and manifest itself as spikes on the chromatogram.

Some degassing occurs by a vacuum filtration of the mobile phase. If more degassing is desired, turning on the vacuum pump and placing a rubber stopper over the top of the flask containing the freshly mixed mobile phase will cause the dissolved air to form tiny bubbles which rise to the surface. Pulling this vacuum for 3–5 minutes is usually sufficient to fully degas a mobile phase. Additionally, placing the flask in an ultrasonic bath will help the emission of gases through this mechanical agitation. However, a sonic bath is not required for complete degassing.

After this process, some manufacturers of low-pressure gradient systems (refer to appropriate sections in Chapters 3 and 7) suggest that helium should be bubbled through the mobile phase during operation of the system. However, bubbling may cause a mobile phase which consists of two solvents to change composition with time. For other solvent delivery systems, "blanketing" of the mobile phase with helium is sometimes recommended. Both approaches often have the added benefit of applying a partial positive pressure which ensures that the mobile phase is "pushed" into the pump, which aids in the smooth delivery of mobile phase and high reproducibility of retention times often result. The degree to which you need to degas depends upon the degree of reproducibility of chromatography which is required. The operator should do what is needed. Certainly, doing all of the degassing activities cannot hurt. However, if some of the degassing activities do not help, and since each step takes up valuable time, why do them? Only the individual can determine what is needed for the situation.

Other Solvent Considerations

While the most important consideration when mixing solvents into mobile phases is their miscibility, other considerations may become important in certain circumstances. For instance, solvent viscosity will influence the back pressure necessary to pump the mobile phase. While this is often not a problem in itself, the operator should be aware that changes in back pressure will occur during gradients. Also, accompanying this viscosity change is a refractive index change which may influence the baseline of the detector. Discussion of these two situations is covered in Chapter 7 (gradients).

In all circumstances, as mentioned earlier, testing the solubility of the aqueous buffer solution in the mobile phase is crucial. Certain buffering salts have very poor solubility with solutions above 80% acetonitrile. When in doubt, test buffer solubility off-line using a small volume in vials and beakers on the bench top. Similarly, the sample should be soluble in the mobile phase. This is especially important if the sample was dissolved in a solvent which is different from the mobile phase. In this case, the only answer is to test the solution by injecting approximately 50 μL of sample solution into a vial containing approximately 1 mL of mobile phase and observing if the sample stays in solution.

Solvent Quality

Generally the quality of solvents that go into a mobile phase should be decided upon by the scientist doing the LC. The type of variability in content and in contaminants in a solvent should be weighted against the cost of the solvent and the repeatability and reproducibility of the chromatography. For instance, tetrahydrofuran (THF) is often stabilized with butyl hydroxytoluene (BHT) to prevent the formation of peroxides. In reverse-phase LC this solvent additive could have a deleterious effect on the chromatography by having BHT coat the column and change the chromatography. However, the degree to which this is a problem needs to be considered with respect to buying a higher grade of THF (BHT free), having to monitor the solvent for peroxide formation, and possibly having to discard solvent when peroxides are detected. In GPC, the presence of BHT is not a problem if peaks are not collected; therefore, using stabilized THF in GPC would be preferable.

What is important in solvent quality is to note the solvent used. When mobile phases are mixed during the development of a method, it is important to note the manufacturer, lot number, grade, type and concentration of the stabilizer(s) if present, and any other attributes of the solvents used, so that other chromatographers can obtain reproducibility using the conditions.

Sometimes the impurities are so low or unknown in a solvent that the chromatographer might not always realize there could be a problem. Methanol is always considered clean, but occasionally it can exhibit a slight "fishy" odor detectable by people with "good" noses. This fishy odor is believed to be due to a small amine contaminant. Some might say that the nose is too sensitive; therefore, this ultratrace contaminant could not cause a problem. The problem, however, would be if a method was developed for an amine-containing sample and only some batches of methanol were contaminated. One methanol mobile phase would "coat" the amine on the column and the separation would be achieved. With clean methanol, the amine coating on the column would wash off and slowly the chromatography of the amine sample would shift to longer and longer elution times. This would be a problem.

Clearly, there is no standard answer for solvent quality. However, by being observant and keeping good records, it is possible to pinpoint problem sources in the solvents.

Water Quality

Because of the very popular reverse-phase mode of LC, water is probably the most popular solvent used in mobile phases. With today's instruments, chemists are now capable of detecting and measuring trace elements down to the level of fractional parts per billion. Typically, however, the thresholds of trace analysis are imposed by problems external to the instrument itself.

"Pure" water can be a major culprit, because it is the starting point for most eluents, standards, buffers, and reagents. Results are obviously meaningless if background contaminants in the water mask the very substances being analyzed.

When impurities were measured in parts per thousand, distilled water was the accepted "standard" of water purity. However, numerous studies have shown that even double- and triple-distilled water contains impurities easily detected with today's instrumentation. As the sole source of water for trace analysis, distillation falls short in many applications. For LC, distilled water can be inappropriate since in some situations even parts-per-billion levels of some organic impurities can be a problem. They can cause baseline shift or background interference, which obscures key components in a sample. Impurities can also concentrate on the column, later bleeding off as random "ghost" peaks. In addition, they can shorten the life of costly columns. Liquid chromatography is used to detect organic and inorganic ions, amines, and carbolyxic acids in such fluids as power-plant feedwater, wastewater, body fluids, and foods. Ionizable impurities in the water used for standards, solvents, and eluting agents can obscure ions of interest or foul the column in this mode of LC.

What is Pure Water?

For all practical purposes, chemically pure water is nonexistent. Purity has meaning mainly in terms of its use. Water may be "pure" enough for bathing, yet unsafe to drink. It may be suitable for human consumption, yet too "foul" for rinsing laboratory glassware.

To resolve such confusion, several professional organizations have recommended water quality standards that are graduated according to classes of use. These specifications enable users to define their needs with more precision. Typical are those of the ASTM, the National Committee for Clinical Laboratory Standards (NCCLS), the College of American Pathologists (CAP), and the American Chemical Society (ACS). These specifications are summarized in Table 6-6. In general, laboratory water quality is defined in terms of its resistivity. The historical focus on resistivity as a measure of water quality presupposes that dissolved minerals are the main concern. Often it is true, but resistivity has been used as the major yardstick for water purity mainly because it has been the most easily measured.

Since minerals form ions in solution, they increase the electrical conductivity of the solution. Conversely, water that is low in dissolved ions has higher resistance to current flow. For example, the calculated resistivity of chemically pure water is 18.3 million Ω (megohms) over a distance of 1 cm at 25°C. In fact, 18 MΩ-cm is the upper limit of current water-purification technology.

TABLE 6-6. Laboratory Reagent Grade Water Specifications[a]

	CAP Type			ASTM Type			
	I	II	III	I	II	III	IV
Specific conductance (microohms/cm)	0.1	0.5	10	0.06	1.0	1.0	5.0
Specific resistance (megohms/cm)	10	2.0	0.1	16.6	1.0	1.0	0.2
Silicate (mg/L)	0.05	0.1	1.0	—	—	—	—
Heavy metals (mg/L)	0.01	0.1	.01	—	—	—	—
Potassium Permanganate Reduction (min)	60	60	60	60	60	10	10
Sodium (mg/L)		0.1	0.1	60.1	—	—	—
Hardness	neg	neg	neg	—	—	—	—
Ammonia	0.1	0.1	0.1	—	—	—	—
Bacterial Growth	10	10^4	—	—	—	—	—
pH	—	—	5.0	—	—	6.2	5.0
	—	—	8.0	—	—	7.5	8.0
CO_2 (mg/L)	3	3	3	—	—	—	—

[a] ACS Specifications similar to CAP Type II. NCCLS specifications similar to CAP. Source of information: Millipore Corp.

Water contains four basic types of contaminants:

Dissolved inorganics
Dissolved organics
Suspended particles
Microorganisms

Inorganic salts and gases freely dissociate in water to form positive and negative ions. These include the minerals (calcium and magnesium) commonly associated with "hard" water and trace amounts of heavy metals.

Organic-free water is a special class of water purity not encompassed by any recognized standard; yet it represents a growing need. New trace-enrichment techniques for HPLC, for example, are now capable of resolving part-per-billion levels of some organic compounds. The mobile phase for work of this type requires organic purity even beyond reagent grade standards.

There are six major water purification methods, each of which has its advantages and disadvantages, as shown in Table 6-7. The requirements of water quality in LC depends upon the type of analysis; and therefore the choice of water purification may be one or more of the methods listed in Table 6-7. For some application, distilled or deionized water may be suitable. For a critical application, it may be necessary to purchase a bench-top or wall-mounted water purification system which includes a reverse osmosis unit, a deionization unit, a carbon adsorption unit, and a final filtration through a 0.22-μm membrane filter. Such systems produce organic-free, reagent-grade water having total organic carbon (TOC) levels less than 50 ppb with 18 MΩ resistivity. Some people prefer to buy bottled water. Each person must make decisions as to the best cost-effective source of water to meet the purity, quality, and chromatographic requirements.

After the water is placed into the HPLC system it is important not to introduce contaminants. For instance, plastic flasks may contain plasticizers which could leach into the mobile phase and interfere with the chromatography. With glass containers, metal ions can leach from the surface and may be a problem in the analysis of ions. Also, there is the problem of microbiological growth in the mobile phase reservoirs. Such growth can be kept in check by frequent cleaning of the reservoirs and using fresh mobile phase. Also, the addition of a bacteriostat to the mobile phase will help if it doesn't affect the chromatography. Acetonitrile is a common reverse-phase solvent which will inhibit microbiological growth at low levels (~5%). When it is used in an aqueous mobile phase, it may obviate the need for an additional bacteriostat.

MOBILE PHASE PREPARATION

When mixing two solvents for preparing a mobile phase, the most important task is to be consistent. For instance, if you are preparing a 50/50 mixture of methanol/water, the mobile phase can be prepared one of two ways.

1. Volume to volume—by measuring 500 mL of methanol and 500 mL of water and pouring each into a large flask.
2. Weight to weight—by weighing the methanol and water in separate containers using the density of each to calculate the appropriate weights. Then mix each into a large flask.

Either method can be used as long as the operator does it consistently. Do not prepare mobile phases one way once and another way the second time. Also do not weigh one solvent and measure another volumetrically. If one method is used and results in too much variability in the chromatography, try the other method. The important point is to use one method, document it, and be consistent.

The clearly wrong way to prepare a mobile phase is to add 500 mL of methanol to a 1000-mL graduated cylinder or a 1000-mL volumetric flask and add water to bring the total volume to 1000 mL. This is not an appropriate way to prepare a 50% water, 50% methanol mobile phase since it is

TABLE 6-7. Six Water Purification Methods[a]

Distillation	
Removes all types of contaminants. Requires only an initial capital investment. Continuously reuseable.	Some contaminants can be carried over into the condensate. Requires careful maintenance to ensure purity. Consumes large amounts of energy.
Deionization	
Removes dissolved inorganics effectively. Relatively inexpensive to operate. Regenerable.	Does not effectively remove particles, organics, pyrogens, or bacteria. Deionization beds can generate resin particles and culture bacteria.
Carbon Adsorption	
Removes dissolved inorganics effectively. Long life capacity.	Carbon fines may be generated. Removal of chlorine may accelerate downstream bacterial build-up.
Microporous Membrane Filtration	
Absolutely removes all particles and microorganisms greater than the pore size. Requires minimal maintenance.	Not regenerable. Potentially high expendable costs. Will not remove dissolved inorganics, organics, pyrogens, or all colloids.
Ultrafiltration	
Effectively removes most particles, pyrogens, microorganisms, and colloids. Produces highest-quality water for least amount of energy. Regenerable.	Will not remove dissolved inorganics.
Reverse Osmosis	
Effectively removes most particles, pyrogens, microorganisms, colloids, and dissolved inorganics. Requires minimal maintenance.	Limited regeneration. Moderately high expendable costs.

[a] *Source:* Millipore Corp.

known that adding 500 mL of methanol to 500 mL of water will result in a total volume less than 1000 mL. Thus, because of this phenomenon of "volume shrinkage," this approach for preparing mobile phases is undesirable.

SOLID-PHASE EXTRACTION

Successful analysis of some samples require a rapid, reliable, precise method for sample preparation and cleanup to remove potential interfering components so accuracy is maximized. It follows then that the goal of sample preparation is to obtain, from the sample, the components of interest in solution in a suitable solvent, free from interfering constituents of the matrix, at a suitable concentration for detection and measurement. Naturally, this preparation needs to be done with minimum time and expense.

An innovative way to perform a sample cleanup that is fast, accurate, and easy, is the use of solid-phase extraction (SPE) columns. In many ways these devices resemble "miniature HPLC columns" upon which a "preseparation" is done. Using these "minicolumns" for sample clean-up eliminates many of the drawbacks associated with traditional liquid–liquid extraction, such as: (1) the use of large amounts of expensive organic solvent, (2) low recovery due to solvent emulsion formation, and (3) large requirements for labor, time, glassware, and bench space.

The procedure to use an SPE column is outlined in Figure 6-16. Essentially, the process of "extraction" works on the basis of selective elution from these miniature LC columns. To successfully do an SPE procedure, four basic steps are involved: conditioning, loading the sample, eluting the undesirable material, and finally isolating the desired material.

Conditioning

Conditioning (Fig. 6-16*a*, Step 1) addresses two purposes. First, any contamination that could be present in the cartridge owing to the packaging and handling processes is washed off. Removing contaminants is essential since they have the potential to elute with and contaminate the sample. Second, conditioning "wets" the packing material and leaves it in a state that is compatible with the initial mobile phase and sample. The function of conditioning is analogous to equilibrating an HPLC column. The first solvent for conditioning should be of equal or stronger eluting strength than the strongest mobile phase to be used. This ensures that all possible contamination that might elute with the analyte(s) is removed from the cartridge prior to sample introduction. Following the first conditioning solvent, the cartridge should be flushed with a second, miscible solvent of the same or close to the same elution strength of the initial mobile phase. If there is any of the strongest conditioning solvent remaining or if the packing is completely dry, poor

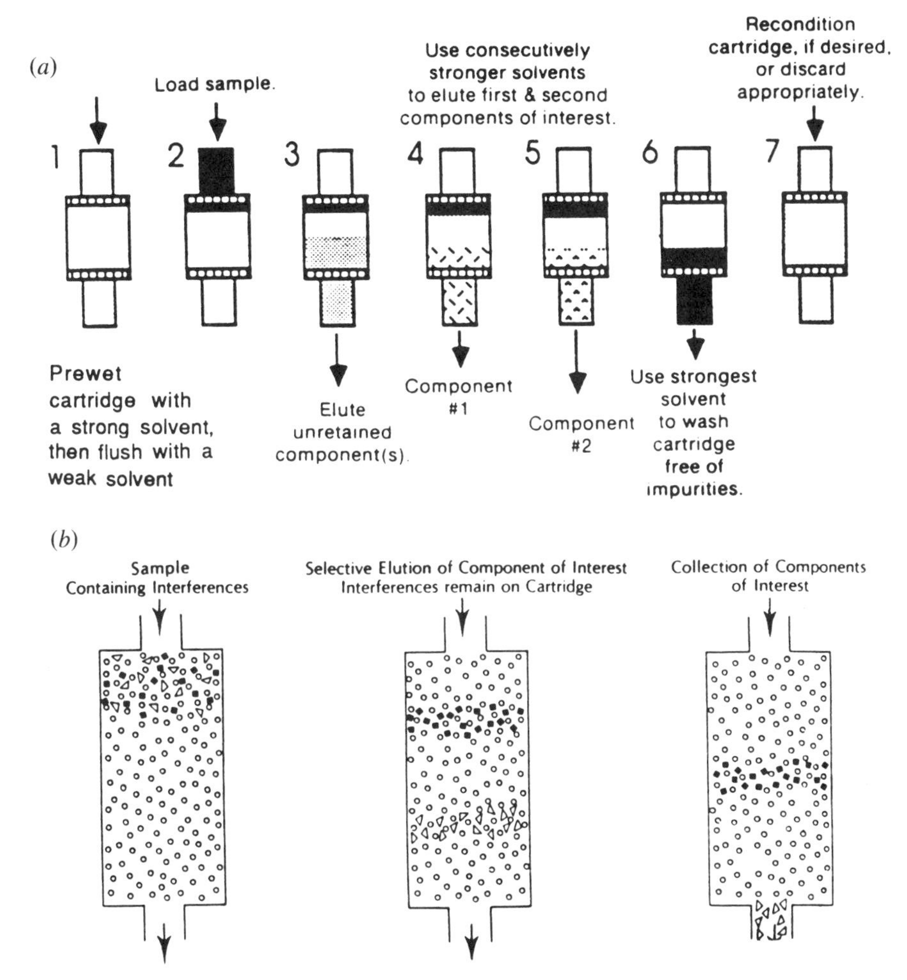

FIGURE 6-16. How a solid-phase extraction cartridge works. (*a*) The general approach to using an SPE cartridge. (*b*) Using an SPE cartridge as a "disposable" device.

retention, poor recoveries, and/or "dirty" isolated analyte(s) may result because of the improper conditioning of the cartridge.

Loading the Sample

The loading step (Fig. 6-16*A*, Step 2) involves placing the sample onto the packing and pumping through the packing, whereupon the analytes of interest are sorbed onto the packing. For the analytes to be retained, the solvent in which they are dissolved (the initial mobile phase) must be a weak solvent for the packing type being used. If the solvent is too strong, the analytes will not be retained and low analyte recoveries will result. For example, for reverse-phase cartridges, the initial mobile phase is usually water or an aqueous buffer. Using the weakest solvent possible will result in the narrowest band width of sample on the packing. A narrow band width is desirable so that only a small amount of stronger mobile phase will be necessary for elution of the analytes from the cartridge.

Eluting the Undesirable Material

After loading the sample, the packing is usually rinsed with a mobile phase to wash off any undesirable sample components (Fig. 6-16*A*, Steps 3 and 4). A mobile phase that is slightly stronger or the same strength as the initial mobile phase (loading solvent) is used as the "rinsing" mobile phase. In this way, unwanted sample components that are early eluters are removed. In some cases, more than one rinse mobile phase may be used. This is especially true for ion-exchange packings or ionic analytes. Solvent pH and ionic strength will determine whether there is a charge on the packing and/or the analyte, thus effecting analyte retention due to charge attraction or repulsion (see Chapter 5). A rinse step also ensures that all of the sample comes in contact with the packing. Small droplets of the loading solvent might adhere to the walls of the tube and a rinse solvent will wash remaining sample onto the cartridge bed. The philosophy of rinsing off early eluters is based upon the assumption that a clean, concentrated sample will result in simpler HPLC chromatograms and extended HPLC column lifetimes.

Isolating the Analytes

After removing the unretained and weakly retained components, it is time to elute the desired components from the cartridge (Fig. 6-16*A*, Step 5) in as narrow a band as possible. Because these cartridges are packed with large particles, they have only approximately 100–300 plates and, as such, are relatively inefficient. This number of plates is similar to that of a TLC plate. Because of the low efficiencies of these cartridges, elution of a component band with a strong mobile phase may require 5–10 column volumes. For the standard irregular shaped, 40-μm particle with 125 Å pores, one bed volume is approximately 60 μL per 100 mg of packing. Therefore, the optimal amount of solvent to elute the analytes from a 500-mg cartridge (typical C_{18} size) will be about 0.3–0.6 mL and for a 1-g cartridge (typical for silica) 0.6–1.2 mL will be required. This, of course, should be experimentally verified and adjusted to the specific situation.

Using too strong a mobile phase will result in the elution of unnecessary sample components that are more strongly retained than the analytes. The use of an optimal mobile phase (solvent) strength would keep these sample components retained on the packing instead of being coeluted with the compound(s) of interest. Using too weak an elution strength will result in too broad an elution volume, which negates the concentrating advantage of SPE cartridges. Just as in HPLC, by blending miscible solvents, the strength of the mobile phase should be tailored for a given application. Once the analyte has been eluted, it may be injected directly into a chromatograph. If it is a very complex problem, or if the eluting solvent is not an appropriate match for the final analysis/separation, the eluent can be concentrated or evaporated to dryness and the extracted analytes dissolved in another solvent for injection or further cleanup.

As was discussed, for HPLC an important solvent consideration is solvent miscibility. Each solvent that is passed through the SPE cartridge should be miscible with the prior solvent. If an immiscible solvent is used, the next solvent may not properly interact with the packing. This will result in poor recoveries and proper sample cleanup will not be the end result. If the use of miscible solvents is not possible, drying of the cartridge between immiscible mobile phases is necessary. Forcing clean nitrogen or clean air through the packing will normally dry the packing bed sufficiently so that the next mobile phase can be added without any miscibility issues.

Reconditioning

In some situations, it is desired to reuse an SPE cartridge. In this case, a final wash with the strongest solvent (Fig. 6-16*A*, Steps 6 and 7) and reconditioning the cartridge back to the starting point (Step 1) is required. Cleaning and reconditioning is more easily accomplished on bonded-phase cartridges which are not as easily contaminated and are more rapidly equilibrated with mobile phase.

Often, however, SPE cartridges are used as shown in Figure 6-16*B*, where the remaining compounds in the sample matrix are left on the packing. In this case, the cartridges are disposed of in a proper manner consistent with the safety policy of the individual's organization.

Application Examples

Figures 6-17 and 6-18 illustrate a simplified sample preparation strategy using an SPE device. Only a fraction of the multicomponent sample shown in Figure 6-17 is of interest. The approach of using a solid-phase cartridge is

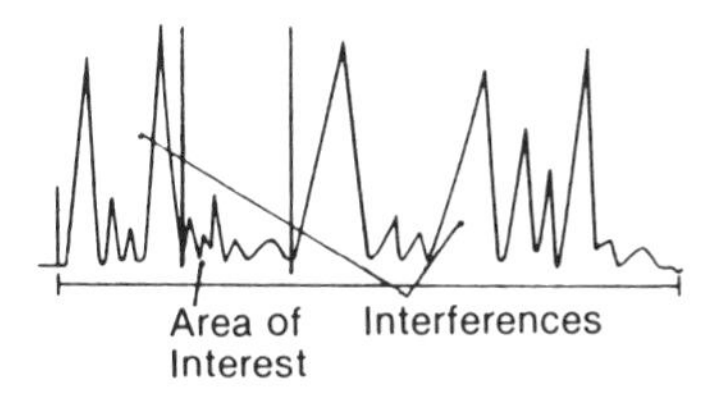

FIGURE 6-17. Chromatogram of a multicomponent complex sample.

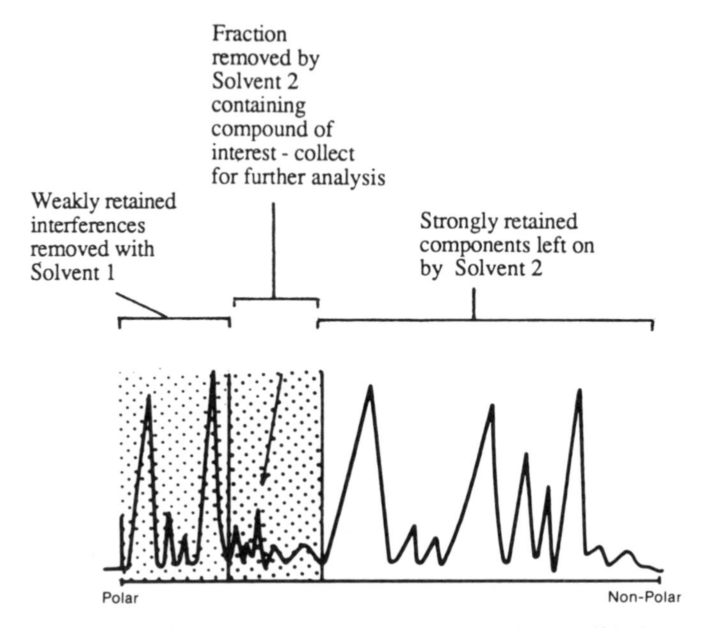

FIGURE 6-18. Simplified sample preparation strategy using a solid-phase cartridge.

shown in Figure 6-18 so that the sample can be fractionated quickly and the unwanted components removed. The component of interest can now be collected and the HPLC analysis optimized as shown in Figure 6-19 without interferences present.

These minicolumns are extremely effective in isolating and selectively eluting a wide range of drugs and metabolites from raw urine. Fractions were taken according to Figure 6-20 and the chromatograms of these individual fractions are shown in Figures 6-21 through 6-23.

Another example is the cleanup of either a multiple vitamin capsule or concentrated vitamin premix for vitamin D analysis. The four-step cleanup scheme is shown in Figure 6-24. Figures 6-25 and 6-26 are the analyses of the final fractions of two samples. If the cleanup was not done, the vitamin D would not have been separable from interferents.

Sample preparation minicolumns can be purchased from a number of manufacturers and suppliers. However, if one wishes, these solid-phase

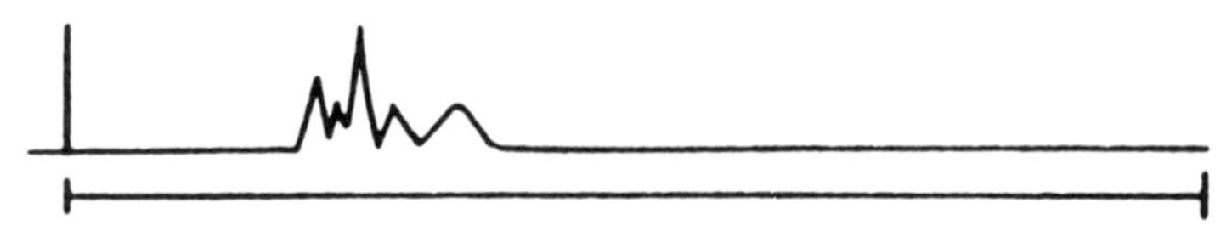

FIGURE 6-19. High performance liquid chromatography analysis of components of interest after solid-phase cleanup.

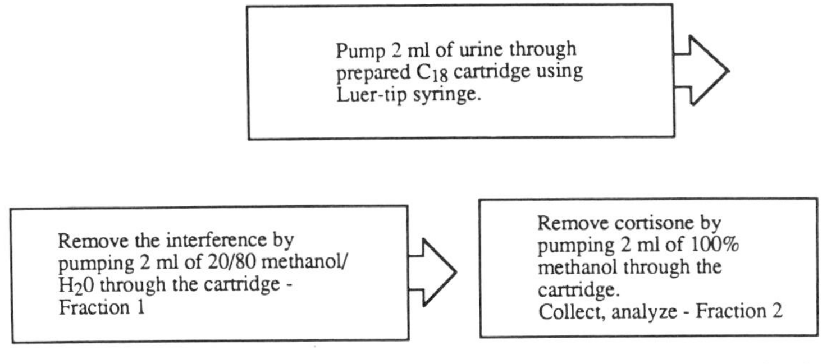

FIGURE 6-20. Cleanup scheme for cortisone from urine using SPE.

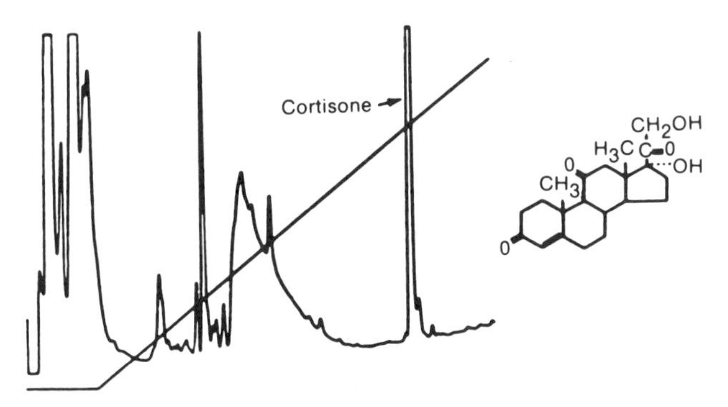

FIGURE 6-21. Early eluting interference. This HPLC chromatogram shows that cortisone analysis could be significantly shortened if the early eluting interferences could be removed.

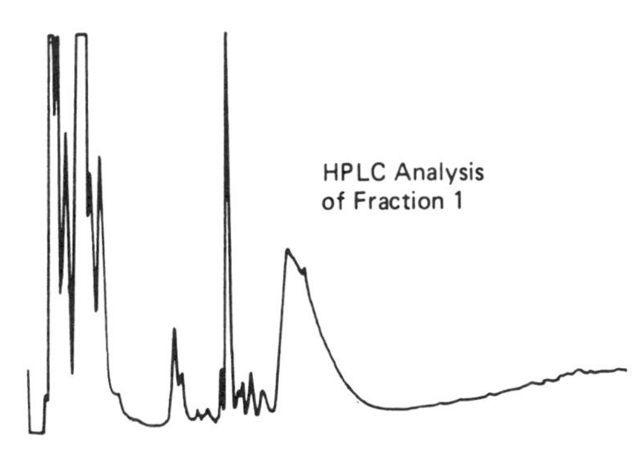

FIGURE 6-22. High performance liquid chromatography, analysis of fraction 1. This chromatogram shows the analysis of the same urine sample after an SPE cleanup. This fraction demonstrates the removal of the interferences while the cortisone remains on the cartridge.

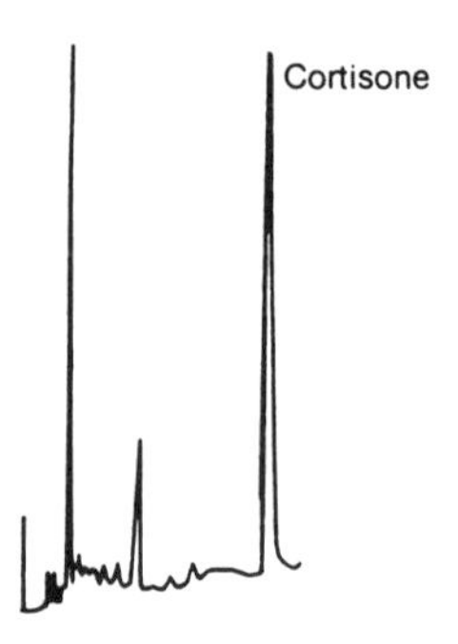

FIGURE 6-23. High performance liquid chromatography analysis of fraction 2. This chromatogram shows that cortisone is eluted quantitatively from the SPE cartridge with 2 mL of 100% methanol. Now conditions can be optimized for a faster HPLC analysis.

extraction columns can be made using a Pasteur pipet, glass wool, and bulk LC packing shown in Figure 6-27. Figure 6-28 shows the insertion of a glass wool plug into the tip of the Pasteur pipet. Figure 6-29 shows the addition of large-particle packing material (37–75 μ) into the pipet. After a known weight (or volume) of packing is added (Fig. 6-30), the minicolumn is wetted with solvent (Fig. 6-31) and is ready for use.

It is too often assumed that because precolumn sample preparation devices, such as solid-phase extraction cartridges, are simple tools, they require relatively little skill or attention to detail for successful use. Cartridges do, however, require attention to detail for successful operation in sample enrichment procedures. Two of the most important parameters to control and understand are flow-rate effects and recovery (or *loadability*) effects.

Proper Operation: Effect of Flow Rate

How often have you observed an analyst taking 10 mL of sample into a syringe and squirting the volume through the solid-phase extraction device? Considering that the flow rate of a "squirt" can be as much as 200 mL/min, it is apparent that the maximum recommended flow rate for applying a sample should be determined.

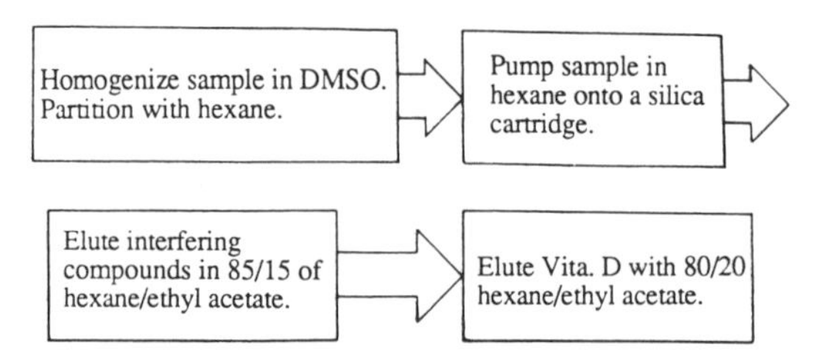

FIGURE 6-24. Cleanup scheme of vitamin D using an SPE silica cartridge.

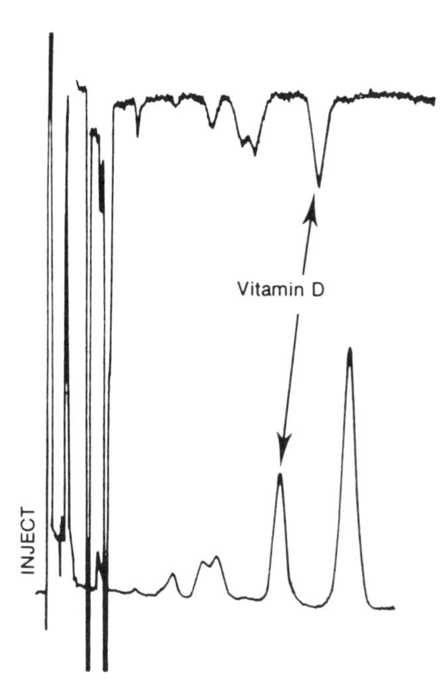

FIGURE 6-25. High performance liquid chromatogram of a multivitamin capsule with a label claim of 400-IU capsule after cleanup with an SPE silica cartridge according to Figure 6-24.

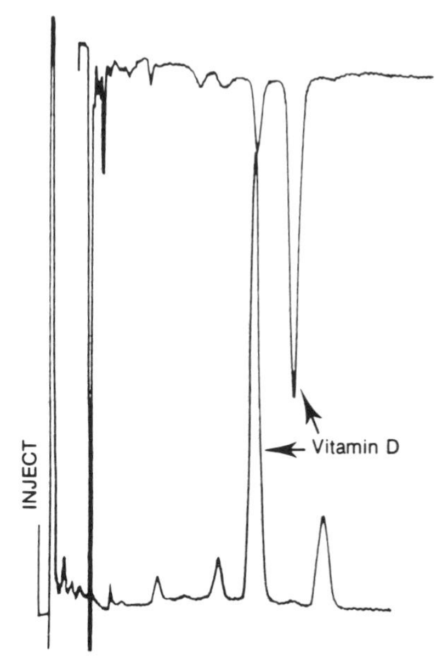

FIGURE 6-26. Chromatogram of a multivitamin oil with a concentration of 40,000 IU/g of vitamin D and 500,000 IU of vitamin A after cleanup according to Figure 6-24.

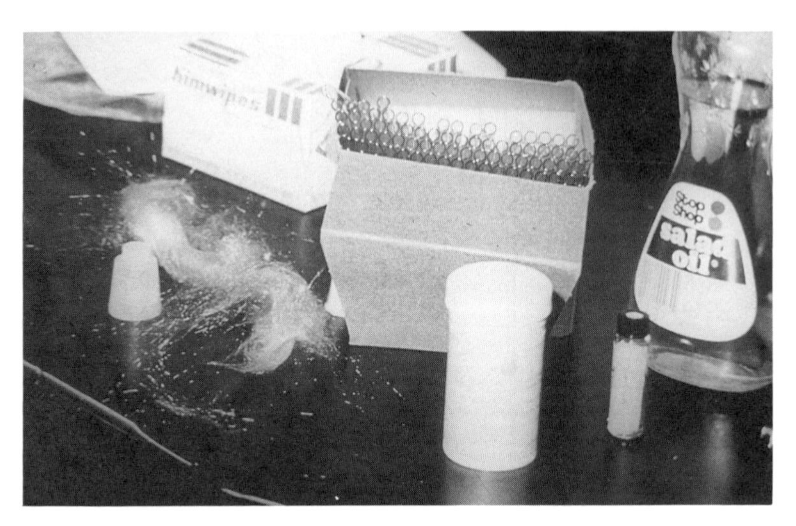

FIGURE 6-27. Materials for making solid-phase extraction columns.

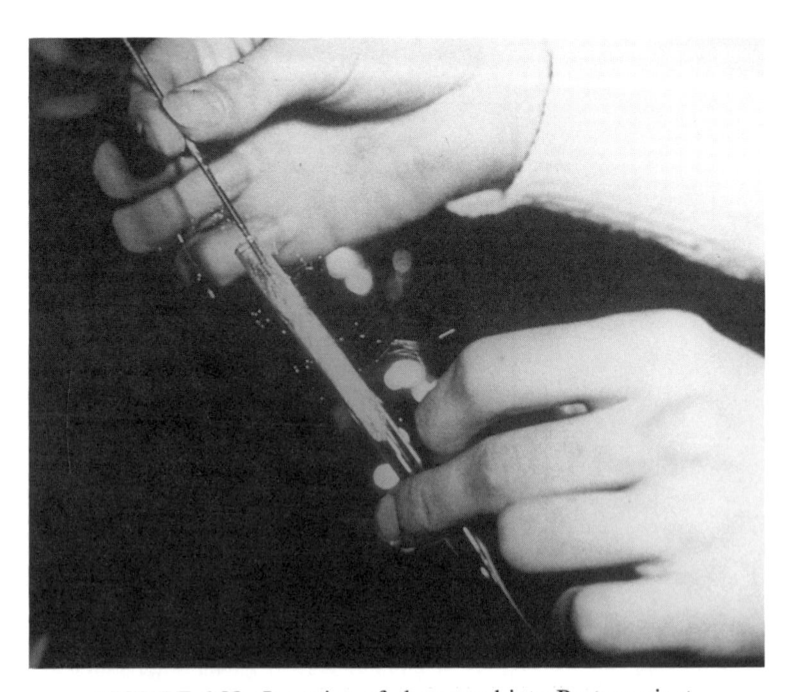

FIGURE 6-28. Insertion of glass wool into Pasteur pipet.

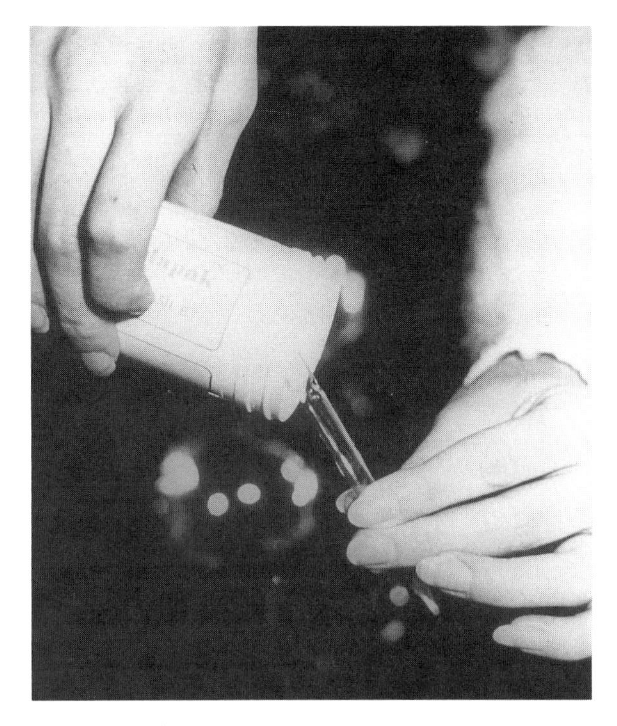

FIGURE 6-29. Addition of packing into Pasteur pipet.

To illustrate potential problems that may occur with real samples, the effect of flow rate was investigated. To demonstrate this effect, riboflavin was ''trace-enriched'' from 10 mL of an aqueous solution. A fresh solid-phase extraction device was properly conditioned and used for each experiment. The surface of the solid phase must be solvated so that the cartridge can be properly conditioned. Directions for solvating the surface can be found in the operating instructions that accompany the cartridge.

The four flow rates chosen for the trace enrichment study were gravity flow (0.3 mL/min), 1 mL/min, 10 mL/min, and 27 mL/min. Note that a ''squirt'' is significantly faster than this last flow rate.

Following trace enrichment, the riboflavin adsorbed by the cartridge was eluted by washing with 2 mL of methanol, and the effluent was then analyzed for riboflavin by LC. The riboflavin adsorbed by the cartridge was eluted by washing with 2 mL of methanol, and the effluent was then analyzed for riboflavin by HPLC. The width of the yellow band of riboflavin during the trace enrichment and the riboflavin recovery at different flow rates are listed in Table 6-8.

These results demonstrated that cartridge performance in trace enrichment is flow-rate dependent and, therefore, is a variable that should be controlled (remember the relationship of plate height versus linear velocity).

TABLE 6-8. Recovery of Riboflavin from a Solid-Phase Extraction Cartridge[a]

Flow Rate (mL/min)	Recovery (%)	Visual Observation of Riboflavin Band on Cartridge
Gravity (.03)	100	Very narrow (≈0.1 cm)
1	100	Narrow (≈0.2 cm)
10	100	Broad (≈0.4 cm)
27	95	Very broad—cartridge totally yellow

[a] Table reprinted from reference 21 with permission.

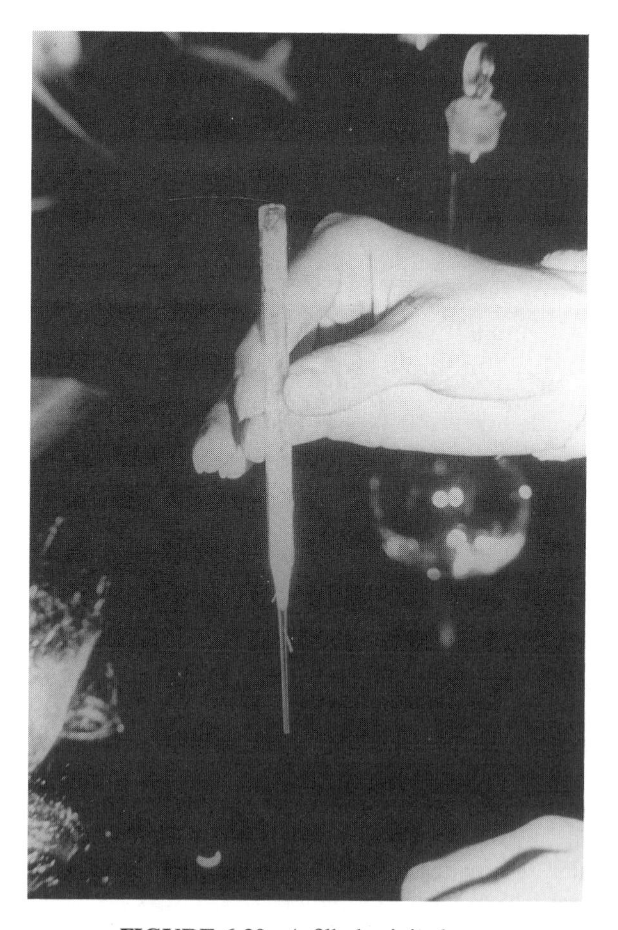

FIGURE 6-30. A filled minicolumn.

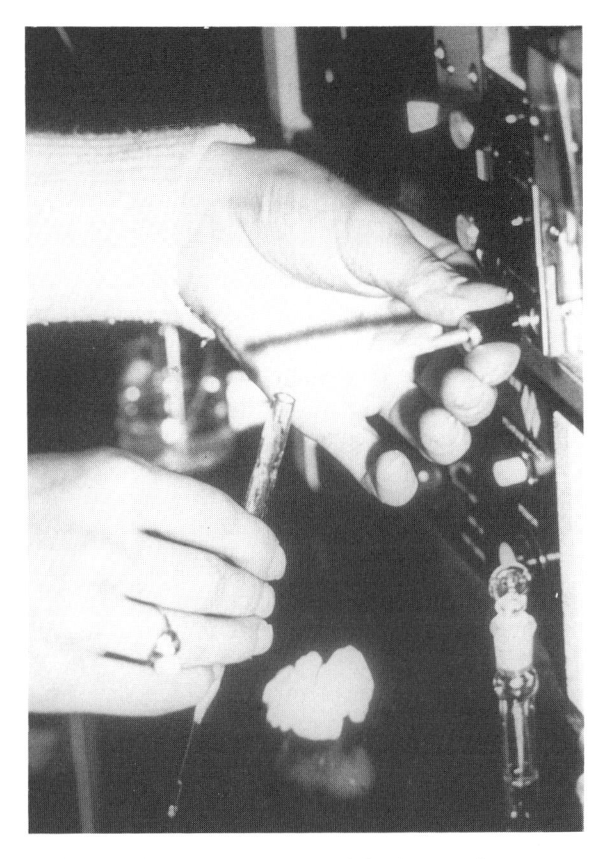

FIGURE 6-31. Wetting the minicolumn prior to use.

The implication for cartridge usage is that results could be nonreproducible or inaccurate if the flow rate is not controlled within a proper range. In this example, flow rates of 1–10 mL/min would be considered proper because they resulted in 100% recovery. Each sample will probably be unique; however, band broadening will always be greater at the higher flow rate (10 mL/min). The recommended flow rate from one manufacturer is up to 30 mL/min, although another manufacturer claims no upper flow-rate limit.

Proper Operation: Loading During Trace Enrichment

The often-asked question, "How much can I load onto a solid-phase extraction cartridge?" is usually answered with, "About 2 mg" or, "About 100 mg" or "You can put on 100 mL easily." Unfortunately, all of these hypothetical answers may be correct because the original question doesn't have an absolute answer for all situations. The answer depends upon the capacity factors of the components on the cartridge and the concentration of the components.

In the following example the assumption was made that two compounds were to be loaded onto a bonded-phase cartridge and, in the eluent that was chosen, peaks A and B had a k' of 10 and 30, respectively (Figure 6-32). These conditions imply that if a continuous stream of liquid containing components A and B is pumped across a cartridge, a frontal chromatogram results, as is illustrated in Figure 6-32.

The results obtained from a trace enrichment would clearly vary depending upon how much volume of sample was passed through the cartridge. Assuming, in this example, that a cartridge with a column volume of 1.0 mL was used, and the sample was applied at the proper flow rate and resulted in a narrow band of sample on the top of the cartridge, the loadability and recovery would be affected for each compound as illustrated in Tables 6-8 and 6-9.

Because $k' = 10$ for compound A, we can load up to 11 mL of sample onto the cartridge and still obtain 100% recovery. If a larger sample—for instance 20 mL—had been loaded, the recovery would have been only 50% because the compounds are eluting from the column after 11 mL. The total amount that could be retained was 1.1×10^{-8} g. For compound B, $k' = 30$, therefore, up to 31 mL of sample can be loaded onto the cartridge with 100% recovery. If we had loaded a volume of 40 mL, the recovery would have been only 77%. The total amount of compound B that could be retained was 3.1×10^{-5} g.

Separation Guidelines for Solid-Phase Extraction

The larger the k' value of the solute, the better the capacity of the cartridge (the ideal k' value is infinity); however, compounds usually have finite k'

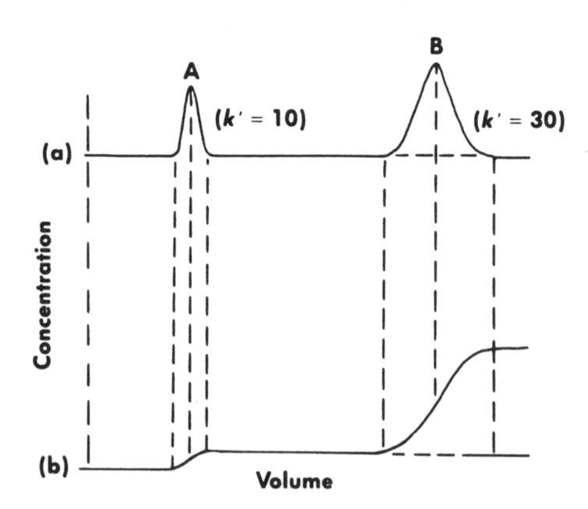

FIGURE 6-32. Comparison of (*a*) analytical (differential) and (*b*) frontal (integral) chromatogram for components A and B.

TABLE 6-9. Loadability and Recovery on a Solid-Phase Extraction Cartridge[a]

Volume Passed Through Cartridge (mL)	Amount Retained (g)	Recovery (%)
	Compound A[b]	
5	5×10^{-9}	100
10	10×10^{-9}	100
11	11×10^{-9}	100
20	11×10^{-9}	55
30	11×10^{-9}	37
50	11×10^{-9}	22
	Compound B[c]	
5	5×10^{-6}	100
20	20×10^{-6}	100
30	30×10^{-6}	100
31	31×10^{-6}	100
40	31×10^{-6}	77
50	31×10^{-6}	62

[a] Table reprinted from reference 21 with permission.
[b] Concentration compound A = 1×10^{-9} g/mL; $k' = 10$.
[c] Concentration compound B = 1×10^{-6} g/mL; $k' = 30$.

values (a value of perhaps 100 or 500). There is, therefore, a finite volume that can be loaded onto a cartridge until breakthrough occurs and, clearly, this is always a consideration that needs to be addressed for the compound of interest for any analysis.

Because solid-phase cartridges are usually used in trace analysis, mass overload is unlikely. It is important, however, to realize that mass overloading can occur on a cartridge, just as it can on an analytical peak (or frontal breakthrough), causing earlier retention times. Therefore, when a cartridge is to be used, if recovery or loading is an important concern, the analyst must verify recovery and breakthrough. The choice of the packing and solvent to produce optimum retention must be made. As in analytical liquid chromatography, retention on a solid-phase extraction cartridge depends upon sample volume and concentration, solvent strength, and packing material as shown in Table 6-10.

PRACTICAL PREPARATIVE LIQUID CHROMATOGRAPHY

As was mentioned in Chapter 1 and demonstrated by the work of R. B. Woodward in Chapter 2, HPLC is an excellent method for rapidly isolating quantities of pure material by diverting the mobile phase to a collection vessel as the peak of interest emerges from the column. When the HPLC is

TABLE 6-10. Separation Guidelines for Solid-Phase Extraction

Chromatographic Mode	Normal Phase (Silica, Florisil®, Alumina, Diol, NH_2, etc.)	Reverse Phase (C_{18}, CN, etc.)	Ion Exchange (NH_2, Anion Exchange, Cation Exchange, etc.)
Packing polarity	High	Low	High
Typical solvent polarity range	Low to medium	High to medium	High
Typical sample loading solvent	Hexane, toluene, CH_2Cl_2	H_2O, buffers	H_2O, buffers
Typical elution solvent	Ethyl acetate, acetone, CH_3CN	H_2O/CH_3OH, H_2O/CH_3CN	Buffers, salt solutions
Sample elution order	Least polar sample components first	Most polar sample components first	Sample components most weakly ionized first
Solvent change required to elute retained compounds	Increase solvent polarity	Decrease solvent polarity	Increase ionic strength or increase pH (AX) or decrease pH (CX)

used to isolate, enrich, or purify one or more components in a given sample, it is referred to as preparative liquid chromatography. This is in contrast to the analytical use of HPLC, which has the objective of accomplishing a quantitative or qualitative measurement. Preparative LC is relatively easy to accomplish by doing a "scale-up" of an analytical separation. If the reader is interested, Chapter 15 (Experiment 8) is a hands-on example for preparative experimentation. When preparative LC is only an occasional requirement in the laboratory and it is not possible to justify a high-performance preparative instrument, it is sufficient to develop an analytical separation and make repeated injections of the largest sample amount that can be introduced without excessive loss of resolution. The standard analytical column (4 mm ID) will suffice when only a few milligrams are required since injection volumes of 100 μL or greater can often be employed on these columns without serious loss of resolution. When larger amounts (>50 mg) are desired, it may be necessary to scale up to a larger column to permit collection of the desired amount of pure material. If preparative LC is to be a frequent requirement in the laboratory, it may be desirable to purchase a special purpose instrument and dedicate an area for this activity. Those who need to do preparative chromatography should consult a dedicated text (22) on this topic for detailed discussions on most areas of preparative LC.

Approaching the Problem

Since the goal of preparative LC is to isolate, enrich, or purify one or more components in a given sample, the approach to solving a preparative should follow a logical sequence of manageable tasks. One such approach is shown in Figure 6-33. Following this flow chart (Fig. 6-33), the first activity is to define the problem answering two fundamental questions:

1. *"How Much Pure Material Is Required?"* A milligram will be the minimum objective of most preparative LC experiments since it is seldom possible to obtain positive identification with less than a milligram. If the compound of interest makes up 10% by weight of the sample and excellent resolution is obtained so that essentially all the compound can be collected on one pass through the column, then 10 mg of total sample must be injected.

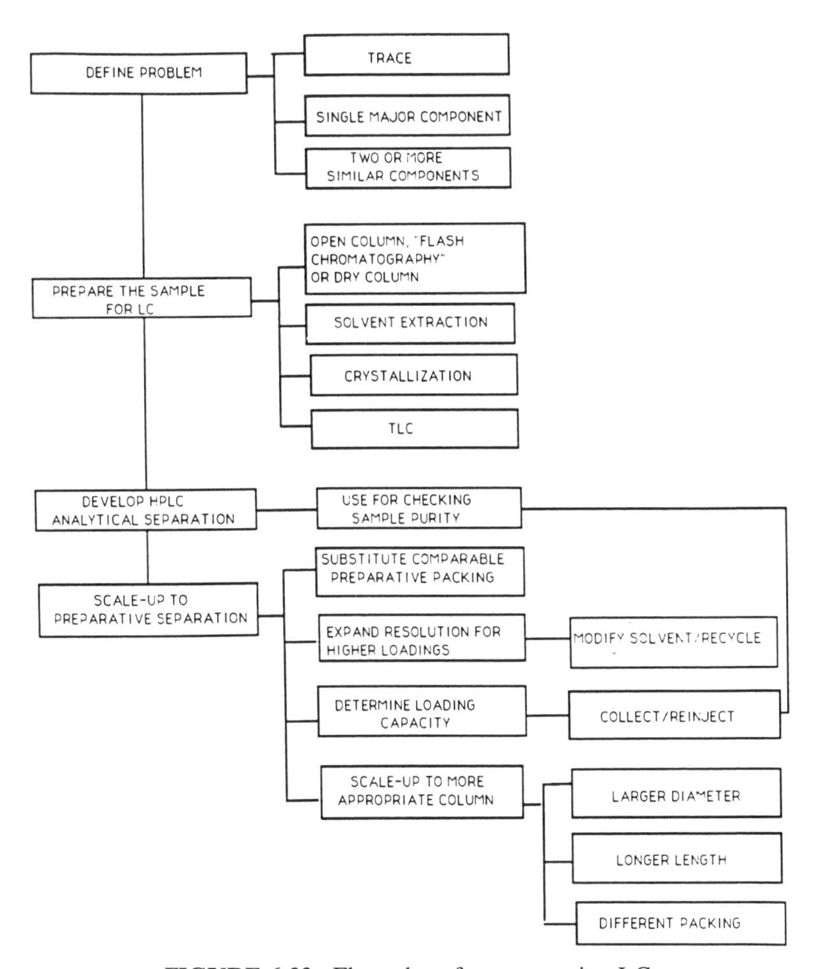

FIGURE 6-33. Flow chart for preparative LC.

Even on typical analytical columns, this can usually be accomplished in one or two injections.

2. *"What Do You Know About the Sample?"* How many other compounds are present and what are their chemical characteristics? Is it necessary to recover all of the components or only part of the material? What will the collected fractions be used for—to test biological activity; do structure elucidation?

Once the answers to these questions are determined, the next activity is to prepare the sample. It may be desirable to preface an LC separation with a simple solvent extraction or class separation on a less expensive packing or open-column system. The high-resolution HPLC column is best suited for separating compounds of a similar chemical nature. If very complex mixtures must be introduced to the high-resolution column, conditions should be adjusted so that most of the unneeded compounds pass through the column unretained. Unneeded material that is more strongly retained than the compound of interest must be regularly cleaned from the column by step elution with a very strong solvent to avoid contamination during subsequent collection experiments. It is valuable to perform a material balance experiment to establish conditions where all compounds are being eluted from the column. Gravimetric and spectrophotometric techniques are commonly used for this purpose.

Preparing the sample (step 2 in Fig. 6-33) is a very important activity in the preparative LC process. Think about the problem and ask how a separation can simplify it. You do not have to solve a separation problem in one step. If the problem involves small amounts or trace materials, the first preparation may only increase the concentration of the desired compound. If the sample contains an unwanted tar or solid material that is very different from the compound of interest, an extraction based on solubility may simplify the process. Filtration is another useful technique. For example, if your compound resides in the bark of a special tree, solvent extraction followed by filtration and crystallization may be required before chromatography is considered. Thin layer, open-column, "flash" (23), or dry-column chromatography (24,25) are all useful sample preparation activities to consider. While TLC often does not have the capacity to prepare large amounts of material; preparative thick-layer plates may be used to isolate sufficient material. Flash chromatography is essentially a speedier version of open column (gravity flow), although flash chromatography is not really very fast compared to HPLC. Dry-column chromatography is a useful technique for scale-up of a TLC separation to approximately 1-g amounts.

In flash chromatography 40–63 μm silica is placed into a glass column fitted at the top with a joint to which a nitrogen tank line can be attached. The sample is layered on the top of the column and then solvent is flushed through the column at a rapid flow rate (as great as 50 mL/min). Flash chromatography is claimed to be fast, cheap, and more efficient than other conventional means of silica gel chromatography. It does not, however, offer the kind of results obtainable with modern HPLC. To separate 1 g of material requires about 1 h to pack the column, load the sample, and elute the components. Loading the sample is the most difficult part. The preferred method of many involves adsorbing the sample onto silica by dissolving the sample in a low boiling solvent. Five times its weight of silica is added and the solvent is removed by rotary evaporation under vacuum. The adsorbed sample is then layered on the top of the column. Separation workup depends upon the number of fractions cut. In a two-component mixture where the difference in R_f values (ΔR_f by TLC) is large (> 0.2), less than five fractions may need to be cut. In a three- or four-component mixture where the ΔR_f is smaller ($\approx$0.1), as many as 10–20 fractions may be cut. Some chromatographers collect fractions in 15-mL test tubes and may have 40 fractions to check. If these fractions are very dilute, sample workup can take an entire afternoon even before the appropriate fractions are combined and the solvent removed (distilled off) under vacuum. In the original paper (23) only 16 plates are reported for efficiency. Therefore, only relatively simple separations (large α's) can easily be achieved on a flash column. For instance, if a ketone is reduced to an alcohol and the reaction mixture contains 95% alcohol and 5% unreacted starting material, these two components have very large α's (different adsorption characteristics) and are easily separated by open-column or flash chromatography.

The dry-column approach which is used for preparative separations is claimed to have equivalent separating capability to TLC and, therefore, appropriate for scaling-up thin layer separations. The adsorbent is packed dry into a flexible, thin-wall nylon tube. The nylon is transparent, flexible, easy to use, can be sterilized, and is resistant to most solvents except for strong mineral and organic acids. The sample is applied to the inlet of the packed tube and the mobile phase moves down the dry column by gravity and capillary action eluting the sample toward the end of the column. After the mobile phase reaches the bottom of the column (usually 15–30 mins) the separation is completed in the same way in which a thin layer plate is eluted. Only one column volume is used. The R_f values obtained on the dry columns are claimed to be similar to R_f values obtained on thin layer plates. To recover the fractions, the column (the nylon tube) is cut into sections and the separated solute in each section is dissolved in an appropriate solvent to extract it from the packing material. The size of the column and the high activity of the packing permits separation of gram quantities of material on a preparative scale.

After the sample is prepared by any or all of the techniques just discussed, the last two steps on the flow chart (Fig. 6-33) involve developing the analytical HPLC separation and scaling it up to the final column upon which the actual preparative separation is to take place. Once the preparative LC separation is accomplished, the purity of the fractions can be determined by analysis using the original HPLC separation. In developing the separation, the general approach is to obtain maximum resolution (α) in the shortest time. Once this is done, the resolution can be expanded by modifying the mobile phase to provide greater spacing between peaks at higher loading levels. Lastly, column loading capacity is determined and, if necessary, a larger column is substituted for higher loading.

Classifying the Preparative Problems

The three separation situations that are likely to be encountered are shown in Figure 6-34. All other preparative problems can often be reduced to one of these three situations by the use of preliminary separation techniques, followed by concentration and reinjection. Each of these situations calls for specific approaches to develop the separation.

Case 1. For isolating a single major component, shown in Figure 6-35, the first step is to develop the analytical separation. Next, in preparation for scale-up, it is necessary to adjust the mobile-phase composition so that resolution is increased, thus providing more room for the peak broadening that will occur with the higher loaded preparative runs. The requirements for peak spacing are not as stringent in this case since the other components are only present in trace quantities. The loading limit is determined by sequentially injecting larger amounts until the broadened peaks merge or overlap. In cases where complete recovery is not important, it is possible to exceed this loading limit and "center-cut" or "heart-cut" the major peak.

Case 2. For two or more main components that elute closely, the same steps as in Case 1 are followed except that a larger α (peak spacing) is required between the main components. When total recovery is not im-

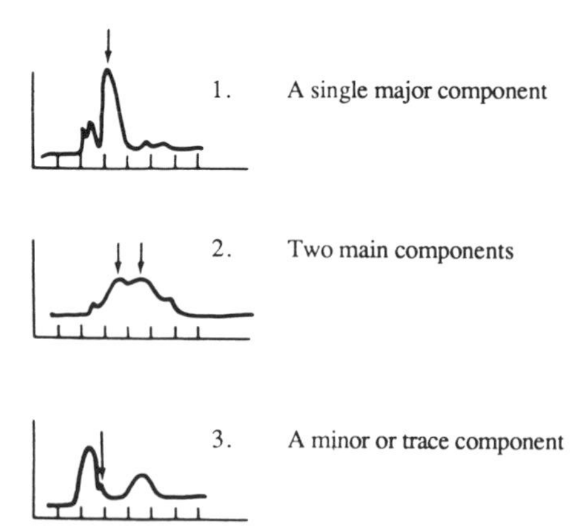

FIGURE 6-34. The three preparative problems. Most separations challenges can be simplified to one or more of these three situations.

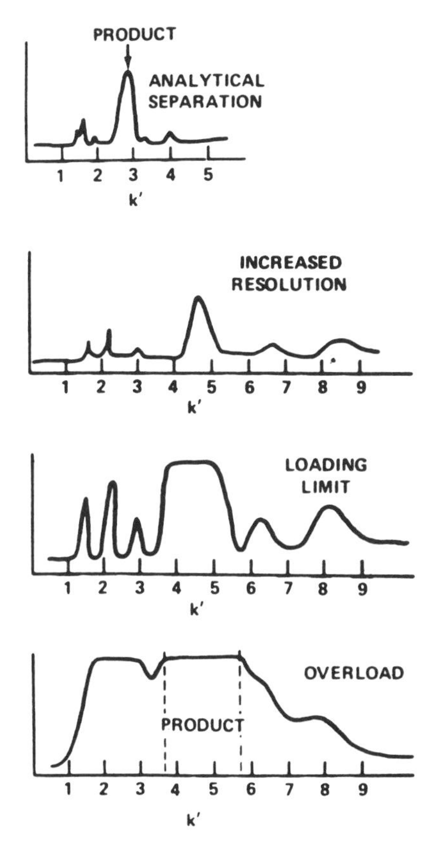

FIGURE 6-35. Developing a preparative separation for a major component (Case 1 of Fig. 6-34).

portant, it is possible to collect the leading edge of the first peak and trailing edge of the second peak. For better recovery, one may "recycle" the overlapped portion, a technique that was described earlier in this chapter.

Case 3. For a minor or trace component, HPLC is used as an "enriching technique." As shown in Figure 6-36, the problem is approached initially as in Case 1. Once the area of retention of the compound is determined for the loading-limited situation, the sample is injected under overload conditions. When this is done, the detector resolution may be lost. Nevertheless, the region of the trace product can be collected on a retention time basis. The injection of sample may be repeated and the collected fractions pooled, concentrated, and reinjected to accomplish a final purification. Now the component of interest is a single major component and should be handled as in Case 1.

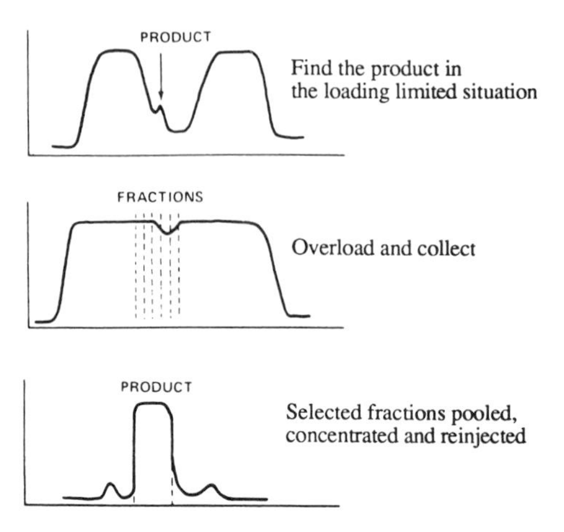

FIGURE 6-36. Preparative scaleup of a minor component (Case 3 of Fig. 6-34).

Peak Shapes in Preparative LC

As columns are overloaded for preparative work, peak shape often deviates from the Gaussian shape typical of analytical work. In preparative work, the peaks can assume a triangular shape because the adsorption isotherm is nonlinear. A typical isotherm is shown in Figure 6-37, where C_M is the concentration of sample in the mobile phase and C_S is the concentration of sample in the stationary phase. At low concentration of sample (C_M) there is a linear adsorption isotherm which results in Gaussian peak shapes. At a point when either the sample adsorption in the stationary phase or the sample solubility in the mobile phase becomes limited, the isotherm becomes nonlinear, assuming either a convex or a concave shape. Convex isotherms are the most common and result in peak tailing. Conversely, concave isotherms cause fronting of the peaks.

At high concentrations of sample, a convex isotherm will result in peak

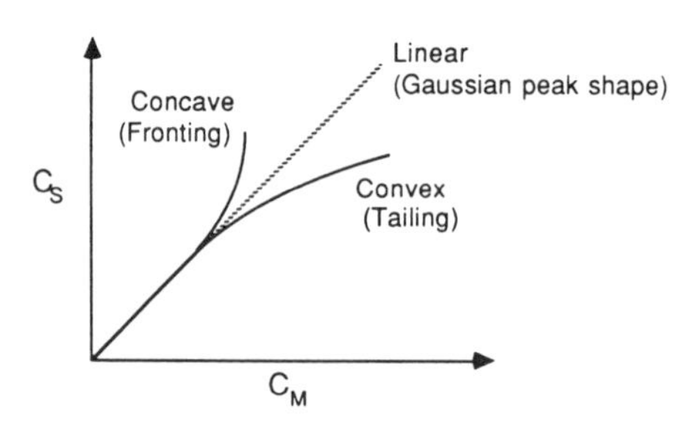

FIGURE 6-37. Plot of Langmuir isotherms (description of corresponding chromatographic peak shape given in parentheses). (Reprinted from reference 22 with permission.)

shapes gradually assuming the form of right triangles. One way to think of this is that the molecules in the band center tend to stay in the mobile phase longer and thus move faster than those at the back side of the peak, which are present at lower concentrations. The result is a steepening of the peak front and broadening of the tail. The vertical leg of the nearly right triangle elutes first at an *apparent* k' much earlier in time than the original k' (obtained under lightly loaded conditions). For a concave isotherm, the shape is reversed with a broad peak front culminating in a steep tail. It is important to note that, in the overloaded situation, peak area, *not peak height,* is proportional to concentration assuming the detector response remains linear.

An example of the scale-up and overload phenomenon is shown in Figure 6-38. As can be seen, the peak shapes become triangular as the loading is increased and the preparative chromatogram does not resemble the original retention, Gaussian peak shape, or resolution, which were obtained in a nonoverloaded situation. Figure 6-38 also exhibits an additional behavior unique to an overloading of two sample components. Notice that for the 40-μL injection, the front peak is sharpened considerably compared to the second peak. This sharpening effect on the first peak results from a "frontal displacement." When a sample component passing through a chromatographic bed is followed closely by a high concentration of a more strongly retained second compound, the molecules of the second will compete with and displace the first component from adsorption sites. When the concentration is high enough, the effect is to sharpen or narrow the band of the first eluting compound and increase the amount of compound that can be collected in pure form from an overlapped mixture. Thus, if possible, the separation should be developed to have the desired component eluting first so that the most amount of pure material can be found in the narrowest volume.

Scaling-up to Larger Columns

As discussed in Chapter 5, preparative LC should be attempted whenever feasible, by liquid–solid (adsorption) chromatography, usually on silica gel. Using the liquid–solid mode provides high capacity and high throughput at moderate cost. With high-capacity silica gel, one can use a mobile phase that has high solubility of the sample (higher loadability) and is easily removed (improves speed of recovery and reduces sample degradation). Silica is relatively inexpensive, available with large surface areas (capacity), and analytical LC or TLC may be used as a method development guide.

Manipulation of the α term is the most effective way to approach a preparative problem. This is true no matter which stationary phase is used. When α is large, the N term need not be large (~500 plates) to accomplish the overloaded preparative separation. Furthermore, this has the advantage of allowing the use of larger particles (30–80 μm). Use of larger particles results in lower pressure drops even at high flow rates and lower instrumentation cost to achieve the needed throughput. Once the analytical separation is

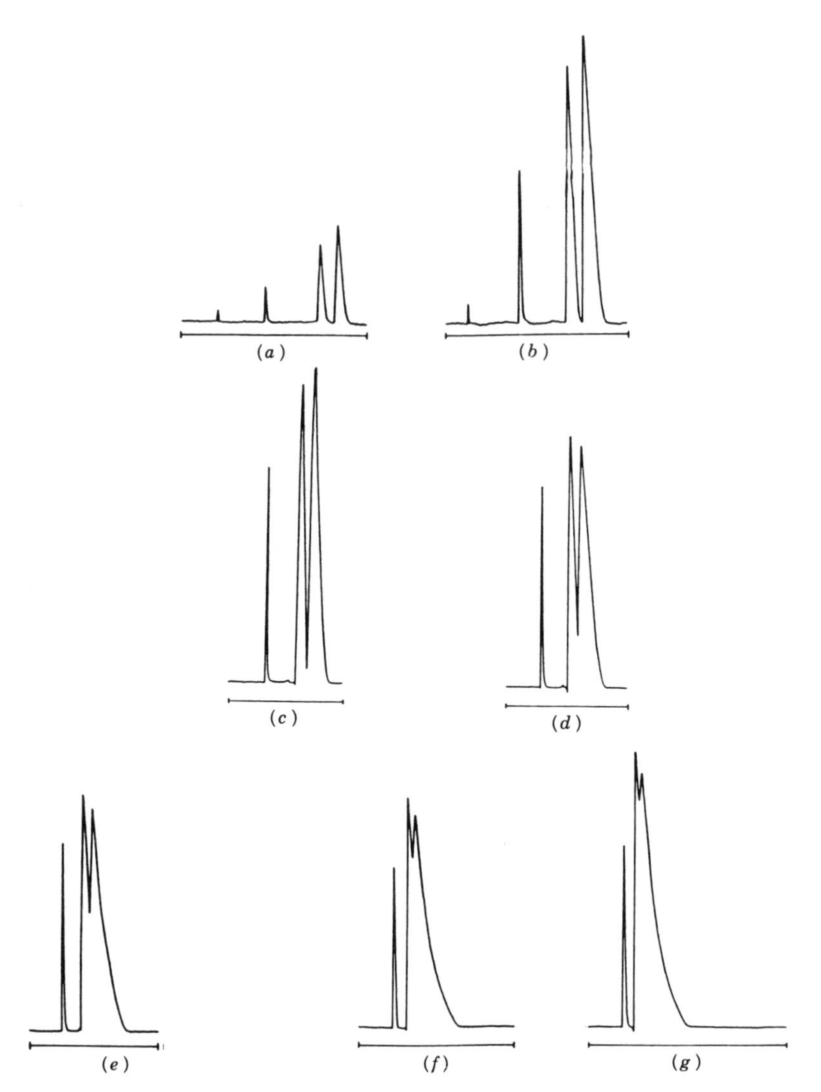

FIGURE 6-38. Scale-up of an analytical separation and RI detector. Flow rate: 3 mL/min. (*a*) 1 μL injected, RI with sensitivity at ×8. (*b*) 5 μL injected, RI with sensitivity at ×8. (*c*) 10 μL injected, RI with sensitivity at ×8. (*d*) 20 μL injected, RI with sensitivity at ×16. (*e*) 40 μL injected, RI with sensitivity at ×32. (*f*) 80 μL injected, RI with sensitivity at ×64. (*g*) 100 μL injected, RI with sensitivity ×64.

developed for the overloaded situation on a particular stationary phase, the scale-up to a larger column follows the guidelines shown in Table 6-11.

Figure 5-43 (Chapter 5) was a good example of following the logical separation development scheme (shown in Fig. 6-33) and of following the guidelines in Table 6-11. Once the fractions were collected (Fig. 5-43C) the purity

TABLE 6-11. Scale-Up Guidelines

Variable	Retention Chromatography				Gel Permeation Chromatography		
Column ID (in.)	1/8	3/8	1	2.4	3/8	1	2.4
Relative internal cross-sectional area	1x	12x	100x	750x	1x	8x	60x
Typical sample load—easy separation just possible by TLC ($\alpha > 1.3$)	40 mg	500 mg	4 g	30 g	2	18	120
Typical sample load—difficult separation ($\alpha < 1.3$)	4 mg	50 mg	400 mL	3 g	1	9	60
Typical injection volume	5–100 μL	0.5–5 mL	4–40 mL	30–300 mL	5–10 mL	45–90 mL	300–600 mL
Typical solvent flow rates (mL/min)	0.3–6	1–10	5–90	60–600	0.5–6	3–30	20–200

of the fractions were determined by analytical HPLC and found to be 99⁺% pure for fractions 2 and 4 respectively. Fractions 1, 3, and 5 were discarded. In this case a preparative instrument was used for the final purification and 10 g of the mixture were separated in 10 minutes. The flow rate was 500 mL/min and this may appear fast; however, the V_0 of the preparative cartridge was 500 mL, so that the flow is equivalent to 3 mL/min on an analytical 3.9 × 30 cm column (i.e., 1 column volume per minute).

Preparative LC separations are generally accomplished (and desired) under isocratic conditions: however, gradient may be needed for the collection of two or more components which have widely different k''s. One technique called "trace enrichment" is sometimes used in preparative work where a very dilute sample is concentrated on the head of the column prior to gradient elution with a stronger mobile phase. Because of the extra cost associated with most gradient-generating devices, it may be desirable to employ step gradient methods. A step gradient can be most conveniently accomplished by placing a low-pressure valve between the pump and reservoirs and switching to the stronger mobile phase at the appropriate time.

Role of the Detectors

Because of the large sample amounts injected in preparative LC, special detector cells may be necessary to avoid "blinding" or saturating the detector. Special flow cells for preparative applications should have large volumes, short path lengths, and large-bore inlet tubing to minimize pressure drops. In addition special stream splitters may be necessary to avoid exceeding the flow-rate/back-pressure capacity of the flow cell. As was discussed in Chapter 3, the response of the RI detector is generally more uniform for compounds of similar structure so that the relative size of peaks is a good indication of the concentration of those components in the original sample. Furthermore, the RI detector is less sensitive than the photometric detector; therefore, it is usually the detector or choice in preparative LC where high sensitivity is not normally a requirement. The RI detector's universal nature allows all components of the sample to be observed and isolated, if desired. Also, because it is less sensitive, the detector is less likely to be blinded (saturated) by very high concentrations of sample. An example of the usefulness of the RI detector is shown in Figure 6-39. At the higher loading, the UV detector is saturated and indicates considerable loss in resolution, due primarily to the high extinction coefficients of the two compounds, whereas the RI shows baseline separation. Collection and analysis of peaks 1 and 2 revealed each to be better than 99% pure. Using only the UV photometric detector, which became saturated with signal, would have given the mistaken impression of column overload and caused the operator to stop short of optimum loading. Also observable in Figure 6-39 is the triangular peak shapes and the change in k' as the sample load is increased.

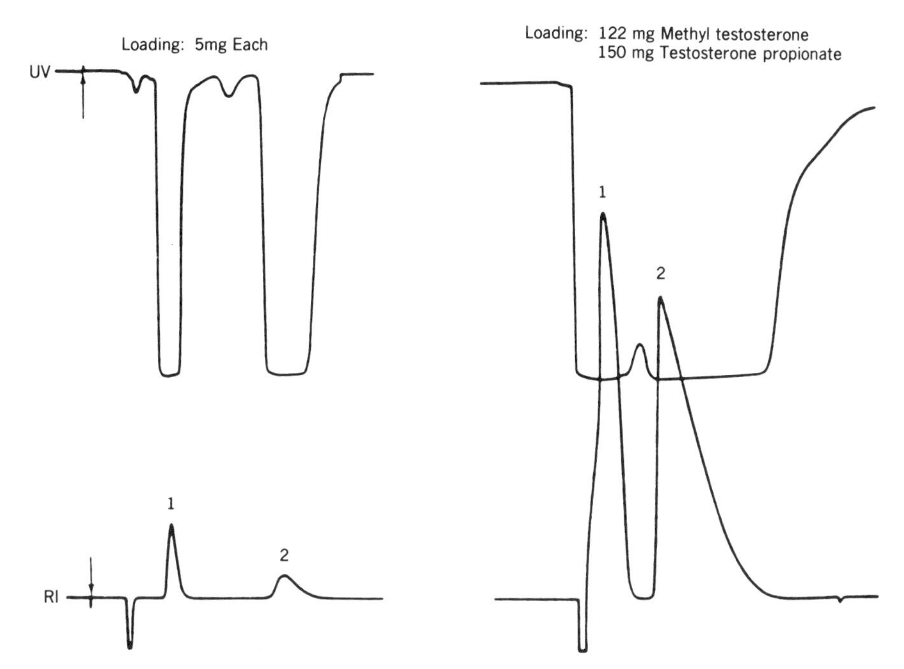

FIGURE 6-39. Comparison of UV and RI in preparative LC. Column: Bondapak C_{18} (37–75 μ) 7.8 mm × 60 cm. Mobile phase: methanol:water (75:25). Sample: (1) methyl testerone and (2) testosterone propionate. At the higher loading, the RI shows baseline separation while the UV photometric detector is saturated owing primarily to the high extinction coefficients of the two compounds. The UV chromatogram suggests considerable loss in resolution. Collection and analysis of peaks 1 and 2 revealed each to be better than 99% pure, which indicates that the RI is responding more appropriately to the actual behavior on the column.

Sometimes, the use of both UV and RI is important. Figure 6-40 shows a separation monitored by both UV and RI detectors and indicates the complementary information that can be obtained by using these detectors in series. The chromatograms show that steroid isomers (peaks 2 and 3) with little UV absorbance elute on either side of a minor impurity (peak 1) that is strongly UV absorbing. Collection of peak 1 in chromatogram A resulted in poor recovery of the total sample injected and low purity. Collection of peaks 2 and 3 in chromatogram B contained high recovery and high purity. Using the RI detector gives a good indication of the compound's concentration in the total sample and, as was stated earlier, this is another example that peak sizes can be misleading when using only a photometric detector. Other detectors such as fluorescence, conductivity, and flame ionization are seldom used in preparative LC because they are very specific and very sensitive.

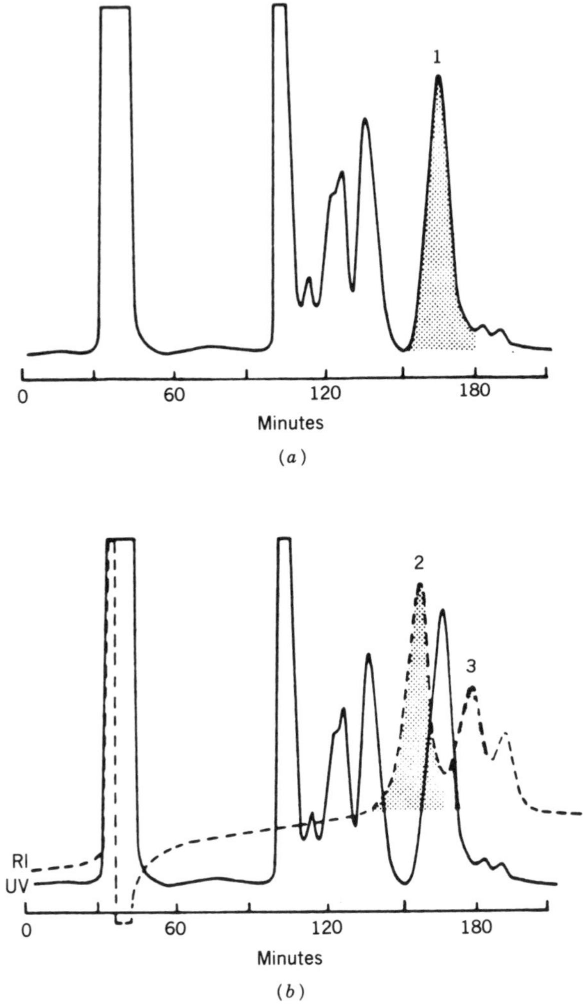

FIGURE 6-40. Preparative steroid separation. (*a*) UV detector only. (*b*) Dual UV/RI detectors as noted on the chromatogram. (Reproduced with permission from reference 26.)

Collection/Recovery

Collection of the purified material is accomplished by diverting flow into an appropriate sized vessel as the peak of interest passes through the detector. Valves are available to facilitate or even automate this operation. Recovery can be accomplished by evaporation (normal phase), by freeze drying (reverse phase), or by extraction from an aqueous phase (reverse phase) into an organic solvent followed by evaporation. Obviously solvent impurities are of concern when recovering the compound of interest since these impurities can be concentrated during the recovery process. In fact, for this reason, a higher-purity solvent may be required in preparative LC than for analytical HPLC.

REFERENCES

1. R. C. George and C. Patel, *Pharm. Tech.*, **1,** 88–99 (1982).
2. J. Kohler and J. J. Kirkland, *J. Chromatogr.*, **385,** 125 (1987).
3. J. Kohler, D. B. Chase, R. D. Farlee, A. J. Vega, and J. J. Kirkland, *J. Chromatogr.*, **352,** 275 (1986).
4. B. A. Bidlingmeyer, J. A. Korpi, and J. Del Rios, *Anal. Chem.*, **54,** 442 (1982).
5. M. J. Walters, *J. Assoc. Off. Anal. Chem.*, **70,** 465 (1987).
6. M. F. Delaney, A. N. Papas, and M. J. Walters, *J. Chromatogr.*, **410,** 31 (1987).
7. K. Karch, I. Sebestian, and I. Halasz, *J. Chromatogr.*, **122,** 3 (1976).
8. J. R. Chretien, B. Walczak, L. Morin-Allory, M. Dreux, and M. Lafosse, *J. Chromatogr.*, **371,** 253 (1986).
9. J. Kohler, D. B. Chase, R. D. Farlee, A. J. Vega, and J. J. Kirkland, *J. Chromatogr.*, **352,** 275 (1986).
10. B. A. Bidlingmeyer and F. V. Warren, *Anal. Chem.*, **56,** 1583A (1984).
11. A. A. Benedetti-Pichler, *Ind. Eng. Chem., Anal. Ed.*, **8,** 373 (1936), Fig. 1.
12. R. P. W. Scott and C. E. Reese, *J. Chromatogr.*, **138,** 283 (1977).
13. R. P. W. Scott and P. M. Kucera, *J. Chromatogr.*, **185,** 27 (1979).
14. T. Wolf, G. T. Fritz, and L. R. Palmer, *J. Chromatogr. Sci.*, **19,** 337 (1981).
15. W. Horwitz, L. R. Kamps, and K. W. Boyer, *J. Assoc. Off. Anal. Chem.*, **63,** 1344 (1980).
16. J. R. Miksic, *Anal. Chem.*, **53,** 2157 (1981).
17. R. F. Adams, R. L. Jones, and P. L. Conway, *J. Chromatogr.*, **336,** 25 (1984).
18. R. E. Lee, Jr., D. Friday, R. Roja, H. James, and J. G. Bause, *J. Liq. Chromatogr.*, **6**(6), 1139 (1983).
19. V. K. Jones and J. G. Tarter, *LC-GC Magazine*, **4,** 1211 (1987).
20. N. B. Godfrey, *Chem. Tech.*, 359 (1972).
21. B. A. Bidlingmeyer, *LC/GC*, **2**(8), 578 (1984).
22. B. A. Bidlingmeyer, Ed., *Preparative Liquid Chromatography*, Elsevier, 1987.
23. W. C. Still, M. Kahn, and A. Mitra, *J. Org. Chem.*, **43,** 2923 (1978).
24. B. Loev and K. M. Snader, *Chem. Ind. (London)*, **1965,** 15.
25. B. Loev and M. M. Goodman, *Chem. Ind. (London)*, **1967,** 2026.
26. J. J. DeStefano and J. J. Kirkland, *Anal. Chem.*, **47,** 1193A and 1103A (1975).

ACKNOWLEDGMENT

Figures 6-4a, 6-5 through 6-10, 6-11b and c, 6-15 a and b, 6-16a, 6-17 through 6-31, 6-34 through 6-36 and 6-39 are courtesy of the Waters Chromatography Division of Millipore.

CHAPTER 7

GRADIENT ELUTION CHROMATOGRAPHY

Hardware contributions to gradient performance
- High-pressure or low-pressure gradient formation
- Mixing devices
- Gradient precision

Developing a gradient separation
- The three basic situations
- Early components bunched, late components bunched
- Early components bunched, late components resolved
- Early components resolved, late components resolved
- Applying a strategy for developing a gradient separation
- Effect of flow rate upon gradient performance
- Applications

Practical considerations
- Viscosity and refractive index effects
- In-line degassing
- Flow programming

Summary
References
Acknowledgment

"Gradient elution" is the term to describe the process by which the composition of the mobile phase is changed during an LC analysis. At the beginning of the analysis, the solvent used is appropriate to elute some of the components but is "weak" in terms of its ability to remove other compounds from the column. As the chromatography proceeds, the "strength" of the mobile phase is changed in order to remove all of the compounds from the column and separate these compounds from one another in a timely fashion. Gradient elution is used when a sample contains dissimiliar components having a wide range of polarities and an isocratic (single) mobile phase does not separate the components in a reasonable time. Therefore, it is necessary to change the mobile phase strength during the run, which has the effect of compressing the run time and bringing together peaks that are too well separated.

An example of such a situation is the separation of oligomers of polystyrene 800 (weight average molecular weight = 800), shown in Figure 7-1. Use of 70% tetrahydrofuran (THF) results in poor resolution of the early eluting peaks. A mobile phase of 60% THF, while resolving the early eluting compounds, does not elute the remaining components from the column in a

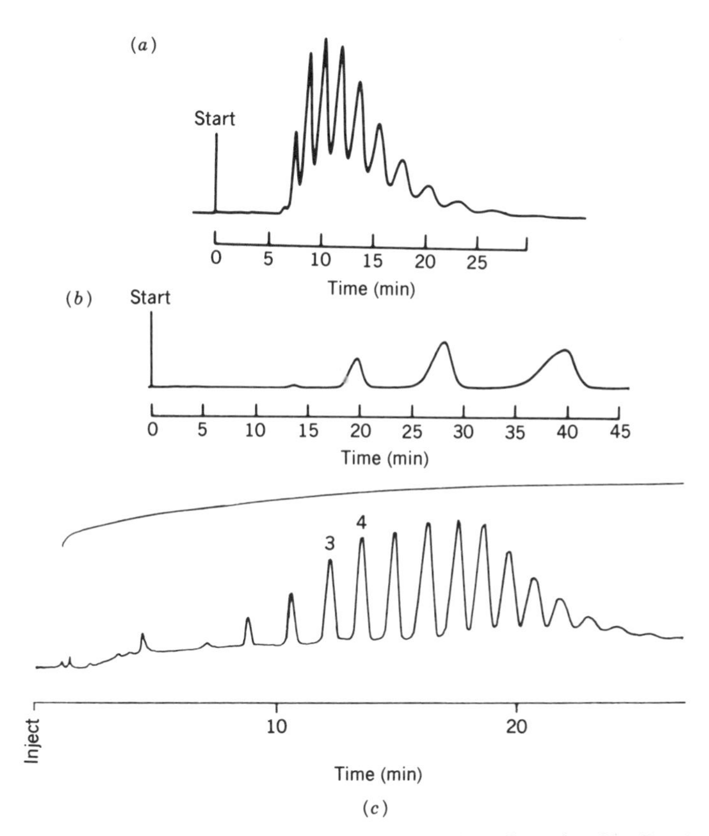

FIGURE 7-1. Isocratic separations of polystyrene 800. Sample: 50 μL of 5% solution in THF. Column: Radial-Pak Resolve C_8 (10 μm), 10 cm ID × 8 mm ID. Flow rate: 1.0 mL/min. Detector: model 40-254 nm. (*a*) 70% THF in water. (*b*) 60% THF in water. (*c*) mobile phase: 50–70% THF in water using gradient curve shown on chromatogram for 20 min. Flow rate: 2.5 mL/min. (Reproduced from reference 1 with permission.)

reasonable length of time. Obtaining adequate resolution in a reasonable length of time is achieved by changing the strength of the mobile phase during the separation, as shown in Figure 7-1*c*. Changing the mobile phase from 50 to 70% THF during a 20-min time interval results in good resolution of all components (1).

It is important to note that there is no standardized nomenclature in gradient elution. One convention is to state the initial mobile phase first, final mobile phase second, and the range of composition over which the variation occurred in terms of the percentage of strong solvent. For the example in Figure 7-1*c*, the gradient was a "water: tetrahydrofuran gradient, 50–70%." (The percent composition is a volume/volume measurement.) Another convention is "50–70% tetrahydrofuran in water." If a composition change is from pure water to acetonitrile it would be written either "water: acetonitrile, 0–100%" or "0–100% acetonitrile in water." The second convention is used in this chapter.

Gradient elution operates on the principle that under the initial mobile phase conditions many of the components have a k' value of essentially infinity in that these components are stopped in a narrow band at the head of the column. As solvent composition is changed, sample components dissolve at a characteristic solvent strength and then migrate down the column leaving the remaining components behind. Changes in the mobile phase composition may be "continuous" with a predetermined set of conditions, or may be done in "steps" of substantial solvent composition changes. Use of step changes in solvent composition is most commonly used in solid-phase extraction and preparative work (refer to Chapter 6), where class separations are desired.

HARDWARE CONTRIBUTIONS TO GRADIENT PERFORMANCE

Gradient chromatography can be accomplished with the use of relatively simple instrumentation. An example of this is shown in Figure 7-2*a* where, using an aqueous acetate–methanol gradient as the mobile phase, more than 50 UV-absorbing constituents have been resolved from homodialysate fluid of uremic patients in a 70-min chromatogram (2). This relatively complex separation was accomplished using one high-pressure solvent delivery system and the gradient-forming device shown in Figure 7-2*b*. The mixing chamber originally contained 50 mL of the aqueous solution and the secondary reservoir contained 19 mL of methanol. At the start of the chromatographic run the stopcock between the two reservoirs was opened. The contents of the mixing chamber were kept uniformly stirred using a Teflon-coated magnetic stirring bar. As the mobile phase was pumped at a flow rate of 1 mL/min from the mixing reservoir into the chromatographic column, the methanol concentration in the mixing chamber increased as shown in the gradient profile in Figure 7-2. The volume of the tubing between

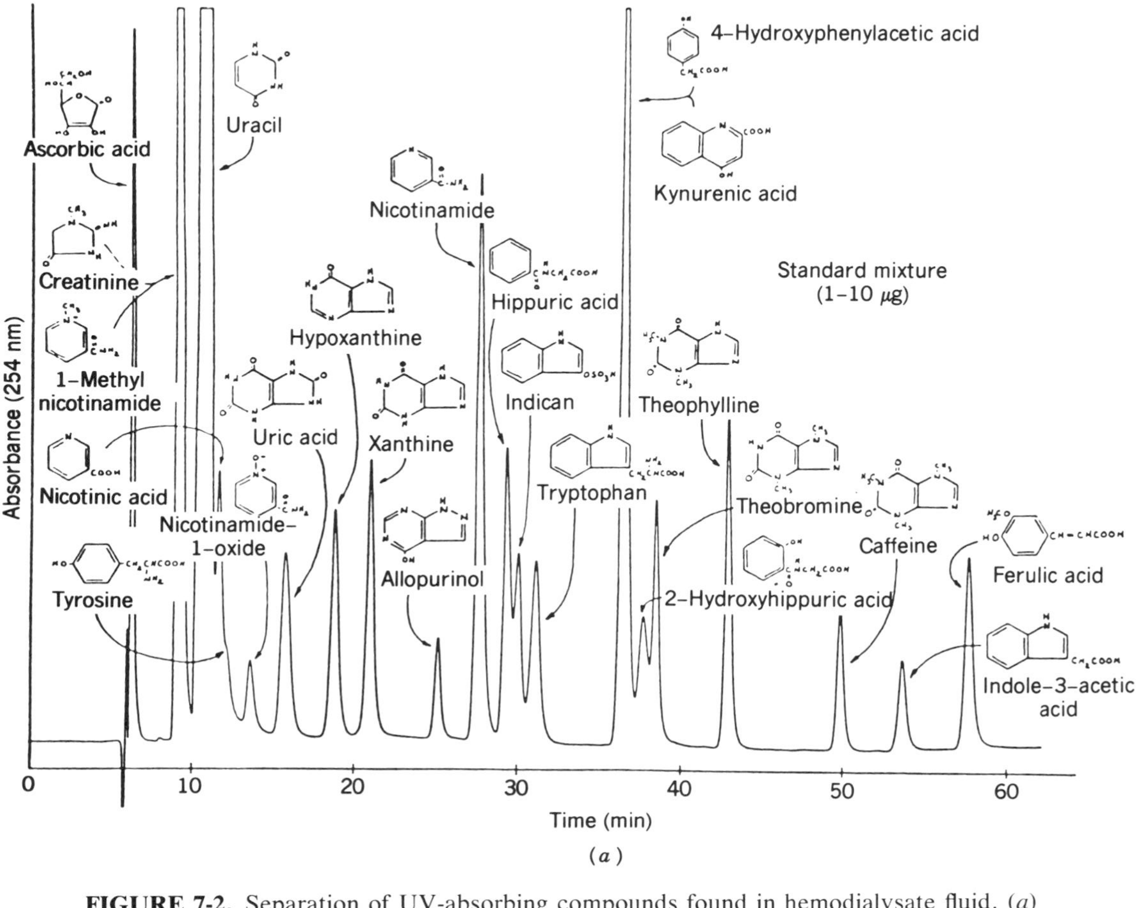

FIGURE 7-2. Separation of UV-absorbing compounds found in hemodialysate fluid. (*a*) Standard compounds. (*b*) Solvent reservoir and mixing chamber. (Reproduced from reference 2 with permission.)

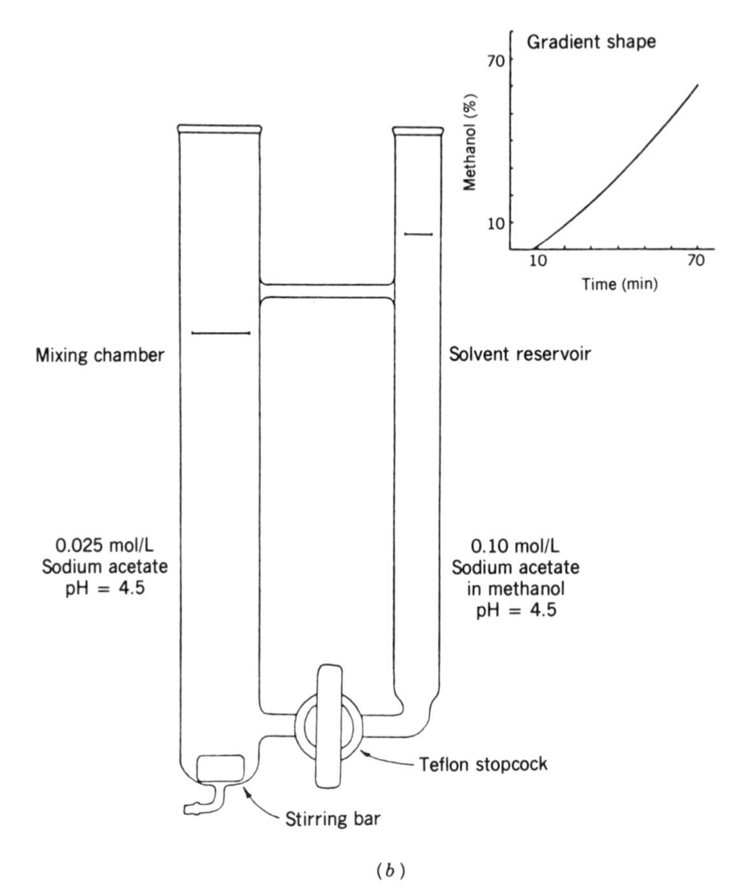

(*b*)

FIGURE 7-2. (*Continued*)

the mixing reservoir and column determined the length of the isocratic period. Modern instrumentation essentially mirrors the process just described and in doing so results in a very convenient, reproducible and precise gradient.

Some individuals do not prefer gradient elution as a quantitative technique because it is more complex than isocratic elution and, hence, more things can *potentially* go wrong. However, with proper control of operating parameters and good instrumentation, it is possible to obtain a separation with excellent quantitative results. Since chromatographic results will only be as precise as the least precise component in the HPLC, it is necessary that all of the components perform at or exceed specifications. This requires that the operator understand the hardware and determine that it is working correctly before attempting a separation. The "ideal" gradient system should be easy to operate, reproducible to provide consistent retention times, versatile to provide capability of generating various concave, convex, and linear gradient shapes, and convenient to provide a rapid turnaround time to initial eluent conditions (equilibration) for fast throughput from analysis to analysis.

High-Pressure or Low-Pressure Gradient Formation

Methods for generating eluent gradients fall into two broad classes shown in Figure 7-3. The first method is referred to as "low-pressure" gradient formation and employs electrically actuated solenoid valves located before a single solvent delivery system (pump). The precision of the gradient depends upon the ability of the solenoid to reproducibly dispense solvents in segments of variable size (volume) depending on the composition desired. If reproducible retention times and stable detector baselines are to be obtained, these "segments of solvent plugs" must be well mixed into a homogeneous mobile phase stream before entering the chromatographic column.

The second method is referred to as "high-pressure" gradient formation and uses a separate solvent delivery device for each solvent, with each being capable of delivering smooth, precise flow rates of as low as a few microliters per minute. Gradients are formed by varying the delivery speeds and simply blending the concurrent solvent streams on the high-pressure side of the pumps. This method is referred to as "high-pressure" gradient formation.

Each method of forming a gradient has supporters who claim that one

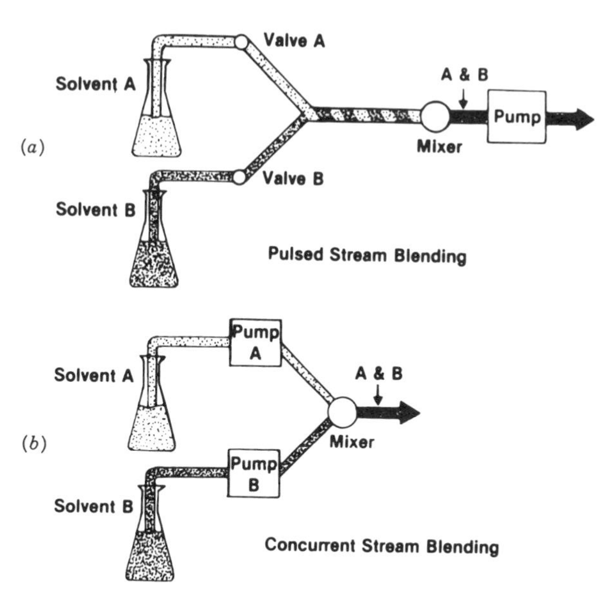

FIGURE 7-3. Methods of gradient formation. (*a*) Single pump with multiple solvents using solenoid valves. (*b*) Multiple pump. (Reproduced from reference 1 with permission.)

method is superior to the other one. However, as in most situations in HPLC, there is no *one* right answer and each gradient method has advantages and disadvantages. In one sense the low-pressure gradient formation is advantageous since it has only one pump (and associated check valves, etc.) rather than two pumps which might foul or "hiccup" during use. However, the low-pressure system has a solenoid to blend the solvents which may fail and the operator has no easy way of knowing if something is wrong. Since low-pressure gradient systems have only one pump, they are often slightly less expensive, but not always so. These systems usually require rigorous degassing of solvents and have a larger mixing volume (than high-pressure systems), so there is a longer lag time (delay volume) in starting the gradient and, hence, in returning to initial conditions before another sample is analyzed. Low-pressure systems can often accommodate three and four solvents for gradient usage. This multiple solvent blending might also be useful for optimization activities in both isocratic and gradient situations.

The high-pressure gradient systems require two solvent delivery systems, which involves twice as many "potential" problems (leaks, etc.). However, there is an advantage when not using a gradient system and that is that the operator has two isocratic systems. In the two-pump gradient system, the solvents should be degassed prior to use but they do not have to be as rigorously or continually degassed as in the case of low-pressure systems. Mixing volumes for high-pressure systems are generally small so these systems give a rapid eluent change and turnaround time to initial conditions. The key differences are summarized in Table 7-1. Individuals deciding on purchasing a gradient system should make their own list of comparisons and assign a personal value to each attribute. The following sections describe other attributes that should be evaluated. Remember that if it is important to you, you should test a system for accuracy, precision, and convenience in your laboratory before the final decision to purchase is made. Both types of gradient systems exist and both work.

TABLE 7-1. Comparison of Gradient Systems

Low-Pressure Mixing	High-Pressure Mixing
One pump with solenoid valve (2, 3, or 4 solvents)	Two pumps
Degassing essential	Degassing desired but not essential
Large volume for mixing	Low volume for mixing
Slow response of solvent change (large system volume)	Fast response of solvent change (small system volume)
Slow return to initial conditions	Fast returns to initial conditions
Lower cost (perhaps)	Higher cost (perhaps)
One isocratic or one gradient system	Multiple isocratic or single gradient system

Mixing Devices

Each gradient formation approach requires mixing of the solvents to form a uniform eluent composition. The three common types of mixing devices are: (1) stirred chamber (often a 2-mL stainless-steel vessel containing a magnetic stirrer); (2) static chamber (an empty tube with about a 2:1 ratio of length to diameter); (3) capillary laminar flow blending (narrow bore tubing, flattened and folded to provide a restricted pathway, with a 1- or 2-mL volume). The "ideal" mixer must provide both homogeneous solvent composition and faithful reproduction of the required gradient profile. While there may be no perfect mixing device, Figure 7-4 shows one comparison of the three mixing devices on a high-pressure gradient system. In this comparison, the capillary laminar flow blending produces rapid response and accurate gradient profile reproduction with minimum delay time. This does not imply that this is the only comparison which should be made, nor does it imply that all systems will give this exact result.

Figure 7-4 emphasizes that if it is important to know the hardware performance of the gradient system, the instrument should be "checked out" in a manner similar to that shown in the figure. For some systems where high sparging rates of solvents will be used, the compound methyl paraben should be used at a low concentration (approximately 0.1% or less) instead of ace-

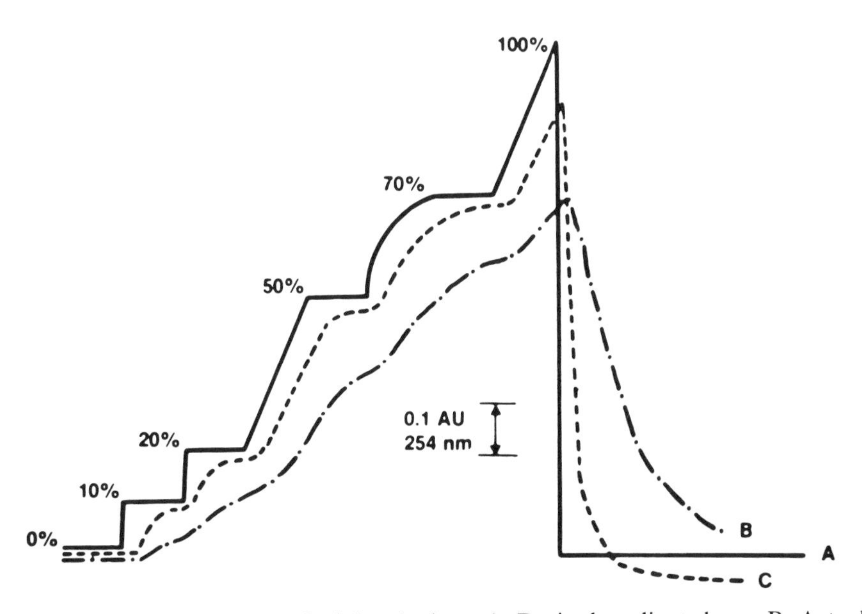

FIGURE 7-4. Comparison of mixing devices. A. Desired gradient shape. B. Actual shape using a stirred mixing chamber. C. Actual shape using capillary laminar flow blending. Solvent A: methanol. Solvent B: 0.5% acetone/methanol. Flow rate: 2.0 mL/min. Run time: 2 min per segment. Solvent delivery: two dual-piston pumps. (Reproduced from reference 1 with permission.)

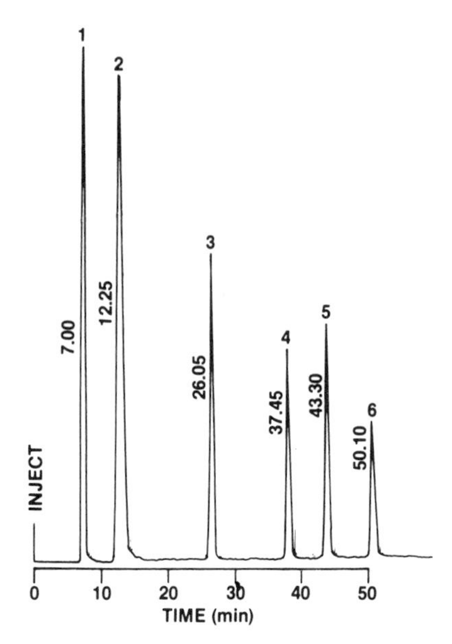

FIGURE 7-5. A 0 to 10% gradient separation of polar compounds. Components (1.0 μg each): (1) uracil, (2) hypoxanthine, (3) 3-methyl xanthine, (4) theobromine, (5) theophylline, and (6) β-hydroxyethyl theophylline. Solvent A: 0.01 M sodium acetate/water. Solvent B: acetonitrile. Flow rate: 2.0 mL/min. Gradient: 0–10% solvent B using a linear shape (top line). Run time: 50 min. Injection volume: 15 μL. Column: Radial-Pak Resolve C_{18} (10 μm) 8 mm ID × 10 cm. Detector: UV at 254 nm, 0.1 AUFS. (Reproduced from reference 1 with permission.)

tone, since acetone can potentially evaporate. By checking out the instrument with the range of gradient concentrations and shapes to be used in the final application, any hardware problems can be identified, and changes made before the instrument is used for analyses.

Gradient Precision

After optimizing a gradient separation qualitatively, one must next consider quantitative aspects of the analysis. After the gradient method is established, the analysis must be run with a known sample mixture to develop the percent standard deviation of the retention times and peak height (and/or area) measurements. One example used a sample mixture that required a gradient from 0 to 10% of acetonitrile, shown in Figure 7-5, and the gradient profile is detailed in Table 7-2 (1). With a flow rate of 1.0 mL/min, this is clearly a demanding test of a gradient system, since the flow rate of one solvent delivery module must go from 0 to 100 μL/min.

TABLE 7-2. Gradient Profile and Shape

Time (min)[a]	%B	Curve Shape
Initial	0	—
50.0	10	Linear
60.0	10	—
60.1	0	Step

[a] Equilibrium time at initial conditions before running the gradient: 10 min.

When the various instrumental and chromatographic factors are set for optimum performance, extremely good quantitation is possible, as shown in Table 7-3. The data represent nine replicate runs of the separation described in Figure 7-5. The values reported are average retention time and peak area along with their corresponding percent relative standard deviation (% RSD). These results are excellent by any standard of performance.

In addition to gradient profile and flow rate, other factors have been reported to possibly influence the analytical precision. Some claim that the initial increments of acetonitrile may be adsorbed onto the column surface until the incoming solvent becomes strong enough to cause desorption. This initial demixing can lead to variable results unless the equilibration of initial conditions is carefully controlled. This situation is believed to be the most serious on highly hydrophobic columns. If you use a gradient, an investigation of the role of time of initial conditions on retention reproducibility needs to be done before the procedure is implemented in routine operation.

DEVELOPING A GRADIENT SEPARATION

A gradient does not improve the overall resolution; it actually destroys or reduces resolution where there is "too much" resolution. A gradient is used to optimize resolution per unit time. A gradient speeds up the separation by

TABLE 7-3. Retention Time and Area Precision (N = 9) for the 0–10% Gradient

Compound	Retention Time (min)	% RSD of Retention Time	Area	% RSD of Area
Uracil	6.98	0.34	21714	0.47
Hypoxanthine	12.24	0.26	30197	0.45
3-Methylxanthine	26.04	0.11	15359	0.48
Theobromine	37.33	0.17	14479	0.39
Theophylline	43.23	0.08	9577	0.32
β-Hydroxyethyl theophylline	49.98	0.15	6331	0.40

reducing excess resolution to a shorter time frame. Therefore, when developing a gradient analysis method, one must have two goals: (1) to obtain adequate resolution of the sample components in minimum time and (2) to ensure high precision and accuracy. The first goal is achieved by careful strategy and suitable equipment, the second goal can be attained by proper control of the hardware contributions. It is assumed that the reader has chosen the most appropriate hardware and that it is operating to specification. The remaining goal, to attain good resolution, involves following a strategy. One strategy follows five fundamental steps: (1) determination of initial and final solvent composition; (2) adjustment of run time; (3) determination of gradient shape (linear, concave, or convex) (refer to Fig. 7-6); (4) adjustment of flow rate to improve resolution; and (5) return of the column to initial conditions. This strategy (or a similar list) should be followed when developing gradient separations.

Developing a gradient separation involves varying the percent change in mobile phase composition per unit volume of mobile phase delivered. This rate of change (ROC) in solvent composition per unit volume is graphically represented in Figure 7-7. As the slope of the line is decreased, the resolution of the separation will improve until the separation is limited by the efficiency of the column. The ROC value can be calculated by:

$$\text{ROC} = \frac{\Delta\ \text{Composition}}{\text{Gradient time}} \times \frac{1}{\text{Flow rate}} = \frac{\Delta\ \text{Composition}}{\text{mL}} \qquad (7\text{-}1)$$

From equation 7-1 it follows that ROC can be decreased by lengthening the run time or by increasing the flow rate. And, of course, ROC can be changed

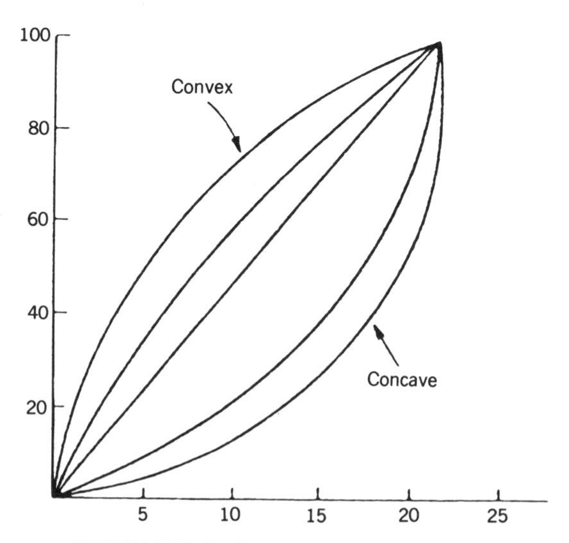

FIGURE 7-6. Common gradient shapes.

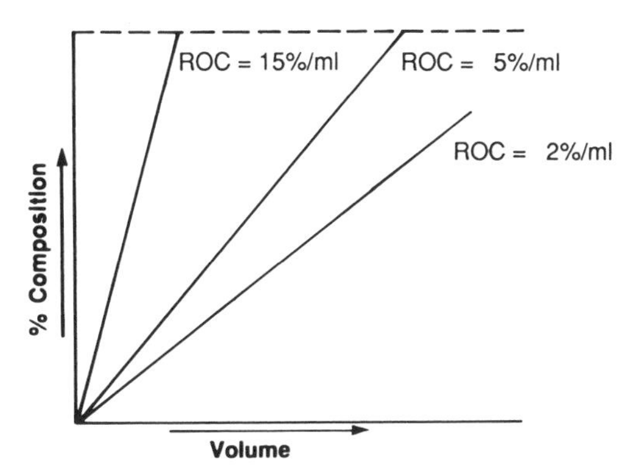

FIGURE 7-7. Rate of gradient change defined by percent composition change per unit volume. (Reproduced from reference 1 with permission.)

by changing the gradient shape. If the separation is "k' limited," and there is inadequate resolution, the ROC value needs to be decreased during the region of inadequate resolution. If slowing the rate of change does not improve the resolution, the separation is not k' limited but must be "plate limited" and improvements in resolution will only come with more plates, for example, a longer or higher-efficiency column. In this case slowing the flow rate to achieve more plates (see Chapter 6) could improve the resolution at the expense of increased analysis time. However, since gradient separations are used for samples with a wide polarity of components, most situations are k' limited. It is important to remember that the gradient separation is the result of the rate of change of mobile phase strength per unit time. All gradient development relies on changing the ROC value in the appropriate fashion to optimize resolution per unit time. However, most chromatographers do not calculate the ROC values per se.

The Three Basic Situations

Most separation problems that require gradient use can be considered to be made up of one or more of the three basic situations shown in Figure 7-8. After the initial gradient choice you should have accomplished your objective and have all components well resolved. This situation does not occur very often on the first try. Most often after the first attempt you will observe either (1) inadequate resolution overall, (2) inadequate resolution at the beginning of the gradient run, or (3) inadequate resolution at the end of the run. As in other areas of chromatography, there is more than one way to attain the desired result; therefore, it is important to understand the choices that can be made and the effect of these choices upon resolution. Developing a gradient separation is a challenging task and it is important to be logical when changing the variables. For the beginner, changing one variable

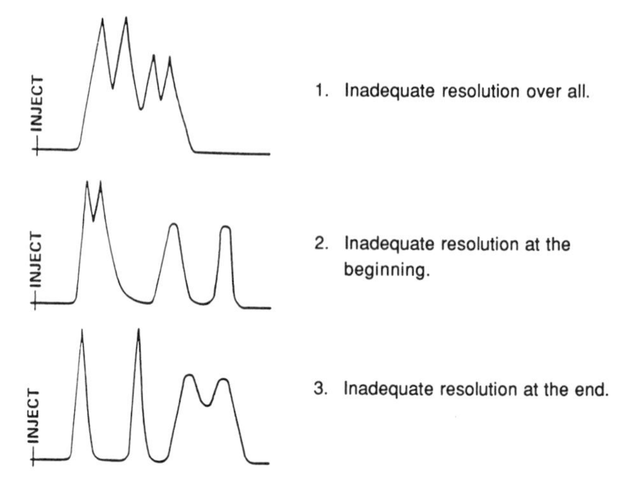

FIGURE 7-8. The logical method of selecting the proper gradient program after one run. Every sample must first be run on a linear gradient. There are three general conditions of inadequate resolution in gradient elution.

at a time can reinforce the "cause-and-effect" behavior in gradient work. When you have become familiar with this behavior and develop an understanding of gradient elution, you can proceed to change two variables at a time.

Early Components Bunched, Late Components Bunched

If the first gradient choice gives inadequate resolution overall as in Case 1 (Fig. 7-8), there are three alternatives to pursue, as shown in Figure 7-9. Reducing the final strength of the solvent to slow down the rate of mobile-phase change should improve the separation. If this does not give the desired separation, the time over which the initial gradient was run can be increased; this also decreases the rate of mobile phase change and should improve the separation. Or, by running the same gradient at a higher flow rate, the result should be a similar separation to the previous two suggested approaches only in a shorter time. This last result is due to larger volume of weak solvent used per unit time, which decreases the rate of solvent composition change per volume of solvent delivered. It is not a result of increasing plates. Sometimes this problem may need to be segmented into two separate problems. In this situation, the problem consists of one or more of the other two examples and may require an S-shaped curve, that is, more concave in the early portion and more convex in the later portion. Optimizing the rates of change in the gradient in the front portion and in the back portion of the chromatogram are the subjects of the next two sections.

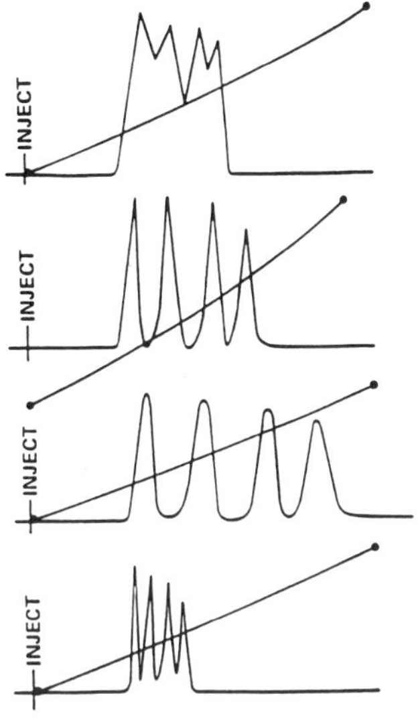

1. Decrease the initial and final conditions. Or if initial is 100%A only decrease final condition.

2. Increase the time.

3. Double the flow rate use original time.

FIGURE 7-9. Approaches to correct inadequate resolution throughout the chromatogram. Refer to text for discussion.

Early Components Bunched, Late Components Resolved

If there is inadequate resolution at the beginning, as in Case 2 (Figure 7-8), the first action, if it is available, is to broaden the range of solvent composition in the programmed run, especially if initial solvent composition is in the 15–30% range. If an initial solvent composition of 15–30% resulted in peaks bunched together, it indicates the solvent composition is initially too strong (components are being eluted together). This problem may often be eliminated by expanding the starting solvent composition, which results in those early components displaying greater differential mobility. The second action to take is to change to a nonlinear, concave gradient as shown in Figure 7-10, which slows the rate of change of solvent composition for the early peaks and increases the rate of change for the later eluting peaks. Determining the correct shape (or rate of change) is a matter of trial and error, as shown in Figure 7-10. However, by proceeding in a logical sequence of steps, this activity does not have to be a time-consuming task. The third action, if needed, is to increase the time span of the solvent program; but this will also increase the analysis time.

Early Components Resolved, Late Components Bunched

This case differs from the previous one in that broadening the range of mobile phase composition will not help; that is, if 70% of the final solvent

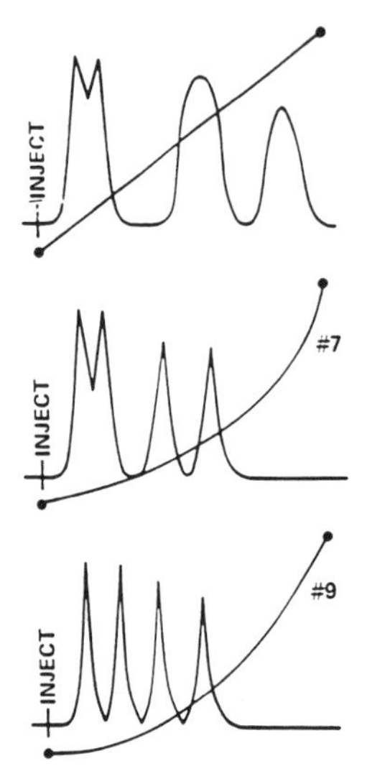

1. Delay the rate at the start and increase the rate toward the end.
2. As above and vary degree.
3. Decrease initial conditions.

FIGURE 7-10. Approaches to correct inadequate resolution at the beginning of the chromatogram. Refer to text for discussion.

gives components bunched together, going to 100% of the final solvent cannot help and could make things worse. If a linear gradient has adequate resolution in the beginning of the chromatogram and results in inadequate resolution at the end of the chromatogram, as in Case 3 (Figure 7-8), it is necessary to decrease the rate of solvent strength change at the end of the

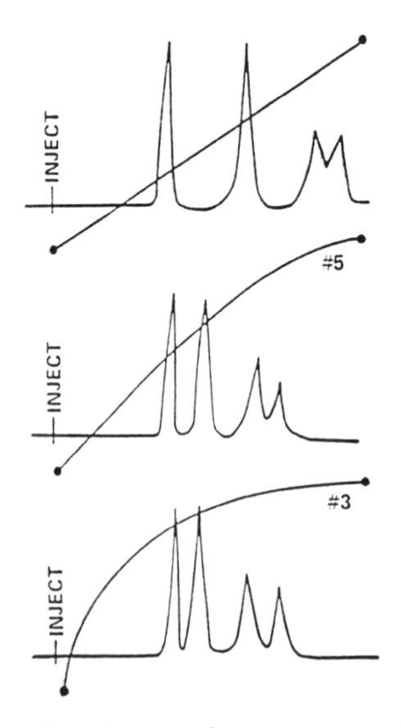

1. Increase the rate at the beginning and decrease the rate toward the end.
2. As above and vary degree.
3. Decrease final condition.

FIGURE 7-11. Approaches to correct inadequate resolution at the end of the chromatogram. Refer to text for discussion.

run by using a convex shape, as shown in Figure 7-11. This will have the effect of improving the resolution for the later eluters. Again, this is a process of trial and error to find the exact curve shape to optimize the resolution per unit time. A second action is to increase the time span of the gradient program, which increases the total run time and should be done only if longer analysis time is acceptable.

Applying a Strategy for Developing a Gradient Separation

Applying the strategy for developing a gradient separation can be illustrated using a sample of polystyrene 800 (see Fig. 7-1); however, the separation for this illustration is on a 30 cm × 3.9 mm reverse-phase C_{18} column rather than the column used in Figure 7-1. Since attempts at isocratic runs failed to give an adequate separation, the investigator first chose a linear gradient from 30 to 100%, shown in Figure 7-12. This was an arbitrary choice of starting and ending mobile phases based upon the observation that nothing eluted at 30% and everything eluted at 100% of the strong eluent. By examining Figure 7-12, it can be seen that resolution is inadequate at the end of this gradient attempt; thus the investigator chose to lower the final conditions to 80% (Fig. 7-13), which improved the separation. Continuing this approach, the final conditions were lowered to 70% (Fig. 7-14), which resulted in separation being complete; but the analysis was long and there was a region at the

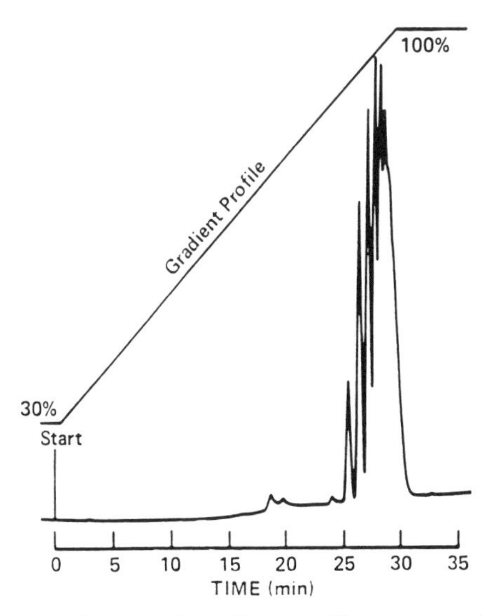

FIGURE 7-12. Effect of solvent and gradient profile upon resolution. Sample: 50 μL 5% polystyrene 800 MW in THF. Gradient conditions: 30–100% THF in water in 30 min. Flow rate: 1 mL/min. Column: μBondapak C_{18} (10 μm), 3.9 mm ID × 30 cm. Detector: UV at 254 nm.

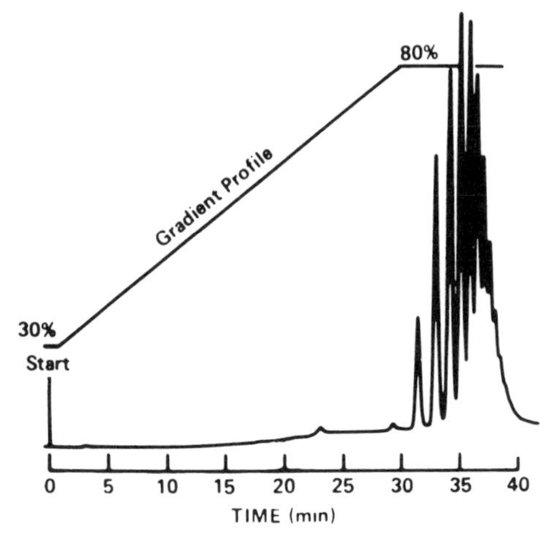

FIGURE 7-13. Second choice: Effect of solvent and gradient profile upon resolution. Sample: 50 μL 5% polystyrene 800 MW in THF. Gradient conditions: 30–80% THF in water in 30 min. Flow rate: 1 mL/min. Column: μBondapak C_{18}, 3.9 mm ID × 30 cm. Detector: UV at 254 nm.

beginning of the run during which time nothing eluted. For some, this separation may be adequate; however, many would consider this an excessive waste of time at the beginning which requires further optimization.

Remember that by changing only one variable at a time, one can observe the cause-and-effect behavior as the individual proceeds to a rapid and logical ending point. In this situation the individual decided to next change the flow rate to determine if the analysis time would be improved. The analysis

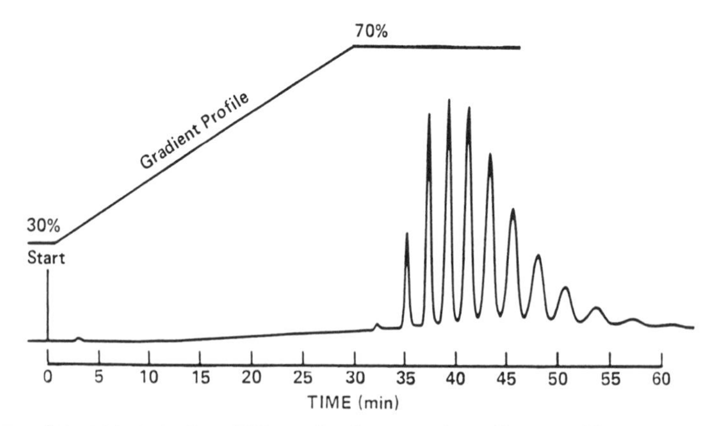

FIGURE 7-14. Third choice: Effect of solvent and gradient profile upon resolution. Sample: 50 μL 5% polystyrene 800 MW in THF. Gradient conditions: 30–70% THF in water in 30 min. Flow rate: 1 mL/min. Column: μBondapak C_{18}, 3.9 mm ID × 30 cm. Detector: UV at 254 nm.

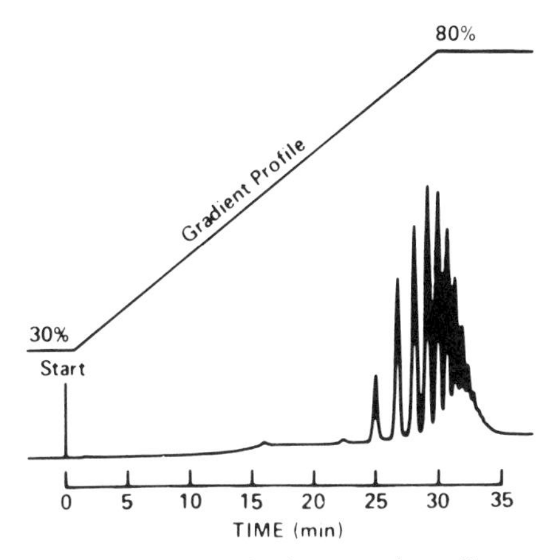

FIGURE 7-15. Fourth choice: Effect of solvent and gradient profile upon resolution. Sample: 50 μL 5% polystyrene 800 MW in THF. Flow rate: 2 mL/min. Column: μBondapak C_{18}, 3.9 mm ID × 30 cm. Detector: UV at 254 nm. Gradient conditions: 30–80% THF in water in 30 min.

time was reduced and the resolution was improved slightly at 2 mL/min. Also, running at a higher flow rate would enable more experiments to be run during a day; hence, the gradient development task would be completed in a faster time frame. Now, at 2 mL/min, the next step was to raise the initial solvent conditions to determine the effect upon reducing the extra time from the beginning of the chromatogram. As the initial conditions were raised from 30 to 40% to 50 to 60% tetrahydrofuran, the separation improved with the first peak moving to earlier elution volumes and the latter peaks having better resolution. This is a direct result of lowering the ROC during the separation. The separations are shown in Figures 7-15 and 7-16.

For many people, this separation would be adequate. However, some individuals might desire to further optimize the separation. In this separation, there is a problem of "excessive" resolution at the beginning and somewhat "inadequate" resolution at the end of the analysis. Thus, it is necessary to increase the rate of solvent composition change during the earlier part of the run and slow the rate during the latter part of the run. To do this a convex gradient profile is chosen (Fig. 7-17) with a run time of 30 min. Then, to attain a little additional resolution at the end of the run, a longer run time of 45 min can be used to slow the rate of change during the end of the run (Fig. 7-18).

Effect of Flow Rate upon Gradient Performance

Running an analysis at a high flow rate is not commonly considered for optimizing gradient HPLC analysis. Therefore, it is appropriate to view the

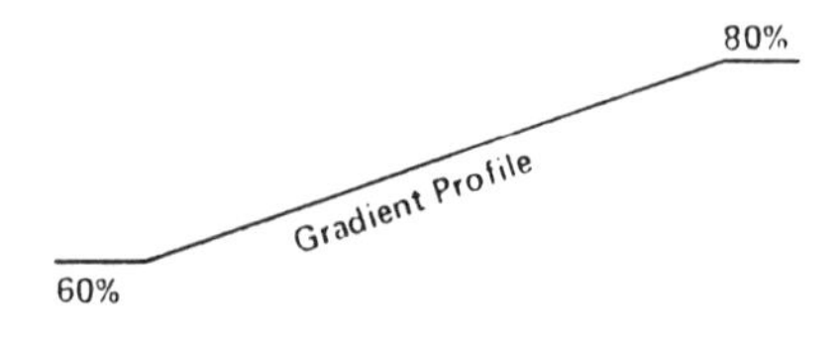

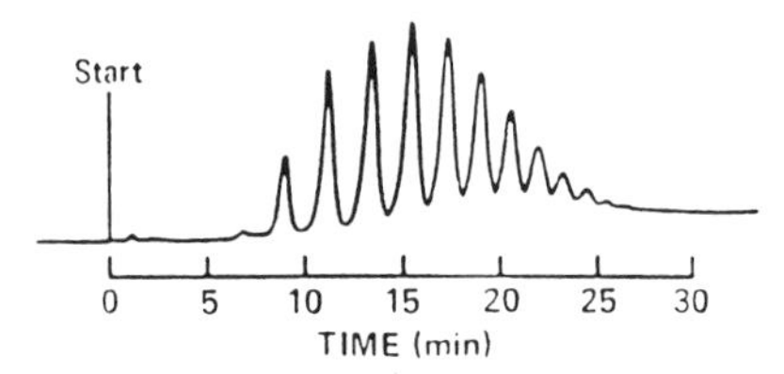

FIGURE 7-16. Fifth choice: Effect of solvent and gradient profile upon resolution. Sample: 50 μL 5% polystyrene 800 MW in THF. Flow rate: 2 mL/min. Column: μBondapak C_{18}, 3.9 mm ID × 30 cm. Detector: UV at 254 nm. Gradient conditions: 60–80% THF in water in 30 min.

role of flow rate in more detail. In Figure 7-19, the effect of flow rate upon resolution in a gradient analysis is shown using a 3.9-mm ID column. By increasing the flow rate while keeping the gradient run time constant, the ROC term is decreased from 1.7 to 0.8, and there is a noticeable improvement in resolution. One reason higher flow rates are not often used is that when using higher flow rates to improve resolution with traditional narrow-

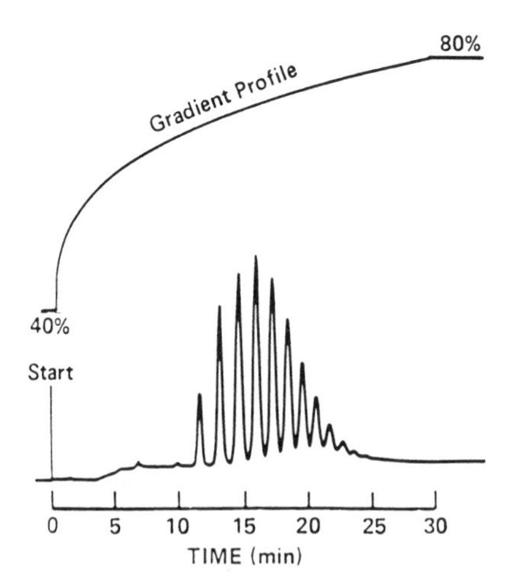

FIGURE 7-17. Use of a convex profile. Sample: 50 μL 5% polystyrene 800 MW in THF. Gradient conditions: 40–80% THF in water in 30 min. Flow rate: 2 mL/min. Column: μBondapak C_{18}, 3.9 mm ID × 30 cm. Detector: UV at 254 nm.

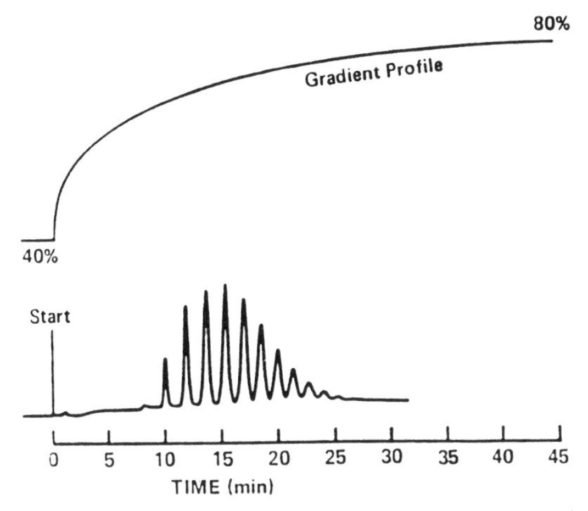

FIGURE 7-18. Final analysis. Sample: 50 μL 5% polystyrene 800 MW in THF. Gradient conditions: 40–80% THF in water in 45 min. Flow rate: 3 mL/min. Column: μBondapak C_{18}, 3.9 mm ID × 30 cm. Detector: UV at 254 nm.

diameter columns there is a backpressure limitation. Many mobile phase/column combinations used in gradients limit the flow to only 2.5–3.0 mL/min (through a nominal 4 mm ID × 30 cm column) due to backpressure. In essence, the analyst pays the price of wasted time developing the separation using traditional narrow-bore, 30-cm length columns because obtaining low ROC values has required long analysis time for the separation. Slow regeneration of initial conditions to avoid column-bed collapse in narrow-bore columns is often recommended in gradient HPLC. This also increases the gradient analysis time.

If available, incorporating a wider diameter, shorter length column into a gradient system enables the use of higher flow rates (up to 10 mL/min) due to lower backpressures. This flow-rate flexibility allows optimization of separation and analysis time. The advantage of very high flow rates is clearly demonstrated in Figure 7-20. Resolution is satisfactory at 2.5 mL/min for a 20-min gradient and ROC = 0.4%/mL. When the flow rate is increased to 5 mL/min and the run time is reduced to 10 min (ROC = 0.4%/mL), the resolution is comparable to a first approximation to Figure 7-20*a*. However, by using a high flow rate the optimized separation (Fig. 7-20*b*) is finished before the unoptimized one has begun (Fig. 7-20*a*). In Figure 20B there was a higher resolution per unit time than in Figure 7-20*a*. (Resolution per unit time is defined by R/t_0 where R is the resolution and t_0 is the time of elution of a nonretained peak.) By using a wide diameter and short column length, backpressure limitations were eliminated and very high flow rates could be used to optimize the separation. The key conclusion to be drawn from this discussion is that gradient elution should be run as fast as is consistent with

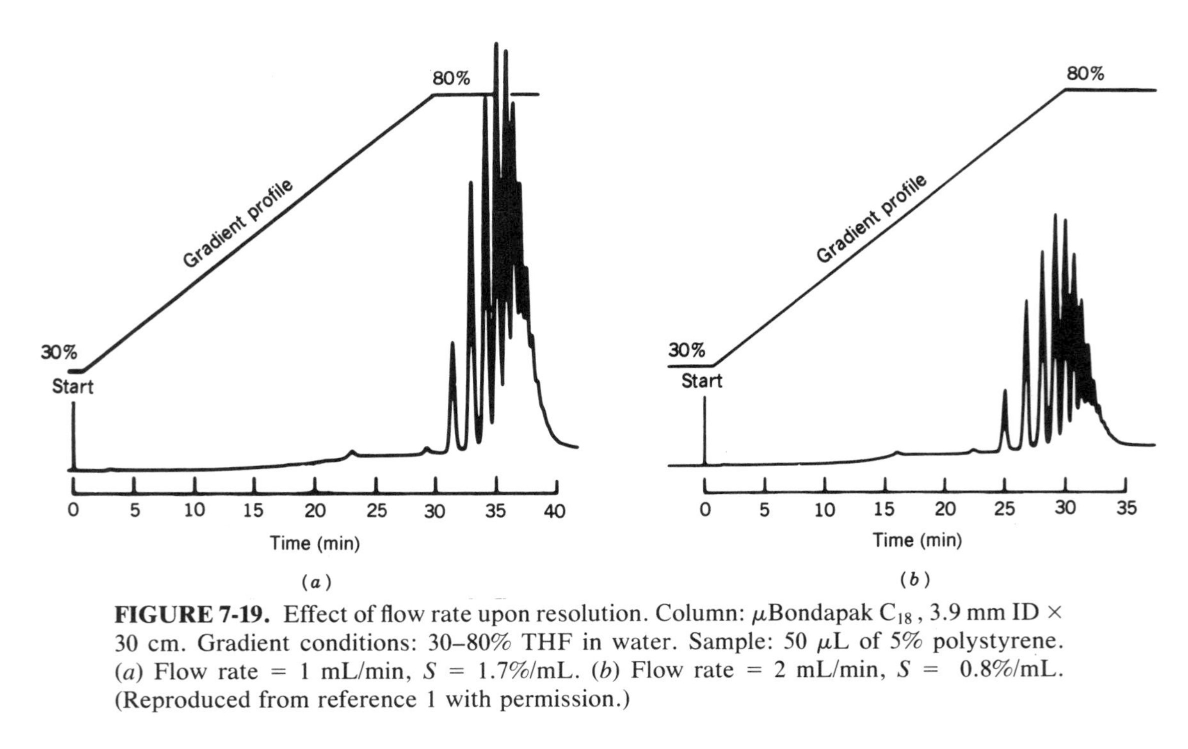

FIGURE 7-19. Effect of flow rate upon resolution. Column: μBondapak C_{18}, 3.9 mm ID × 30 cm. Gradient conditions: 30–80% THF in water. Sample: 50 μL of 5% polystyrene. (*a*) Flow rate = 1 mL/min, S = 1.7%/mL. (*b*) Flow rate = 2 mL/min, S = 0.8%/mL. (Reproduced from reference 1 with permission.)

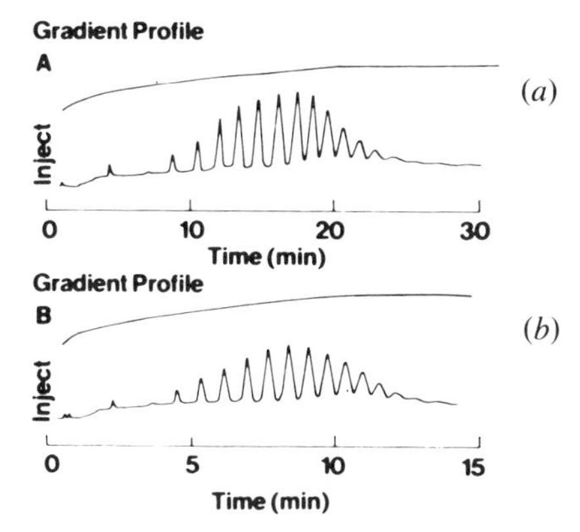

FIGURE 7-20. High-speed gradient analysis. Column: Radial-Pak Resolve C_8, (10 μm), 8 mm ID × 10 cm. Gradient conditions: 50–70% THF/H_2O; using convex curve. Detector: UV at 254 nm, 0.5 AUFS. Sample: polystyrene MW800. (*a*) Flow rate: 2.5 mL/min; run time: 30 min; backpressure: 750 psi; chart speed: 1 cm/min. (*b*) Flow rate: 5 mL/min; run time: 15 min; backpressure: 1500 psi; chart speed: 2 cm/min. (Reproduced from reference 1 with permission.)

required results to minimize run time no matter what column dimensions are used.

Applications

Gradient HPLC helps solve what has been referred to as the "general elution problem" and sometimes called the "general resolution problem" of having early eluting compounds not well resolved while the late eluting compounds are retained too long and exhibit considerable peak broadening (Fig. 7-21*a*). As shown, peaks 1 and 2 are not resolved, peaks 5 and 6 have excessive retention, and the total analysis time is too long. In the first attempt (Fig. 7-21*b*) to improve this separation and reduce the total analysis time, a linear gradient resulted in excess resolution of the early peaks, consolidated the late peaks, and did not reduce the total time appreciably. It should be noted that during a gradient run, pressure readings vary because of the pronounced viscosity changes as the water–methanol ratio is varied, but the solvent delivery system maintains a constant reproducible flow rate throughout the run. Viscosity changes are addressed in the next section. Satisfactory resolution of all components and a significant reduction in analysis time was achieved by starting with a stronger solvent and using a convex gradient (Fig. 7-21*c*). If a separation involves fewer components, further reductions in analysis time are feasible by adjusting initial and final solvent strength and gradient curvature. Note that this separation was done on a larger particle packing and an adequate separation was attained.

It should be no surprise that the gradient approach gives high resolution

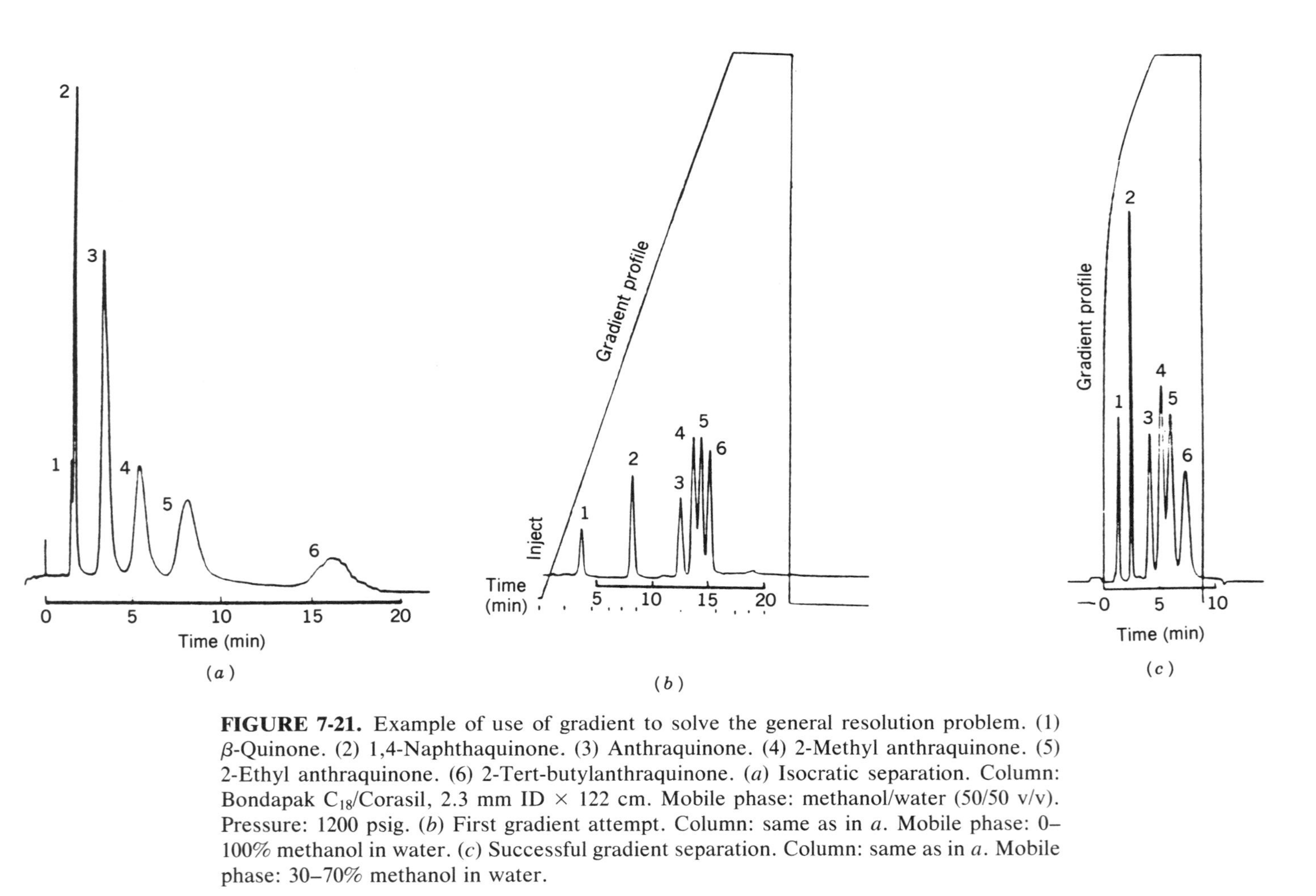

FIGURE 7-21. Example of use of gradient to solve the general resolution problem. (1) β-Quinone. (2) 1,4-Naphthaquinone. (3) Anthraquinone. (4) 2-Methyl anthraquinone. (5) 2-Ethyl anthraquinone. (6) 2-Tert-butylanthraquinone. (*a*) Isocratic separation. Column: Bondapak C_{18}/Corasil, 2.3 mm ID × 122 cm. Mobile phase: methanol/water (50/50 v/v). Pressure: 1200 psig. (*b*) First gradient attempt. Column: same as in *a*. Mobile phase: 0–100% methanol in water. (*c*) Successful gradient separation. Column: same as in *a*. Mobile phase: 30–70% methanol in water.

even when using columns containing large particles. This is because the separation is not limited by the number of theoretical plates and the spacing of the peaks is adjusted using the changing mobile-phase concentration. In fact, when using gradient elution on columns with particle sizes greater than 10 μm, the separation may actually result in narrower peaks than that obtained for that column in an isocratic mode.

Gradient steepness (ROC) has an important effect on resolution and peak height in gradient separations. The steeper the gradient (ROC), the higher the peak height and the closer the peaks will elute. Figure 7-22 shows this behavior for the separation of seven antihistamines on a C_{18} column. The gradient times were changed from 5 to 10 to 20 to 40 min while the starting and ending mobile-phase compositions remained constant. The chromatogram with the highest ROC resulted in the sharpest peak profile and lowest retention of the peaks. As the ROC is slowed by increasing the run time, retention, and resolution increase. A nice separation was achieved in Figure 7-22*b*.

A more complex example of gradient usage is the separation of an 11-component herbicide sample shown in Figure 7-23. Figure 7-23*a* shows the initial chromatogram in which all of the 11 peaks are resolved, but the resolution between peaks 7 and 8 (the 12- to 13-min time region) is only R = 0.8. While the chromatogram is 15 min, the total run time is 20 min, which includes the gradient run, plus the additional time to return to initial conditions and the time required to reequilibrate the stationary phase with the starting mobile phase. Because of an interest in improving the resolution of the critical pair of peaks (peaks 7 and 8), it was necessary to reduce the ROC during the elution of these peaks. Two longer run times were chosen to investigate the effect of a lower ROC upon the separation (Figs. 7-23*b* and *c*). The result showed improved resolution at the expense of run time.

Figure 7-23*c* was the best resolution obtained, but it was clear that this separation can be optimized further by eliminating the "dead time" at the beginning of the chromatogram. One approach to improve this separation is to increase the starting percentage of the organic component of the mobile phase. However, it is also desirable to keep the same relative peak spacing while reducing the dead time. To accomplish both goals, it is necessary to shorten the run time of the gradient and maintain the some ROC. For instance, a 10% increase in the starting organic concentration will require a decrease in the gradient run time of 10 min to maintain a comparable ROC to the previous chromatogram. Figures 7-24*a*–*c* show the results of progressively increasing the percentage of starting organic concentration while appropriately decreasing the gradient run time. Depending upon your personal preference, chromatograms *b* or *c* are both quite good, however, chromatogram *c* has the fastest analysis time. The example shown in Figure 7-24 is a good illustration of the point mentioned earlier, that as your experience with gradient optimization increases, the development process can be enhanced by changing more than one variable at a time.

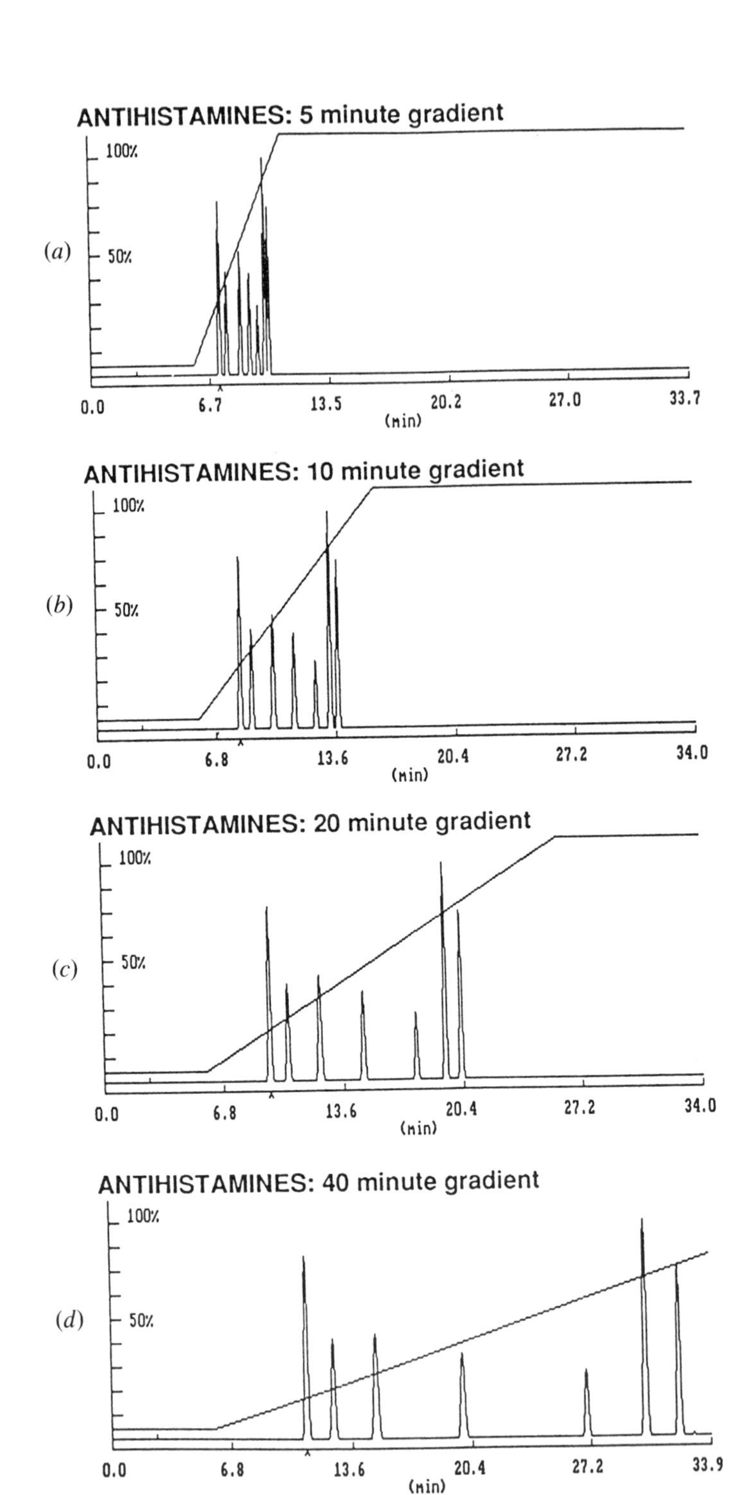

FIGURE 7-22. Effect of gradient steepness upon resolution. The retention depends upon the steepness. The steeper the gradient, the lower the retention and the sharper the peaks. (Reproduced with permission from LC Resources, Inc.)

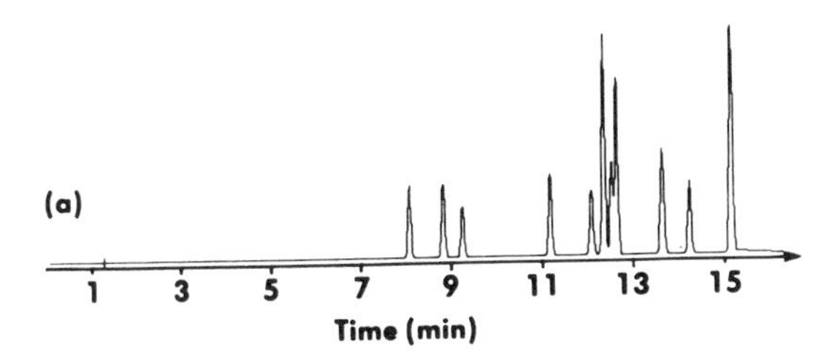

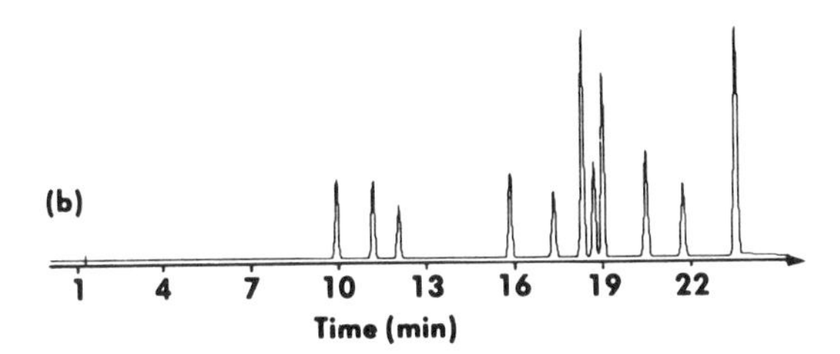

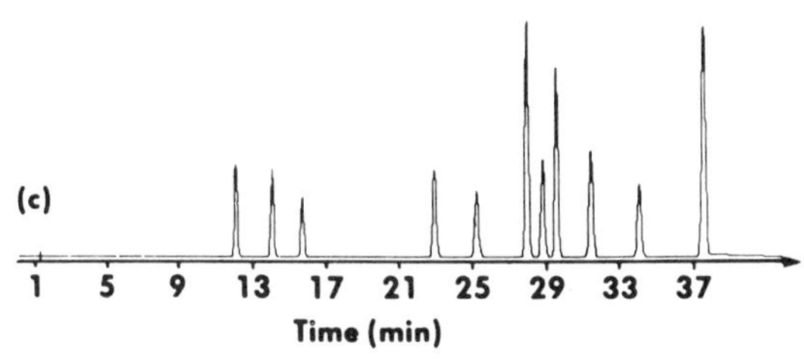

FIGURE 7-23. Separation of an 11-component herbicide sample as a function of gradient time (TG). (*a*) TG = 20 min, R = 0.8. (*b*) TG = 40 min, R = 1.5. (*c*) TG = 80 min, R = 2.2. Column: Zorbax C_8 (5 μm), 4.6 mm × 25 cm. Gradient: linear, 5–100% acetonitrile/water. Flow rate: 2 mL/min. Temperature: 35°C. (Reproduced from reference 3 with permission.)

PRACTICAL CONSIDERATIONS

As in isocratic work, the HPLC hardware must be functioning correctly. For the best precision in gradient work, flow rates must be reproducible, dead volumes should be minimal, and columns must be reequilibrated before each run. Unless simple precautions are taken, gradient elution can be a difficult technique. For instance, when developing gradients that contain buffers in the mobile phase, make sure the buffer will not precipitate out of solution as the concentration of the solvents is changed. Buffers and buffer solubility were discussed in Chapter 6. If the buffers precipitate in the HPLC, such column blockage is very difficult to remove. Once solvent compatibility is

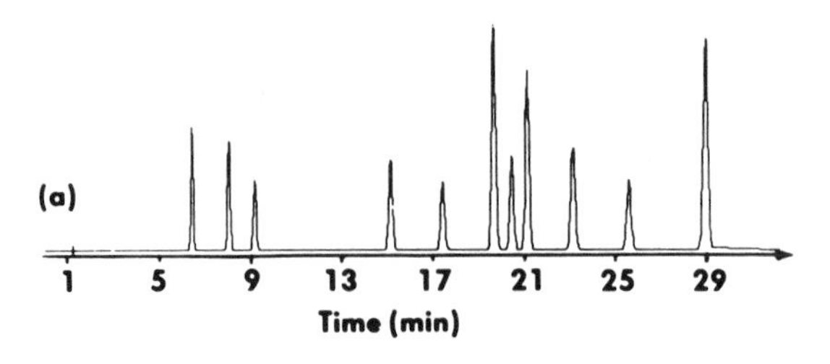

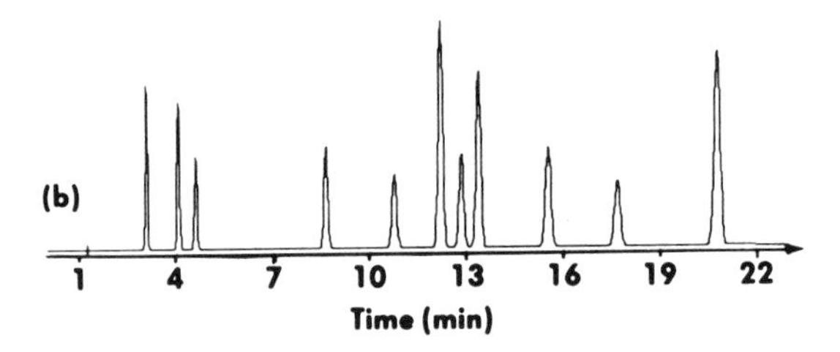

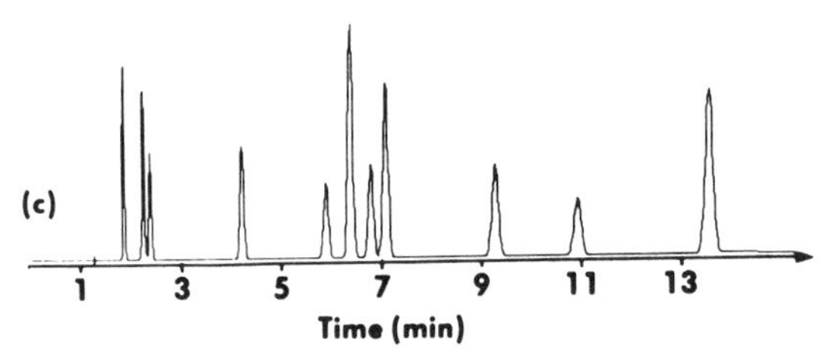

FIGURE 7-24. Separation of an 11-component herbicide sample as a function of the initial organic in the mobile phase. (*a*) Gradient: 15–100% acetonitrile/water over 70 min, R = 2.1. (*b*) Gradient: 25–100% acetonitrile/water over 60 min, R = 1.9. (*c*) Gradient: 35–100% acetonitrile/water over 50 min, R = 1.5. Gradient slope and other conditions the same as in Figure 7-23. (Reproduced with permission from reference 3.)

known, it is also important to run a "blank" gradient as a record of the background detector response during the generation of the gradient. Baselines often drift or show interference during gradient elution. A good baseline should look like that shown in Figure 7-25. Another influence that will affect gradient elution exists if the initial and final solvent have a different background absorbance. This situation will result in a changing baseline. An example of the type of change that can occur is shown in Figure 7-26.

Viscosity and Refractive Index Effects

Another factor to consider with gradients is that, as the composition of the mobile phase changes, mobile-phase viscosity and refractive index change.

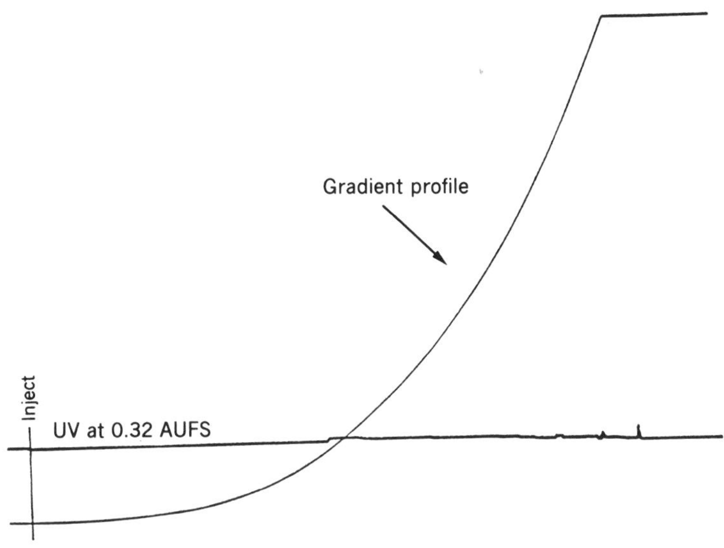

FIGURE 7-25. Blank gradient baseline. Column: μBondapak C_{18}, 3.9 mm ID × 30 cm. Solvent: A-water; B-acetonitrile; O-100% acetonitrile in water using a convex curve for 1 hr. Flow rate: 1.0 mL/min. Detector: UV at 254 nm.

For instance, as shown in Figure 7-27*a*, the viscosity of methanol/water mixtures is higher than either water or methanol alone. Similarly, this is the case for acetonitrile/water mixtures (Fig. 7-27*b*). The viscosity change manifests itself in a changing pump pressure as the gradient proceeds from water to the organic solvent (Fig. 7-28); therefore, if pressure is being monitored it

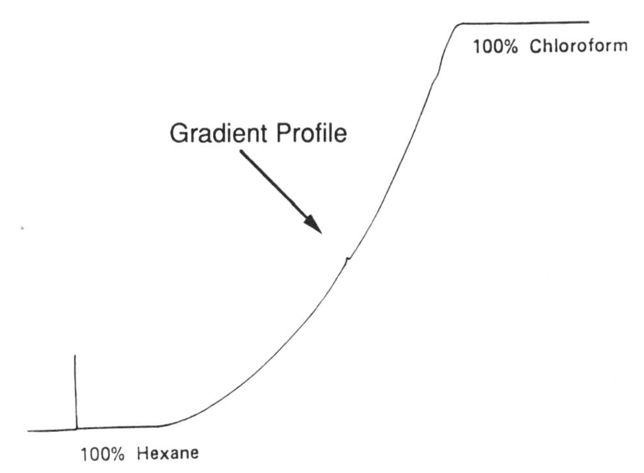

FIGURE 7-26. Baseline drift caused by solvent change during a gradient. Column: μPorasil (silica, 10 μm), 3.9 mm ID × 30 cm. Gradient conditions—solvent: A-hexane; B-$CHCl_3$. Convex curve: O-to 100% CH_3Cl in hexane over 10 min. Flow rate: 4 mL/min. Detector UV at 254 nm.

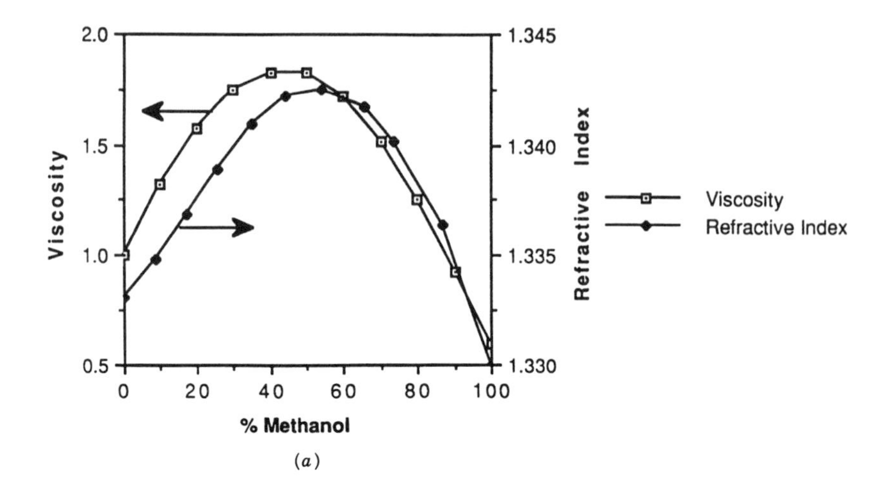

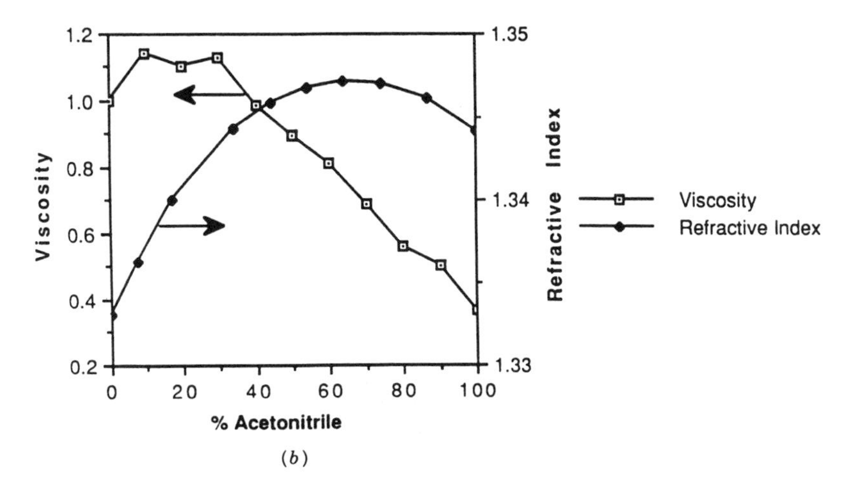

FIGURE 7-27. Mobile phase viscosities and refractive indices. (*a*) Methanol/water mixtures at 20°C. (*b*) Acetonitrile/water mixtures at 20°C.

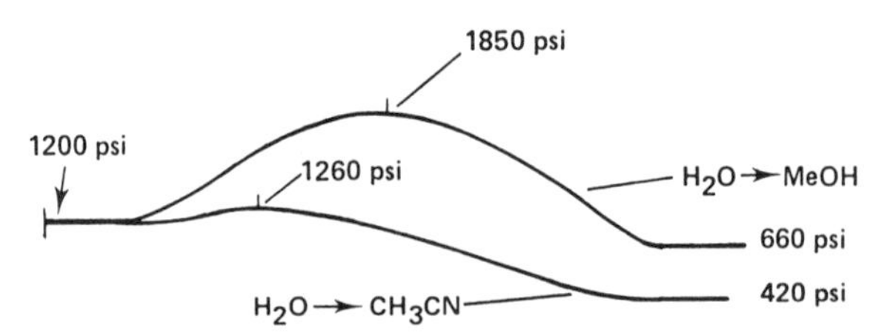

FIGURE 7-28. Pressure effects during gradient run. Column: μBondapak C_{18}, 3.9 mm ID × 30 cm. Gradient conditions: 0–100% solvent B in water over 10 min (linear). Flow rate: 2.5 mL/min.

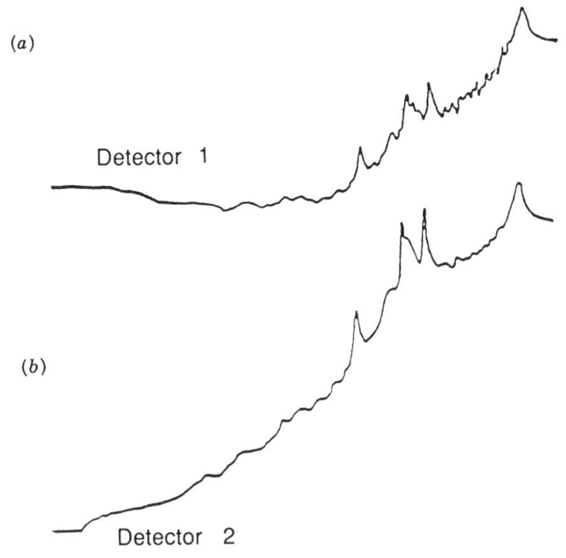

FIGURE 7-29. Refractive index effects of water–methanol gradient. Column: μBondapak C_{18}, 3.9 mm ID × 30 cm. Mobile phase: 0–100% methanol in water with a linear change over 30 min. Flow rate: 3 mL/min. Detector: (*a*) UV at 254 nm, 0.08 AUFS; (*b*) UV at 254 nm, 0.05 AUFS.

will change during a gradient analysis. However, if the pressure change is dramatically different for duplicate gradient runs, some maintenance may be necessary.

The refractive index affects the baseline changes in apparent "absorbance" during a gradient. Since the RI of a solvent changes with composition and pressure, the light transmitted to the photosensor changes and because the photosensor is in a spectrophotometer, this change in light contributes to the apparent "absorbance." An example of this is shown in Figure 7-29. Different detectors will have different UV absorbance profiles depending upon their sensitivity to the RI effect.

In-line Degassing

If quantities of air are dissolved in solvent A (i.e., water) and have much lower solubility in solvent B, then this dissolved air will be released during the course of the gradient. This is a physical trait of certain solvents and is commonly referred to as "out-gassing" during gradient elution. The out-gassing is minimized if the two solvents are mixed at high pressures since the dissolved air will remain dissolved. Since all the pressure drop in HPLC is across the column, as the pressure decreases the out-gassing may reappear. As this gas exists the column, bubbles can become trapped in the detector cell and appear as spike on the chromatogram. These spikes are the result of

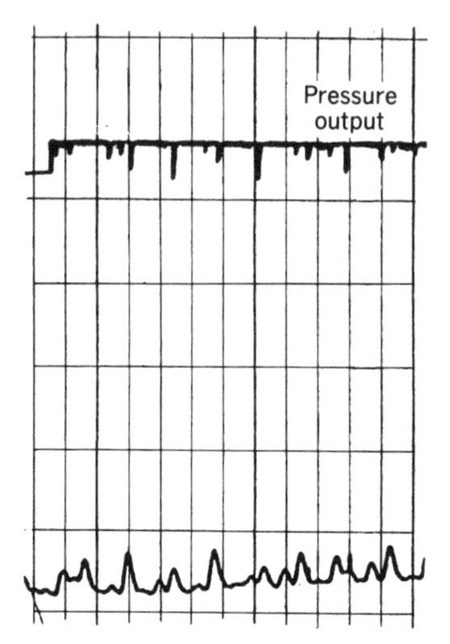

FIGURE 7-30. Effects of undegassed mobile phase. Column: μBondapak C_{18}, 3.9 mm ID × 30 cm. Mobile phase: 50:50 MeOH : water. Flow rate: 2.0 mL/min. Detector: UV at 254 nm, 0.01 AUFS.

air bubbles forming in and/or while leaving the cell (Fig. 7-30). As discussed in Chapter 6, degassing solvents thoroughly before use is one approach to removing these baseline spikes, but sometimes this does not completely eliminate the problem. Another approach is to add pressure resistance to the effluent line using a short (1- or 2-in.) length of standard 0.009-in. tubing. While this is appropriate for most photometers (however, check the manual or manufacturer before doing it), remember most refractive index detectors will not take pressures above 100 psi and should be removed from the system.

Flow Programming

In certain situations a complement to solvent gradient is flow programming. With this technique the flow rate is gradually increased during the run, thus speeding elution of the peaks. As a practical matter, compounds that do not have a UV absorbance are not easily monitored during a gradient separation. Flow programming using the refractometer for detection, on the other hand, greatly extends the separation range of non-UV absorbing compounds and permits fast turnaround time between samples because the column is always in equilibrium with a single solvent and no column–mobile phase reequilibration is required.

The saccharides in the sample shown in Figure 7-31 were produced by hydrolyzing cornstarch, resulting in a mixture of sugars containing from 1 to 10 simple saccharide repeating units. To provide adequate resolution of this complex sample with reasonable retention times, the flow rate was gradually

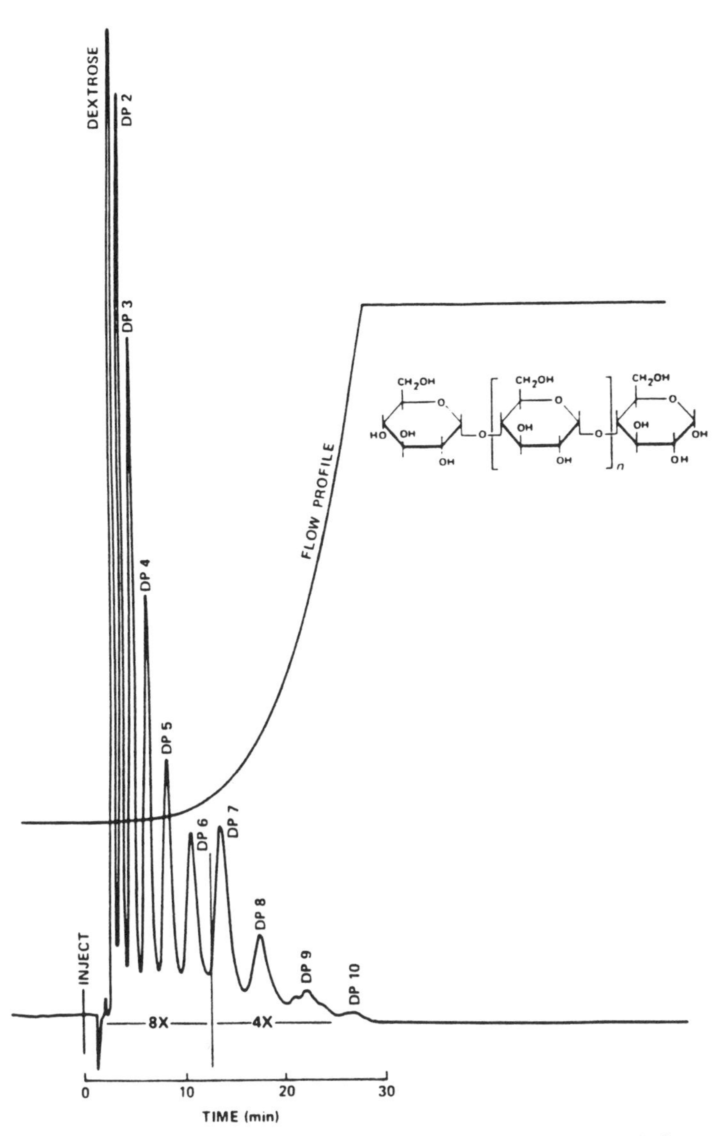

FIGURE 7-31. Separation of saccharides using flow programming. Column: carbohydrate analysis, 3.9 mm ID × 30 cm. Mobile phase: H_2O/CH_3CN (35/65). Flow rate: flow programmed from 2 to 4 mL/min. Detector: RI at a sensitivity of 8× and 4× as noted on the chromatogram.

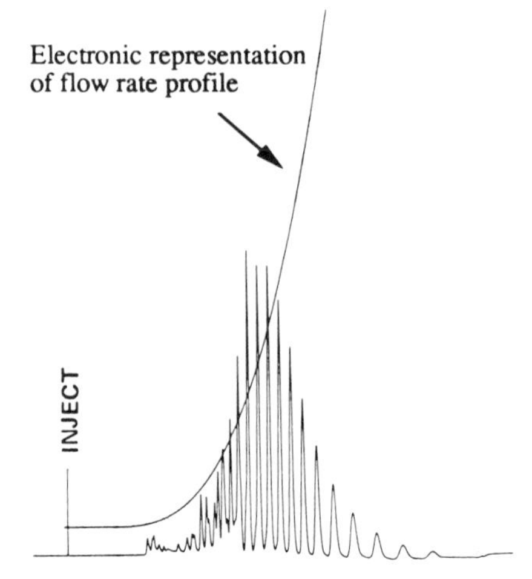

FIGURE 7-32. Flow programming. Sample: Triton X-100. Column: μPorasil (silica, 10 μm), 3.9 mm ID × 30 cm. Mobile phase: DMSO:$CHCl_3$:heptane (3:10:87). Flow rate: 0.5–8.0 mL/min; convex curve for 20 min. Detector: UV at 280 nm.

increased during the separation from 2.0 to 4.0 mL/min. When sample components of a wide polarity and no UV response are separated, the refractive index detector and flow programming are particularly useful to shorten the analysis time.

As mentioned earlier, flow programming may be used with UV detection. In Figure 7-32, a sample of Triton X-100 is separated using flow programming and monitoring the UV absorbance. The resolution obtained on this sample compares very favorably with that obtained using a solvent gradient.

SUMMARY

Gradient elution LC requires control of many operating parameters. The choice of high performance chromatographic modules is critical for good gradient performance. In the development of the separation, an initial flow rate is chosen, the gradient shape is determined, and finally, flow rate and run time are optimized. In the regeneration of initial column conditions, a step return to the initial eluent conditions may be possible using wide diameter columns with minimal risk of column bed collapse, and equilibration at high flow rates may be possible. The isocratic mode has been the method of choice for quantitative LC because the number of variables is kept to a minimum. The information presented here demonstrates that the separating power and convenience of gradient elution, as well as the ability to obtain excellent quantitative information, makes this technique attractive for the routine analysis of samples.

REFERENCES

1. J. Korpi and B. A. Bidlingmeyer, *American Laboratory,* **6,** 110 (1981).
2. F. C. Senftleber, A. G. Halline, H. Veening, and D. A. Dayton, *Clinical Chemistry* **22**(9), 1522 (1976).
3. L. Snyder and J. Dolan, *LC.GC,* **5,** 970 (1987).

ACKNOWLEDGMENT

Figures 7-6, 7-8 through 7-18, 7-21, 7-25, 7-26, 7-28 through 7-32 are courtesy of the Waters Chromatography Division of Millipore.

CHAPTER 8

EXPERIMENT 1: DEMONSTRATING THE FUNDAMENTALS

Objective
Background
Additional materials
Safety and disposal
Experiments
  A. Isocratic separation—C_{18} cartridge
  B. Isocratic separation—silica cartridge
Optional experiments
  C. Isocratic separation—silica cartridge
  D. Step gradient separation—C_{18} cartridge
  E. Step gradient separation—silica cartridge
Evaluation
  Retention
  Capacity Factor
  Selectivity
  Efficiency
Discussion
Self-help questions
Additional experiments
References

OBJECTIVE

Demonstrate the fundamentals of LC in a vivid manner so that the experimenter is introduced to the key equations and relationships using familiar materials.

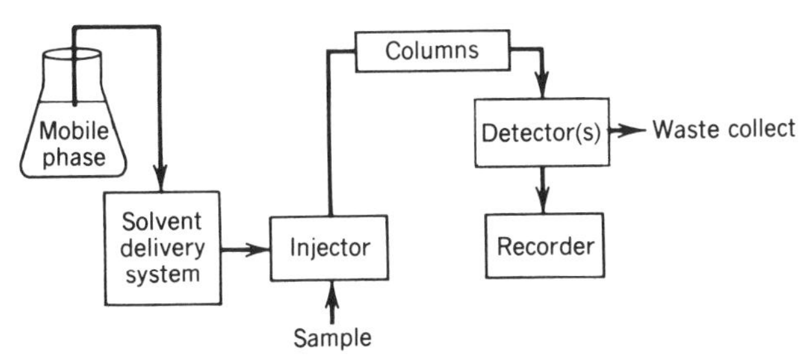

FIGURE 8-1. Block diagram showing the components of an isocratic HPLC instrument.

BACKGROUND

Figure 8-1 shows the components of a simple HPLC instrument which is capable of isocratic (single-eluent) separations. Each component of Figure 8-1 is represented in this introductory experiment. A 10-mL (disposable) syringe will serve as the "eluent reservoir" and "solvent delivery system." The "injector" is a smaller (disposable) syringe. The "LC column" is packed with silica gel or silica modified by having a nonpolar alkyl chain (C_{18}) bonded to it. The mobile phase (eluent) consists of isopropanol mixed with water or with household, distilled white vinegar (dilute acetic acid). The samples are highly colored dyes found in a popular grape drink and the human eye provides an effective "detector." Finally, the "recorder" is the experimenter's recording of the experiment done manually with pen and laboratory notebook.

The reverse-phase mode is used for all the separations performed in this experiment. "Reverse phase" is the term used when the stationary phase is more nonpolar than the mobile phase with regard to the polarity of the sample. The isopropanol/water and isopropanol/vinegar mobile phases are typical of reverse-phase mobile phases, which generally are composed of water mixed with polar organic modifiers. The bonded C_{18} column used is a very nonpolar surface and is the most popular stationary phase for reverse-phase HPLC. In this experiment, the silica column when used in the reverse-phase mode provides a very weak "nonpolar" surface in comparison to C_{18}. Silica is normally thought of as a highly polar surface and is most commonly used in the normal-phase mode. The use of silica in the normal-phase mode, with a nonpolar mobile phase is the subject of Chapter 9 (Experiment 2).

For additional information on the use and preparation of solid-phase extraction devices, read the appropriate section in Chapter 6.

ADDITIONAL MATERIALS

5 mini columns. These five columns are conveniently available as Sep-Pak® cartridges (Waters, Milford, MA) of two varieties, silica and C_{18}.

While the experiment has been tested on the SepPak® brand, equivalent solid-phase extraction devices should also work in this experiment. As an alternative, homemade columns can be constructed from Pasteur pipets plugged with glass wool at the outlet tip and dry-packed with bulk packing material (refer to Chapter 6 for details). Silica gel (200 mesh) can be purchased from many supply houses. Preparative C_{18} packing of 50 to 100-μm particle size is available from many chromatography suppliers.

1 100-mL volumetric flask

2 10-mL graduated cylinders

1 100-mL graduated cylinder

Unsweetened grape-flavored Kool-Aid® powder (obtained locally) and dissolved in purified water at approximately the manufacturer's suggested concentration (0.3 g/100 mL) in a volumetric flask. (Note: This solution is not intended for food use).

FD&C Blue 1 and Red 40 available from several suppliers including H. Kohnstamm, New York, NY [optional experiment for spectroscopic determination (quantitation) of the dyes].

Distilled water.

Distilled white vinegar, household (or 5% glacial acetic acid).

Isopropanol (or household rubbing alcohol, nominally 70% isopropanol in water).

1 3-mL syringe used for sample injection.

1 10-mL syringe (Luertip) used to pump solvent.

Note: If the use of syringes is undesirable, small laboratory wash bottles can be substituted for the 3- and 10-mL syringes.

SAFETY AND DISPOSAL

This experiment does not purport to address the safety issues associated with its use. It is the responsibility of the user to establish appropriate safety and health practices and to determine the applicability of regulatory limitations prior to use. All chemicals should be handled and disposed of in an appropriate manner consistent with the safety policy of the experimenter's company, school, or organization.

EXPERIMENTS

Five LC separations are described and suggestions for further experimentation are provided. For the three isocratic separations (Sections A, B, and C), elution volume data can be collected to allow calculation of resolution (R), selectivity (α) and efficiency (N). The most accurate data will be obtained by collecting the column effluent in a 10-mL graduated cylinder. The volumes corresponding to the beginning and end of each colored band should be recorded (see Table 8-1 for sample data).

For some experiments there may not be a perfect separation between the blue- and red-colored bands. In this case, data for the beginning and end of the intermediate purple band (overlap of red and blue) should be recorded. The center of the purple band will serve as the end of the first band and beginning of the last.

Each separation requires 5–10 mL of mobile phase (eluent) and can be performed in a few minutes. If time permits, students should repeat each experiment three times so that the reproducibility of the method can be checked. Calculation of a mean and standard deviation for selected data and/or calculated values would be a useful exercise. The cartridges will not perform properly if the required pretreatment and between-injection washings are ignored. This is particularly true for a new C_{18} column, which will show little retentivity toward the dyes unless thoroughly prewetted with isopropanol.

A. Isocratic Separation—C_{18} Cartridge

Activity	Comments
1. Prepare 18% (v/v) isopropanol as the mobile phase.	1. Add 18 mL of pure isopropanol to 82 mL of water since 100-mL amount is sufficient. If rubbing alcohol (70% isopropanol) is used, then 25.7 mL of rubbing alcohol is added to 3 mL of distilled water.
2. Remove the piston from the syringe barrel. Push the longer end of the cartridge onto the syringe luer tip. Hold assembly upright and add the appropriate solution into the syringe barrel. Insert piston and "pump" contents through the cartridge.	2. This is the general procedure for operating the syringe as a pump.
3. Pretreat the C_{18} cartridge by "pumping" through 10 ml of isopropanol at 5–10 ml/min followed by 10 ml of water at the same flow rate. If rubbing	3. If using a Sep-Pak® cartridge, it may be helpful to cut off the exit tube at the point where it meets the body of the cartridge. This will eliminate the

alcohol is used, pump through 10 ml of the rubbing alcohol.

remixing of closely eluting bands in the tube, which could complicate the determination of the end of one band and the beginning of another.

4. Remove the cartridge from the syringe tip, then remove piston from the syringe. Reassemble according to Step 2 for additional pumping activities.

4. This is the general procedure for changing mobile phases.

5. Slowly inject 1 ml of sample (Kool-Aid® powder in solution) onto the cartridge. (Discard the column effluent).

5. This solution was prepared in a 100 ml volumetric flask at a concentration of 0.3 g/100 ml. (Note: this solution is not intended for food use.)

6. Fill the pump with mobile phase and elute the dyes by pumping at a steady flow rate of 5 ml/min.

7. Now begin to collect the column effluent in a 10-ml graduated cylinder. As each colored band elutes from the column, stop the flow momentarily and record volumes for the beginning and end of the band (i.e., the first and last colored drops).

7. These results will be used to fill in Table 8-1.

8. Repeat the experiment

8. Between injections, wash the column with 10 ml of water at 5–10 ml/min. Should colored material build up on the column after several injections, repeat the pretreatment procedure to clean the cartridge.

9. Repeat the experiment
10. Average the results

B. Isocratic Separation—Silica Cartridge

Activity	Comments
1. Repeat Activity A-1 and prepare 18% (v/v) isopropanol as the mobile phase.	1. Refer to Comment A-1.

2. Pretreat the silica cartridge by pumping 10 mL of water through the cartridge at 5–10 mL/min.

3. Slowly inject 1 mL of the sample prepared in Activity A-5 onto the silica cartridge.

4. Elute the dyes by pumping the mobile phase at approximately 1 mL/min; collect each dye band in a 10-mL graduated cylinder as in Activity A-7.
5. Record data as in Activity A-7.

6. Wash the column with 10 mL of water between injections.
7. Repeat the experiment.
8. Repeat the experiment.
9. Average the results.

2. Repeat Activity A-2 in order to properly insert the silica cartridge onto the syringe and pump the mobile phase. Remember to use water in this step.
3. Slow injection is particularly important in this separation to minimize elution of red dye during the injection step.
4. Approximately 20 drops is equivalent to 1 mL. Once the red dye is collected, the blue band may be eluted at 5 mL/min.
5. Construct Table 8-1 based upon your experimental parameters.

TABLE 8-1. Suggested Worksheet for the Isocratic Separations with Sample Data

	Red Color (mL)	Blue Color (mL)	Experimental System (To be filled in)
V_{RS} (start)	1.0	1.3	______
V_{RE} (end)	1.3	2.9	______
$W = V_{RE}$ (end)-V_{RS} (start)	0.3	1.6	______
$V_R = V_{RS}$ (start) + 1/2 W	1.2	2.1	______
L (cm)	1.25[a]	1.25[a]	______
r (cm)	0.5[a]	0.5[a]	______
$V_M = 0.5\,\pi r^2 L$	0.49[a]	0.49[a]	______
$k' = \dfrac{V_R - V_M}{V_M}$			
	1.4	3.3	______
$\alpha = \dfrac{k'_2}{k'_1}$			
$N = 16\left(\dfrac{V_R}{W}\right)^2$		2.4	______
$R = \dfrac{V_{R1} - V_{R2}}{1/2(W_1 + W_2)}$		27	______
		0.95	______

[a] These are the dimensions for a Sep Pak C$_{18}$® cartridge. Refer to the section titled Evaluation for a discussion of this topic.

OPTIONAL EXPERIMENTS

C. Isocratic Separation—Silica Cartridge

Activity	Comments
1. Repeat Activity A-1 and prepare 18% isopropyl alcohol in vinegar as the mobile phase.	1. If using rubbing alcohol (70% IPA) make the appropriate volume adjustment (see Comment A-1) and use vinegar instead of water.
2. Pretreat the silica cartridge by pumping 10 mL of vinegar through the cartridge at 5–10 mL/min.	2. Refer to Activity A-2 to properly insert the cartridge into the syringe.
3. Slowly inject 1 mL of sample.	
4. Elute the dyes by pumping the mobile phase at no faster than 5 mL/min and measure the band volume in a graduated cylinder.	4. Refer to Activity A-7 and Comment A-7. If the blue dye does not elute before the red dye, repeat the pretreatment with vinegar. Also, once a cartridge is treated with vinegar, it cannot be used for Sections B or E.
5. Record the data.	5. This is to be able to fill in Table 8-1.
6. Wash the column with 10 mL of vinegar between injections.	

D. Step Gradient Separation—C_{18} Cartridge

Activity	Comments
1. Prepare two eluents: 5% isopropanol in water and 25% isopropanol in water	1. If using rubbing alcohol (70% IPA) make the appropriate volume adjustments (refer to Activity A-1 and Comment A-1).
2. Pretreat the C_{18} cartridge as in Activity A-3, with the isopropanol in water. The last wash should be 10 mL of water.	
3. Slowly inject 1 mL of sample.	
4. Elute the red dye with the 5% isopropanol (in water) mobile phase.	4. Note that large volumes of this eluent can be pumped without eluting the blue dye.
5. Use the 25% isopropanol (in water) mobile phase to elute the blue dye.	5. See Activity D-7 regarding data collection.

6. Wash the column with 10 mL of water before making any further injections.
7. Data need not be collected for this experiment.

E. Step Gradient Separation—Silica Cartridge

Activity	Comments
1. Prepare a 15% isopropanol (in water) mobile phase.	1. If using rubbing alcohol (70% IPA) make the appropriate volume adjustments (see Activity A-1 and Comment A-1).
2. Pretreat the silica cartridge with 10 mL of water.	2. Refer to Activity A-2.
3. Slowly inject 1 mL of sample.	
4. Elute the red dye with the water mobile phase at 1 mL/min.	4. Approximately 20 drops is equivalent to 1 mL.
5. Elute the blue dye with the 15% isopropanol (in water) mobile phase at 5 mL/min.	
6. Wash the column with 10 mL of water.	
7. Data need not be collected	7. Refer to Activity D-7.

EVALUATION

The objective of all chromatographic separation is resolution. This experiment illustrates resolution and the factors that affect it. As discussed in Chapters 1 and 3 resolution cannot occur if the components are not partially retained or slowed down (retarded) by the column. Therefore, before calculating resolution, it is important to use the results of the experiment to calculate the fundamental chromatographic parameters of retention, capacity factor, selectivity, and efficiency.

Retention

The equation which defines retention for a chromatographic process is

$$V_R = V_M + KV_S \tag{8-1}$$

where V_R is retention volume, V_M is mobile phase volume, V_S is the stationary phase volume, and K is the equilibrium constant for the concentration distribution between the two phases [K = (solute concentration in stationary phase)/(solute concentration in mobile phase)]. The most common term used to describe retention is the capacity factor k':

$$k' = K\left(\frac{V_S}{V_M}\right) \tag{8-2}$$

The capacity factor takes into account the fact that the observed retention will be determined by the equilibrium distribution constant corrected for the relative volumes of the two phases.

By combining equations (8-1) and (8-2) the more common relationship appears

$$V_R = \frac{V_M(1 + k')}{V_M} \tag{8-3}$$

According to equation (8-3), the retention volume for a solute is equal to $(1 + k')$ multiples of the columns' mobile phase volume V_M. The k' value is a unitless relative retention measurement which can be calculated from experimental values. Rearrangement of equation (8-3) yields

$$k' = \frac{V_R - V_M}{V_M} \tag{8-4}$$

Equation (8-3) is used to determine k' values for experiments A–C, as indicated in the table. The V_R values for the dyes in this experiment are the volumes corresponding to the centers of the red and blue bands. No value is measured for V_M, however. The best method for obtaining a true value of V_M is to inject on the column a compound that is very similar in chemical retentivity to the mobile phase and that responds to the detector. Such a compound would be unretained by the stationary phase and would elute after passing through the volume V_M which is occupied by the mobile phase within the column. Due to the limitations of the apparatus used in these experiments, we will approximate V_M. For a column filled with porous packing, the value of V_M represents about 50% of the total empty column volume and can be estimated by the equation

$$V_M = 0.5\ \pi r^2 L \tag{8-5}$$

where r is column radius and L is column length, both expressed in centimeters. A typical value of r for Sep-Pak® cartridges is 0.5 cm. Typical L values are 2.50 cm for silica and 1.25 cm for C_{18} cartridges.

Once a value for V_M is calculated using equation (8-5), it is a straightforward matter to determine k'. The value of V_R for each dye is taken to be the volume corresponding to the center of the band. (This approximation method is justified since peaks usually elute with symmetrical shapes.) Table 8-1 provides a convenient format for recording values of L, r, V_M and k'.

Selectivity

When k' values have been determined, the concept of selectivity can be introduced and a separation factor calculated. The degree of separation between two band centers after elution is called the selectivity or separation factor. It is the ratio of the k' values defined by equation (8-6):

$$\alpha = \frac{k_2'}{k_1'} \tag{8-6}$$

The definition requires selectivity to be calculated for only two components and the larger k' value is always in the numerator. Thus, α is always unity or greater. For example, an α value of 1.1 indicates that the column exhibits a 10% greater retentivity for the later eluting component. The silica and C_{18} columns can be compared with respect to their selectivity on the basis of calculated values of α. One word of caution is appropriate: For early-eluting solutes k' may approach zero, especially when V_M is approximated roughly as in equation (8-5). In this case α may be very large, even though the separation of the colored bands is not dramatic. In practice chromatographers usually select a mobile phase which gives k' values between 2 and 10.

Efficiency

While the band centers are separating due to the relative selectivity of the column, band broadening occurs simultaneously. Band broadening is detrimental to the resolution in the separation and is minimized for columns of high efficiency. Martin and Synge (1) first introduced the concept of efficiency or theoretical plates into chromatography. In their work they derived the concept of LC from partition theory and random statistics which is similar to separation concepts developed for extraction and distillation. In essence, they assumed that the column could be divided into a number of equal sections where an equilibrium distribution of the solutes occurred between the two phases. These equilibrium stages were called theoretical plates. Thus, a separation was modeled, like an extraction process, to occur when the mobile phase is transferred to the next stage (or plate) where a new equilibrium is established. For a chromatographic column the number of theoretical plates, N, can be calculated by the equation

$$N = 16\left(\frac{V_R}{W}\right)^2 \tag{8-7}$$

where W is the band width.

For experiments in Sections A–C, both V_R and W can be measured. By collecting the column effluent and measuring the beginning and end of each band, the band width W can be approximated (see the table). The retention volume V_R of the band can be obtained by measuring the volume of solvent eluted before the band collection and assuming that the band profile is symmetrical so that the band maximum occurs in the center of the band (i.e., volume of band/2 = center of band). Thus it is straightforward to calculate the N value. It is important to note that the calculation of N should be based on the longer-retained dye. Since this experiment uses a cartridge and not a high performance LC column, the number of theoretical plates will be in the range of 20 to 200.

Resolution

Now that the fundamental parameters of LC have been defined and calculated, the focus of the experiment can be directed to the primary objective: resolution. Resolution is the measure of how well two compounds are separated by the LC column. The quantitative measurement of resolution between the two bands of color takes into account the separation of the band centers as well as the width of each colored band according to the equation

$$R = \frac{V_{R1} - V_{R2}}{1/2(W_1 + W_2)} = \frac{\Delta V_R}{W_{Ave}} \quad (8\text{-}8)$$

The numerator in the resolution equation is the distance between bands, which is related to the selectivity, and the denominator is the average band width, which is proportional to the efficiency of the column. For experiments A–C resolution can be calculated from the already measured values and compared for both column types.

Table 8-1 contains sample data for this experiment. You should create your own Table 8-1. It should be filled in with your experimental results and the various terms calculated. One table should be made with the data you collected for each experiment performed (e.g., if you did Sections A and B, you should construct two tables, if you did A, B, and C you should have three tables). Do not prepare a Table 8-1 for the gradient experiments.

DISCUSSION

The five LC experiments (Sections A–E) introduce several key features of chromatography. Sections A and B illustrate the relative retentivity of two different columns. In Section B it is clear that the silica stationary phase is less retentive than the C_{18} phase. The red dye is barely retained in Experiment B, with the result that this dye may begin to elute during the injection. The red dye can actually be eluted without the addition of any isopropanol

(Section E). A "stronger" eluent (more isopropanol) is required when the C_{18} cartridge is used, indicating the difference in retentivity between the two stationary phases. Calculation of R_s values for silica and C_{18} may reflect the observed difference in chromatographic performance.

Sections B and C indicate the power of mobile-phase modification in achieving a separation. With the isopropanol/vinegar eluent the blue dye elutes before the red dye, unlike any of the other separations. Comparison of experimental results from Sections A–C and D–E will highlight the differences between isocratic elution and gradient elution in LC separations. By the proper selection of two solvents, each dye could be eluted in a narrow band by the step gradient. In HPLC instruments, continuous gradients rather than step gradients are used, but the same advantage of narrow elution bands and shorter analysis time is achieved.

SELF-HELP QUESTIONS

1. Which column gave the best separation performance? Why? Did you make any unusual observations? What were they?

2. If you carefully smell the original Kool Aid® sample and carefully smell the collected fractions, where does the grape smell elute? Does it elute? The smell is due to another organic compound, methyl anthranilate, whose structure is shown below.

$C_6H_4(COOCH_3)(NH_2)$

Did the methyl anthranilate always elute in the same location on the different stationary phases? If not, how can you explain this?

3. What is the effect of the "strength" of the mobile phase?

ADDITIONAL EXPERIMENTS

A. *Quantitative Determinations*. One of the three isocratic separations could also be used for a quantitative determination of the amounts of FD&C Blue 1 and Red 40 present in the powdered drink mix. A spectrophotometer would be needed. After scanning to determine the best wavelength for the analysis of each dye, measurements on standard solutions could be used to construct a calibration curve. The isocratic separation could then be performed, and the entire band containing each dye could be collected and diluted accurately to an appropriate volume. With the volume of the eluted bands known, the mass of dye can be calculated.

B. A published report (4) describes a method for the rapid separation of

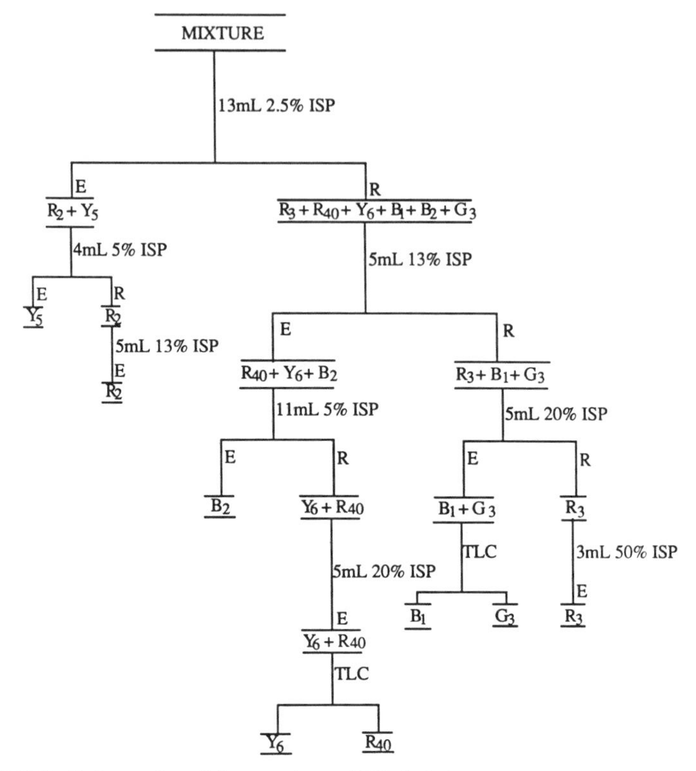

FIGURE 8-2. Schematic of Separation of FD&C dyes on a solid-phase extraction device. Isopropanol (ISP): E, elutes and R, retains. R_2, Red = 2; Y_5, Yellow = 5; Y_6, Yellow = 6; R_3, Red = 3; R_{40}, Red = 40; B_1, Blue = 1; B_2, Blue = 2; G_3, Green = 3. (Reproduced from reference 4 with permission.)

seven permitted FD&C dyes (Red = 3, Red = 40, Blue = 1, Blue = 2, Yellow = 5, Yellow = 6, and Green = 3) and one presently banned dye, Red = 2. The procedural scheme is shown in Figure 8-2. Samples were dissolved, acidified, and filtered prior to injection. Most dyes can be isolated by the use of step-gradient procedures using isopropanol/water mobile phases on a reverse-phase solid-phase extraction cartridge (e.g., SepPak C_{18} cartridge). In the present scheme, Yellow = 6 and Red = 40 are separated by the application of TLC procedures. In all cases, if quantitation is desired, it is accomplished using spectrophotometry. This procedure has been successfully applied to more than 200 food samples including candy, beverage syrup, gum balls, pudding and pie fillings, and cologne. Sugars and flavorings in the food products were generally found to be unretained, giving no interference with the analysis.

As an investigative project, the student should verify the literature procedure, select several samples, and determine which FD&C dyes are present.

If quantitation is desired, consult the original literature and incorporate the appropriate steps in this supplemental experiment.

C. *Separation of Charged Solutes*. Another popular mode of reverse-phase LC might be applicable to the separation of the two FD&C dyes, both of which have negatively charged sulfonate groups in aqueous solution. Paired-ion chromatography has become very popular for the separation of charged solutes (2). In this technique, an ionic surfactant is added to the mobile phase at a concentration of 0.1–5 mM. Cetylpyridinium chloride (CPC) has been used for this purpose (3). Since this compound is the active ingredient of Cepacol® mouthwash, Cepacol might serve as a convenient reagent for further experimentation (CPC is available from Pfaltz and Bauer, Stamford, CT). The retention behavior of both red and blue food dyes can be altered by injecting 1 mL of Cepacol® mouthwash and washing with 5 mL of water prior to the injection of Kool-Aid® solution in Section A. The dyes could not be eluted with the 18% isopropanol mobile phase. With higher concentrations of isopropanol, the dyes were eluted but, unfortunately, not separated. It might be possible to prepare an eluent which contains an appropriate concentration of cetylpyridinium chloride to affect a separation.

REFERENCES

1. A. J. P. Martin, and R. L. M. Synge, *Biochem. J.,* **35,** 91, 1358 (1941).
2. B. A. Bidlingmeyer, *J. Chromatogr. Sci.,* **18,** 525 (1980).
3. B. A. Bidlingmeyer and F. V. Warren Jr., *Anal Chem.,* **54,** 2351 (1982).
4. M. L. Young, *J. Assoc. Offic. Anal. Chem.,* **67,** 1022 (1984).

CHAPTER 9

EXPERIMENT 2: NORMAL-PHASE CHROMATOGRAPHY

Objective
Background
Additional materials
Safety and disposal
Experiments
  A. Preparation of instrument
  B. Sample preparation
  C. Chromatographic analysis
Evaluation
Discussion
Self-help questions
Additional experiments

OBJECTIVE

To demonstrate the approach used in developing a normal-phase separation on silica by observing the effect upon the solutes of changing the mobile-phase composition.

BACKGROUND

The term "normal phase" refers to a chromatographic system that utilizes a polar stationary phase and a nonpolar mobile phase. Normal-phase chromatography usually succeeds in the separation of compounds that differ in the number or chemical nature of their polar groups. Geometrical isomers are often separated successfully by normal-phase chromatography. In general, normal-phase chromatography is not used with polar samples. The most frequently used stationary phase for normal-phase chromatography is silica. Typical carrier solvents are hydrocarbons and chlorinated hydrocarbons. The mechanism of separation occurs through liquid–solid adsorption. Thus, the most important consideration is the selection of the optimum mobile phase.

As a general rule, a mobile phase should be chosen which is slightly less polar than the sample (solute). Since retention occurs as a result of competition between the mobile phase and the sample for the silica surface, the mobile phase for a given separation may have to be a mixture of two solvents. There are other secondary interactions that occur in the separation process: (1) between the mobile phase and the sample and (2) between the mobile phase and the adsorbent. In cases where mobile phase–sample interactions are encountered, changing from one solvent to another with similar polarity but different structure dramatically changes the separation.

Thus, the approach in developing a separation in normal-phase chromatography is to choose a mobile-phase solvent that is slightly less polar than the sample. If the retention times are too long, a higher-polarity solvent should be used as the eluent. As the retention time decreases, the analysis time should decrease. If resolution is reduced as the eluent is adjusted, a change to a solvent of different structure with a similar polarity can be utilized. There are several qualitative tables of solvents which may be useful to the chromatographer. The most commonly used is the Eluotropic Series. Other tables are the Hildebrand Solubility Parameters and the Snyder Eluent Strength Parameters. (Use of these tables is discussed in the text.) For additional information, read the appropriate sections in Chapters 3, 4, and 5.

ADDITIONAL MATERIALS

1 Silica gel column. The following columns have been tested in this experiment: a 2-mm ID × 60 cm Porasil A column (35–75 μm) (or a hand-packed column); a 3.9 mm ID × 15 cm μPorasil column (10 μm); a 3.9 mm × 15 cm Resolve Silica column (5 μm); and a 3.9 mm ID × 30 cm μPorasil column (10 μm). Other silica-gel columns should work in this experiment; however, they have not been tested.

4 10-mL (stoppered) flasks.

25-μL syringe.

Benzene (any supplier).

Biphenyl (any supplier).

Naphthalene (any supplier).

Carbon tetrachloride (spectrograde).

Isooctane, filtered through a 0.45-μ filter using a solvent-clarification device.

Methylene chloride filtered through a 0.45-μ filter using a solvent clarification device.

Filter paper (any supplier).

Molecular sieves (any supplier).

SAFETY AND DISPOSAL

Some of the chemicals used in this experiment are considered hazardous. This experiment does not purport to address the safety issues associated with its use. It is the responsibility of the user of this experiment to establish appropriate safety and health practices and to determine the applicability of regulatory limitations prior to use. All chemicals should be handled and disposed of in an appropriate manner consistent with the safety policy of the experimenter's company, school, or organization.

EXPERIMENTS

A. Preparation of Instrument*

Activity	Comments
1. Clear (purge) the instrument of any solvent that is immiscible with isooctane.	1. Refer to Chapter 6 and observe the miscibility rules.
2. Prepare 1 L of "water-saturated" isooctane.	2. This is accomplished by adding 2 mL of water to 1000 mL of isooctane with vigorous stirring for 4 h. Allow the water to settle and decant through filter paper to remove the excess water. This is the "water-saturated" isooctane.
3. Prepare 1 L of dry isooctane.	3. Drying the solvent may be done in many ways. One way is to add activated molecular sieves to a bottle of solvent

* This may be done by the instructor before the lab period.

and allow to sit for 2 hours with occasional swirling of the flask. (Molecular sieves may be activated by drying overnight at 200°C. They should be cooled in a dessicator before adding to the iso-octane.)

4. Prepare 1 L of 50% water-saturated isooctane by mixing 500 ml of water-saturated solvent with 500 ml of dry solvent. This is the first mobile phase and should be placed in the reservoir.

4. It is difficult to reproduce water content on a silica column when using "dry" solvent directly from a supplier's bottle. Therefore, water-saturated solvents are useful in attaining an equilibrium with the silica surface while also eliminating very active sites. This will reduce tailing of solute peaks. A 50% water-saturated solvent will make retention times reproducible.

5. The injector should be in the "inject" position.

5. This guarantees that the sample holding loop will be sufficiently flushed and cleaned to avoid contamination from the previous solvent.

6. Set the spectrophotometric detector sensitivity to 1.0 AUFS (254nm).
7. Prime the pump and begin flowing at 1 ml/min. into an appropriate collection vessel.

7. Refer to Pump manual, if necessary.

8. Stop the flow rate and place a silica column in the instrument.
9. Pump 30 column volumes* across the column before the first sample.

9. This ensures that the column will be equilibrated. Proper column equilibration is important in normal-phase LC; otherwise, nonreproducible retention times may result. This is especially true if the mobile phase composition is changing with time (e.g., adsorbing water

* The calculation of the column volume is to be alone by the student.

Activity	Comments
	from the air). It is a good practice when doing any type of chromatography to inject the sample a minimum of three times to check for retention time reproducibility, which implies that the column has reached equilibrium with the mobile phase.
10. While the instrument is equilibrating, prepare the next mobile phase of 5% methylene chloride in isooctane (50% water-saturated) by adding 25 ml of methylene chloride into 475 ml of isooctane from Step A-2.	10. This mobile phase will be used later.

B. Sample Preparation

Activity	Comments
1. Weigh 20 mg of naphthalene into a 10-mL flask and add 10 mL of carbon tetrachloride. Stopper the flask.	
2. Weigh 10 mg of biphenyl into a 10-mL flask (stoppered) and add 10 mL of carbon tetrachloride. Stopper the flask.	
3. Place 0.5 mL of benzene into a 10-mL flask and add 9.5 mL of isooctane (50% saturated) from Activity A-4. Stopper the flask.	3. This sample should be diluted to 19.5 mL with isooctane if a high-efficiency column is used.
4. Using a pipet, measure 2 ml of the naphthalene solution and 1 mL each of the biphenyl and benzene solutions into separate 10-mL flasks (stoppered) and mix well.	4. This is a three-component mixture in carbon tetrachloride.

C. Chromatographic Analysis

Activity	Comments
1. Set recorder chart speed at 0.5 cm/min or equivalent.	
2. Using 50% water-saturated isooctane as the mobile phase, inject 10 μL of the three-component mixture from Activity B-4.	2. Carbon tetrachloride is an unretained peak. An example chromatogram is shown in Figure 9-1.
3. After the three components have eluted, inject 5 μL of the naphthalene solution from Activity B-1.	3. This is necessary to match the retention time of a known compound with the corresponding peak in the mixture.
4. After the component has eluted, inject 5 μL of the biphenyl solution from Activity B-2.	4. See Comment C-3.
5. After the component has eluted, inject 5 μL of the benzene solution from Activity B-3.	5. See Comment C-3.

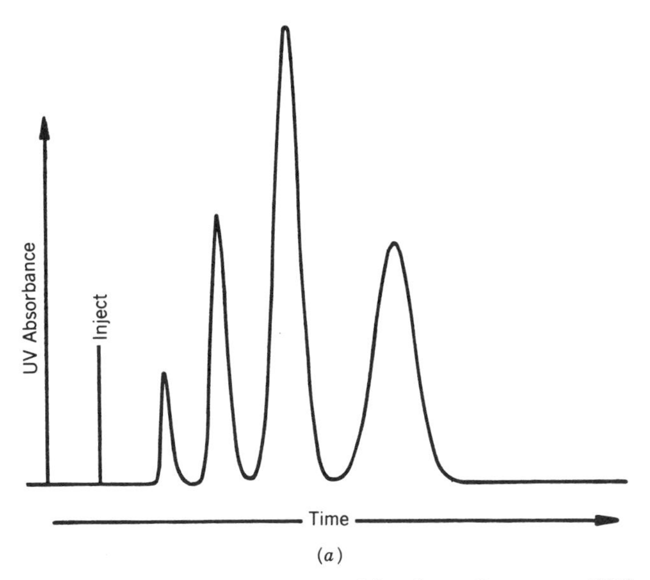

FIGURE 9-1. Example chromatograms. Mobile phase: Isooctane (50% water-saturated). Flow rate: 1.0 mL/min. Detector: UV, 254 nm. Sample: mixture of aromatics, 10 μL. Recorder: 0.2 in./min. Column: (*a*) Porasil A, 2 mm ID × 61 cm (low plate count); (*b*) Porasil A, 2 mm ID × 61 cm (high plate count); (*c*) Resolve silica, 3.9 mm ID × 15 cm; (*d*) μPorasil, 3.9 mm ID × 30 cm; (*e*) μPorasil, 3.9 mm ID × 15 cm. (Note: Actual separation will depend upon the quality of the mobile phase and column packing.)

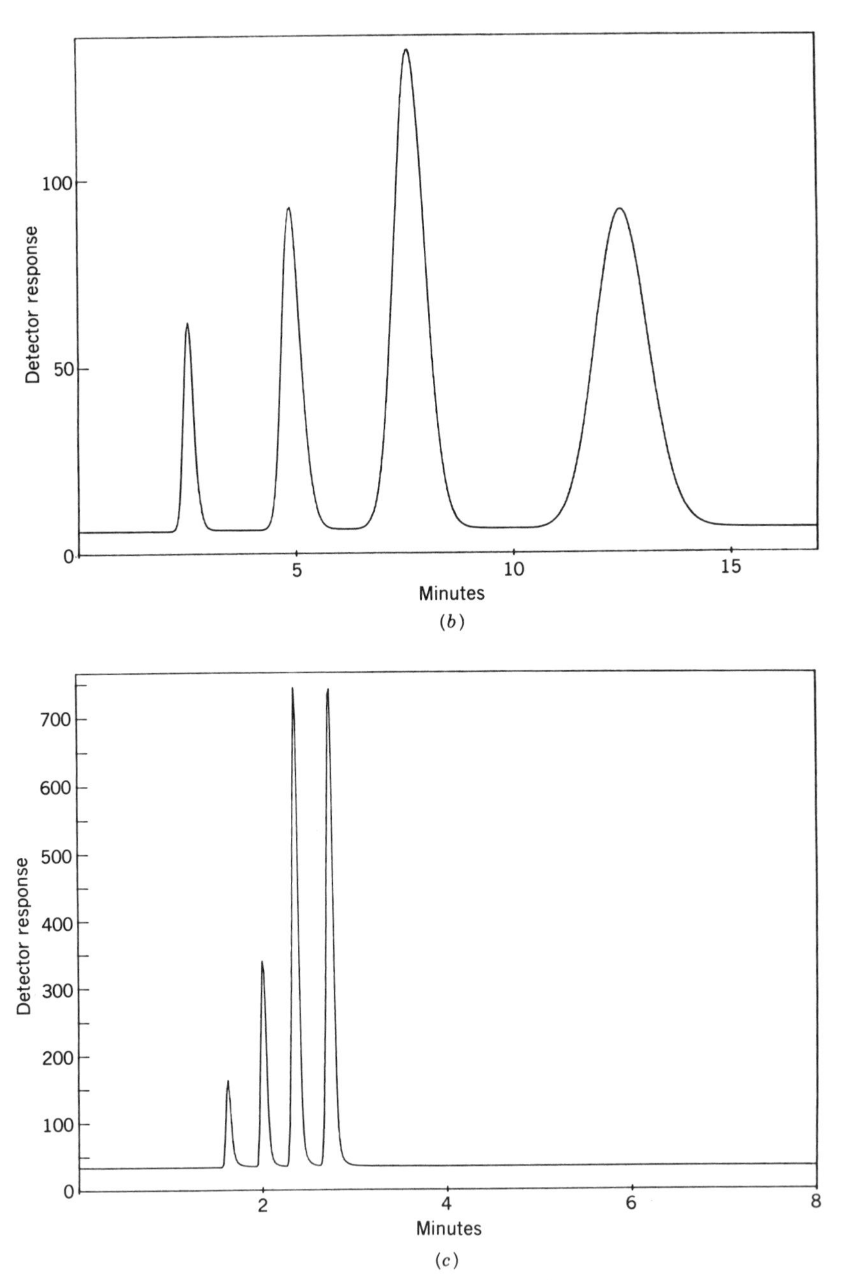

FIGURE 9-1. *(Continued)*

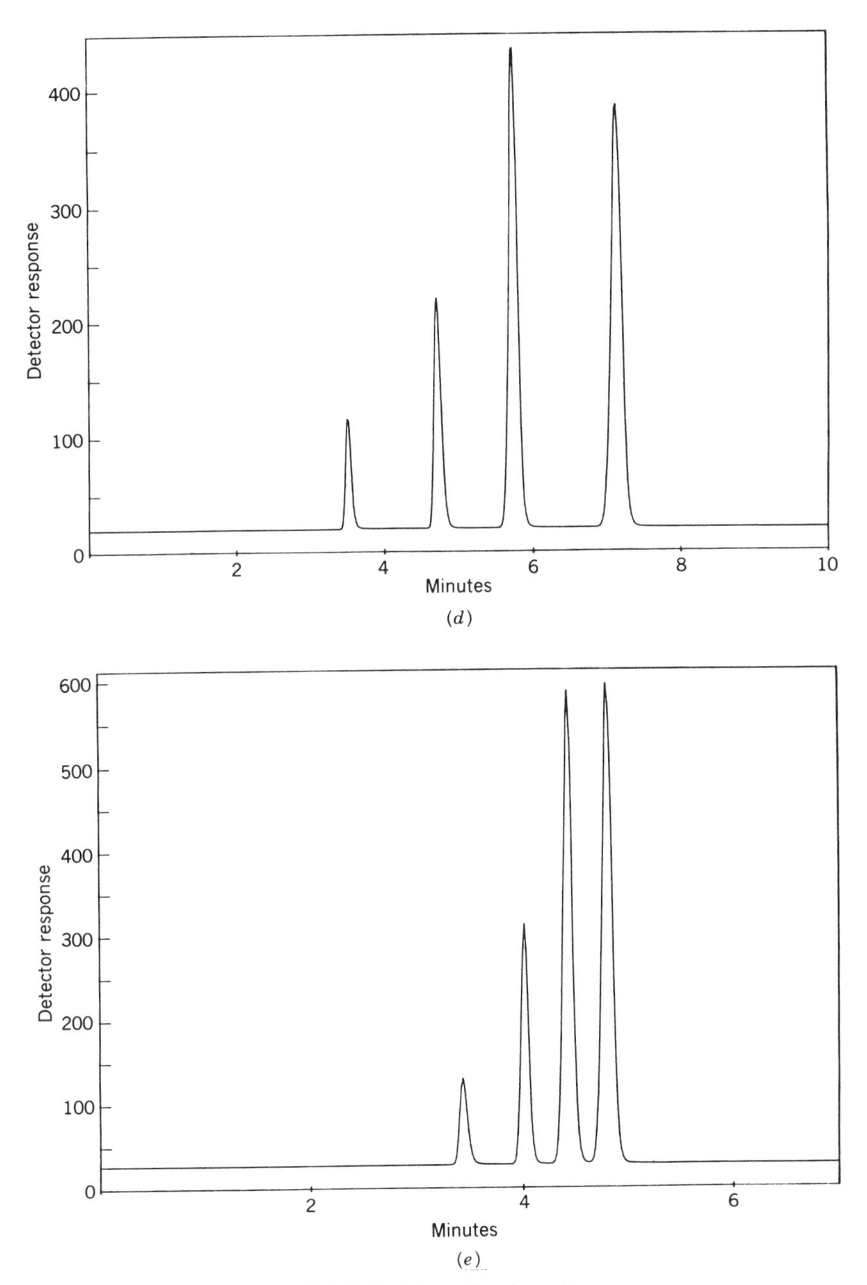

FIGURE 9-1. *(Continued)*

6. Change solvent composition to 5% methylene chloride in isooctane (50% water saturated). Wait 30 column volumes to equilibrate the system. (How can you determine if the system has equilibrated?)

6. The student should calculate a column volume.

7. Inject 10 μL of the three-component mixture (from Activity B-4).

7. An example chromatogram is shown in Figure 9-2. If time permits, do Steps C-3 through C-5.

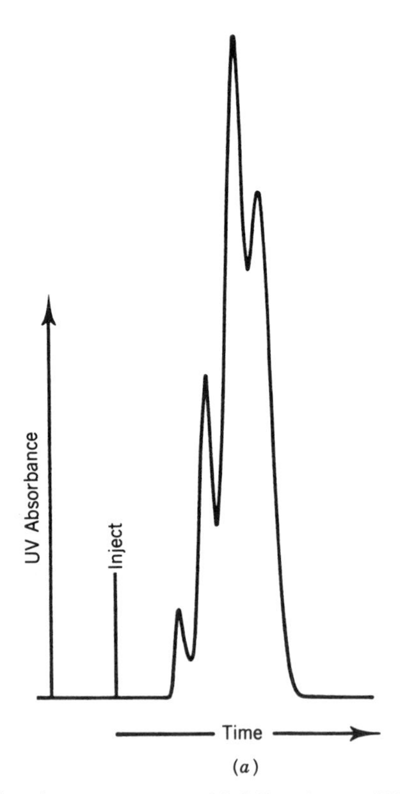

FIGURE 9-2. Example chromatograms. Mobile phase: 5% CH_2Cl_2 in isooctane (50% water saturated). Flow rate: 1.0 mL/min. Detector: UV. Sample: mixture of aromatics, 10 μL. Recorder: 0.2 in./min. Column: (*a*) Porasil A, 2 mm ID × 61 cm (low plate count); (*b*) Porasil A, 2 mm ID × 61 cm (high plate count); (*c*) Resolve silica, 3.9 mm ID × 15 cm; (*d*) μPorasil, 3.9 mm ID × 30 cm; (*e*) μPorasil 3.9 mm ID × 15 cm. (Note: Actual separation will depend upon the quality of the mobile phase and column packing).

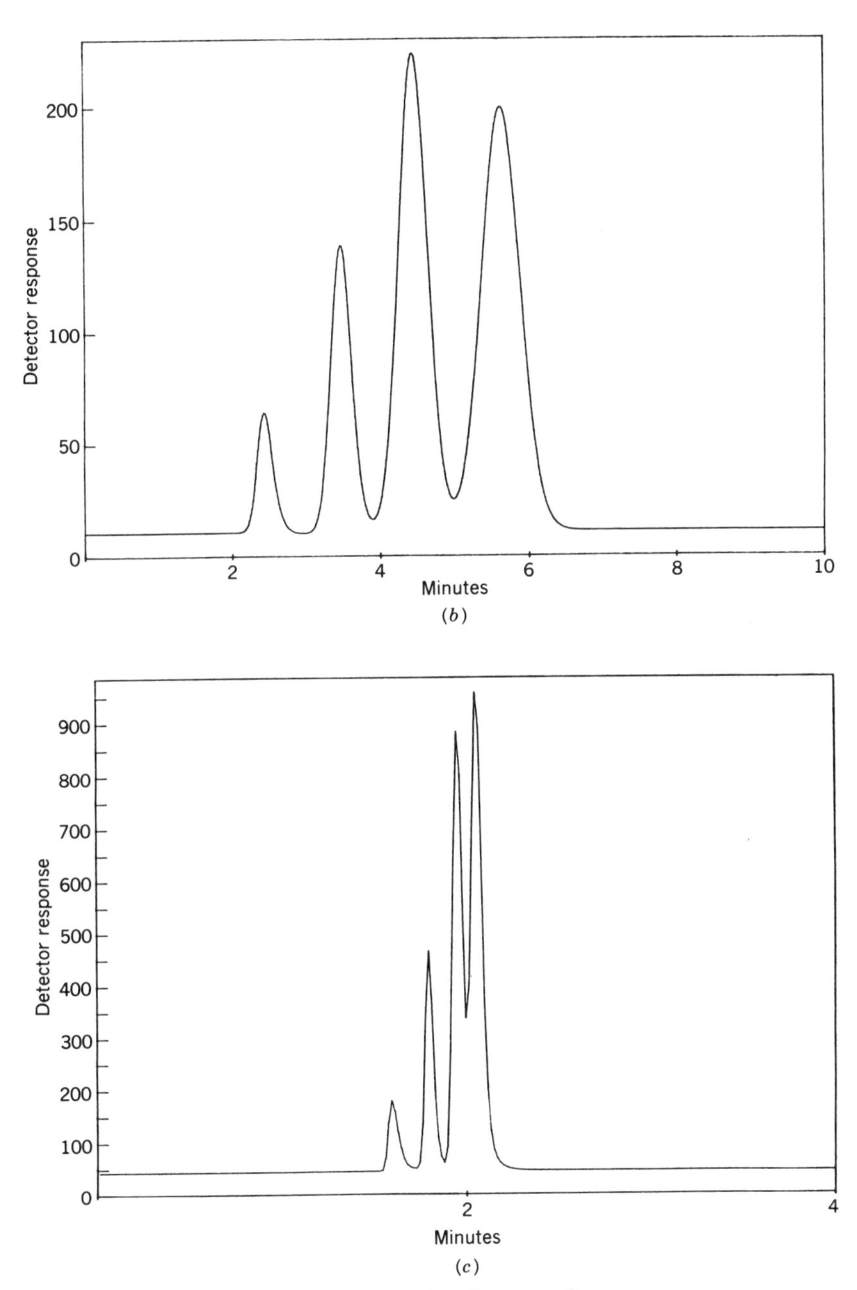

FIGURE 9-2. (*Continued*)

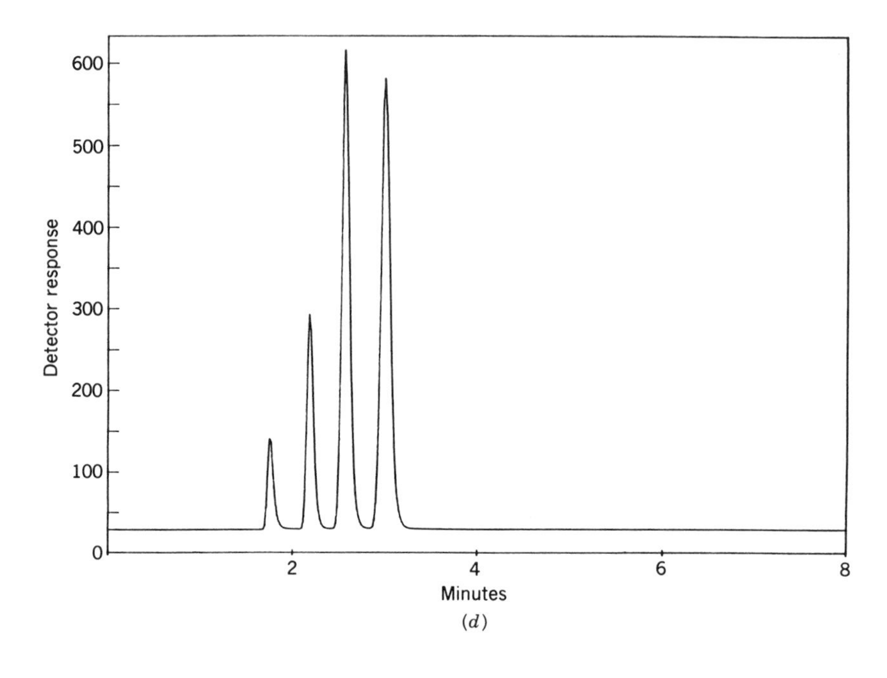

(*d*)

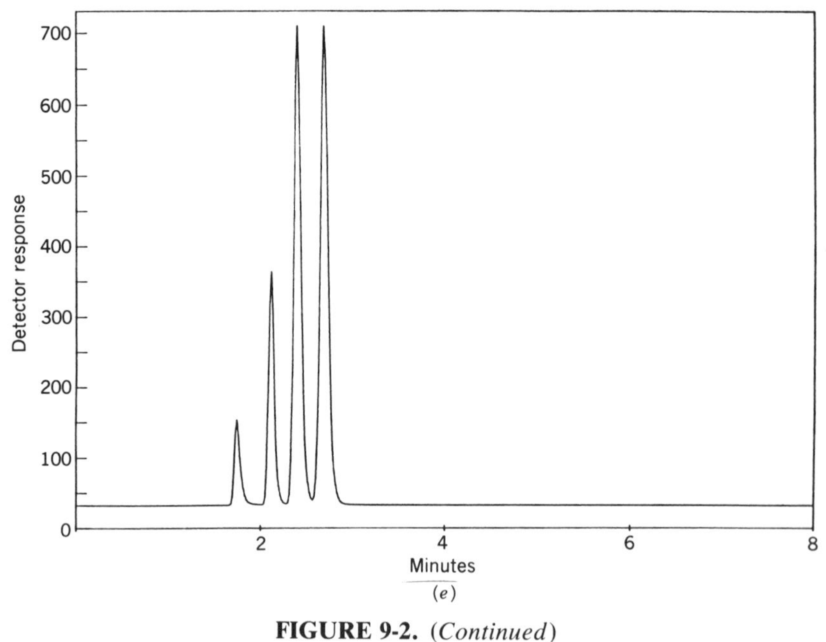

(*e*)

FIGURE 9-2. *(Continued)*

TABLE 9-1. Performance—Solvent: Isooctane (50% Water Saturated)

	Mixture				Individual	
Sample	k'	N	α	R	k'	N
Carbon tetrachloride	___	___	___	___	___	___
Benzene	___	___	___	___	___	___
Naphthalene	___	___	___	___	___	___
Biphenyl	___	___	___	___	___	___

EVALUATION

A. All chromatograms should be properly labeled and attached to the last sheet in this experiment. Proper labeling requires the following:

Sample: Components, approximate amount of each and injection volume.
Detector: Type of detector and sensitivity.
Column: Type of column used, length, and diameter (ID).
Solvent: Mobile-phase composition and flow rate.
Recorder: Recorder chart speed.

B. Example results for the chromatographic separations of benzene, naphthalene, and biphenyl for specified conditions are presented in Figures 9-1 and 9-2. Your chromatograms should be used for all subsequent calculations of performance parameters.

C. On a separate sheet of paper construct Tables 9-1 and 9-2 and record the appropriate performance parameters. Is carbon tetrachloride an unretained peak? How do you support this conclusion?

DISCUSSION

Adsorption of the sample molecules to the stationary phase is the basis for retention. As the mobile phase becomes more like the stationary phase in

TABLE 9-2. Performance—Solvent: Isooctane (50% Water Saturated): CH_2Cl_2

	Mixture				Individual	
Sample	k'	N	α	R	k'	N
Carbon tetrachloride	___	___	___	___	___	___
Benzene	___	___	___	___	___	___
Naphthalene	___	___	___	___	___	___
Biphenyl	___	___	___	___	___	___

terms of polarity, the attraction of the sample to the stationary phase is reduced and the sample elutes more quickly. In this experiment, methylene chloride is more polar than isooctane and, therefore, the 5% methylene chloride in isooctane solvent gave less retention than pure isooctane.

SELF-HELP QUESTIONS

1. What is k' a measure of?
2. What is N a measure of?
3. Do the data in Tables 9-1 and 9-2 support your definitions? In what way?
4. From the data recorded in Tables 9-1 and 9-2, what is the relation of plate number (N) to k'?
5. From the data recorded in Tables 9-1 and 9-2, do the number of plates (N) and k' depend upon whether the sample is injected as an individual sample or as a mixture?
6. What can be said about the relation of selectivity (α) to solvent strength, to k', and to R?

The resolution obtained with a given column and solvent system is a function of column length. Resolution can be improved by lengthening the column or by utilizing "recycle" (Chapter 6). It is important to remember that the resolution increases directly with the square root of the length (to increase resolution by 2X requires 4X column length). This topic is discussed in Chapter 10 (Experiment 3).

ADDITIONAL EXPERIMENTS

The following is a suggested experiment which will further increase your understanding of the use of normal-phase chromatography:

A. Repeat the experiment described above using a mobile phase containing 0.2% isopropanol in isooctane and observe the effect that trace quantities of alcohols (more polar solvent) can have on silica separations.

B. Repeat the experiment described above using a mobile phase containing 0.1% acetonitrile in isooctane. What is the effect?

CHAPTER 10

EXPERIMENT 3: EFFECT OF COLUMN LENGTH AND RECYCLE

Objective
Background
Equipment and reagents
Safety and disposal
Experiments
  A. Preparation of instrument
  B. Sample preparation
  C. Chromatographic analysis
  D. Recycle chromatography
Evaluation
Discussion
Self-help questions
Additional experiments

OBJECTIVE

To determine the influence of column length upon the number of theoretical plates and upon resolution and to demonstrate the ability of increased resolution through utilization of recycle chromatography.

BACKGROUND

The goal of the chromatographic process is the separation of one component of a mixture from another. The measurement of the degree of separation between two components is resolution, defined as the distance between the peak centers of two peaks divided by the average base width of the peaks. There are two ways to improve the resolution between two peaks—selectivity and efficiency considerations. In Chapter 9 (Experiment 2) the approach was to change the selectivity through the chemistry of interactions between the mobile phase–solute–stationary phase. Because the mobile-phase strength influences both the capacity factor (k') and the selectivity (α), it is difficult to predict the exact effect of the solvent changes upon resolution. However, increasing the number of theoretical plates will improve the resolution in a straightforward way and that is the subject of this experiment.

The number of theoretical plates is a measurement of band spreading of a peak after it traveled through the chromatographic system. The smaller the band spreading is, the higher the number of theoretical plates. Four major factors affect the number of theoretical plates in a column: the size of the particles, the distribution of particle sizes, whether the particles are fully porous or pellicular, and how well the column is packed.

The resolution of a separation will increase by the square root of the increase in the number of theoretical plates. So, in principle, any separation could be attained by simply obtaining a column(s) with "enough" plates. In practice, however, this is not possible because as the length of column is increased, so is the pressure required to provide flow. In addition, the cost of adding many columns may become excessive. One effective way to increase column length without the additional burden of column cost and without an increase of pressure drop across the system, is to use the same column(s) more than once by utilizing "recycle." In recycle, the column effluent from the detector, instead of going to waste, is diverted to the inlet of the solvent delivery system and sent through the column again. In this technique we can use the column system again and again to achieve higher resolution. By utilizing recycle, we reduce the investment in columns and we achieve high efficiencies without the subsequent cost of high pressure to achieve the flow rate desired. Recycle is discussed in Chapter 6.

Equipment for recycle operations differs from conventional HPLC equipment. For a recycle system to be useful, the extracolumn band spreading must be small relative to the band spreading of the column. This involves the solvent delivery system, transport tubing, and detector(s). Also, because a recycle system is a closed system with a finite volume, the operator must be aware that fast-moving materials could eventually overtake slower-moving materials and remix. To prevent peak overlap, a means must be provided to allow the operator to remove a portion of the sample components before overlap can occur.

This experiment is intended to first demonstrate the additivity of plates between columns. Recycle is then investigated from the theoretical viewpoint and from the practical standpoint of "shaving" to collect fractions.

For additional background information refer to the appropriate sections of Chapter 6.

EQUIPMENT AND REAGENTS

2 Silica gel columns. The following columns have been tested in this experiment: two 2 mm ID × 61 cm Porasil A columns (35–75 μm) (or hand-packed columns); two 3.9 mm ID × 15 cm μPorasil columns (10 μm); two 3.9 mm ID × 15 cm Resolve Silica columns (5 or 10 μm). Other silica gel columns should work in this experiment; however, they have not been tested. A pellicular column (packed or purchased) may also be used (e.g., 2 - 2 mm ID × 61 cm Corasil II columns).

4 10-mL flasks (stoppered).

25-μL syringe.

Benzene (any supplier).

Biphenyl (any supplier).

Naphthalene (any supplier).

Carbon tetrachloride (any supplier).

Isooctane, filtered through a 0.45-μ filter using a solvent clarification device.

Methylene chloride, filtered through a 0.45-μ filter using a solvent clarification device.

SAFETY AND DISPOSAL

Some of the chemicals used in this experiment are considered hazardous. This experiment does not purport to address the safety issues associated with its use. It is the responsibility of the user to establish appropriate safety and health practices and to determine the applicability of regulatory limitations prior to use. All chemicals should be handled and disposed of in an appropriate manner consistent with the safety policy of the experimenter's company, school, or organization.

EXPERIMENTAL

A. Preparation of Instrument*

Activity	Comments
1. Clear (purge) the instrument of any solvent that is immiscible with isooctane.	1. Refer to Chapter 6 and observe miscibility rules.

* This may be done by the instructor before the lab period.

Activity	Comments
2. Prepare 1 L of 50% water-saturated isooctane for the mobile phase.	2. If there is a need to remake this mobile phase refer to Chapter 9 (Experiment 2), Activity A-2 through A-4.
3. The injector handle should be in the "inject" position.	3. An autosampler should not be used in this experiment if the recycle section is attempted. Also, 0.009-ID tubing should be used throughout the system, using the shortest possible lengths for optimum results.
4. Set the spectrophotometric detector sensitivity to 1.0 AUFS (254 nm).	
5. Prime the solvent delivery system (pump) if necessary, and begin pumping at a flow rate of 2 mL/min. Sufficient mobile phase should be "flushed" through the system to insure a stable baseline on the photometric detector.	5. Refer to the manufacturer's solvent delivery (pump) manual, if necessary.
6. Stop the flow rate and place two silica columns in the instrument.	
7. Pump 30 column volumes* across the column before injecting the first sample.†	7. This ensures that the column will equilibrate. This will result in reproducible retention times of the sample components.

B. Sample Preparation

Activity	Comments
1. Weigh 20 mg of naphthalene into a 10-mL flask. Add 10 mL of carbon tetrachloride and stopper the flask.	

* The calculation of a column volume is to be done by the student.

† This can be done at 4 mL/min. However, the solvent system flow should gradually be brought up to 4.0 mL/min over several minutes when *any* column is used. This will insure that an initially high flow rate will not cause the column bed to collapse. Remember to return to 2.0 mL/min before beginning the experiment. If the column has become "contaminated," flush the system with 30 column volumes each of: 50% methanol in methylene chloride, methylene chloride, and isooctane (50% water saturated). This should clean the column and bring its activation level back to its original value.

Activity	Comments
2. Weigh 10 mg of biphenyl into a 10-mL flask. Add 10 mL of carbon tetrachloride and stopper the flask.	
3. Place 0.5 mL of benzene into a 10-mL flask. Add 9.5 mL of isooctane and stopper the flask.	
4. Using the pipet, measure 2 mL of the naphthalene solution and 1 mL each of the biphenyl and benzene solutions into a separate 10-mL flask (stoppered). Mix well.	4. This is a three-component mixture in carbon tetrachloride.
5. Using the pipet, measure 2 mL of the naphthalene solution and 1 mL of the biphenyl solution into a 10-mL flask (stoppered). Mix well.	5. This is a two-component mixture in carbon tetrachloride.

C. Chromatographic Analysis

Activity	Comments
1. Set recorder chart speed at 0.5 cm/min or equivalent.	
2. Using 5% methylene chloride in isooctane (50% water saturated) isooctane as the mobile phase, inject 10 μL of the three-component mixture from Activity B-4.	2. Carbon tetrachloride is an unretained peak. An example chromatogram is shown in Figure 10-1.
3. After the components have eluted, repeat Activity C-2.	
4. Remove one of the columns and reconnect the remaining column in the system. Cap the column, which was removed, as it will be used again in Activity C-7.	4. This experiment will determine the efficiency of one of the columns.
5. Inject 10 μL of the mixture of sample solution from Activity B-4.	5. A sample chromatogram is shown in Figure 10-1.
6. After the components have eluted, repeat the 10-μL injection of the mixture of sample solution from Activity B-4.	

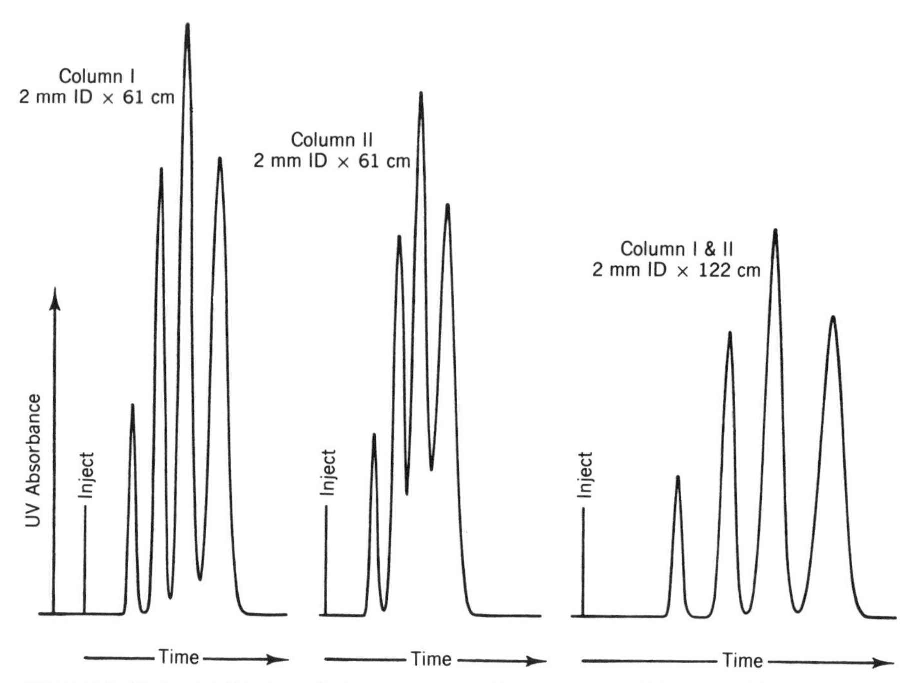

FIGURE 10-1. Additivity of plates on a totally porous packing. Mobile phase: 5% CH_2Cl_2 in isooctane (50% water saturated). Flow rate: 2 mL/min. Detector: UV, 254 nm, 0.5 AUFS. Sample: Mixture of aromatics, 10 μL. Column: Porasil A, 2 mm ID × 122 cm. (Note: Actual separation will depend upon quality of mobile phase and column packing.)

7. Replace the present column with the one removed in Activity C-4.

   This experiment will measure the efficiency of the second column. An example chromatogram is shown in Figure 10-1.

8. Inject 10-μL injection of the mixture of sample solution from Activity B-4.
9. Repeat the 10-μL injection of the mixture of sample solution from Activity B-4.
10. Place the two 2 mm ID × 61 cm columns into the instrument.

    This is the column to be used in the following section.

D. Recycle Chromatography

Activity	Comments
1. Change mobile phase to 10% methylene chloride in isooctane (50% water saturated) from Activity A-2.	1. This mobile phase may have to be adjusted to give an R value of about 0.7 since the plate number of columns will vary. Check with your instructor.
2. Pump 30 column volumes of mobile phase across the column before injecting the first sample.	2. This equilibrates the column.
3. Inject 10 μL of the solution from Activity B-5.	3. This will be used in the recycle of two peaks with low α values. Refer to Figure 10-2 for an example chromatogram.
4. After 15 sec, rotate the injector to the "load" position.	4. This will insure minimum spreading in the recycle mode.
5. After the peak from the carbon tetrachloride reaches baseline, wait 15 sec and turn the recycle knob to the "recycle" position on the instrument or solvent delivery system.	5. The 15 sec ensures that no carbon tetrachloride will be recycled.
6. After the second recycle, turn the recycle knob to the "waste" position on the instrument or solvent delivery system.	6. This will give three recycles through the column or the equivalent of four times the number of plates in the 122-cm (4-ft) section. Refer to Figure 10-2.
7. When a stable baseline is again attained, inject 10 μL of the three-component mixture from Activity B-4.	7. This experiment will use the recycle mode to shave the peaks for purity as one would do in a preparative situation.
8. After the injection, wait 15 sec and return the injector to the "load" position.	8. This ensures minimum spreading in the recycle mode. Refer to Figure 10-3 for an example chromatogram.
9. Turn the recycle valve knob to the collect position.	9. This causes solvent to drop from the collect port.
10. When the carbon tetrachloride peak elutes, collect the peak.	10. Refer to Figure 10-3.
11. After the carbon tetrachloride (first peak) reaches baseline, wait 15 sec and turn the recy-	11. The 15 sec ensures that no carbon tetrachloride will be recycled. The order of elution

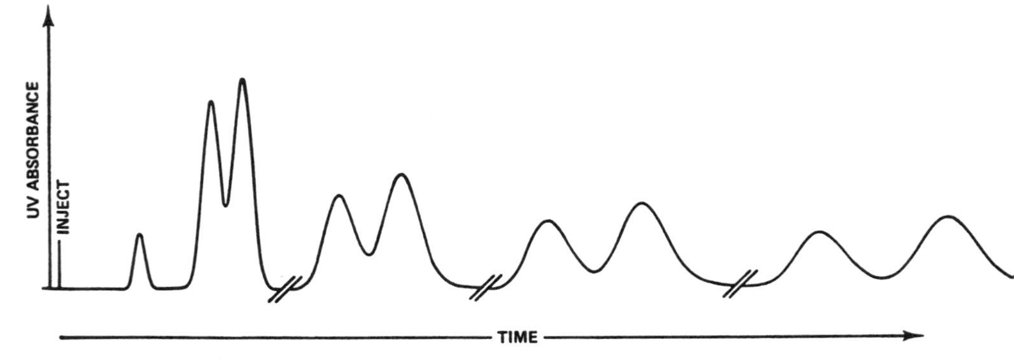

FIGURE 10-2. Recycle of two peaks on a totally porous packing. Mobile phase: 10% CH_2Cl_2 in isooctane (50% water saturated). Flow rate: 2.0 mL/min. Detector: UV, 254 nm, 1.0 AUFS. Sample: naphthalene and biphenyl. Column: Porasil A, 2 mm ID × 122 cm. (Note: Actual separation will depend upon quality of mobile phase and column packing.)

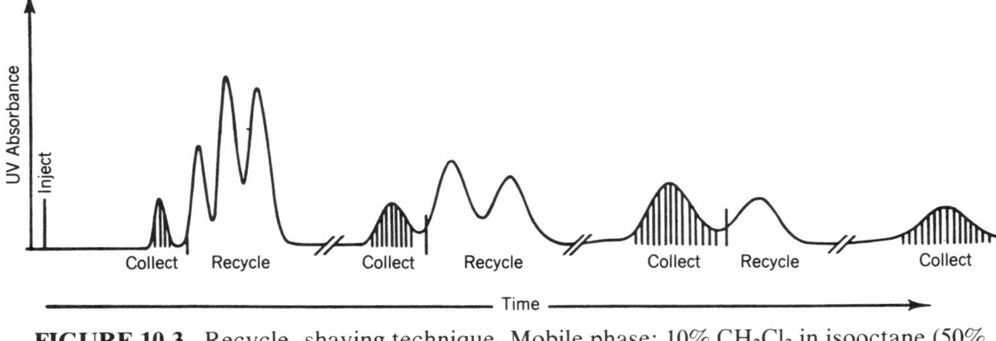

FIGURE 10-3. Recycle–shaving technique. Mobile phase: 10% CH_2Cl_2 in isooctane (50% water saturated). Flow rate: 2 mL/min. Detector: UV, 254 nm, 1.0 AUFS. Sample: mixture of aromatics. Column: Porasil A, 2 mm ID × 122 cm. (Note: Actual separation will depend upon quality of mobile phase and column packing.)

cle switch to the "recycle" position on the valve.

will be carbon tetrachloride, benzene, naphthalene, and biphenyl.

12. After the last peak elutes, turn the recycle switch to the "collect" position.

12. Since the shaving technique is being used, this prepares the instrument for collection of the next peak. Refer to Figure 10-3.

13. As the benzene peak elutes, collect it.
14. After the recorder pen starts forming the next peak, wait 15 sec and turn the recycle valve knob to the "recycle" mode.

14. The 15 sec is the "lag" volume between the recorder response and the recycle switch.

15. After the last peak elutes, turn the recycle valve knob to collect.
16. As the naphthalene peak elutes, collect it.

16. Refer to Figure 10-3.

17. After the recorder pen starts forming the next peak, wait 15 sec and turn the recycle knob to the "recycle" position.
18. As the final peak elutes, collect it.
19. Set the spectrophotometric detector to 0.05 AUFS (254 nm).
20. Inject 100 μL of the collected fraction from Activity D-10.

20. This will check the "purity" of the cut. Clean the syringe by rinsing the syringe three times with the mobile phase to ensure that no contamination occurs between samples.

21. After the component elutes, inject 100 μL from Activity D-13.

21. Refer to Comment D-20.

22. After the component elutes, inject 100 μL from Activity D-16.

22. Refer to Comment D-20.

23. After the component elutes, inject 100 μL from Activity D-18.

23. Refer to Comment D-20.

TABLE 10-1. Number of Plates per Column for Four-Component Mixture

Compound	Column I	Column II	Columns I & II
Carbon tetrachloride			
Benzene			
Naphthalene			
Biphenyl			

EVALUATION

A. All chromatograms should be properly labeled and attached to the last sheet in this experiment. Proper labeling requires the following:

Sample: Components and injection volume.
Detector: Type of detector and attenuation.
Column: Type of column used, length, and diameter (ID).
Solvent: Mobile-phase composition and flow rate.
Recorder: Recorder chart speed.

B. Example chromatograms of the separations of benzene, naphthalene, and biphenyl for the additivity of plates experiment are presented in Figure 10-1. Your chromatogram should be used for all subsequent calculations of performance parameters.

C. On a separate sheet of paper construct Table 10-1 and record the number of plates (N).

D. The example of recycle for naphthalene and biphenyl is shown in Figure 10-2. For your recycle chromatogram, calculate the number of plates in each pass and place the data in a table constructed under the format of Table 10-2.

E. Plot the number of plates (N) versus cycle number from Table 10-2. Also calculate R for the naphthalene–biphenyl pair and plot R versus N. Use an average N value.

F. The example of recycle shaving of the four-component mixture is shown in Figure 10-3. Compare your chromatogram to this figure.

TABLE 10-2. Number of Plates per Recycle of Naphthalene and Biphenyl

	Cycle Through Column			
Compound	1st	2nd	3rd	4th
Naphthalene				
Biphenyl				

TABLE 10-3. Cost Savings when Using Recycle[a]

		Relative Resolution	
Column Length (cm)	Column Cost ($)	Single Pass	Eight Recycles
122	200	1	2.8
244	400	1.4	4
488	800	2	5.6
976	1600	2.8	Generally unnecessary
1952	3200	4	Generally unnecessary

[a] The same resolution obtained with 1952 cm of column can be achieved by recycling 8 times through 244 cm of column at considerably less investment in columns.

DISCUSSION

The number of theoretical plates (N) is an empirical measure of the efficiency of the chromatographic column. It can be easily measured and its influence upon the resolution of a chromatographic separation can be predicted. Since the number of theoretical plates (N) varies linearly with column length, a straightforward approach to improving resolution is to increase column length. However, with increase in column length, backpressure increases as will the cost. An alternative to adding longer and longer columns is to recycle the sample through the chromatographic column. In this way, the number of theoretical plates will increase without the inconvenience of high backpressure and increased cost.

Considering a modest cost of columns to be $50.00 per foot, the advantages of recycle are shown in Table 10-3.

SELF-HELP QUESTIONS

1. Were the plates from columns I and II additive (refer to Table 10-1)? What sources of error are there in the measurement of plates?
2. What is the relationship between the number of plates and the resolution of two peaks?
3. Is the "shaving–recycle" technique useful in preparing pure fractions? Refer to your data.
4. Why is it desirable to have a large k' when doing recycle?
5. Is recycle a useful technique? Explain.

ADDITIONAL EXPERIMENTS

The following suggested supplemental experiment will increase your understanding of the use and characteristics of high-resolution LC. This may be done under the operating conditions previously established, depending on the lab time available and course requirements.

If you have bulk quantity of silica gel available (37–75 μm), pack a 2 mm ID × 61 cm column and measure its efficiency. Compare this number to the number of plates for the columns used earlier in the experiment.

CHAPTER 11

EXPERIMENT 4: GEL PERMEATION CHROMATOGRAPHY USING DUAL DETECTORS (UV AND RI)

Objective
Background
Additional materials
Safety and disposal
Experiments
A. Preparation of instrument
B. Sample preparation
C. Chromatographic analysis
Evaluation
Discussion
Self-help questions
Additional experiments

OBJECTIVE

The objective of this experiment is to calibrate a gel permeation chromatographic column and to separate a series of small molecules by GPC. Additionally, this experiment will demonstrate the usefulness of a spectrophotometric and RI detector in GPC preparative separations.

BACKGROUND

Gel permeation chromatography can be considered as a mechanical sorting of sample molecules according to differences in their "effective size" in solution. As the sample (solute) molecules travel through the column they penetrate the pores of the packing. The degree to which they penetrate the pores depends upon the size of the solute molecules relative to the size of the pores in the three-dimensional column packing matrix. The smaller molecules in the sample migrate into more of the smaller pores of the cross-linked polymer gel than the larger molecules in the mixture. Consequently, the larger molecules elute first, followed by successively smaller molecules until all are eluted at the total pore volume of the column. Because the sample and mobile phase are chosen so that the sample has no attraction to the stationary phase (i.e., the packing is inert with respect to the sample), the maximum amount of solvent required for complete sample elution from the column will equal the liquid volume of the column. In other words, all sample components will elute between V_i (volume of liquid between the particles, often called "interstitial volume") and V_o (total column volume). This makes GPC very predictable for data interpretation and methods development.

In the case of a polymer, the largest molecular chains may be excluded completely from the pores in the packing material and move only through the interstitial spaces between the packing particles. For smaller molecular chains, the path through the column is more tortuous since these molecules enter the pores of the packing material and are retarded more than the larger molecules. The result is a chromatogram that represents a distribution of molecular sizes in the polymer. Gel permeation chromatography is also an extremely useful method for the separation of low-molecular weight species as this experiment demonstrates.

Gel permeation chromatography columns should be calibrated to determine the values for V_i and V_o for the particular column(s) being used and to determine the range of resolving power for the column(s). (Note that sometimes in the GPC literature the term V_t is used in place of V_o.) When V_R (retention volume of a component) equals V_i, all sample component molecules are being excluded. This results in no separation. To achieve separation in this situation, the largest pore diameter of the packing must be increased. When V_R equals V_o, all the sample components are small enough to enter all accessible pores and no separation occurs. A column that has smaller pores should be added.

Since the calibration of a GPC is dependent upon the *effective size in solution* of the sample molecules, the type (structure) of molecules used for the calibration is important. The ideal case is to calibrate with a "standard" sample(s) of the material of interest. However, this is not always possible. In those instances, arbitrary standards are chosen. The arbitrary standards are used to construct a "size" calibration where the molecular size is calculated from the standard. For polymer analysis, these standards are often polystyrene of narrow molecular weight distribution. These standards may be purchased from a variety of suppliers.

In GPC it is possible to have two compounds of identical molecular weights but with different shapes. Since the shape determines the effective size in solution, the compounds may separate quite well on a gel permeation column. Also, it is possible to have two molecules with very different weights but essentially identical shapes in solution. This coelution results in no separation. This is why calibration of the GPC column should be done with compounds identical to or similar to the compounds to be separated.

If the sample is readily soluble in the mobile phase, GPC is unmatched by any other mode of chromatography for simplicity, since the entire analysis is accomplished in a column volume. The time and effort required to develop a separation is less than any other mode of HPLC. It can be of immense value in the purification or organic and inorganic synthesis reaction mixtures, purification of natural products extracts, and for the rapid clean-up of extracts (from plants, insects, soil, etc.) prior to the assay of small molecules. Aqueous size separation is referred to as gel filtration chromatography and is very useful for protein separations and the analysis of water-soluble polymers.

For additional background information on GPC refer to the appropriate sections in Chapters 3 and 4.

ADDITIONAL MATERIALS

1 High-efficiency (> 4500 plates) GPC column. The following columns have been tested for this experiment: a 7.8 mm ID × 30 cm 100-Å μStyragel column and a 7.8 mm ID × 30 cm 100 Å Ultrastyragel column. Other high-efficiency GPC columns (100 Å or equivalent) should be suitable; however, they have not been tested in this experiment.

6 100-mL volumetric flasks.

Dimethyl phthalate (any supplier).

Dibutyl phthalate (any supplier).

Dioctyl phthalate (any supplier).

Benzene (any supplier).

Tetrahydrofuran (THF) filtered through a 0.45-μ filter using a solvent clarification device.

1-mL pipet (graduated in 0.1-mL increments).

25-μL syringe.

100-μL syringe.

Note: Substitute compounds could be used for this experiment (e.g., any homologous series). These could be chosen according to the interest of the students and instructor. The molecules selected for calibration are to some extent determined by the type of detector(s) available. If an RI detector is available, any compound may be used.

SAFETY AND DISPOSAL

Some of the chemicals used in this experiment are considered hazardous. This experiment does not purport to address the safety issues associated with its use. It is the responsibility of the user of this experiment to establish appropriate safety and health practices and to determine the applicability of regulatory limitations prior to use. All chemicals should be handled and disposed of in an appropriate manner consistent with the safety policy of the experimenter's company, school, or organization.

EXPERIMENTS

A. Preparation of Instrument*

Activity	Comments
1. Clear (purge) the HPLC system (with no column) with the THF to be used as the mobile phase.	1. The mobile phase already in the instrument must be compatible with the mobile phase to be used in this experiment. Refer to Chapter 6 and observe miscibility rules.
2. Place 700 mL of THF in the mobile phase reservoir.	2. For use on some HPLC hardware this mobile phase may need to be degassed.
3. Prime the solvent delivery system (pump) and start the flow at 2.0 mL/min.	3. Refer to manufacturers' solvent delivery system (pump) manual. Since no column is installed, either collect flow from the injector into a suitably sized container or jumper the injector to the detector in order to flush the entire system to waste.
4. Stop the flow rate, install one GPC 100-Å column in the instrument and return the flow rate to 1.0 mL/min. (Read the column use instructions carefully.)**	4. If a recommended column is used, read the procedure for purging and setting the flow rate for the μStyragel column in the Gel Permeation Chromatography Column Care and Use Manual (CU84037-μStyragel). If another column is used refer to the corresponding Care and Use Manual.

* This may be done by the instructor before the lab period.

** The GPC column must be a "good" column. Therefore, if there is any question about the condition of the column refer to the Care and Use Manual for testing procedures.

Activity	Comments
5. Set the spectrophotometric detector sensitivity setting to 0.5 AUFS (254 nm).	5. If an RI detector is available, set it at an appropriate sensitivity.
6. Allow the system to equilibrate under these conditions while the samples are being prepared.	6. An equilibrated system is best indicated by a stable baseline on the detector.

B. Sample Preparation

Activity	Comments
1. Standards: Prepare individual solutions of a 1.25 mg/mL concentration of dioctyl, dibutyl, and dimethyl esters of phthalic acid by weighing 0.125 g of each ester into an individual 100-mL volumetric flask and diluting to volume with THF from the same lot as used in the solvent reservoir. Place 0.1 mL of benzene in another 100-mL volumetric flask and dilute to the mark with THF as before.	1. Density values for the samples are: dioctyl phthalate, $d = 0.981$ g/mL; dibutyl phthalate, $d = 1.043$ g/mL; and dimethyl phthalate, $d = 1.190$ g/mL. There should be four standard solutions prepared in this activity.
2. Standard mix: Pipet 0.5 mL of dioctyl, dibutyl, and dimethyl phthalate solutions from Activity B-1 into a vial and mix well.	
3. Preparative mix: Prepare a solution containing 20 mg/mL of dioctyl, dibutyl, and dimethyl phthalate by weighing 0.200 g of each ester into a 10-mL volumetric flask and bringing it to volume with THF.	

C. Chromatographic Analysis

Activity	Comments
1. Set recorder to 0.5 cm/min or equivalent.	
2. Inject 10 μL of the individual phthalate standards from Activity B-1 and analyze sequentially.	2. Clean syringe thoroughly between injections. If a peak is off scale, inject less volume (e.g., 1

Record the elution times to the centers of the peaks. Inject 25 μL of the benzene solution prepared in Activity B-1.

μL) and/or make the detector less sensitive.

3. Inject 50 μL of the standard mix (Activity B-2) of the esters.

3. This is to familiarize the operator with the complex chromatogram. Note the beginning and the end of the individual peaks in the mixture.

4. Inject 100 μL of ester standard mix (Activity B-2).

4. Be prepared to collect fractions.

5. When the peak begins, collect the mobile phase for 30 sec.

5. This fraction should contain no sample as it represents the "lag" volume in the connecting tube. Some solvent delivery systems may contain a valve for collecting mobile phase effluent. If so, simply turn this "selector" valve to the "collect" position.

6. Collect fractions at 30-sec intervals until the recorder is at baseline for at least 30 sec.
7. Inject 20 μL of each collected fraction and record elution times and estimated purity of the major component peak

EVALUATION

A. All chromatograms should be labeled. Proper labeling requires the following:

Sample:	Components, approximate amount of each and injection volume.
Detector:	Type of detector and sensitivity.
Column:	Type of column used, length, diameter (ID) and pore size.
Solvent:	Mobile phase composition and flow rate.
Recorder:	Recorder chart speed.

B. Typical chromatograms for the separation of dimethyl, dibutyl, and dioctyl phthalates on one 100-Å μStyragel column are shown in Figure 11-1 and for the 100-Å Ultrastyragel in Figure 11-2.

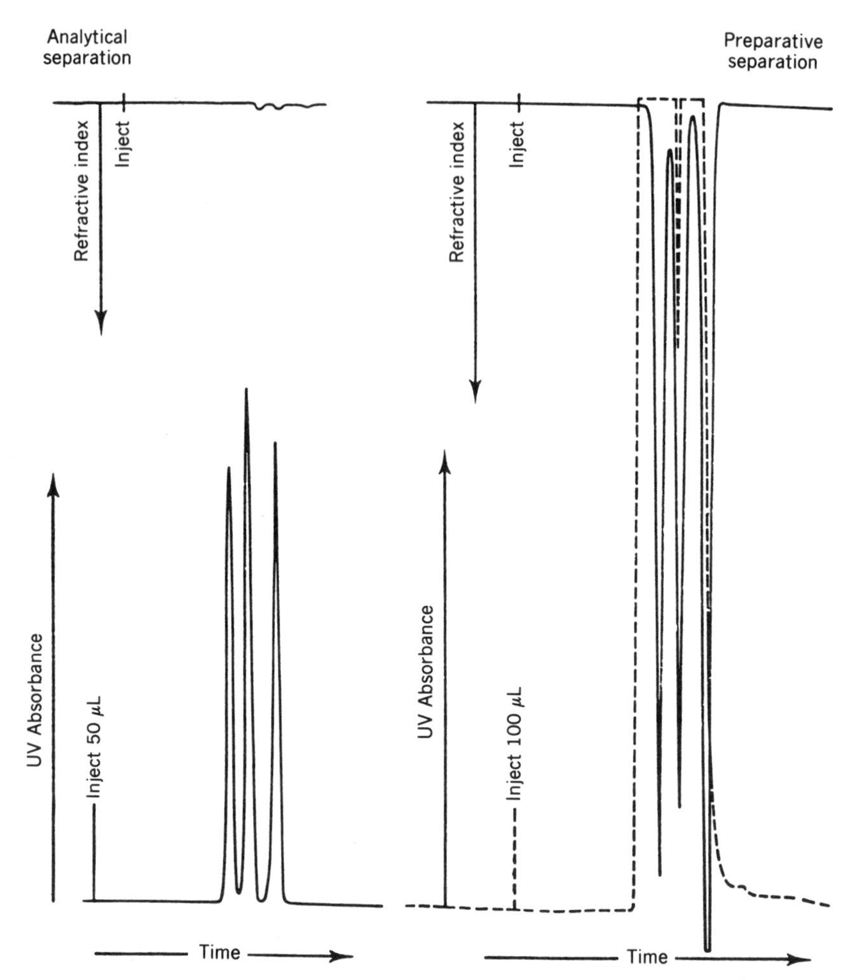

FIGURE 11-1. Separation of mixture of phthalates by GPC on a 100-Å μStyragel column. Mobile phase: THF. Flow rate: 1 mL/min. Detector: UV at 254 nm, 0.5 AUFS, RI at X32. Sample: dioctyl, dibutyl, and dimethyl phthalate; sample size noted on chromatogram. (Note: Actual separation will depend upon the quality of the mobile phase and column packing.)

FIGURE 11-2. Separation of mixture of phthalates by GPC on a 100-Å Ultrastyragel column. Analytical loadings are shown in (*a*) and preparative loadings are shown in (*b*). Mobile phase: THF. Flow rate: 1 mL/min. Detector: UV at 254 nm, 0.5 AUFS in (*a*) and 2.0 AUFS in (*b*); RI at X8 in (*a*) and X64 in (*b*). Sample: dioctyl, dibutyl, and dimethyl phthalates with 50 μL injection in (*a*) and 100-μL injection in (*b*). (Note: Actual separation will depend upon the quality of the mobile phase and column packing).

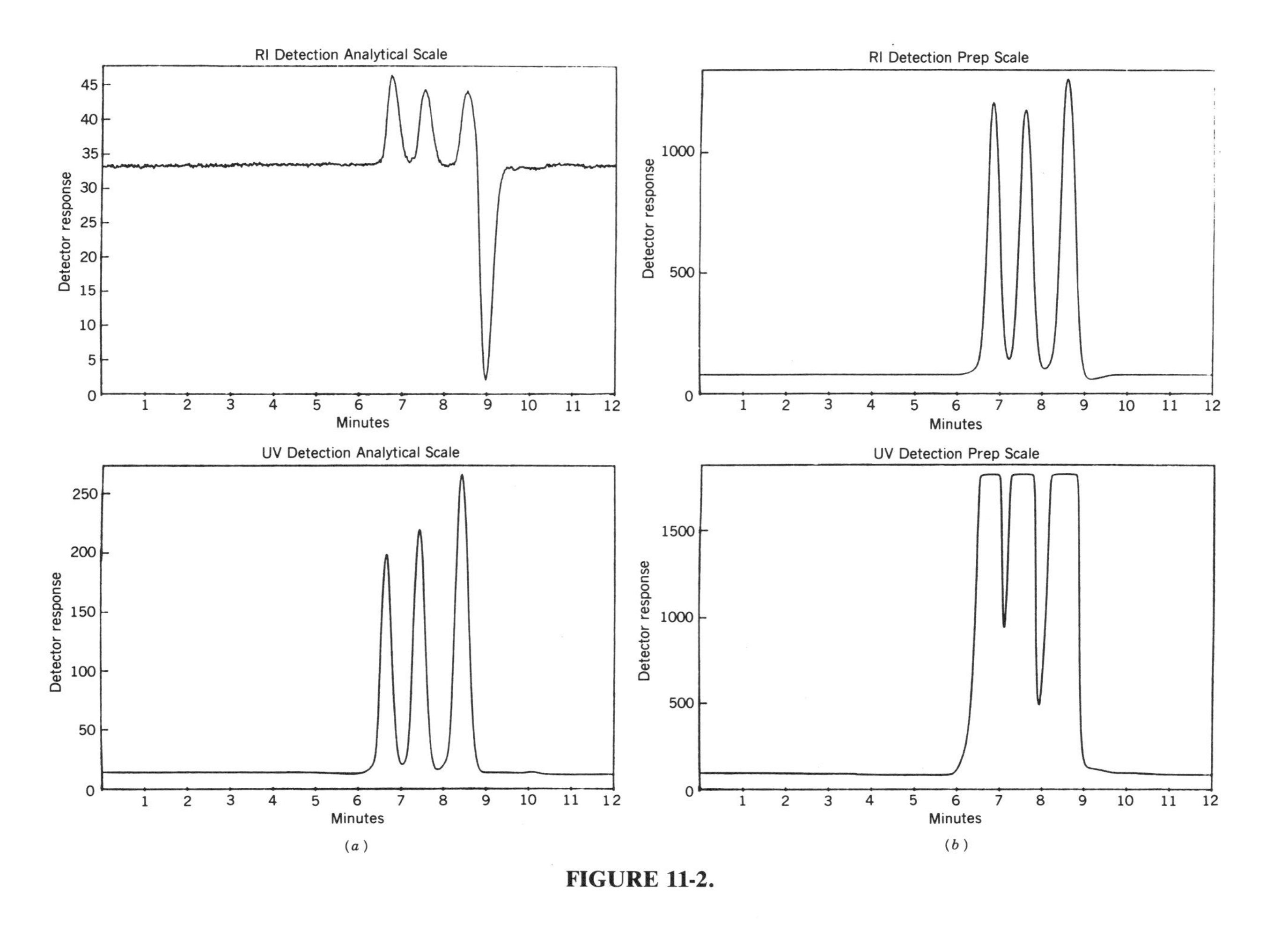
RI Detection Analytical Scale
RI Detection Prep Scale
UV Detection Analytical Scale
UV Detection Prep Scale
Detector response
Minutes
(a)
(b)

FIGURE 11-2.

C. Record the molecular weight of the known compounds used in the calibration and also record the elution volume from your chromatograms for the center of each peak measured from the point of injection. Place this information into a tabular format. This table will be referred to as Table 11-1 later in this experiment.

D. The calibration curve should be constructed by plotting the log of the molecular weight for the test compounds versus their individual elution volume in milliliters.

DISCUSSION

Gel permeation chromatography separates the sample components in a highly predictable manner. The larger molecules always elute from the column first. Also, the maximum amount of mobile phase required to elute all sample components is equivalent to one column volume. Furthermore, the elution volume of a molecule can be predicted from a calibration curve for a given set of columns.

In deciding when to utilize the gel permeation technique, consideration *must* be given to the configurational differences of the molecules to be separated. This is often in the same ratio as their molecular weight differences. However, compounds with small differences in molecular weight occasionally may have vast differences in molecular conformation in solution. It is important to realize that the separation is based upon molecular *size* not *molecular weight.*

One example that illustrates this is the GPC separation of aspirin and propyl paraben. These two compounds have identical molecular weights (MW = 180) but can be separated using GPC (e.g., on a 100-Å μStyragel columns). From the structures (Fig. 11-3) it can be hypothesized that propyl paraben acts as a "larger" molecule in solution than does aspirin. Since the side chain on propyl paraben would give the molecules a larger end-to-end size than the more compact aspirin molecule, propyl paraben should elute before aspirin.

As can be seen from this and other experiments in this book, the spectrophotometric detector is excellent for analytical scale detection of separated molecules that contain a chromophoric group. If these molecules have moderate or large molar absorbtivities at the operating wavelength, excellent sensitivity can be achieved. However, under these same conditions, the spectrophotometer is much less useful for preparative scale chromatographic separations because it tends to become optically "blinded" when relatively large amounts of materials are injected (i.e., all light entering the detector cell is absorbed and no separation can be seen because of the off-scale readout). This is the reason for the difference in appearance of the chromatogram in Figure 11-1. Also, spectrophotometers will be "blind" to

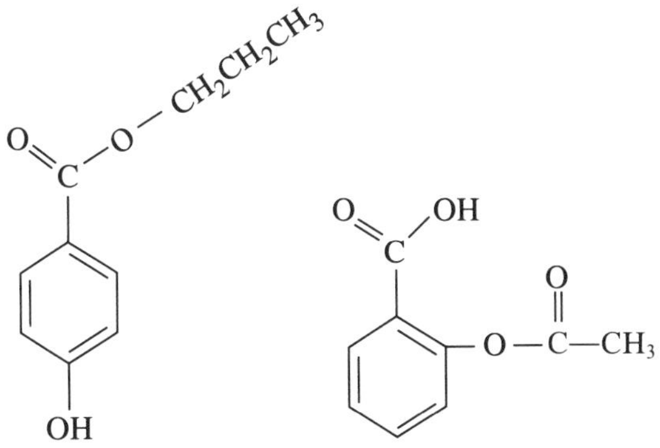

Propyl paraben (MW = 180)

Aspirin (MW = 180)

FIGURE 11-3. Structures of propyl paraben and aspirin.

molecules that have no chromophore in their structure. The differential refractometer (RI) is the preferred detector for preparative LC because generally it is difficult to overload the RI detector, and it responds to all compounds with a refractive index that is different from the mobile phase. The RI is essentially a universal detector which sees any component in the sample, not just those with chromophores.

SELF-HELP QUESTIONS

1. What was the elution order of your samples? Is this what was predicted?
2. Since the separation is based upon size in solution, why were you able to calibrate in terms of log molecular weight versus elution volume?
3. Where would dihexylphthalate elute?
4. Refer to your Table 11-1 (which you prepared) and comment as to the purity of the collected fractions. Is collecting "blind" a disadvantage?
5. What amount of material was being preparatively separated in the experiment?

ADDITIONAL EXPERIMENTS

The following are suggested supplemental experiments which will further increase your understanding of the use and characteristics of GPC. They may be done under the operating conditions previously established.

A. For one GPC column (100 Å):
   1. Vary fraction collection time and investigate the effect of these changes in collection time on the purity of each fraction.
   2. Using one of the phthalate samples from the original experiment, vary the injection volume, that is, use injection volumes greater than and less than 100 μL. Note the effect of these changes upon the location of peak maximum and width. At what injection volume does the peak maximum shift?
   3. Cut a piece of Tygon® tubing (12 in. in length) into quarters and extract with 50 mL of THF. Inject the extract (10 μL) into the HPLC (adjust injection size if necessary). Are any plasticizers present? Do the peak maxima correspond to those in the experiment? Are there any other peaks present? What are they?

B. One of the most important variables for improving an LC separation is column length. Increased column length gives increased resolution because it increases the number of theoretical plates. If two or more GPC columns (100 Å size) are available, it is instructive to repeat this experiment with multiple columns in series and to compare the results.

C. A rule of thumb in small-molecule GPC is that it is possible to analyze two compounds whose molecular weight differs by 10% or more by the judicious selection of the mobile phase. For instance, the separation of tolnaftate (the active ingredient of an antifungal preparation) (MW = 302) and BHT (which is present as an antioxidant) (MW = 220) can be accomplished using methylene chloride, a nonhydrogen bonding solvent as shown in Figure 11-4. For this assay, sample prep-

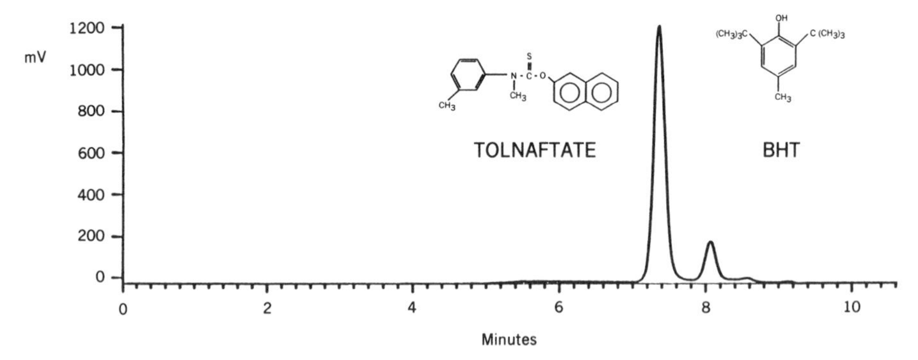

FIGURE 11-4. Separation of an antifungal cream. Mobile phase: methylene chloride. Flow rate: 1 mL/min. Detector: 280 nm. Column: Ultrastyragel (100 Å). Sample size: 5-μL injection of 0.843 g of cream dissolved in 5 mL of THF. Active ingredient tolnaftate is claimed on the label to be 1%. (Note: Actual separation will depend upon the quality of the mobile phase and column packing.)

aration is dissolving the sample in THF (approximately 1 g/5 mL) and filtering through an appropriate, solvent compatible filtration device, and injection of a 5-μL aliquot. The overall analysis takes only 12 min. As an experiment, the student can calibrate the detector and analyze several over-the-counter creams which are commercially available. Detection at 254 nm is possible; however, the response intensity will be reduced by approximately 50%, which should not pose any difficulty since the concentration of active ingredient is high. The student will have to calculate the concentration of tolnaftate in the ointment from the label, choose the concentration range for the calibration curve, calibrate for the active ingredient, and implement the final analyses.

CHAPTER 12

EXPERIMENT 5: DEVELOPING A REVERSE-PHASE CHROMATOGRAPHIC SEPARATION

Objective
Background
Additional materials
Safety and disposal
Experiments
  A. Preparation of instrument
  B. Sample preparation
  C. Chromatography
Evaluation
Discussion
Self-help questions
Additional experiments

OBJECTIVE

To demonstrate the approach used in developing a reverse-phase separation by observing the effect upon the solutes of changing the composition of the mobile phase.

BACKGROUND

The term reverse-phase is used to describe a chromatographic system where the stationary phase (column packing material) is nonpolar and the mobile phase is polar. The most commonly used stationary phase for reverse-phase

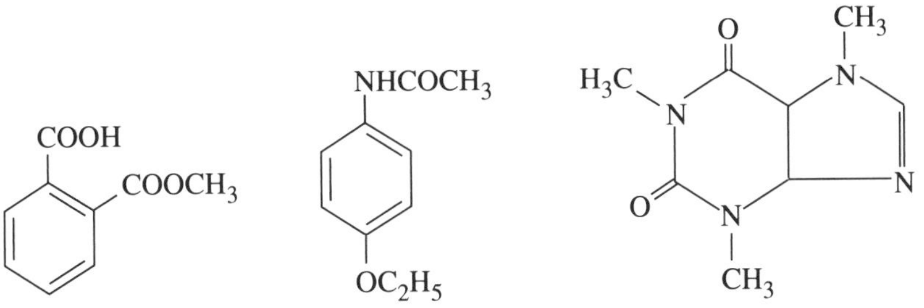

FIGURE 12-1. Structures of sample molecules used in this experiment.

separations is an octadecyl alkyl hydrocarbon chain (C_{18}) which is chemically bonded to the silica substrate. Bonding the stationary phase eliminates the problem of "bleed," which would occur if the hydrocarbon was physically coated on the silica. Typical mobile phases are various ratios of methanol/water, acetonitrile/water, or tetrahydrofuran/water. Reverse-phase systems are often chosen for samples that are too nonpolar to adsorb on silica (normal phase) or are so polar that elution from the silica does not occur within a reasonable period of time and results in unsymmetrical peaks. Reverse phase is an extremely versatile mode of chromatography and can be employed successfully with compounds soluble in solvents ranging from chloroform to water. Refer to Chapter 5 for additional information.

For reverse-phase chromatography, the mobile phase is made by choosing one solvent in which the sample is very soluble and another solvent in which the sample is less soluble. One can then prepare a mobile phase by adjusting the amount of the "strong" and "weak" solvents to a ratio where the attraction of the solutes to the packing is in a competitive equilibrium with the attraction (solubility) of the solutes to the mobile phase. The equilibrium of the solutes in the mobile phase relative to the bonded phase determines the retention time and effects the separation.

"Method development" is simply finding an appropriate mobile phase (the blend of the good and poor solvents) in which the sample solutes are retained and separated. For this experiment the structures of sample components are shown in Figure 12-1, methanol is the strong solvent and water is the poor solvent. The water contains acetic acid which adjusts the pH so that the aspirin is not ionized. This ionization control is necessary to have adequate retention and good peak shape for the aspirin.

ADDITIONAL MATERIALS

1 C_{18} bonded-phase column. The following columns have been tested in this experiment: a 2 mm ID × 61 cm Bondapak C_{18}/Porasil B (37–75 μm) column; a 3.9 mm ID × 15 cm μBondapak C_{18} (10 μm) column; a

3.9 mm ID × 30 cm μBondapak C_{18} (10 μm) column; and a 3.9 mm ID × 15 cm Resolve C_{18} (5 or 10 μm) column. Other C_{18} bonded-phase columns should work in this experiment; however, they have not been tested.

2 150-mL Erlenmeyer flasks, glass stoppered.

1 100-mL graduated cylinder.

Magnetic stirring plate.

Stirring bar (Teflon® or glass).

Glass funnel.

25-μL syringe.

APC tablet (aspirin, phenacetin, caffeine, any supplier). If these tablets are not available, a laboratory mixture may be prepared by mixing 226.8 mg of acetylsalicylic acid (aspirin), 32.4 mg of caffeine, and 162 mg of phenacetin for this experiment. This laboratory mixture will suffice as an APC "tablet."

No. 2 filter paper (any supplier).

Methanol, filtered through a 0.45-μ filter using a solvent clarification device.

Distilled water, filtered through a 0.45-μ filter using a solvent clarification device.

Glacial acetic acid.

SAFETY AND DISPOSAL

This experiment does not purport to address the safety issues associated with its use. It is the responsibility of the user of this experiment to establish appropriate safety and health practices and to determine the applicability of regulatory limitations prior to use. All chemicals should be handled and disposed of in an appropriate manner consistent with the safety policy of the experimenter's company, school, or organization.

EXPERIMENTS

A. Preparation of Instrument*

Activity	Comments
1. Clear (purge) the instrument of any solvent that is immiscible with water.	1. Refer to Chapter 6 and observe miscibility rules.

* This may be done by the instructor before the lab period.

Activity	Comments
2. Prepare a 2-L stock solution of 4% (v/v) acetic acid in distilled water.	2. This solution will be used to prepare the aqueous methanol mobile phase used for elution.
3. Prepare a 500-mL reservoir of methanol.	
4. Prepare a 50:50 (v/v) methanol:acetic acid solution.	4. Acetic acid is used in the mobile phase to eliminate tailing of peaks due to interaction with the residual silica active sites. This technique is referred to as "ion suppression" since it also suppresses the ionization of the sample compounds (refer to Chapter 5).
5. Prepare a 35:65 (v/v) methanol:acetic acid solution.	
6. Prepare 25:75 (v/v) methanol:acetic acid solution.	
7. Place the solution of methanol from Activity A-3 (100% methanol) in the mobile phase reservoir.	
8. The injector handle should be in the "inject" position.	8. This guarantees that the sample holding loop will be sufficiently cleaned to avoid contamination from the previous solvent.
9. Prime the solvent delivery system and start flowing at 2 mL/min and pump until a stable baseline is attained on the detector (approximately 30 mL).	9. Refer to operator's manual necessary.
10. Stop the flow and install a single column in the instrument and resume flow to 1 mL/min.	
11. Set spectrophotometer sensitivity to 1.0 AUFS (254 nm).	
12. Allow the instrument to equilibrate under these conditions while preparing the sample.	12. This ensures that the detector baseline is stabilized before the experiment is begun.

B. Sample Preparation

Activity	Comments
1. Crush an APC tablet with a glass stirring rod. Transfer the	1. Since the purpose is to develop a separation, the tablet was not

Activity	Comments
crushed tablet to a 150-mL Erlenmeyer flask. Add 100 mL of methanol. Insert a glass or teflon-coated magnetic stirring bar. Stir for 5 min on a magnetic stirring plate to dissolve the contents.	weighed. The substitute laboratory mixture may be used instead of the tablet as described in the section entitled Additional Materials. If the 5- or 10-μm column (15 or 30 cm length) is used, the sample should be diluted with 200 mL of methanol.
2. Filter through No. 2 filter paper into another 150-mL Erlenmeyer flask. If the substitute laboratory mixture is used there is no need to filter.	2. This ensures that the excipient "floaters" will not plug the syringe needle.

C. Chromatography

Activity	Comments
1. Set recorder chart speed to 0.5 cm/min or equivalent.	
2. Inject 10 μL of the tablet solution.	2. The mobile phase is methanol and the sample components should have little or no retention; therefore, a suitable separation will not occur.
3. Change the mobile phase composition to 50:50 (v/v) methanol:acetic acid from Activity A-4.	
4. The injector handle should be in the "inject" position. Reequilibrate the system by passing 30 mL of mobile phase through the column. This may be done at a higher flow rate to save time.	4. This ensures that the detector baseline is stabilized before the next injection is made.
5. Change the flow rate to 1 mL/min. Allow system to pump at this flow rate for another 5 min. Return injector to "load" position.	5. This ensures that a stable baseline is attained.
6. Inject 10 μL of the tablet solution from Activity B-2.	6. The mobile phase may still be too strong and a separation may not be apparent.

7. Change mobile phase composition to 35% MeOH:65% acetic acid (v/v), and repeat Activity C-4 through C-5.

7. At this solvent composition the peaks should be partially separated. The elution order is caffeine, aspirin, and phenacetin. A 5-μL injection might be used if a column containing 5 or 10 μm particles (15 or 30 cm length) is used and the peak heights are too high.

8. At 1 mL/min flow rate for the mobile phase, inject 10 μL of the tablet solution (Activity B-2).
9. Change mobile phase composition to 25% MeOH:75% acetic acid (v/v), and repeat Activity C-4 and C-5.
10. At 1 mL/min flow rate for the mobile phase, inject 10 μL of the tablet solution (Activity B-2).

EVALUATION

A. All chromatograms should be properly labeled and attached to the last sheet in this experiment. Proper labeling requires the following:

Sample:	Components, approximate amount of each and injection volume.
Detector:	Type of detector and sensitivity.
Column:	Type of column used, length, and diameter (ID).
Solvent:	Mobile phase composition and flow rate.
Recorder:	Recorder chart speed.

B. Example results of the separation of the APC tablet (or mixture) under the conditions of the experiments are presented in Figures 12-2 to 12-7. These chromatograms are used for all subsequent sample calculations of performance parameters.

C. On a separate sheet of paper construct Table 12-1. Calculate the performance parameters and fill in the tablet with the appropriate values.

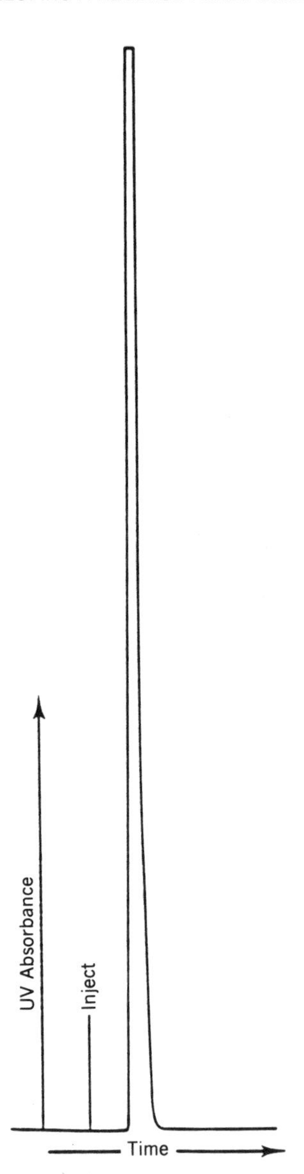

FIGURE 12-2. Example of reverse-phase separation of an APC tablet. Mobile phase: MeOH. Flow rate: 1 mL/min. Detector: UV at 254 nm, 1.0 AUFS. Sample: APC tablet in 100 mL MeOH, 10-μL injection. Column: Bondapak C_{18}/Porasil B, 2 mm ID × 61 cm. (Note: Actual separation will depend upon the quality of the mobile phase and column packing.)

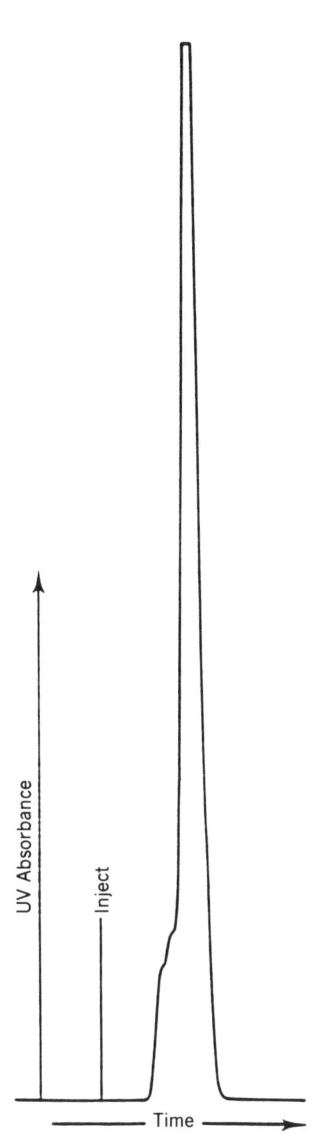

FIGURE 12-3. Example of reverse-phase separation of an APC tablet. Mobile phase: MeOH/H_2O (4% HOAc), 50/50. Flow rate: 1 mL/min. Detector: UV at 254 nm, 1.0 AUFS. Sample: APC tablet in 100 mL MeOH, 10-μL injection. Column: Bondapak C_{18}/Porasil B, 2 mm ID × 61 cm. (Note: Actual separation will depend upon quality of mobile phase and column packing.)

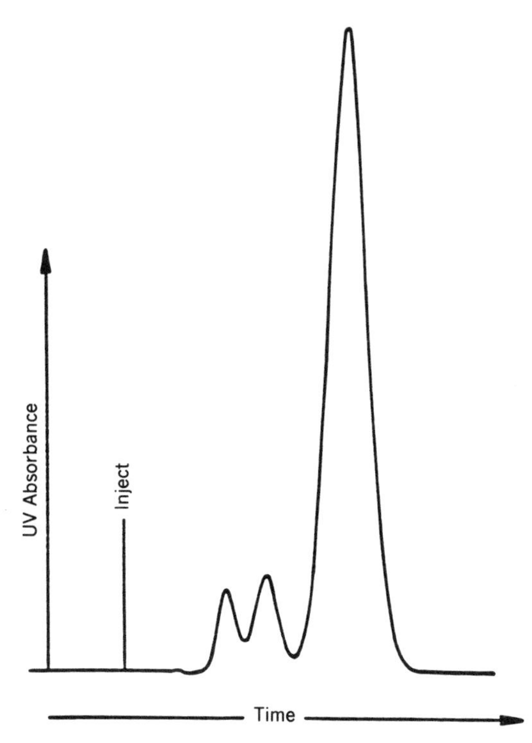

FIGURE 12-4. Example of reverse-phase separation of an APC tablet. Mobile phase: MeOH/H_2O (4% HOAc) 35/65. Flow rate: 1 mL/min. Detector: UV at 254 nm, 1.0 AUFS. Sample: APC tablet in 100 mL MeOH, 10-μL injection. Column: Bondapak C_{18}/Porasil B, 2 mm ID × 61 cm. (Note: Actual separation will depend upon the quality of the mobile phase and column packing.)

DISCUSSION

Aspirin, caffeine, and phenacetin are very polar compounds. Normal-phase separations of polar compounds like these are generally not successful. Reverse-phase chromatography is usually a better method.

The "solubility" of the sample components in the mobile phase relative to the packing determines the retention time on the reverse-phase column. In this case, increasing the amount of aqueous solution alters the solubility of the sample and the retention is increased.

SELF-HELP QUESTIONS

1. Are the aspirin, phenacetin, and caffeine more soluble in methanol or water? With this in mind can you explain the increase in k', α, and R as the ratio of methanol to water (4% HOAC) decreases?

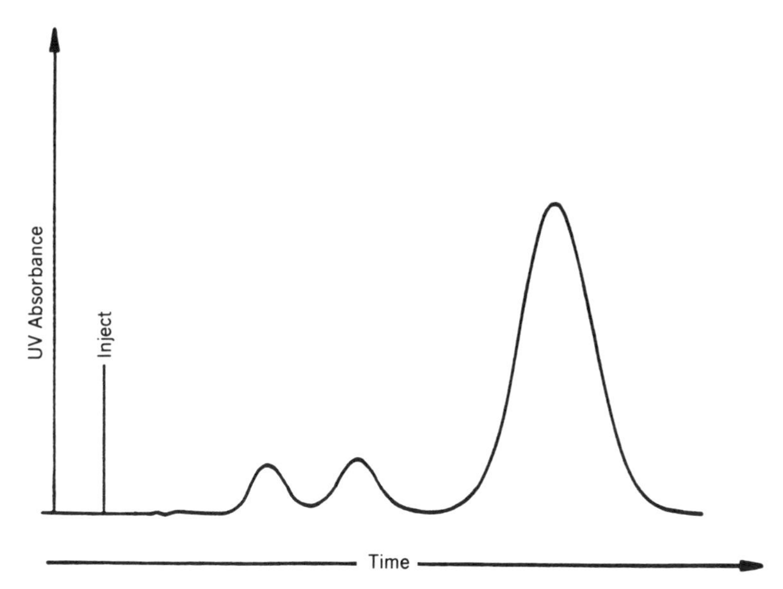

FIGURE 12-5. Example of reverse-phase separation of an APC tablet. Mobile phase: MeOH/H_2O (4% HOAc) 25/75. Flow rate: 1 mL/min. Detector: UV at 254 nm, 1.0 AUFS. Sample: APC tablet in 100 mL MeOH, 10-μL injection volume. Column: Bondapak C_{18}/Porasil B, 2 mm ID × 61 cm. (Note: Actual separation will depend upon the quality of the mobile phase and column packing.)

2. Comment on the relationship between k' and N at the various mobile phase concentrations. Is this similar to the relationship observed in Chapter 9, Experiment 2?
3. What mobile-phase composition gave the "best" separation of all three components? Why is this separation the best?
4. If you wish to measure the amount of aspirin in a tablet on a routine basis which mobile-phase composition would you choose? Why? Which mobile phase would be suitable for monitoring phenacetin in a tablet on a routine basis? Why?

ADDITIONAL EXPERIMENTS

The following are suggested supplemental experiments which will further increase your understanding of the use and characteristics of reverse-phase chromatography. They may be done under the operating conditions previously established, depending on the lab time available and course requirements.

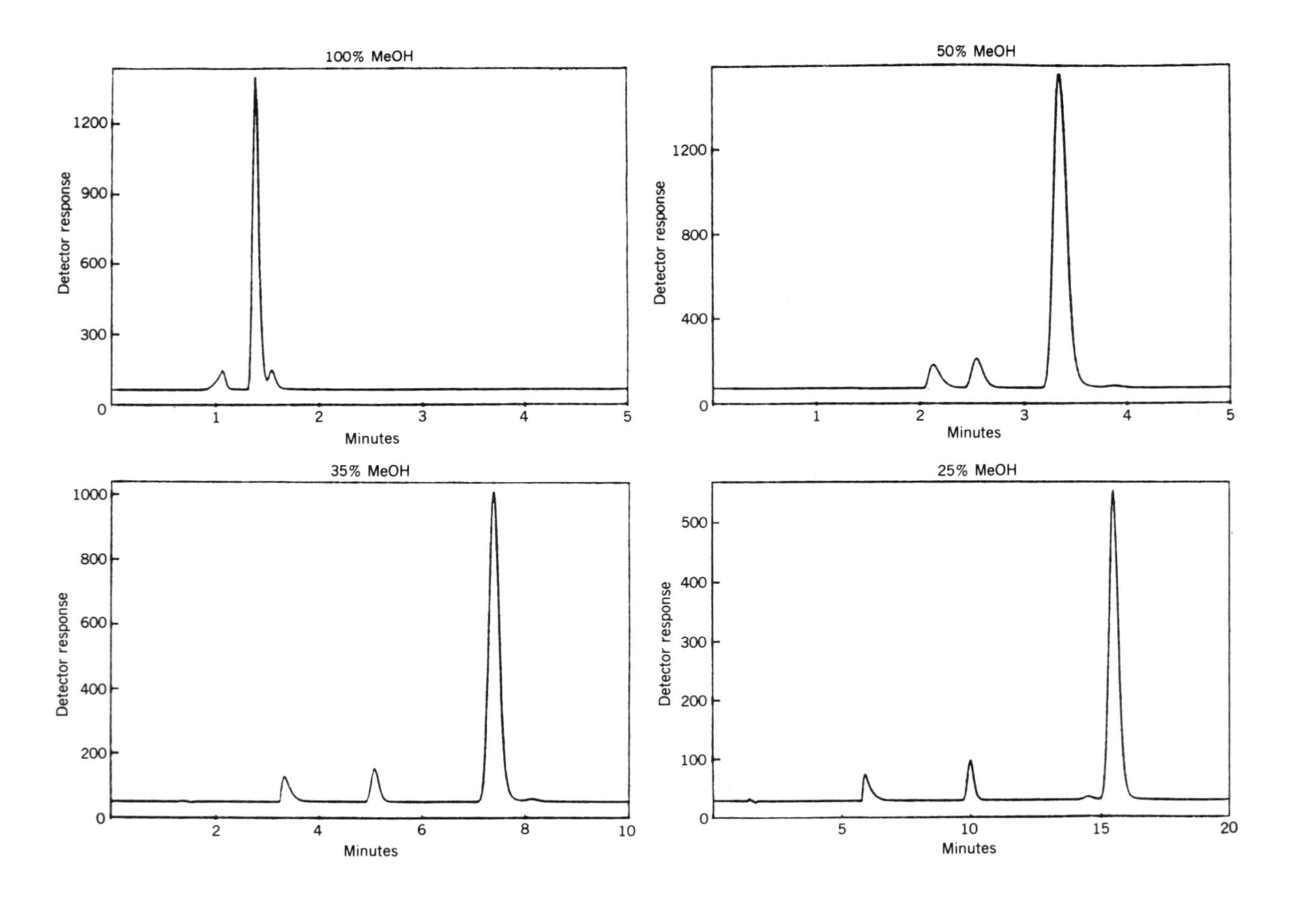
100% MeOH
Detector response
Minutes
50% MeOH
Detector response
Minutes
35% MeOH
Detector response
Minutes
25% MeOH
Detector response
Minutes

TABLE 12-1. Calculated Performance Parameters

Mobile Phase Composition		Performance Parameters			
MeOH:water (4% HOAc)		k'	N	α	R
50%:50%	Caffeine	___	___	×	×
	Aspirin	___	___	×	×
	Phenacetin	___	___	×	×
	Caffeine/Aspirin	×	×	___	___
	Aspirin/Phenacetin	×	×	___	___
35%:65%	Caffeine	___	___	×	×
	Aspirin	___	___	×	×
	Phenacetin	___	___	×	×
	Caffeine/Aspirin	×	×	___	___
	Aspirin/Phenacetin	×	×	___	___
25%:75%	Caffeine	___	___	×	×
	Aspirin	___	___	×	×
	Phenacetin	___	___	×	×
	Caffeine/Aspirin	×	×	___	___
	Aspirin/Phenacetin	×	×	___	___

←

FIGURE 12-6. Example of reverse-phase separation of an APC tablet. Column: Resolve C_{18} column (5 μm) 3.9 mm × 15 cm. Mobile phase: percent methanol noted on chromatogram and remaining percentage is 4% aqueous acetic acid. Injection volume: 5 μL. Flow rate: 1 mL/min. Detector: 254 nm, 1.0 AUFS. (Note: Actual separation will depend upon the quality of the mobile phase and column packing.)

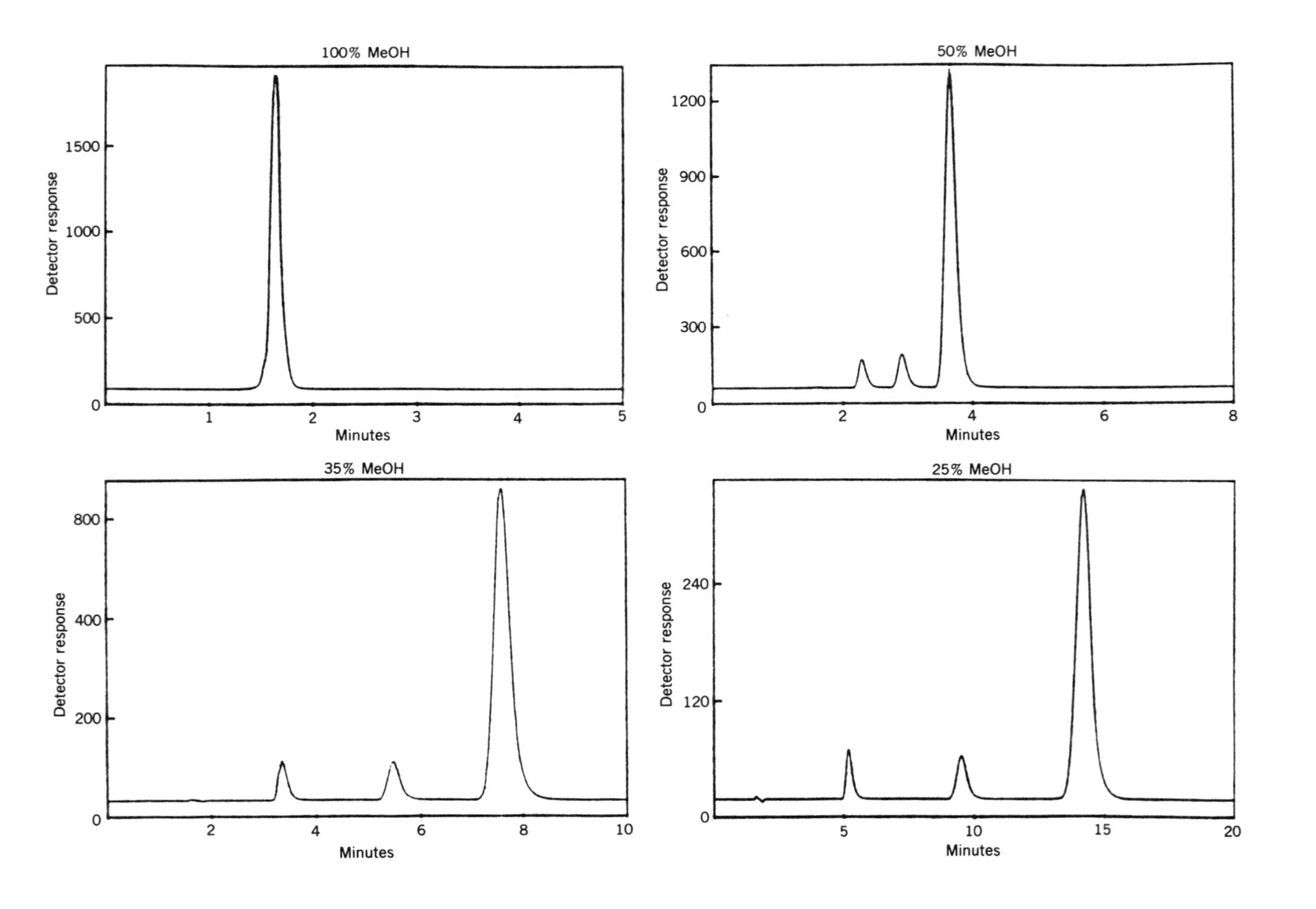

100% MeOH
Detector response
Minutes
50% MeOH
Detector response
Minutes
35% MeOH
Detector response
Minutes
25% MeOH
Detector response
Minutes

FIGURE 12-7. Example of reverse-phase separation of an APC tablet. Column: μBondapak C_{18} column (10 μm) 3.9 mm × 15 cm. Mobile phase: percent methanol noted on chromatogram and remaining percentage is 4% aqueous acetic acid. Injection volume: 5 μL. Flow rate: 1 mL/min. Detector: 254 nm, 1.0 AUFS. (Note: Actual separation will depend upon the quality of the mobile phase and column packing.)

A. Repeat the experiment described above using acetonitrile in place of methanol in the mobile phase in the same proportions. Alter the solvent concentration to optimize the resolution.

B. If sufficient columns are available, investigate the effect of increasing column length and add another 60-cm section of Bondapak C_{18}/Porasil B. Use the separation achieved with a mobile-phase composition of 50:50 (v/v) MeOH:H_2O (Fig. 12-3). Calculate the appropriate performance parameters and compare results.

CHAPTER 13

EXPERIMENT 6: QUANTITATION

Objective
Background
Additional materials
Safety and disposal
Experiments
A. Preparation of instrument for APC and aspirin analysis
B. Sample preparation for APC and aspirin analysis
C. Chromatographic analysis for the APC tablet
D. Quality control of aspirin
E. Preparation of instrument for analysis of caffeine, saccharin, benzoate, and aspartame from beverages
F. Sample preparation for beverage analysis
G. Chromatographic analysis of caffeine, saccharin, benzoate, and aspartame
Evaluation
Discussion
Self-help questions
Additional experiments
References

OBJECTIVE

Demonstrate a quantitative analysis by LC, measuring the precision of the method by generating and using a calibration curve and performing an actual quality-control assay.

BACKGROUND

Quantitative analysis of a chromatogram is the process of measuring the peak response from the detector and comparing the response to that of a known calibration curve. If the chromatographic system is operating correctly, the precision of the analysis will depend upon the reproducibility of the injection. The precision may be determined by repetitive injections of the sample and by statistical averaging of the results. For minimum operator interaction, one usually chooses a computer or an integrator. Other methods are, however, quite suitable. For a symmetrical peak, measurement of peak height is a rapid and convenient method to obtain good quantitative data.

Once the precision of an injection of a standard compound is known, quantitation of unknown samples for the components of interest is determined from a calibration curve. A calibration curve is a plot of detector response (peak area height) versus the mass injected. This curve is generated by injecting the same volume from a series of known solutions which differ in concentration. However, in certain instances, a standard curve may be obtained by injecting different volumes (amounts) of a single solution by partially filling the injector loop. This is possible only if the sample injection volume has a negligible effect upon peak spreading, and the mass injected does not overload the column. The latter approach is certainly the simplest and, therefore, is used in this experiment. The compounds used in Sections A–E of Experiments are the same ones used in Chapter 12, Experiment 5.

A typical case of quantitation routinely performed using an LC is the analysis of a pharmaceutical product where the final tablet must contain ±10% of the amount of active ingredients stated on the label (100% ± 10%) (YES–NO quantitation). Often the quality-control procedure requires that a specified number of tablets selected at random be tested, and that a percentage of these tablets be within specification. In this case, standards are prepared that correspond to 90 and 110% of the label statement, and these are injected into the liquid chromatograph. This is followed by the injection of the samples prepared from randomly selected tablets. After the last tablet has been injected, the standards are again injected. Connecting the 90% of label standards (drawing a line between the top of the peaks) and connecting the 110% of the label standards will give an "acceptance window" which will quickly show the number of tablets that do not meet specifications.

Also included in this experiment is a procedure for the analysis of direct and indirect additives in beverages and soft drinks (Sections E–G). This analysis may be substituted for the APC procedure if the instructor desires. High performance LC is a popular approach to analyzing beverages for caffeine, saccharin, benzoate, and other additives. Reverse-phase methods (1–5) like the one described in this experiment have been used to determine caffeine levels in coffee (2,3,5) tea (4), and soft drinks (1,5,6), with most methods including the simultaneous separation and analysis of saccharin and sodium benzoate, with minimal sample preparation required (6). With the introduction of caffeine-free beverages, this experiment takes on new relevance. Also, the experimental conditions can be modified slightly to allow the simultaneous analysis of aspartame, which is increasingly being used to sweeten dietary beverages. Both separations can be applied to the analysis of soft drinks, artificial sweeteners, fruit juices, and coffee.

ADDITIONAL MATERIALS

1 C_{18} bonded-phase column. The following columns have been tested in this experiment: 2 mm ID × 61 cm Bondapak C_{18}/Porasil B (37–75 μm); a 3.9 mm ID × 15 cm μBondapak C_{18} (10 μm); a 3.9 mm ID × 30 cm μBondapak C_{18} (10 μm); and a 3.9 mm ID × 15 cm Resolve C_{18} (5 and 10 μm). Other C_{18} bonded-phase columns should work in this experiment; however, they have not been tested.

6 100-mL volumetric flasks (optional section requires 7 additional flasks).

Phenacetin (any supplier).

Aspirin (any supplier).

Aspirin, phenacetin, caffeine (APC) tablet.

Aspirin tablet (any brand name).

Methanol, filtered through a 0.45-μ filter using a solvent clarification kit.

Water, filtered through a 0.45-μ filter using a solvent clarification kit.

Glacial acetic acid.

25-μL syringe.

15 10-mL vials with screw caps.

For determination of caffeine, saccharin, benzoate, and aspartame: 1 C_{18} bonded-phase column. Only a 3.9 mm × 15 cm μBondapak C_{18} (10 μ) has been tested. Resolve C_{18} (10 μ) (3.9 mm × 15 cm) may have a tailed caffeine peak.

Caffeine (Eastman USP or equivalent supplier).

Sodium saccharin (Mallinckrott USP or equivalent supplier).

Benzoic acid (Fisher ACS or equivalent supplier).

Aspartame (any supplier).

1 100-mL volumetric flask.

5 10-mL volumetric flasks for samples.

SAFETY AND DISPOSAL

This experiment does not purport to address the safety issues associated with its use. It is the responsibility of the user of this experiment to establish appropriate safety and health practices and to determine the applicability of

regulatory limitations prior to use. All chemicals should be handled and disposed of in an appropriate manner consistent with the safety policy of the experimenter's company, school, or organization.

EXPERIMENTS

For this experiment decide which section(s) are to be completed before beginning.

A. Preparation of Instrument for APC and Aspirin Analysis*

Activity	Comments
1. Mix 25 mL of concentrated acetic acid with 625 mL of distilled water; to this mixture add 350 mL of methanol and mix thoroughly.	1. Since the stationary phase in this case is quite nonpolar, the mobile phase must be polar to assure retention.
2. Place this solution (65% of 4% acetic acid:35% methanol) from Activity A-1 in the solvent reservoir.	2. Acetic acid is used in the mobile phase to eliminate tailing of the peaks due to interaction with the residual silica active sites. This technique is referred to as "ionic suppression" since it also suppresses the ionization of the sample components.
3. Prime the solvent delivery system, if necessary and begin flowing at 1 mL/min.	3. Refer to manufacturers manual, if necessary.
4. The injector handle should be in the "inject" position.	4. This ensures that the sample holding loop is thoroughly cleaned.
5. Stop the flow and install a column into the instrument. Restart flow at 1 mL/min.	
6. Set the spectrophotometric detector sensitivity at 1.0 AUFS (254 nm).	
7. Allow the instrument to equilibrate under these conditions while preparing the samples.	7. This ensures that the detector baseline is stabilized before the experiment is begun.

* This may be done by instructor before the lab period.

B. Sample Preparation for APC and Aspirin Analysis

For the APC analysis:

Activity	Comments
1. Weigh accurately 150 mg of pure phenacetin. Transfer quantitatively to a 100-mL volumetric flask. Dilute to mark with methanol.	1. This solution concentration will be 1.5 μg per μL.
2. Place one APC tablet in each of five 100-mL volumetric flasks. Add 75 mL of methanol to each flask, dissolve the active ingredients by vigorous agitation for 3 mins. Dilute to mark with methanol and allow the insoluble binders (excipients) to settle.	2. If these tablets are not available, a laboratory tablet may be prepared. (See Chapter 12, Experiment 5 for the preparation).

Optional—for the quality control of aspirin (Section D):

Activity	Comments
3. Weigh 4.5 and 5.5 grains of aspirin (acetyl salicylic acid) and place each into separate 100-mL volumetric flasks. Dilute to mark with methanol.	3. Aspirin label claims are in terms of "grains." 1 grain = 60 mg.
4. Into five separate 100-mL volumetric flasks place one aspirin tablet. Add 75 mL of methanol to each flask, dissolve the active ingredients, and allow excipients to settle. Dilute to mark with methanol. Mix thoroughly and allow to settle again. Filter a representative amount (approximately 10 mL) of each solution through a sample clarification device or filter (0.45 μ) into a 10-mL vial with a screw cap or another suitable container. Label each container as to its contents.	4. A new sample clarification device should be used for each sample.

C. Chromatographic Analysis for the APC Tablet

Activity	Comments
1. Set recorder chart speed to 0.5 cm/min or equivalent.	
2. Rotate the injector handle to "load" position.	
3. Inject 10 μL of the standard phenacetin solution (from Activity B-1) using the 25-μL syringe.	
4. After the peak has eluted, repeat Activity C-2 and C-3 for four more analyses.	4. This is to determine the precision of injections.
5. Inject 2 μL of the standard phenacetin solution.	5. These results will be used to construct a standard response curve for phenacetin. Since precision is a function of the volume injected, there will be a small variability as the sample size is changed. This approach is used for economy of time, and the error should not be significant.
6. Inject 5 μL of the standard phenacetin solution.	
7. Inject 10 μL of the standard phenacetin solution.	
8. Inject 15 μL of the standard phenacetin solution.	
9. Inject 10 μL of each of the APC tablet solutions.	

D. Quality Control of Aspirin

Activity	Comments
1. Make two separate 10-μL injections, each for the 4.5 and 5.5 grain 100-mL standard solutions of aspirin. These were the solutions prepared in Activity B-3. The spectrophotometric detector should be set to 0.2 AUFS (254 nm).	1. This sets the -10% and $+10\%$ limits. Care should be taken to inject a particle-free solution. This may mean that an aliquot must be filtered through a sample-clarification device (0.45 μ) for each solution.

Activity	Comments
2. Inject 10 μL from one of the aspirin solutions from Activity B-4.	
3. Repeat above Activity D-2 for the remaining four solutions.	
4. Make two separate 10-μL injections for the 4.5 grain/100 mL and 5.5 grain/100 mL standard solutions of aspirin from Activity B-3.	4. This checks the −10% and +10% limits to determine if any chromatographic variation has occurred during the analyses.

E. Preparation of Instrument for Analysis of Caffeine, Saccharin, Benzoate, and Aspartame from Beverages*

Activity	Comments
1. If it is desired to analyze aspartame as part of this experiment, the mobile phase should contain 20% methanol and 80% of 1-M acetic acid adjusted to pH = 4.2. This pH adjustment is made by adding concentrated sodium hydroxide dropwise to 1000 mL of 1-M acetic acid while continuously stirring until the pH meter gives a constant value of 4.2.	1. Since the stationary phase is quite nonpolar, the mobile phase must be polar to assure retention. Acetic acid is used in the mobile phase to eliminate tailing of the peaks due to interaction with the residual silica active sites. This technique is referred to as "ion suppression" since it also suppresses the ionization of the sample components.
2. If it has been determined that only caffeine, saccharin, and benzoate are to be analyzed, mix 34 mL of concentrated acetic acid with 766 mL of distilled water; to this mixture add 200 mL of methanol and mix thoroughly.	2. This mobile phase does not require pH adjustment and aspartame will coelute with caffeine.
3. Place the appropriate mobile phase from either Activity E-1 or E-2 into the solvent reservoir.	3. This decision is based upon the compounds to be analyzed.
4. Prime the solvent delivery system, if necessary, and begin flowing at 1.5 mL/min.	4. Refer to manufacturers' manuals, if necessary.
5. The injector handle should be in the "inject" position.	5. This ensures that the sample holding loop is thoroughly cleaned.

* This may be done by instructor before the lab period.

Activity	Comments
6. If the column is not already installed in the instrument, stop the flow and install a column. Restart flow at 1.5 mL/min.	6. The μBondapak C_{18} is fully porous silica particles that have the surface coated with permanently bonded C_{18} functional groups and is "endcapped." Resolve C_{18} is fully coated with the C_{18} functional group and is not endcapped.
7. Set the spectrophotometric detector sensitivity at 0.5 AUFS (254 nm).	
8. Allow instrument to equilibrate under these conditions while preparing the samples.	8. This ensures that the detector baseline is stabilized before the experiment is begun.

F. Sample Preparation for Beverage Analysis

Activity	Comments
1. Weigh accurately 20 mg of each standard (caffeine, saccharin, and benzoic acid). Transfer quantitatively to a 100-mL volumetric flask. Dilute to mark with the mobile phase from Section E. For the analysis involving aspartame, 20 mg of aspartame was transferred to a 100-mL volumetric and was diluted to the mark with the previously prepared standard solution. This procedure results in two standard solutions. One solution contains all four standards and the other contains only three known compounds.	1. This solution concentration will be 0.2 μg/μL of caffeine, saccharin, and benzoic acid. The concentration of aspartame will be 2 μg/μL.
2. Take 10 mL of a beverage (e.g., coffee, tea, or soda) and place it in a small vial. Place this solution into an ultrasonic bath for 3 min.	2. This is to degas the sample.
3. Take 7 mL of the degassed beverage (Activity F-2) and place it into a 10-mL volumetric flask and dilute to the mark with deionized water.	

4. Filter through a 0.45-μ filter.
5. For each beverage to be analyzed, prepare the sample following Activities F-2 through F-4.

G. Chromatographic Analysis of Caffeine, Saccharin, Benzoate, and Aspartame

Activity	Comments
1. Set recorder chart speed to 0.5 cm/min or equivalent.	
2. Rotate the injector handle to "load" position.	
3. Inject 10 μL of the three-component standard solution (from Activity F-1) using the 25-μL syringe. After observing the resulting chromatogram, and if aspartame is to be analyzed, inject the four-component standard solution (from Activity F-1) using the 25-μL syringe.	3. Your chromatogram should resemble either Figure 13-5 or 13-6. If the retention time of the last peak is too long or the resolution is inappropriate, adjust the mobile phase appropriately by increasing or decreasing the percentage of methanol. If there is a coelution, changing the pH may improve the separation. (Refer to Fig. 13-7 for pH effects on μBondapak C_{18}, 3.9 mm × 15 cm.) The standards are already made as described in Section F and these activities do not have to be repeated. If aspartame is to be analyzed, compare the chromatograms with Figures 13-5 and 13-6 and with each other and identify each peak.
4. After the peaks have eluted, repeat Activity G-2 and G-3 for four more analyses of the appropriate standard solution.	4. This is to determine the precision of injections. Either use the three- or four-component standard solution.
5. Inject 2 μL of the standard solution (from Activity F-1).	5. These results will be used to construct a standard response curve for caffeine, saccharin, and benzoate. Since precision is a function of the volume

injected, there will be a small variability as the sample size is changed. This approach is used for economy of time, and the error should not be significant.

6. Inject 5 μL of the standard solution.
7. Inject 10 μL of the standard solution.
8. Inject 15 μL of the standard solution.
9. Inject 20 μL of the standard solution.
10. Inject 10 μL of each of the beverage solutions.

10. This is for the analysis of the beverage additives. Duplicate injections are appropriate, if time permits.

EVALUATION

For the analysis of an APC tablet, do items A–E.

A. Figure 13-1 shows an example chromatogram for the separation of aspirin, phenacetin, and caffeine on a low-efficiency column. Figure 13-2 shows the example chromatograms used in the construction of the phenacetin calibration curve (Fig. 13-3).
B. Figure 13-4 shows an example of the phenacetin standards and APC tablet analysis on a high-efficiency column.
C. All chromatograms should be properly labeled and attached to the last sheet in this experiment. Proper labeling requires the following:
   - Sample: Components, approximate amount of each and injection volume.
   - Detector: Type of detector and sensitivity.
   - Column: Type of column used, length, and diameter (ID).
   - Solvent: Mobile-phase composition and flow rate.
   - Recorder: Recorder chart speed.
D. On a separate sheet of paper construct Table 13-1 by making the appropriate measurements and recording the data.
E. On a separate sheet of paper construct Table 13-2 by making the appropriate measurements and record the data. An example of typical results are shown in Table 13-3.

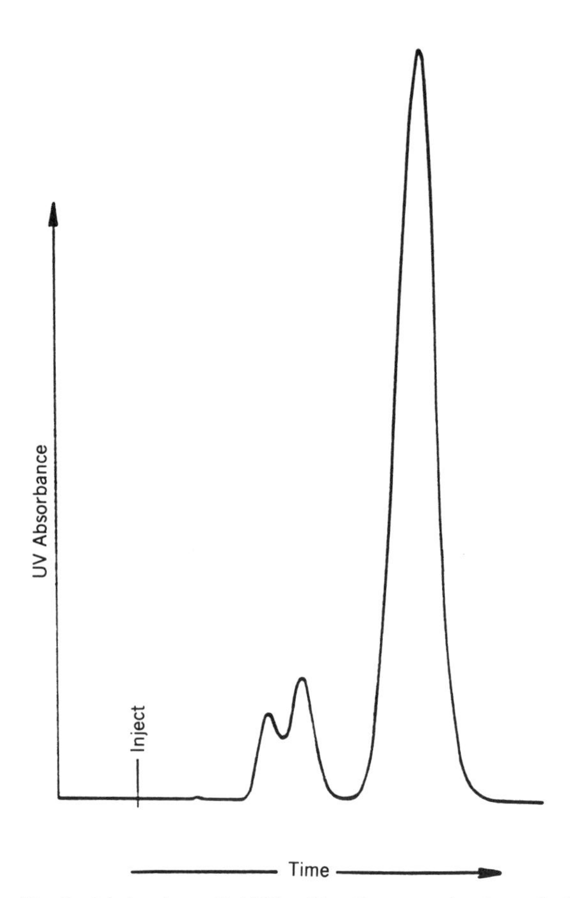

FIGURE 13-1. Typical injection of APC tablet for quantitation of phenacetin. Mobile phase: MeOH/H_2O (4% HOAc) 35/65. Flow rate: 1 mL/min. Detector: UV at 254 nm, 1.0 AUFS. Sample: APC tablet in 100 mL MeOH, 10 μL. Column: Bondapak C_{18}/Porasil B, 2 mm ID × 60 cm. (Note: Actual separation will depend upon the quality of the mobile phase and column packing.)

For the quality control of aspirin tablets, do items F and G.

F. On a separate sheet of paper construct Table 13-4 by making the appropriate measurements and recording the data.
G. Construct a quality-control chart from the data used in Table 13-4.

For the analysis of beverage additivies, do items H–K.

H. Figure 13-5 shows an example chromatogram for the separation of caffeine, saccharin, and benzoate. Figure 13-6 shows an example

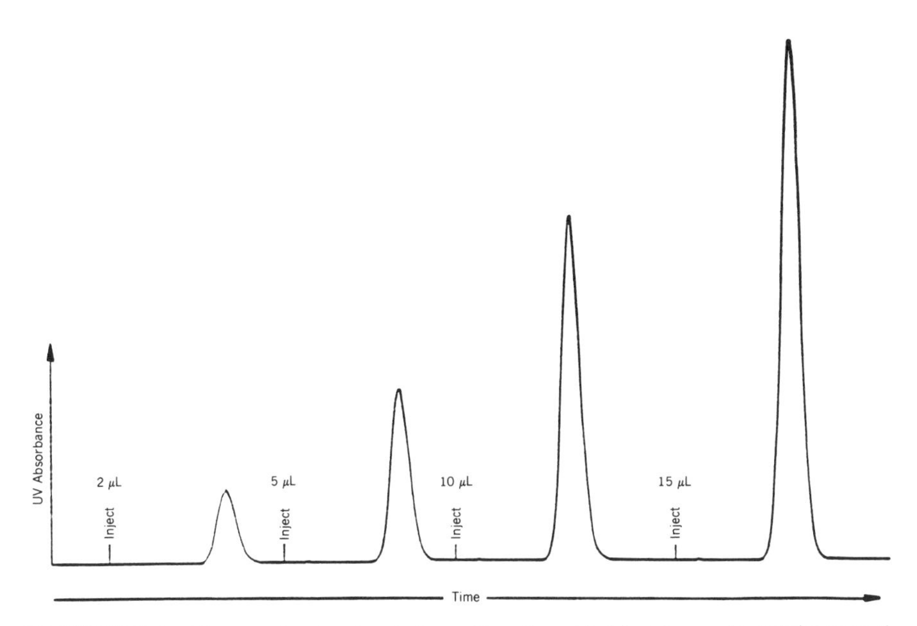

FIGURE 13-2. Phenacetin standards for calibration. Mobile Phase: MeOH/H_2O (4% HOAc) 35/65. Flow rate: 1 mL/min. Detector: UV: 254 nm, 1.0 AUFS. Sample: phenacetin standards, injection size variable. Column: Bondapak C_{18}/Porasil B, 2 mm ID × 60 cm. (Note: Actual separation will depend upon quality of mobile phase and column packing).

TABLE 13-1. Phenacetin Results

Sample	Peak Retention (cm)	Peak Height (cm)	Average Peak Retention, (cm)	Average Peak Height (cm)	Standard Deviation of Peak Height[a]	Standard Deviation of Peak Retention
Repetitive 10-μL injections of the phenacetin standard solution	1					
	2					
	3					
	4					

[a] Standard deviation is calculated by the formula $S = \sqrt{(X - \overline{X})^2/N - 1}$ where X is the individual value and $\overline{X}$ is the average value.

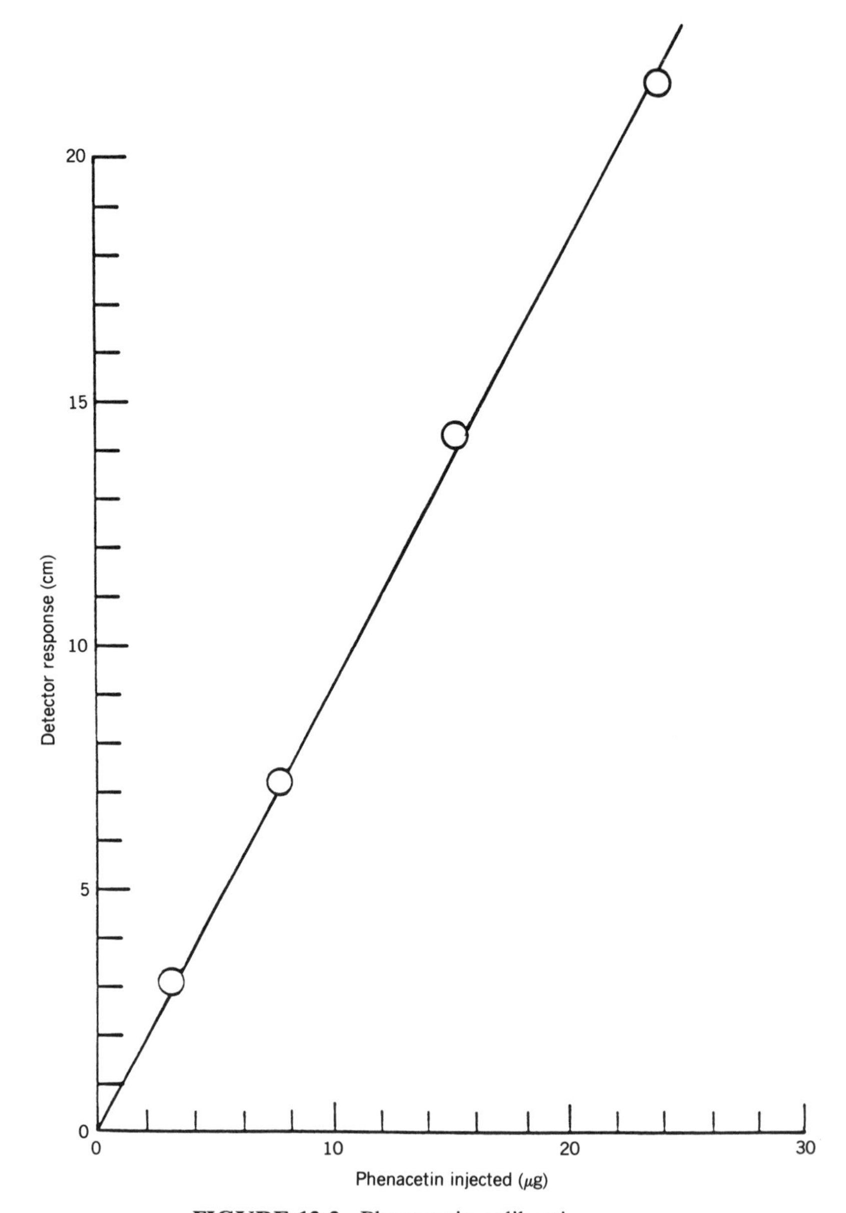

FIGURE 13-3. Phenacetin calibration curve.

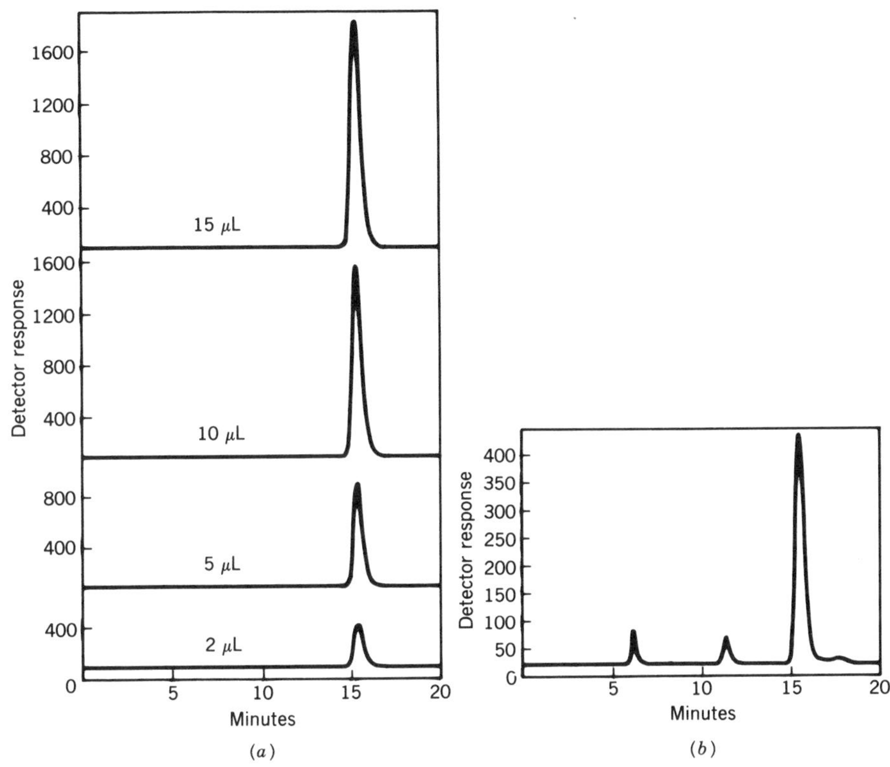

FIGURE 13-4. Phenacetin standards for calibration (*a*) and injection of APC tablet for phenacetin quantitation (*b*). Mobile phase: MeOH/H_2O (4% HOAc) 35/65. Flow rate: 1 mL/min. Detector: UV at 254 nm, 1.0 AUFS. Sample: (*a*) Phenacetin standards, injection size variable as shown on chromatograms; (*b*) APC tablet. Column: 3.9 mm × 15 cm μBondapak C_{18} (10 μ). (Note: Actual separation will depend upon the quality of the mobile phase and column packing.)

chromatogram of the analysis of caffeine, saccharin, benzoate, and aspartame.

I. All chromatograms should be properly labeled and attached to the last sheet in this experiment. Proper labeling requires the following:

Sample:	Components, approximate amount of each and injection volume.
Detector:	Type of detector and sensitivity.
Column:	Type of column used, length and diameter (ID).
Solvent:	Mobile-phase composition and flow rate.
Recorder:	Recorder chart speed.

J. On a separate sheet of paper construct Table 13-5 by making the appropriate measurements and recording the data.

TABLE 13-2. Calibration and Analysis Data

Sample	Distance from Injection to Peak Center[a] (cm)	Individual Peak Height (cm)	Total Phenacetin Injected (μg)
Standard			
2 μL			
5 μL			
10 μL			
15 μL			
Tablet			
No. 1			
No. 2			
No. 3			
No. 4			
No. 5			

[a] The phenacetin peak in the APC chromatograms may be identified from the elution time of the phenacetin standard.

K. On a separate sheet of paper construct Table 13-6 by making the appropriate measurements and recording the data. Typical results are shown in Table 13-7. There are approximately 355 mL per 12-oz can. For various carbonated beverages (e.g., colas), the range of caffeine was 10.9–17.8 mg per can, the range of benzoate was 15.9–35 mg per can, and the range of aspartame when present was 79–93 mg per can. Your results may vary.

TABLE 13-3. Example Table for Phenacetin Results

Sample	Distance from Injection to Peak Center[a] (cm)	Individual Peak Height (cm)	Total Phenacetin Injected (μg)
Standard			
2 μL	5.20	3.1	3.0
5 μL	5.24	7.2	7.5
10 μL	5.18	14.4	15.0
15 μL	5.20	21.6	22.5
			Total Phenacetin Found (μg)
Tablet			
No. 1	5.20	150	16.0 μg
No. 2	5.20	150	16.0 μg
No. 3	5.22	152	16.2 μg

TABLE 13-4. Tabular Summary of Results of Tablet Uniformity[a]

Sample	Injection Number	Peak Retention Distance from Injection to Peak Center (cm)	Peak Height (cm)
Low-limit aspirin (90% of label)	1		
	2		
	3		
	4		
Upper-limit aspirin (110% of label)	1		
	2		
	3		
	4		
Individual tablets	1		
	2		
	3		
	4		
	5		

[a] This table is optional.

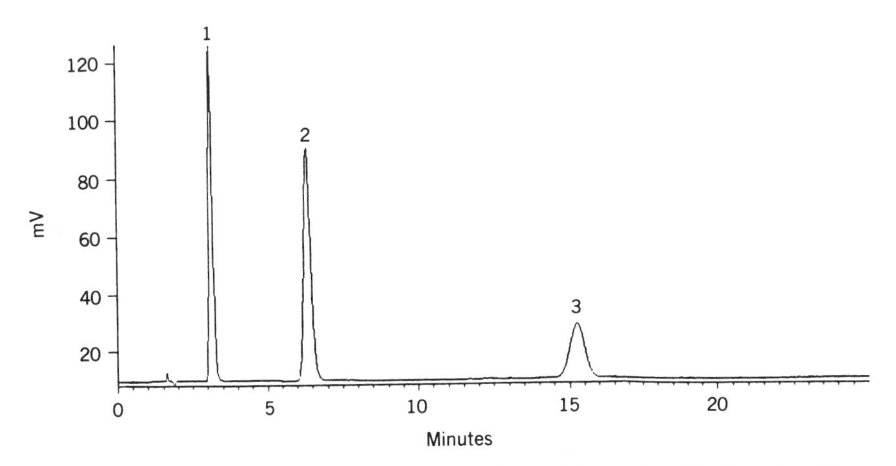

FIGURE 13-5. Analysis of caffeine, saccharin, and benzoic acid. Sample: (1) saccharin, (2) caffeine, and (3) benzoate. Column: μBondapak C_{18} (10 μm) 3.9 ID mm × 150 mm. Flow rate: 1.0 mL/min. Mobile phase: 20% MeOH/80% 1 M acetic acid, pH = 2.4. (Note: Actual separation will depend upon the quality of the mobile phase and column packing.)

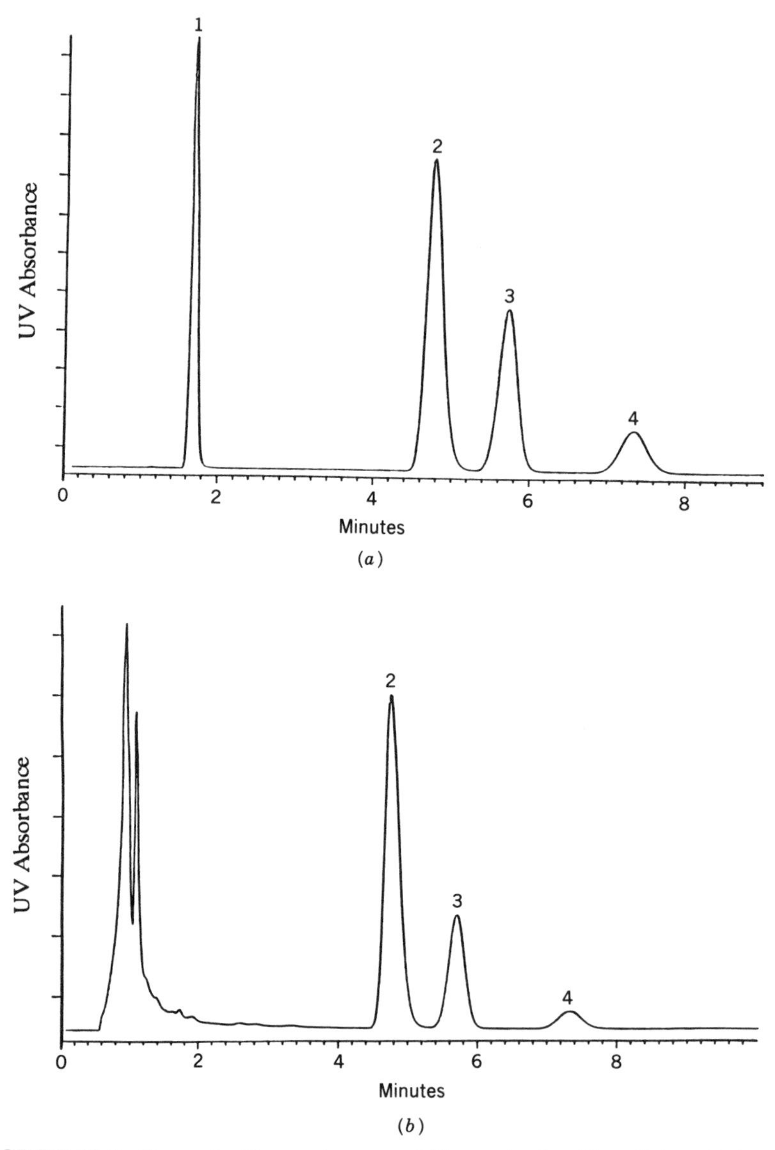

FIGURE 13-6. Example Chromatogram of (1) saccharin, (2) caffeine, (3) benzoate, and (4) aspartame. Column: μBondapak C_{18}, 3.9 mm ID × 150 mm. Mobile phase: 20% methanol/80% 1 M acetic acid; pH 4.2. Flow rate: 1.5 mL/min. Detection: UV at 254 nm (0.1 AUFS). (*a*) Standards. (*B*) A carbonated beverage. (Note: Actual separation will depend upon the quality of the mobile phase and column packing.) (Reproduced from reference 6 with permission.)

TABLE 13-5. Caffeine Standard Results for Beverage Analysis[a]

Sample	Peak Retention (cm)	Peak Height (cm)	Average Peak Retention (cm)	Average Peak Height (cm)	Standard Deviation of Peak Height[b]	Standard Deviation of Peak Retention
Repetitive 10-μL injections of the standard solution	1					
	2					
	3					
	4					

[a] Construct a similar table for saccharin and benzoate (if these are to be measured).
[b] Standard deviation is calculated by the formula $S = \sqrt{(X - \overline{X})^2/N - 1}$ where X is the individual value and $\overline{X}$ is the average value.

DISCUSSION

Since the response of the detector (and the separation) is a function of a flow rate, it is essential that the standard response curve be determined at the same flow rate as the tablet assay. If retention times differ significantly from the runs of the standards, there is a need to troubleshoot the HPLC to determine where the problem resides. Refer to Chapter 3 for a discussion of retention time precision.

TABLE 13-6. Calibration and Analysis Data for Caffeine in a Beverage[a]

Sample	Distance from Injection to Peak Center[b] (cm)	Individual Peak Height (cm)	Total Caffeine Injected (μg)
Standard			
2 μL			
5 μL			
10 μL			
15 μL			
20 μL			
Beverage			
No. 1			
No. 2			
No. 3			
No. 4			
No. 5			

[a] Construct a similar table for saccharin and benzoate (if these are to be measured).
[b] To be done for each compound which is being analyzed.

TABLE 13-7. Example Table for Caffeine Results

Sample	Distance from Injection to Peak Center (cm)	Individual Peak Height (cm)	Total Caffeine Injected (μg)
Standard			
2 μL	5.71	3.20	0.40
5 μL	5.69	8.00	1.00
10 μL	5.68	16.00	2.00
15 μL	5.65	24.00	3.00
			Total Caffeine Found (μg)
Beverage			
No. 1	5.68	10.90	1.36
No. 2	5.66	15.00	1.88
No. 3	5.67	12.20	1.53

SELF-HELP QUESTIONS

These are to be answered with respect to the experiments performed.

1. What is the standard deviation of your analysis? Is this good?
2. What effect would a variable rate of solvent delivery have upon the precision of analysis?
3. What effect would the purity of the standard have upon the final quantitation?
4. How could a higher purity standard be prepared?
5. What other sources of error are there in this experiment and how could they be minimized?
6. Assume you are using a higher-efficiency reverse-phase column than you used in this experiment. Comment as to how you could use this column to "improve" your analysis.
7. If you did the optional section, comment on the "quality" of the tested aspirin (see Table 13-3).
8. For those who analyzed the APC tablet, if available, refer to *Experiments for Instrumental Methods*, C. N. Reilley and D. T. Sawyer, McGraw Hill, New York, 1961, pp. 185–190, and compare the spectrophotometric method for determining phenacetin in an APC tablet to the chromatographic method which you just completed.
9. As shown in Figure 13-7, the pH of the mobile phase influences the retention of the various compounds in the sample. What does this imply about the qualitative nature of retention time in HPLC? Is HPLC a quantitative or qualitative tool?

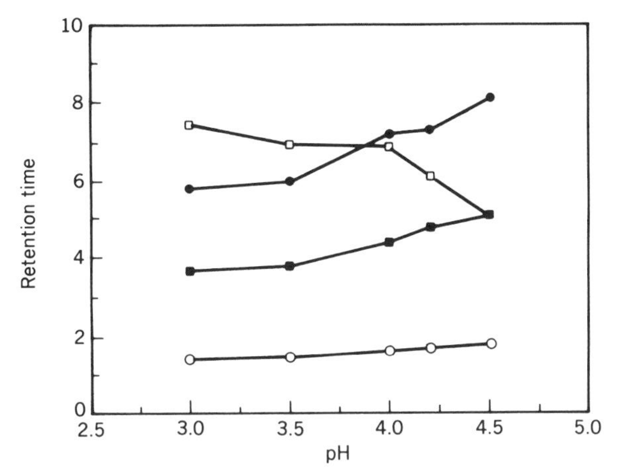

FIGURE 13-7. Effect of pH on the retention times of the beverage additives: □, benzoic acid; ●, aspartame; ■, caffeine; ○, saccharin. Conditions are the same as Figure 13-6 with the pH of the acetic-acid component adjusted with 50% sodium hydroxide to the desired pH. (Note: Actual separation will depend upon the quality of the mobile phase and column packing.) (Reproduced from reference 6 with permission.)

ADDITIONAL EXPERIMENTS

The following suggested supplemental experiments will increase your understanding of the use and characteristics of LC. They may be done under the operating conditions previously established, depending on the lab time available and course requirements.

A. Repeat the experiment A or G using a flow rate of 3 mL/min. Calculate the appropriate values and compare results. Which flow rate is best—the original or 3 mL/min?

B. Repeat the experiment D but carry out the quantitation for the aspirin and/or caffeine in the APC tablets. What sources of error are present?

C. Repeat the experiment using three different groups of five aspirin tablets from three different suppliers.

D. If your standards are not pure, you can inject a large amount on the column and collect "pure" standards while letting the impurities go to waste. The solvent could then be evaporated and the standards used for maximum accuracy.

REFERENCES

1. D. S. Smyly, B. B. Woodward, and E. C. Conrad, *J. Assoc. Offic. Anal. Chem.*, **59,** 14 (1976).

2. B. L. Madison, J. W. Kozarek, and C. P. Damo, *J. Assoc. Offic. Anal. Chem.*, **59,** 1258 (1976).
3. M. Attina and G. Ciranni, *Farm. Ed. Prac.*, **31,** 650 (1976), through *Anal. Abs.*, **33,** 1F36 (1977).
4. A. C. Hoeffler and P. Coggon, *J. Chromatogr.*, **129,** 460 (1976).
5. B. B. Woodward, G. P. Heffelfinger, and D. I. Ruggles, *J. Assoc. Offic. Anal. Chem.*, **62,** 1011 (1979).
6. S. Schmitz, F. V. Warren and B. A. Bidlingmeyer, *J. Chem. Ed.*, **68,** A195 (1991).

CHAPTER 14

EXPERIMENT 7: MONITORING KINETICS

Objective
Background
Additional materials
Safety and disposal
Experiments
A. Preparation of instrument
B. Preparation of samples
C. Chromatography
Evaluation
Discussion
Self-help questions
Additional experiments

OBJECTIVE

To monitor the kinetics of a chemical reaction using HPLC.

BACKGROUND

If the components of a chemical reaction can be separated by LC, monitoring the course of the reaction is easily accomplished by injecting an aliquot of the mixture into the chromatograph at specified times. Since each component in the mixture can be separated, both the rate of decrease of the reactant(s) as well as the rate of increase of the product(s) can be monitored by measuring the peak heights for the respective compounds. If the separation is significantly faster than the reaction, insignificant changes occur between sampling time and analysis. This is the situation in this experiment which studies the hydrolysis of aspirin into salicylic acid and acetic acid. The reaction is shown in Figure 14-1. The separation is by reverse-phase using a spectrophotometric detector. Aspirin (acetylsalicylic acid) and salicylic acid have large molar absorptivities at 254 nm, which makes their presence easy to detect while acetic acid, which has no absorption at this wavelength, will not be detected.

A reaction may be classified according to the number of molecules whose concentrations affect the rate of the reaction. This determines the order of the reaction. Aspirin decomposes according to the reaction in Figure 14-1 and the rate constant of this hydrolysis is directly proportional to the concentration of aspirin. No other dependencies of the rate are observed under the conditions of this experiment so that the reaction may be called pseudo-first order and may be described by the equation

$$-\frac{dC}{dt} = k_{\mathrm{D}}C \qquad (14\text{-}1)$$

where C is the concentration of aspirin, t is the time, and k_{D} is the rate constant of disappearance of aspirin.

If C_0 represents the original concentration of aspirin and C_t represents the amount of aspirin a time t_i, equation (14-1) may be expressed as:

$$k_{\mathrm{D}} = \frac{2.303}{t} \log \frac{C_0}{C_t} \qquad (14\text{-}2)$$

Thus, by following the concentration (C_t) of aspirin at various times (t_i), k_{D} may be determined.

ASPIRIN + H_2O (WATER) ⟶ SALICYLIC ACID + ACETIC ACID

FIGURE 14-1. Hydrolysis of aspirin to form salicylic acid and acetic acid.

ADDITIONAL MATERIALS

Use the column that was used in Chapter 12 (Experiment 5).

2 150-mL Erlenmeyer flasks, glass-stoppered.
1000-mL beaker.
100-mL graduated cylinder.
Stirring hot plate.
No. 2 filter paper.
Glass funnel.
Stirring bars (Teflon® or glass).
25-μL syringe.
Aspirin tablet (any supplier).
Methanol, filtered through a 0.45-μ filter using a solvent-clarification device.
Distilled water, filtered through a 0.45-μ filter using a solvent-clarification device.
Glacial acetic acid.

SAFETY AND DISPOSAL

This experiment does not purport to address the safety issues associated with its use. It is the responsibility of the user of this experiment to establish appropriate safety and health practices and to determine the applicability of regulatory limitations prior to use. All chemicals should be handled and disposed of in an appropriate manner consistent with the safety policy of the experimenter's company, school, or organization.

EXPERIMENTS

A. Preparation of Instrument*

Activity	Comments
1. Mix 25 mL of concentrated acetic acid with 625 mL of distilled water; to this mixture, add 350 mL of methanol. Mix thoroughly.	1. This is a mobile phase of 35% methanol: 65% water with 4% acetic acid.

* This may be done by the instructor before the lab period.

Activity	Comments
2. Place the mobile phase from Activity A-1 in the mobile phase reservoir	
3. Prime the solvent delivery system (pump) and deliver flow at 2 mL/min.	3. Consult the manufacturer's manual, if necessary.
4. The injector handle should be in the "inject" position.	4. This ensures that the sample holding loop is thoroughly cleaned. If an autosampler is used in this experiment, there must not be an excessive delay between the time a sample is removed from the reaction and the time when it is injected into the LC.
5. Stop the flow rate and install a column into the instrument. Restart the flow rate to 1 mL/min.	5. Use the column that was used in Chapter 11 (Experiment 4).
6. Set the spectrophotometric detector sensitivity to 0.5 AUFS (254 nm).	
7. Allow the system to equilibrate under these conditions while the samples are being prepared.	7. Allow at least 10 column volumes to flow through the column. This may be done at a higher flow rate (3 mL/min) if it is compatible with the instrument. Set flow to 1 mL/min and ensure that the detector baseline is stabilized before the experiment is begun.

B. Preparation of Samples

Activity	Comments
1. Place an empty 150 mL Erlenmeyer flask in a 1000-mL beaker. Add water to the beaker to cover the 150-mL mark on the flask. Refer to Figure 14-2.	1. This ensures that the reaction will be accommodated easily in the heating bath.
2. Remove the flask and place the beaker of water on a stirring hot plate and bring the water to a boil.	2. The magnetic stirring will be used later in the experiment.
3. Crush the tablet with a glass	3. The solution may still be cloudy

stirring rod. Transfer the crushed tablet to a 150-mL Erlenmeyer flask. Add 100 mL of distilled water. Stir to dissolve aspirin.

at this point due to excipients present in the tablet.

4. Filter through No. 2 filter paper into another 150-mL Erlenmeyer flask (stoppered).

4. The narrow neck and stopper of this flask are necessary to minimize evaporation during the experiment. The filtration ensures that the excipients ("floaters") will not plug the syringe needle.

5. Place a small (appropriate size) magnetic stirring bar in the flask.

5. This will be used later to magnetically stir the solution when it is being heated in the boiling water.

C. Chromatography

Activity	Comments
1. Set recorder chart speed at 12 in/h or equivalent.	
2. Carefully place the 150-mL Erlenmeyer flask containing the dissolved tablet (from Activity B-4) into the beaker containing the boiling water. Your apparatus should resemble that shown in Figure 14-2. Immedi-	2. This injection is considered to be the zero time reference of the experiment. If for some reason the peak height is off scale, inject a smaller amount, 5 μL, to bring the peak height onto the recording device. Note: if a

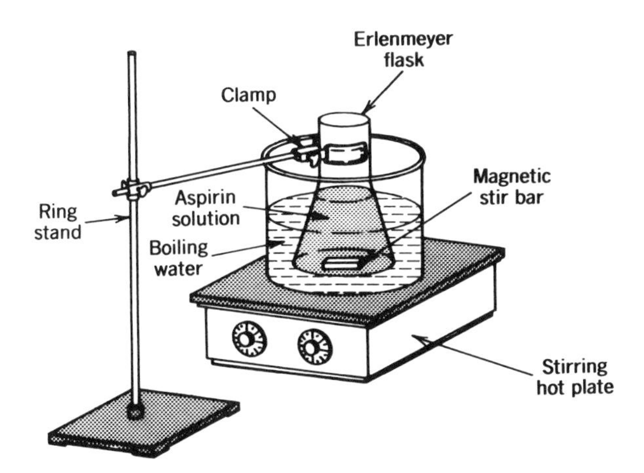

FIGURE 14-2. Experimental set-up for hydrolysis of aspirin.

ately withdraw a 10-μL aliquot from the flask and inject 10 μL of the aspirin solution into the HPLC.

smaller sample size is used, use that sample size (e.g., 5 μL) throughout the rest of the experiment.

3. Stir contents of flask gently with the magnetic stirrer.
4. Fill and dispense the syringe 5 times with distilled water.
5. Reinject 10 μL of the aspirin solution every 15 min* for a total elapsed reaction time of 105 min. Clean syringe between injections as described in Activity C-4.

3. This ensures that the reaction will proceed uniformly.
4. This prevents syringe contamination and "carryover" from the initial injection.
5. "Quenching" of the reaction occurs (for all intents and purposes) when the heat is removed. This occurs when the sample is removed from the reaction flask and injected into the LC.

EVALUATION

A. All chromatograms should be properly labeled and attached to the last sheet in this experiment. Proper labeling requires the following:

Sample:	Reaction mixture and injection volume.
Detector:	Type of detector and sensitivity.
Column:	Type of column used, length, and diameter (ID).
Solvent:	Mobile phase composition and flow rate employed.
Recorder:	Recorder chart speed.

B. Results of the separation of the aspirin reaction mixture under the conditions of this experiment are presented in the series of chromatograms in Figure 14-3A through 14-3H for the separation on a low-efficiency column. Figure 14-4 shows example chromatograms of the reaction products at various times during the reaction on a high-efficiency column.

C. On a separate sheet of paper construct Table 14-1. Calculate the appropriate performance parameters and record the data using the data from your chromatogram obtained after 30 min into the reaction.

D. On a separate sheet of paper construct Table 14-2. Make the appropri-

* A sample is removed from the reaction mixture every 15 min, however, in some cases the chromatographic run time may be 20 min. If this is the case, the sample may be held in a syringe until the injection. Alternatively, the run time can be decreased by using a flow rate of 2.0 mL/min. This increased flow decreases retention time and should not have a deleterious effect on the resolution. It is cautioned, however, that the backpressure of the column should not be exceeded, as suggested by the column care and use manual of the manufacturer.

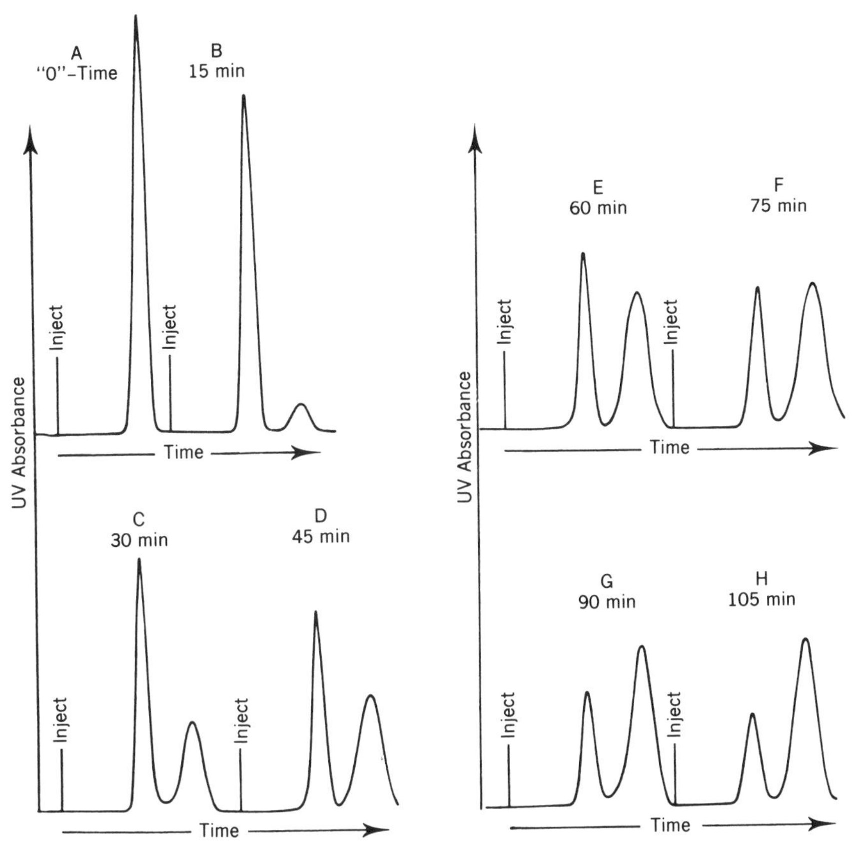

FIGURE 14-3. Separation of aspirin reaction mixture on a low-efficiency column. Column: Bondapak C18/Porasil B (37–75 μm) 2 mm ID × 61 cm. A–H are times noted on the chromatogram. Mobile phase: MeOH/4% HOAc in H_2O (35/75). Flow rate: 1 mL/min. Detector: UV at 254 nm, 0.2 AUFS. Sample: 10 μL of aspirin solution. (Note: Actual separation will depend upon the quality of the mobile phase and column packing).

TABLE 14.1 Performance Parameter at Time of 30 Minutes

k' (Aspirin)	k' (Salicylic acid)	α	N (Aspirin)	N (Salicylic acid)	R

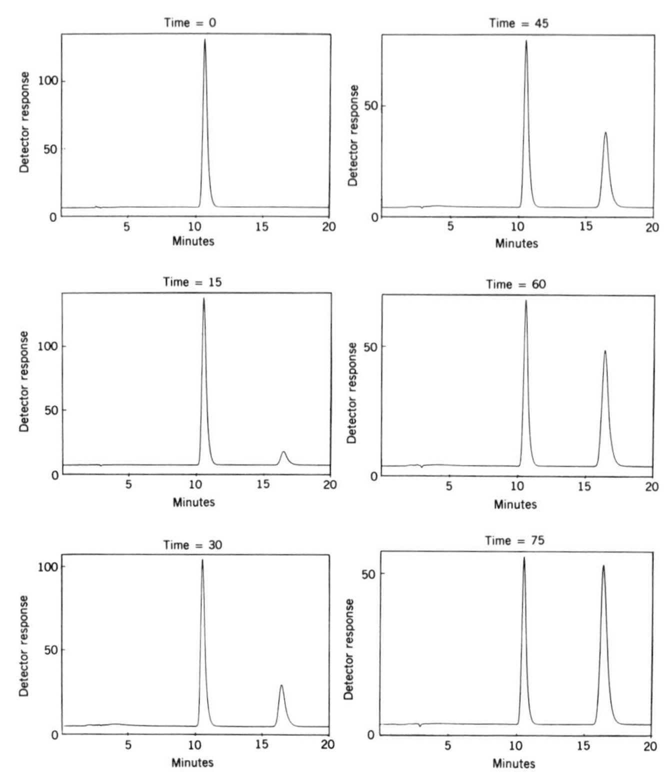

FIGURE 14-4. Separation of aspirin reaction mixture on a high-efficiency column. Reaction times are noted on the chromatograms. Column: μBondapak C_{18} (10 μm) 3.9 mm ID × 15 cm. Flow: 1 mL/min. Mobile phase: MeOH/4% HOAc in H_2O (37/75). Detector: UV at 254 nm; Sample: 10 μL of aspirin solution. (Note: Actual separation will depend upon the quality of the mobile phase and column packing.)

TABLE 14-2. Tabular Summary of Experimental Results

Condition after Hydrolysis is Started (Min)	Peak Height (mm)	
	Aspirin-1	Salicylic Acid-2
0	____	____
15	____	____
30	____	____
45	____	____
60	____	____
75	____	____
90	____	____
105	____	____

TABLE 14-3. Example Table From Figures 14-3, Chromatograms A–H

Condition after Hydrolysis is Started (Min)	Peak Height (mm)	
	Aspirin-1	Salicylic Acid-2
0	155.5	0
15	135.5	12.5
30	102.0	36.0
45	80.0	46.5
60	66.5	54.5
75	55.0	57.5
90	45.0	63.5
105	36.5	66.5

ate measurements and record the data. Refer to Table 14-3 for the example.

E. Construct a plot of the log of the peak height (in mm) for aspirin versus the elapsed time for the decomposition. Calculate the rate constant for this reaction. Remember from the section entitled Background, that the reaction rate is directly proportional to the concentration of aspirin as denoted by equation (14-3):

$$k_D = \frac{2.303}{t} \times \log \frac{C_o}{C_t} \tag{14-3}$$

If log C_t is plotted against t_i, then the ordinate intercept is equal to log C_o, and it is also true that

$$\text{Slope} = \frac{-k_D}{2.303} \tag{14-4}$$

DISCUSSION

Liquid chromatography can be used in real time to aid in kinetic studies. All that is required is that the reactant(s) and product(s) be separable. If the reaction is fast, the reaction must be quenched before the HPLC analysis or the analysis time and sampling time must be accounted for.

SELF-HELP QUESTIONS

1. What sources of error are there in this experiment?
2. What other analytical technique might you employ to study the reaction?

ADDITIONAL EXPERIMENTS

The following suggested supplemental experiments will further increase your understanding of the use and characteristics of high-resolution chromatography. They may be done under the operating conditions previously established, depending upon the time available and course requirements.

If a temperature bath is available, run the experiment at a temperature of 80°C. Compare the results to that run at 100°C. If an appropriate fluid is available for the heating bath, run the reaction at 130°C. Compare the results to those obtained at the other temperatures.

CHAPTER 15

EXPERIMENT 8: PREPARATIVE LIQUID CHROMATOGRAPHY

Objective
Background
Additional materials
Safety and disposal
Experiments
- A. Preparation of instrument
- B. Sample preparation
- C. Chromatography

Evaluation
Discussion
Self-help questions
Supplemental experiments
Reference

OBJECTIVE

To develop an understanding of the relationship between sample loading and resolution for preparative separations; and in doing so to investigate the use of a spectrophotometric detector and an RI detector in preparative separations.

BACKGROUND

Preparative LC is similar to analytical LC in that the same basic chromatographic relationships between V_0, k', α, R, and N still apply, but each technique is used to obtain different objectives. Maximum resolution of the sample components in a minimum period of time is of primary importance in analytical LC. In preparative chromatography, the main objective is the maximum recovery of pure compounds per unit time. This is often referred to as throughput. A thorough description of preparative LC is given elsewhere (1) and the reader who wants to know more should read this text.

Before attempting a preparative separation, the separation should be optimized. Once this step is accomplished, it is simply a matter of increasing the sample load until the desired level of overload is attained. Often separations are developed on a high-efficiency analytical column and then transferred to a packing material with a higher surface area and larger particle diameter. The use of large-diameter columns, longer column length, or the recycle technique permit the chromatographer to optimize his separation with respect to the preparative sample amounts which he needs. (Refer to Chapter 10, Experiment 2 for the effect of additional column length and recycle upon separations).

Another approach, which is used in this experiment, is to develop the analytical separation on the high-surface-area packing and increase the amount injected into the column to determine the loading level for preparative work. A common "problem" in preparative LC is detector "saturation." Detector saturation occurs when the concentration of sample eluting from the column is so high that the detection system is electronically overloaded. The result of detector saturation is loss of the ability to observe the peaks. This is demonstrated in this experiment when the spectrophotometer is saturated, and the refractive index detector is not.

ADDITIONAL MATERIALS

1 Silica gel column. The following columns have been tested in this experiment: a 2 mm ID × 61 cm Porasil A (35–75 μm) column (or a hand-packed column), and 3.9 mm × 15 cm μPorasil (10 μm) column, a 3.9 mm ID × 15 cm Resolve Silica (5 μm) column, and a 3.9 mm ID × 30 cm μPorasil (10 μm) column. Other silica gel columns should work in this experiment; however, they have not been tested.

3 20-mL flasks (or vials), stoppered.

25-μL syringe.

Diethyl phthalate (any supplier).

Dibutyl phthalate (any supplier).

Dioctyl phthalate (any supplier).

Methylene chloride, filtered through a 0.45-μ filter using a solvent clarification device.

Acetonitrile (HPLC grade), filtered through a 0.45-μ filter using a solvent clarification device.

Carbon tetrachloride (spectrograde).

SAFETY AND DISPOSAL

Some of the chemicals used in this experiment may be considered hazardous. This experiment does not purport to address the safety issues associated with its use. It is the responsibility of the user of this experiment to establish appropriate safety and health practices and to determine the applicability of regulatory limitations prior to use. All chemicals should be handled and disposed of in an appropriate manner consistent with the safety policy of the experimenter's company, school, or organization.

EXPERIMENTS

A. Preparation of Instrument*

Activity	Comments
1. Clear (purge) the instrument of any solvent that is immiscible with methylene chloride.	1. Refer to Chapter 6 and observe the miscibility rules.
2. Prepare 1000 mL of an 0.1% (v/v) acetonitrile in methylene chloride solution and place in the reservoir.	2. The acetonitrile will reduce tailing in the separation of the phthlates.
3. Prime the solvent delivery system (pump) and start flow rate at 4.0 mL/min.	3. Refer to manufacturer's solvent delivery system (pump) manual, if necessary.
4. Rotate injector handle to the "inject" position.	4. This ensures that the sample holding loop is thoroughly cleaned.
5. Stop the flow rate and place the column in the instrument. Reset the flow rate to 4 mL/min.	5. Consult manufacturer's manual, if necessary, to determine if a flow-rate limitation exists for a column or instrument. If 4 mL/min exceeds this limit use the highest flow available (not greater than 4 mL/min.)
6. Set the spectrophotometric detector sensitivity at 0.5 AUFS, 254 nm.	
7. Allow the instrument to equilibrate under these conditions while preparing the samples. Pumping thirty column volumes through the column should be sufficient.	7. This ensures that the detector baseline is stabilized before the experiment is begun.

* This may be done by the instructor before the lab period.

B. Sample Preparation

Activity	Comments
1. Weigh 500 mg each of diethyl phthalate, dibutyl phthalate, and dioctyl phthalate into a 20-mL flask (or vial) and add 10 mL of mobile phase. Stopper the flask.	1. Density values for the samples are: diethyl phthalate (d = 1.118 g/mL), dibutyl phthalate (d = 1.043 g/mL), and dioctyl phthalate (d = 0.981 g/mL). This solution will be the concentrated sample solution.
2. Pipet 1.0 mL of the solution from Activity B-1 into a 20-mL flask (or vial) and add 10 mL of mobile phase. Stopper the flask.	2. This is the diluted sample solution.
3. Place approximately 5 mL of carbon tetrachloride into a 20-mL flask. Stopper the flask.	3. Carbon tetrachloride will be injected into the column during Section C.

C. Chromatography

Activity	Comments
1. Set recorder chart speed to 0.5 cm/min or equivalent.	
2. Inject 10 μL of carbon tetrachloride.	2. Carbon tetrachloride will measure the V_0 value.
3. After the component has eluted from the column, change attenuation of the spectrophotometer to 0.1 AUFS (254 nm).	3. If you have a refractometer, set the sensitivity to 8X or an equivalent midrange setting.
4. Inject 1 μL of the solution from Activity B-2.	4. This injection is from the dilute sample.
5. After the components have eluted from the column, change attenuation on the spectrophotometer to 0.2 AUFS (254 nm).	
6. Inject 5 μL of the sample solution from Activity B-2.	6. This injection is from the dilute sample.
7. After the components have eluted from the column, change the attenuation on the spectrophotometer to 0.5 AUFS (254 nm).	
8. Inject 10 μL of the solution from Activity B-2.	8. This injection is from the dilute sample.

9. After the components have eluted from the column, change the attenuation on the spectrophotometer to 1.0 AUFS (254 nm).
10. Inject 10 μL of the solution from Activity B-1.

    10. This injection is from the concentrated sample.
11. After the components have eluted from the column, change the attenuation on the spectrophotometer to 2.0 AUFS (254 nm).
12. Inject 25 μL of solution from Activity B-1.

    12. This injection is from the concentrated sample. This response will saturate the detector. However, an estimate of retention volume and peak width is possible.

Note: At this point, the spectrophotometric detector is probably saturated or "blinded." If you have a refractometer you may continue with the remaining steps if time and the lab instructor permit.

13. Change refractometer setting to 16X or the next less-sensitive setting and inject 50 μl and 100 μl of solution from Activity B-1.
14. After the components have eluted from the column, change the refractometer to 32X or the next less-sensitive setting and inject 200 μL of solution from Activity B-1.

EVALUATION

A. All chromatograms should be properly labeled and attached to the last sheet in this experiment. Proper labeling required the following:

Sample:	Components, approximate amount of each and injection volume.
Detector:	Type of detector and sensitivity.
Column:	type of column used, length, and diameter (ID).
Solvent:	Mobile phase composition and flow rate employed.
Recorder:	Recorder chart speed.

TABLE 15-1. Tabular Summary of Experimental Results

Injection Volume (μL)	Solution	Load Sample Weight	k'[a]			N			α		R	
			1	2	3	1	2	3	1 & 2	2 & 3	1 & 2	2 & 3
1	B-2	____	__	__	__	__	__	__	____	____	____	____
5	B-2	____	__	__	__	__	__	__	____	____	____	____
10	B-2	____	__	__	__	__	__	__	____	____	____	____
10	B-1	____	__	__	__	__	__	__	____	____	____	____
25	B-1	____	__	__	__	__	__	__	____	____	____	____
50	B-1	____	__	__	__	__	__	__	____	____	____	____
Optional												
100	B-1	____	__	__	__	__	__	__	____	____	____	____
200	B-1	____	__	__	__	__	__	__	____	____	____	____

[a] k' calculations require V_0, which is obtained from the retention of carbon tetrachloride.

B. On a separate sheet of paper, construct Table 15-1. Make the appropriate measurements and calculations and record the data. Since some of these peaks may be tailed, attempt to draw the best tangents to the side of the peaks for base width determination. Refer to Figures 15-1*a* and *b*.

DISCUSSION

As the amount of sample injected is increased (increased load) on a chromatographic column, the surface area of column packing material available for interaction decreases. As a result, the resolution of the peaks will decrease. However, in preparative chromatography this is not bad since resolution must be compromised with the load. The extent of this compromise depends upon the individual chromatographer.

As can be seen from this experiment, the spectrophotometric detector is excellent for analytical scale detection of the phthalates which have a good molar absorptivity at 254 nm. However, as was observed in this experiment, the spectrophotometer is much less useful for preparative scale chromatographic separations because it tends to become optically saturated when relatively large amounts of materials are injected, that is, all light entering the detector cell is absorbed and no separation can be seen because of the off-scale readout. Also, spectrophotometric detectors will be ''blind'' to molecules that have no chromophore in their structure. The differential refractometer is the preferred detector for preparative LC because generally

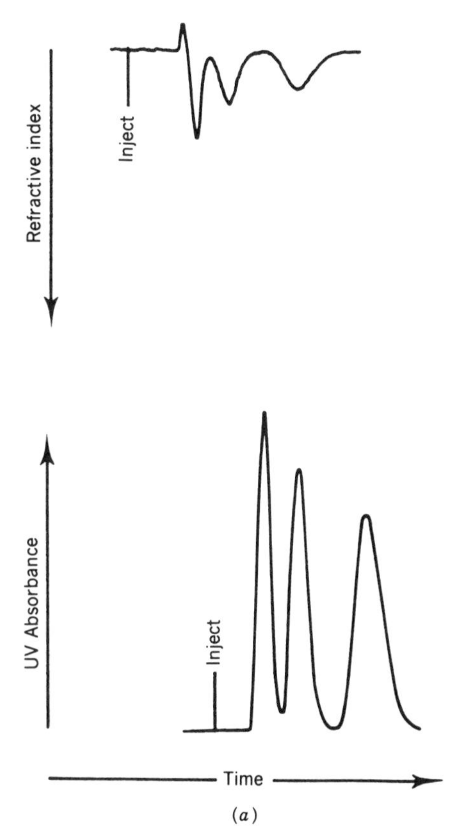

FIGURE 15-1. (*a*) Example chromatogram on a low-efficiency column. Column: Porasil A (37–75 μm) 2 mm ID × 61 cm. Solvent: 0.1% CH_3CN in CH_2Cl_2. Flow rate: 4 mL/min. Detector: UV at 254 nm, 1.0 AUFS. RI: 8X. Sample: 10 μL solution of B-2 containing 0.5 mg each of dioctyl, dibutyl, and diethyl phthalate. Recorder: 0.2 in./min. Top trace: Refractive index. Bottom trace: UV at 254 nm. (Note: Actual separation will depend upon the quality of the mobile phase and column packing.) (*b*) Example chromatogram on a high-efficiency column. Column: μPorasil (10 μm) 3.9 mm ID × 15 cm. Solvent: 0.1% CH_3CN in CH_2Cl_2. Flow rate: 4 mL/min. Detector: UV at 254 nm, 1.0 AUFS. RI: 8X. Sample: 10 μL solution of B-2 containing 0.5 mg each of dioctyl, dibutyl, and diethyl phthalate. Recorder: 0.2 in./min. Top trace: Refractometer. Bottom trace: UV at 254 nm. (Note: Actual separation will depend upon the quality of the mobile phase and column packing.)

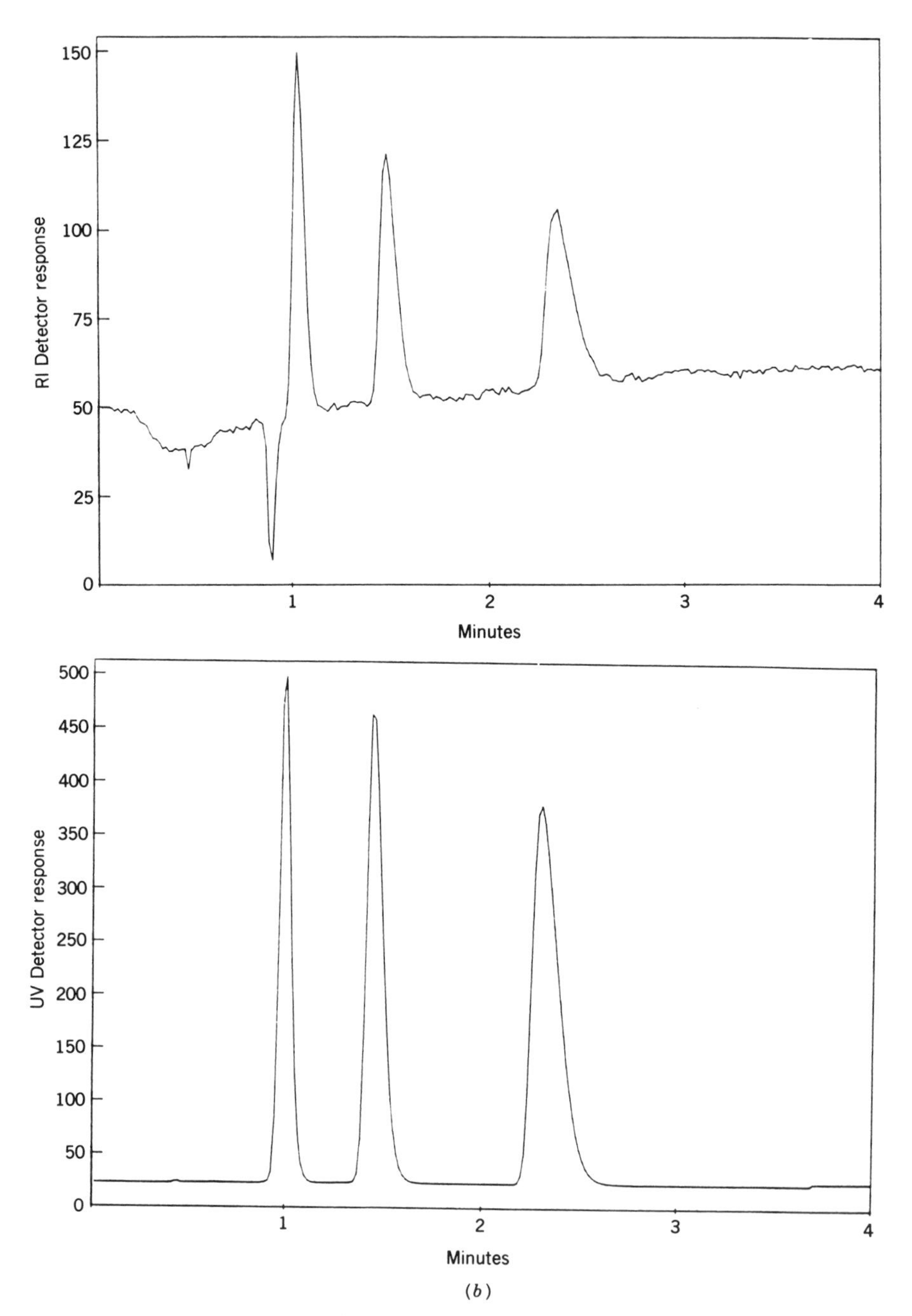

(*b*)

FIGURE 15-1. (*Continued*)

it is a detector that is difficult to saturate. It responds to all compounds with a refractive index different from the mobile phase. The RI detector is essentially a universal detector that sees any component of the sample.

SELF-HELP QUESTIONS

1. From your data in Table 15-1:
   Do k' values decrease as sample load increases?
   Do plates (N) values decrease as sample load increases?
2. The goal in preparative experiments is to maximize load with respect to resolution. What do your data in Table 15-1 tell you about this goal?
3. Discuss your approach to the following two preparative problems:
   a. You have 10 mg of extract which is 80% pure and you wish to purify it by chromatography. How do you do it?
   b. You have 10 g of material which is 80% compound A and 20% compound B. How do you chromatographically approach the separation?
4. From your plot in Section B and Table 15-1, does overloading occur faster with large or small values of k'? Can you suggest an explanation?
5. Porasil A has a surface area of 450 m^2/g. Porasil B has a surface area of 200 m^2/g. What would you expect the overloading level to be on Porasil B compared to Porasil A?
6. If your spectrophotometer had not been saturated, what do you think would have occurred with resolution as sample load increased? For those who used a refractometer in this experiment, did your observation support your concept?

SUPPLEMENTAL EXPERIMENTS

The following are suggested supplemental experiments that will further increase your understanding of the use and characteristics of high-resolution LC. This may be done under the operating conditions previously established.

A. Determine if an analytical separation can be developed on a 2 mm ID × 61 cm pellicular column (Corasil II). Transfer this separation to the large particle, totally porous column (Porasil A). Compare the plate number and k' values from the two columns. Comment as to the similarities and differences.

B. Using the RI detector, load the injected sample to a level where the resolution is equal to 0.7. Collect fractions at 30-sec intervals across the peak and determine the relative purity of each. How much overlap is there between peaks with $R = 0.7$?

C. Install two 2 mm ID × 61 cm large-particle, totally porous columns (Porasil A) and inject a sample load so that you obtain an R value equal to 0.7. Using the refractometer as a guide, shave the first section of the peak and recycle the overlapped portion. Continue the shave–recycle sequence for purity. [Note: Before attempting this experiment, Chapter 10 (Experiment 3) should be performed to develop the technique of recycle.]

REFERENCE

1. B. A. Bidlingmeyer, Ed., *Preparative Liquid Chromatography,* Elsevier, Amsterdam 1987.

CHAPTER 16

EXPERIMENT 9: ANALYSIS OF ESSENTIAL OILS (STEAM DISTILLATES)

Objective
Background
Additional materials
Safety and disposal
Experiments
- A. Preparation of the instrument
- B. Steam distillation of oil of cloves
- C. Optional: Additional steam-distillation extracts
- D. Sample preparation
- E. Chromatography

Evaluation
Discussion
Self-help questions
References

OBJECTIVE

To demonstrate the use of reverse-phase liquid chromatographic separation in analyzing the major essential oil from a natural product by steam distillation.

BACKGROUND

Steam distillation is an effective method of separating volatile compounds from a complex mixture (1–3). The process occurs at 100°C, rather than at the boiling points of the various components of the mixture. Since many compounds decompose at their boiling points, steam distillation provides a simple way of isolating reasonably pure samples. The process is popular in the isolation of essential oils from natural products (4–8). As an example, cinnamaldehyde boils at 246°C at atmospheric pressure with significant decomposition, but can be readily separated from cinnamon bark by steam distillation.

This experiment is intended to illustrate the complementary nature of analytical LC with steam distillation by measuring the composition of a steam-distilled extract of a spice.

ADDITIONAL MATERIALS

1 C_{18} bonded-phase column. The following columns have been tested in this experiment: a 2 mm ID × 81 cm Bondapak C_{18}/Porasil B column (35–75 μm), a 3.9 × 15 cm μBondapak C_{18} column (10 μm), a 3.9 mm ID × 30 cm μBondapak C_{18} column (10 μm) and a 3.9 mm ID × 15 cm Resolve C_{18} column (5 or 10 μm). Other C_{18} bonded columns (5 or 10 μm) should work in this experiment; however, they have not been tested.

Steam distillation apparatus (refer to any one of many laboratory texts for organic chemistry which describes the apparatus and set-up, e.g., references 1–3).

Spices: cloves. Optional: allspice, aniseed, fennel, cinnamon, and cumin seed.

Methanol (LC grade) filtered through a 0.45-μ filter using a solvent clarification device.

Methylene chloride.

Essential oil standards: eugenol, anethole, cinnamaldehyde, and cuminaldehyde (available from supermarkets, drug, specialty stores or various chemical supply houses) for optional parts of experiment.

1 100-mL separatory funnel.

1 100-mL Erlenmeyer flask, stoppered (one for each distillate and standard).

1 25-μL syringe.

SAFETY AND DISPOSAL

This experiment does not purport to address the safety issues associated with its use. It is the responsibility of the user of this experiment to establish appropriate safety and health practices and to determine the applicability of

regulatory limitations prior to use. All chemicals should be handled and disposed of in an appropriate manner consistent with the safety policy of the experimenter's company, school, or organization.

EXPERIMENTAL

A. Preparation of the Instrument*

Activity	Comments
1. Clear (purge) the instrument of any solvent that is immiscible with water.	1. Refer to Chapter 6 and observe miscibility rules.
2. Prepare the mobile phase by adding 850 mL of methanol and 150 mL of water together in a container of appropriate size.	2. This mobile phase should be adequate for equilibration of the column and for the separations.
3. Place the methanol:water (60:40) (v/v) mobile phase onto the instrument.	
4. The injector should be in the "inject" position.	4. This guarantees that the sample holding loop will be sufficiently flushed and cleaned to avoid contamination from the previous solvent.
5. Prime the solvent delivery system (pump) (if necessary) and flow at 1 mL/min.	5. Since no column is installed either collect flow out of the injector into a suitably sized container or jumper the injector to the detector in order to flush the entire system to waste.
6. Stop the flow.	
7. Install the column into the instrument.	
8. Set the spectrophotometric detector to 1.0 AUFS at 25 nm.	
9. Allow the instrument to equilibrate with the mobile phase while preparing the sample (Sections B and D).	9. A flow rate of 1.0 mL/min is adequate. Faster flows may be used to speed up the process if these higher flows (and backpressures) are compatible with your instrument.

* This may be done by the lab instructor before the lab period.

B. Steam Distillation of Oil of Cloves

Activity	Comments
1. Assemble a flask, condenser, and receiver for distillation.	1. Refer to any one of many organic chemistry laboratory texts for description of apparatus and assembly for steam distillation (references 1–3).
2. Put approximately 200 mL of water into the flask and add a small amount of dry cloves, which should occupy approximately one quarter of the volume of the flask to be used for steam distillation.	2. Weigh the amount of cloves used.
3. Heat the flask and distill the clove oil and water.	3. It may be necessary to interrupt the distillation, add more water, and then distill further until droplets of oil no longer appear.
4. Collect the distillate.	4. Observe to determine when no more oil appears to be coming over.
5. Stop the distillation process.	
6. Add 25 mL of methylene chloride to this distillate, shake and pour off the water layer.	6. The water layer is the upper layer.
7. Repeat Activity B-6 using another 25-mL aliquot of methylene chloride	
8. Combine the two 25-mL portions of methylene chloride from Activity B-6 and B-7.	
9. Wash the methylene chloride with saturated sodium chloride solution.	9. Discard the aqueous salt solution.
10. Dry the methylene chloride with anhydrous sodium sulfate.	10. Magnesium sulfate may be substituted for sodium sulfate in the drying step. If desired, the methylene chloride may be filtered after the drying.
11. The clear extract may be evaporated which results in oil of cloves.	11. Oil of cloves is about 80% eugenol, 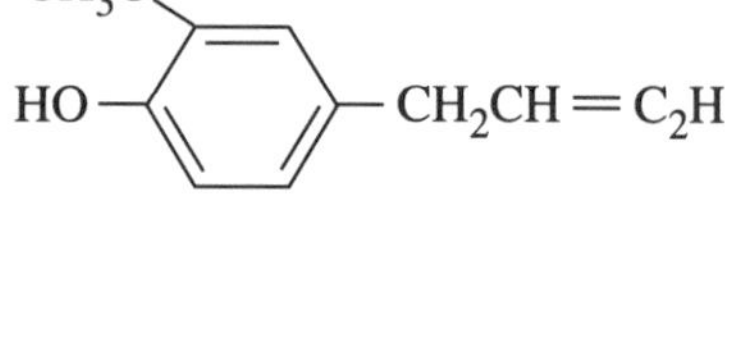

CH_3O

$HO-C_6H_3-CH_2CH{=}C_2H$

C. Optional: Additional Steam Distillation Extracts

A number of other spices may be used for sources of other oils, shown in Table 16-1. It should be noted that the amount of collection of oils may vary depending upon the age of the spice. Older spices may have already lost most of their volatiles. If any of these oils are desired, prepare (extract) the oils from the spices using the activities listed in Section B.

TABLE 16-1. Composition of Some Common Essential Oils

Spice	Major Essential Oil	Structure
Cloves	Eugenol (80%)	CH_3O; HO–C_6H_3–$CH_2CH{=}C_2H$
Allspice	Eugenol	
Aniseed (Anise)	Anethole	CH_3O–C_6H_4–C(H)=C(H)–CH_3
	p-Allylanisole (Trace)	CH_3O–C_6H_4–CH_2–C(H)$=CH_2$
Fennel	Anethole	
Cinnamon (bark)	Cinnamaldehyde (50–65%)	C_6H_5–CH=CHCHO
Cumin Seed	Cuminaldehyde (30–40%)	$(CH_3)_2$–CH–C_6H_4–CHO

D. Sample Preparation

Activity	Comments
1. Dissolve 25 μL of the extracted oil from Section B in 10 mL of methanol.	
2. Filter through a 0.45-μ filter or suitable sample clarification device into a flask.	2. This filtration is to ensure that none of the insoluble material is introduced into the HPLC.
3. Dissolve 25 μL of the standard into 10 mL of methanol.	3. Eugenol is the standard for oil of cloves.
4. Filter through a 0.45-μ filter or suitable sample-clarification device into a flask.	
5. If other steam distilled extracts are to be analyzed, prepare the extract as in Activity D-1 and D-2 and prepare the standard as in Activity D-3 and D-4.	5. Refer to Table 16-1.

E. Chromatography

Activity	Comments
1. Set recorder chart speed to 0.5 cm/min and set flow rate to 1.0 mL/min.	
2. Rotate the injector to the "load" position.	
3. Inject 10 μL of the standards of interest (see Activity D-4 and D-5).	3. Analysis time is approximately 5 min on a 3.9 mm ID × 15 cm μBondapak C_{18} 10 min on a 3.9 mm ID × 30 cm μBondapak C_{18} column. (Refer to Figure 16-1*a*.) If the mobile phase is too weak or too strong to obtain an adequate separation, develop the separation by adjusting the methanol : water composition according to the strategy and tactics discussed in Chapter 5.
4. Repeat Activity E-3.	4. Retention time of the standard should be reproducible.
5. Label the chromatograms.	
6. Inject 10 μL of all of the essential oils obtained from the steam distillations (Sections B and C).	
7. Label the chromatograms.	

EVALUATION

A. Figure 16-1 shows an example chromatogram for several of the essential oils and standards.

B. All chromatograms should be properly labeled and attached to the last sheet in this experiment. Proper labeling requires the following:

Sample:	Components, approximate amount of each and injection volume.
Detector:	Type of detector and sensitivity
Column:	Type of column used, length, and diameter (ID).
Solvent:	Mobile-phase composition and flow rate.
Recorder:	Recorder chart speed.

C. If the V_0 value is known (see Chapter 12, Experiment 5), on a separate sheet construct a table of compounds and the k' values.

DISCUSSION

Clove oil may be obtained from cloves by steam distillation. A number of other spices may be used for extracting other oils (9,10) whose structures are shown in Table 16-1. The essential oils are rich in one main compound and since all components are UV absorbing, the fixed wavelength detector at 254 nm is adequate for detection. Liquid chromatography is a rapid analytical tool in analyzing the essential oils.

SELF-HELP QUESTIONS

1. Did the chromatograms of the essential oil standards agree with those of the steam distillates? If not, what were the differences and why would such differences exist?
2. Does the order of retention of the main components of the essential oils correlate with the structural differences in the main components? For instance, anethole (main component in anise oil) is retained longer than is eugenol (main component in clove oil); explain this in terms of reverse-phase chromatography. If you cannot explain why the retention is as it occurs, refer to Chapter 4 and 5 or to the results of Chapter 12 (Experiment 5: Reverse-Phase Chromatography) for suggestions.
3. (Optional) The major essential oil of dill is (+) carvone and the main constituent of spearmint oil is (−) carvone. These are optical isomers. Would the column used in this experiment separate these optical isomers? Discuss how you could affect a separation of these two optical isomers using HPLC.

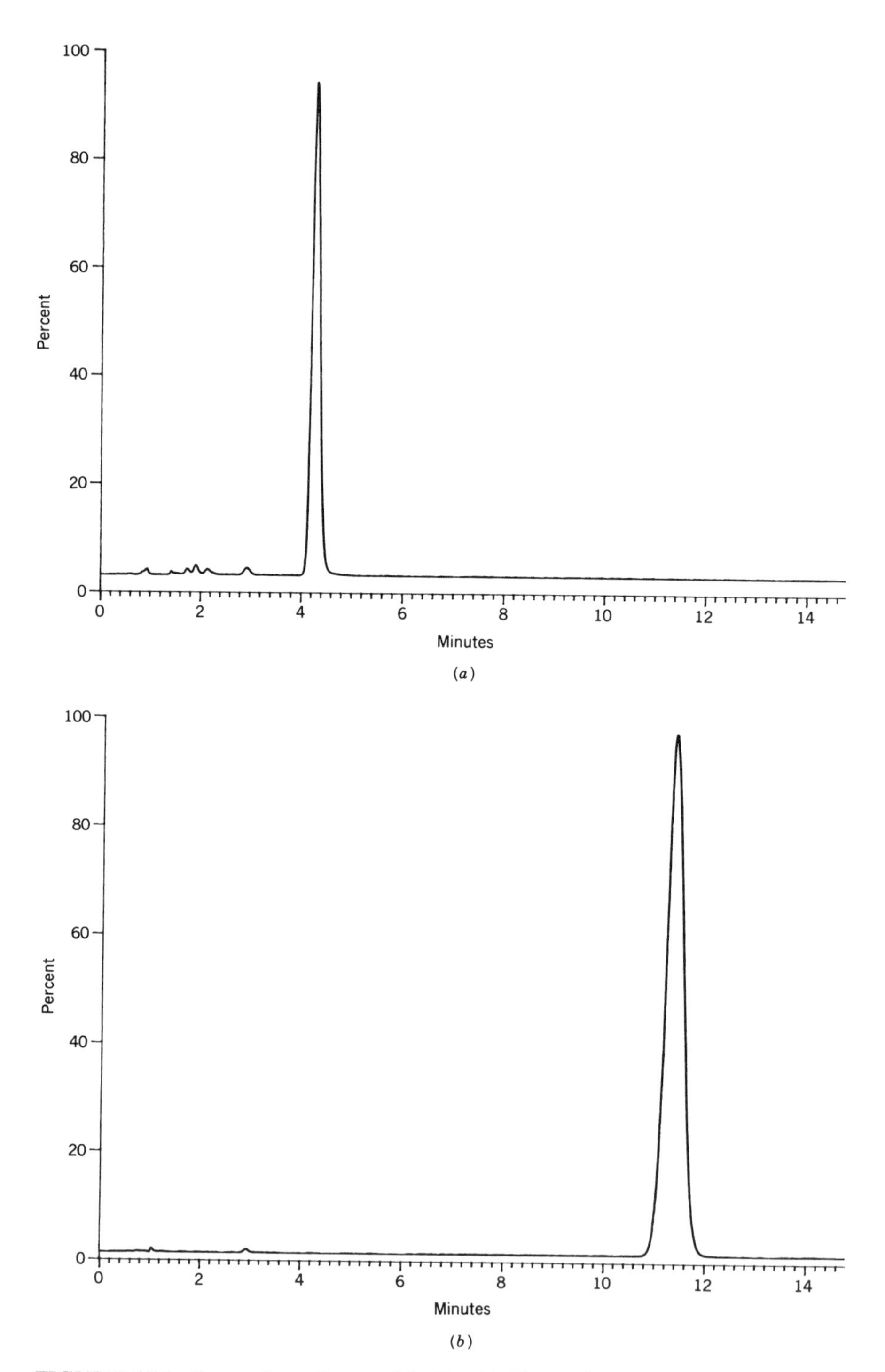

FIGURE 16-1. Separation of essential oils. (*a*) Eugenol. (*b*) Anethole. (*c*) *p*-Benzaldelhyde. (*d*) Cinnamaldehyde. Column: μBondapak C_{18} (10 μm), 3.9 mm ID × 15 cm. Mobile phase: 60% MeOH: 40% Water (v/v). Flow rate: 1.0 mL/min. Detection: 254 nm. Concentration: 1 mg/mL. Injection volume: 10 μL.

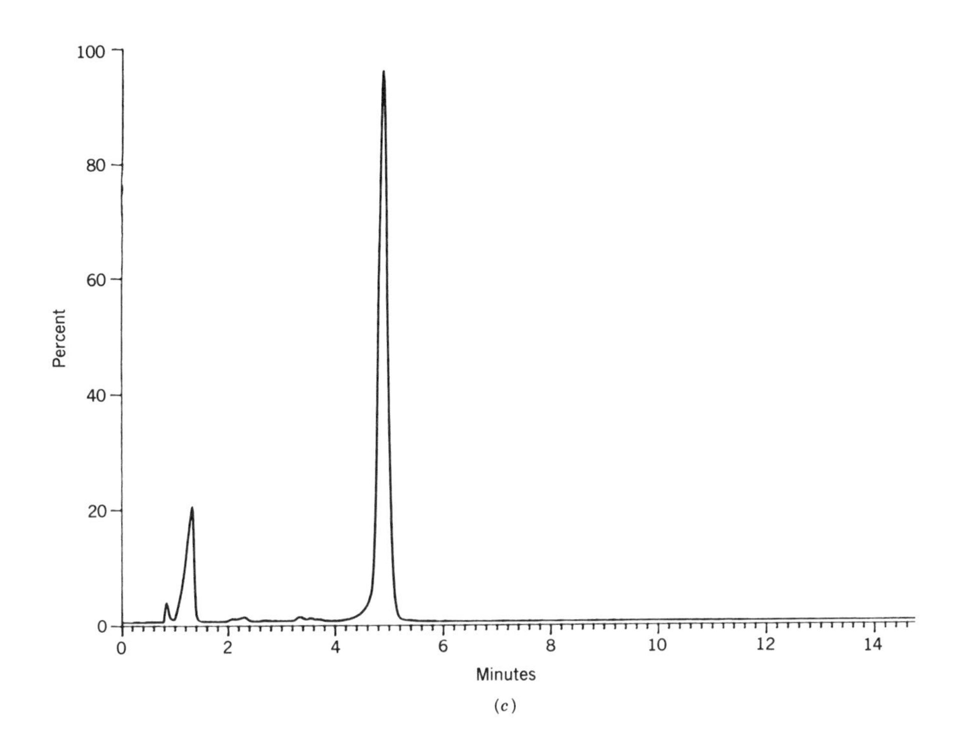

(*c*)

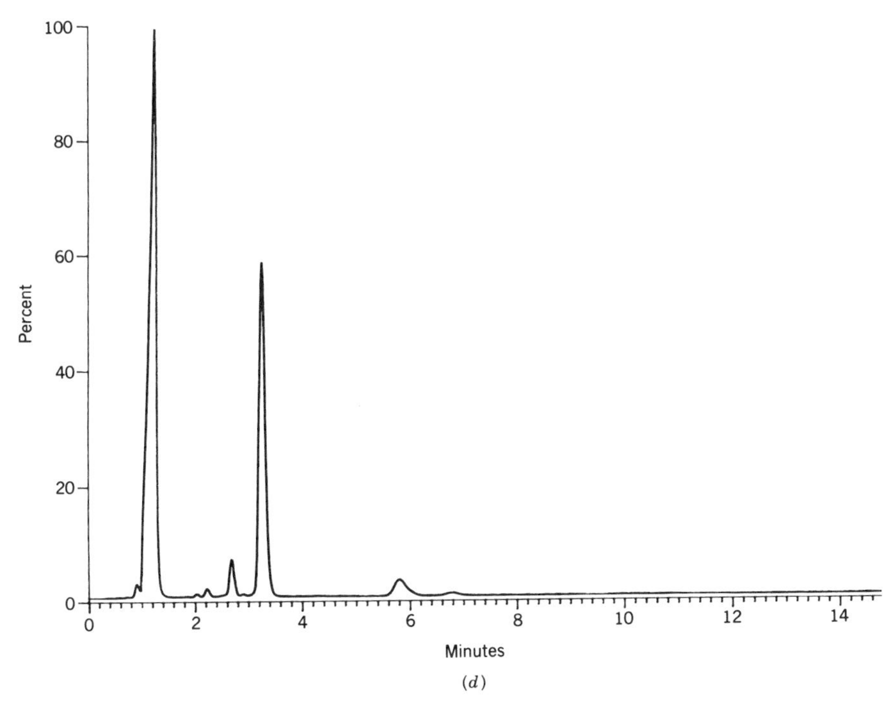

(*d*)

FIGURE 16-1. (*Continued*)

REFERENCES

1. R. J. Fessenden and J. S. Fessenden, *Laboratory Techniques for Organic Chemistry,* Brooks/Cole Publishing Co., Pacific Grove, CA, 1984, 99–103.
2. J. W. Zubrick, *The Organic Chemistry Survival Manual,* Wiley, New York, 1988, p. 174–178.
3. D. L. Pavia and G. M. Lapman, *Introduction to Organic Laboratory Techniques,* 3rd ed., Saunders College Publishing, Philadelphia, PA, 1988, p. 582–587.
4. F. H. Greenburg, *J. Chem. Educ.,* **45,** 537 (1968).
5. O. Runquist, *J. Chem. Educ.,* **46,** 846 (1969).
6. S. L. Murov and M. Pickering, *J. Chem. Educ.,* **50,** 74 (1973).
7. D. L. Garvin, *J. Chem. Educ.,* **53,** 105 (1976).
8. M. S. Ntamila and A. Hassanali, *J. Chem. Educ.,* **53,** 263 (1976).
9. H. B. Heath, *Flavor Technology: Profiles, Products, Applications,* Av., Publishing Co., Westport, CT, 1978.
10. E. Guenther, G. Gilbertson, and R. T. Koenig, *Anal. Chem.,* **49** (Annual Review), 83R–98R (1977).

CHAPTER 17

EXPERIMENT 10: GRADIENT ELUTION

Objective
Background
Additional materials
Safety and disposal
Experimental
A. Preparation of instrument
B. Sample preparation
C. Chromatography
Evaluation
Discussion
Self-help questions
Supplemental experiment

OBJECTIVE

To demonstrate the benefits and limitations of changing solvent composition during the course of the chromatographic separation of a complex sample.

BACKGROUND

There are times when a sample will contain components that have a wide range of polarities. In this situation, the peaks will have a wide range of retention times in an isocratic system and, often, the early eluting peaks will be sharp and the late eluting peaks will be quite broad. The general disadvantage is that a long analysis time is required for the entire sample. It may also be difficult to quantitate the broad, late-eluting peaks. Gradient elution (mobile phase programming) makes the separation of very dissimilar components more manageable. The mobile-phase composition can be programmed from a weak to strong solvent during the course of an analysis. Thus, the separation occurs in less time as peaks are brought together which otherwise would have been too well separated. Refer to Chapter 7 for additional information on gradient elution.

ADDITIONAL MATERIALS

1 C_{18} Bonded-phase column. The following columns have been tested in this experiment: a 2 mm ID × 60 cm Bondapak C_{18}/Porasil B (37–75 μm) (or hand-packed columns), a 3.9 mm ID × 15 cm μBondapak C_{18} (10 μm), a 3.9 mm ID × 30 cm μBondapak C_{18} (10 μm) and a 3.9 mm ID × 15 cm Resolve C_{18} (5 and 10 μm). Other C_{18} columns should work in this experiment; however, they have not been tested.

2 250-mL Erlenmeyer flasks.

2 Rubber stoppers, each of which contains 2 holes (refer to Figure 17-1), to fit the Erlenmeyer flasks.

4 9-in. lengths of $\frac{1}{8}$ in. OD stainless-steel tubing (glass may be substituted).

3 Short sections of Teflon tubing $\frac{1}{8}$ in. OD.

2 Stainless-steel solvent reservoir filters.

2 Magnetic stirrers.

2 Teflon or glass-coated magnetic stirring bars.

4 150-mL Erlenmeyer flasks, glass stoppered.

250-mL graduated cylinder.

25-μL syringe.

Orange peel (cut into 0.5-in. squares).

Grapefruit peel (cut into 0.5-in. squares).

Sample clarification devices (0.45-μ filter) for filtering samples.

Methanol, filtered through a 0.45-μ filter using solvent clarification device.

Water, filtered through a 0.45-μ filter using solvent clarification device.

SAFETY AND DISPOSAL

This experiment does not purport to address the safety issues associated with its use. It is the responsibility of the user of this experiment to establish appropriate safety and health practices and to determine the applicability of

regulatory limitations prior to use. All chemicals should be handled and disposed of in an appropriate manner consistent with the safety policy of the experimenter's company, school, or organization.

EXPERIMENTS

A. Preparation of Instrument*

[Note: This experiment is most easily performed using a one-pump configuration using the technique described below. The experiment was tested using a Waters Model 6000A Solvent Delivery System and using a Waters Model 510 Solvent Delivery System. Other single solvent delivery devices should work but have not been tested. As an alternative to the procedure used in this experiment, a standard automatic (high pressure) gradient system may be used if a second pump and a gradient controller are available (see Evaluation (*B*).]

Activity	Comments
1. Obtain two 250-mL Erlenmeyer flasks and two rubber stoppers to fit them. Also obtain four 9-in. pieces of $\frac{1}{8}$ in. OD stainless-steel tubing (or glass, note: glass is fragile) and approximately 2 ft of $\frac{1}{8}$ in. ID Teflon tubing. Clearly label one flask A and the other flask B.	1. This simple equipment will be used to construct the gradient forming device. Refer to Figure 17-1 for visualization of the apparatus.
2. In each of the stoppers bore two holes that are large enough to just barely admit the stainless-steel tubing. (Caution: if glass is being substituted, please observe appropriate safety and handling procedures when inserting the tubes.)	2. It is extremely important to keep these holes as tight around the tubing as possible. If this is not done, a good syphon will not be formed and the gradient will not work correctly.
3. For the flask A stopper, push one of the stainless-steel tubes through a hole to a depth approximately 2 in. above the bottom of the flask. Attach a reservoir filter to this tube using a short piece of Teflon tub-	3. The reservoir filters are used to prevent suspended particles in the solvent from getting to the column packing material and causing a high operating back pressure.

* This may be done by the instructor before the lab period.

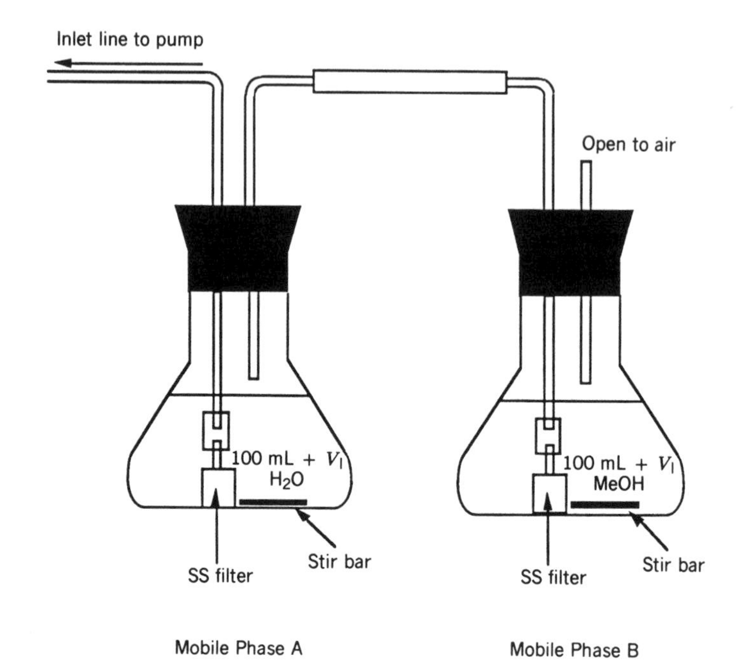

FIGURE 17-1. Schematic of the gradient apparatus.

ing. When the stopper is in place, the reservoir filter should rest on the bottom of the flask. Repeat this procedure for flask B.

4. Push the third piece of stainless-steel tubing through the second hole in stopper A so that it extends just below the bottom of the stopper. (The long extending portion of this tube can be cut to a more convenient length if desired). With the fourth piece of stainless-steel tubing do the same for flask B.
5. Measure the volume of the inlet line that runs from flask A to the pump. To do this, fill flask A with 200 mL of water. Place the stopper apparatus into the flask so that the filter is submerged in the water.

5. Measurement of the inlet line volume allows you to determine when the gradient actually starts. A certain "lag" volume must flow through the system before the concentration change actually begins.

Now open the draw valve and using the syringe fill the inlet line with water.

6. Once the line is filled, remove the stopper apparatus so that the filter is no longer in the water. Draw out all of the water in the line with the syringe and record this volume in milliliters. V_I (Volume inlet line) = —— mL.

   Comment: 6. Flask A should contain a teflon coated magnetic stir bar and be placed on a magnetic stirrer to ensure adequate mixing of the two solvents. This will provide a smooth gradient profile. Fill in the V_I volume in Activity A-6.

7. Attach a piece of Teflon tubing to the stainless-steel tube that has the reservoir filter attached on Stopper B, of sufficient length to reach the stainless-steel tube on the Stopper A that does not have the reservoir filter attached. AT THIS POINT DO NOT ATTACH THE TEFLON TUBING TO STOPPER A.
8. Place 100 mL of methanol in flask B and stopper firmly with Stopper B. Elevate flask B a few inches above flask A.
9. Install a single column in the instrument.
10. The injector handle should be in the "inject" position.

    Comment: 10. This ensures that the sample holding loop is thoroughly cleaned.

11. Using flask A (which contains the remainder of the 200 mL of water) as a reservoir, prime the pump and let the system run at 2.0 mL/min while the sample is being prepared.

    Comment: 11. This allows the system to purge itself and equilibrate.

B. Sample Preparation

Activity	Comments
1. Peel one average size orange and cut the peel into approximately 0.5-in. squares.	

2. Place the chopped orange peel into a 150-mL Erlenmeyer flask and add 100 mL of methanol at room temperature. Shake occasionally for a 10-min time period.
3. Filter 20 mL of methanol extract through the sample clarification device into another 150-mL Erlenmeyer flask.

   3. This will remove any particulates from the solution.
4. Repeat Activity B-1 through B-3 for the grapefruit.

C. Chromatography

1. Stop the pump. Empty out the contents of flask A. Fill flask A with 100 mL of filtered distilled water plus the volume of the inlet line that was previously measured. Stopper firmly. (For example: if the volume of the inlet line V_I was measured to be 10 mL, then add 110 mL of filtered distilled water to flask A.)

   1. This enables you to create a gradient that is almost linear. If the volume to the inlet line is not considered, then a larger amount of methanol will be added to a smaller amount of water and the concentration shift will bias the resulting elutions.
2. Open the solvent draw-off valve and pull liquid to the entrance of the pump.
3. Attach the Teflon line from flask B containing methanol to flask A containing water.
4. Place the injector handle in the "load" position.
5. Set the spectrophotometer sensitivity at 0.1 AUFS, 254 nm.
6. Start the solvent delivery system (pump) flowing at 2.0 mL/min. Open the solvent draw-off valve and using the syringe withdraw just enough water so that the methanol from flask B just begins to drip into flask A. Remove the syringe and close the solvent draw-off valve.

7. Turn the injector handle to the "inject" position.
8. Allow the blank gradient to develop for 50 min.
9. After 50 min, stop the pump, clean out all flasks and lines, and return to original conditions. To do this fill flask A with water. Prime the pump and let the system run at 2.0 mL/min with the injector handle in the inject position. Watch for the UV baseline to reestablish its initial position.
10. Repeat Activity C-1 to C-4.
11. Load 25 μL of the orange peel extract into the injector.
12. Repeat Activity C-6 and C-7 and allow the gradient to develop for 50 min.
13. Repeat Activity C-9 to C-11 using the grapefruit extract.

7. Since this is a "blank" run, no sample is injected.
8. A blank gradient is required to distinguish sample components from contaminants in the solvent which might elute during the analysis.
9. If possible, inject 2 ml of MeOH onto column (see Fig. 17-3 and 17-4) at end of run to insure column is "cleaned."

EVALUATION

A. All chromatograms should be properly labeled and attached to the last sheet in this experiment. Proper labeling requires the following:

Sample:	Components and injection volume.
Detector:	Type of detector and attenuation.
Column:	Type of column used, length, and diameter (ID).
Solvent:	Mobile-phase composition and flow rate.
Recorder:	Recorder chart speed.

B. Example results of the blank gradient, and the separation of orange peel extract and grapefruit peel extract under the conditions of this experiment on a low-efficiency column are given in Figures 17-2–17-4. Example results for a high-efficiency column using the gradient described in this experiment are shown in Figure 17-5. Example results for a high-efficieny column in an instrumental (high-pressure) gradient system are shown in Figures 17-6 and 17-7. Compare your results for the orange peel with the grapefruit extract.

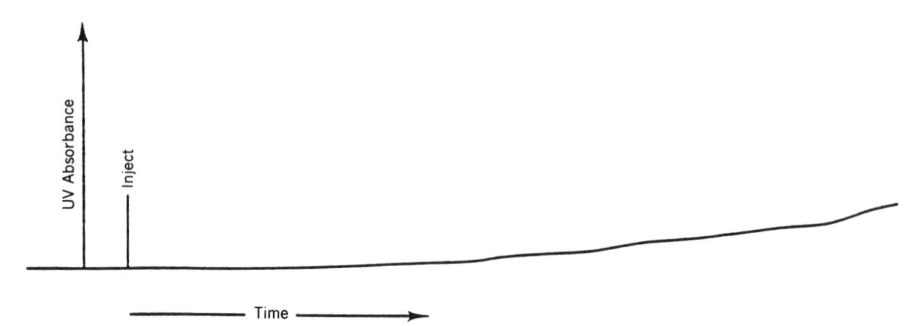

FIGURE 17-2. Gradient chromatography. Mobile phase: H_2O to MeOH gradient. Flow rate: 2 mL/min. Detector: UV at 254 nm, 0.1 AUFS. Sample: Blank. Recorder: 0.2 in./min. Column: Bondapak C_{18}/Porasil B, 2 mm ID × 60 cm. (Note: Actual separation will depend upon the quality of the mobile phase and column packing.)

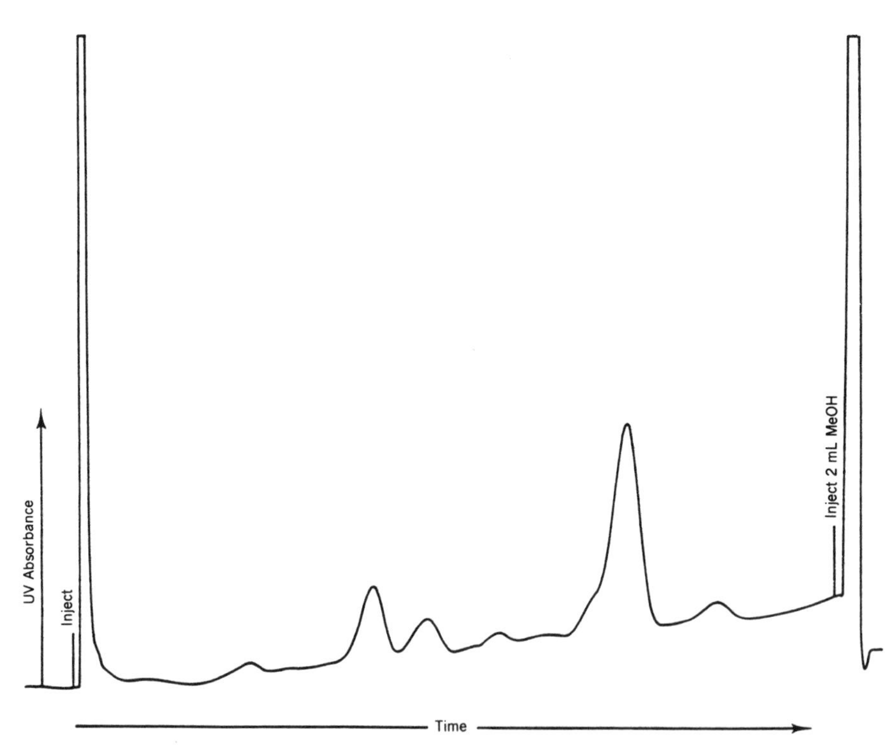

FIGURE 17-3. Gradient chromatography. Mobile Phase: H_2O to MeOH gradient. Flow rate: 2 mL/min. Detector: UV at 254 nm, 0.1 AUFS. Sample: orange peel extract, 25 μL. Recorder: 0.2 in./min. Column: Bondapak C_{18}/Porasil B, 2 mm ID × 60 cm. (Note: Actual separation will depend upon the quality of the mobile phase and column packing.)

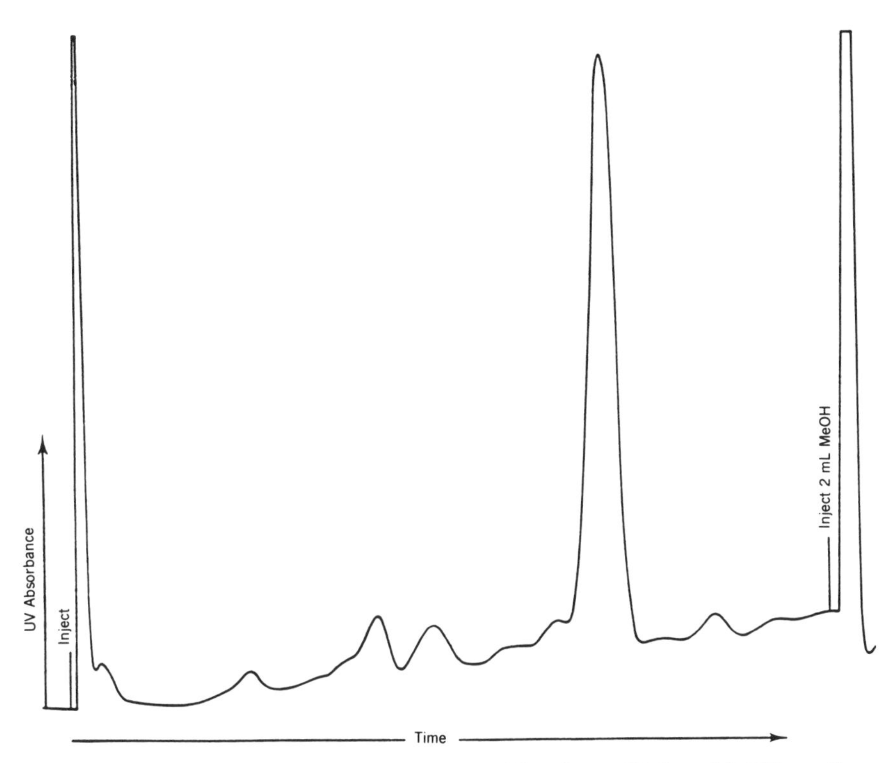

FIGURE 17-4. Gradient chromatography. Mobile phase: H_2O to MeOH gradient. Flow rate: 2 mL/min. Detector: UV at 254 nm, 0.1 AUFS. Sample: grapefruit peel extract, 25 μL. Recorder: 0.2 in./min. Column: Bondapak C_{18}/Porasil B, 2 mm ID × 60 cm. (Note: Actual separation will depend upon the quality of the mobile phase and column packing.)

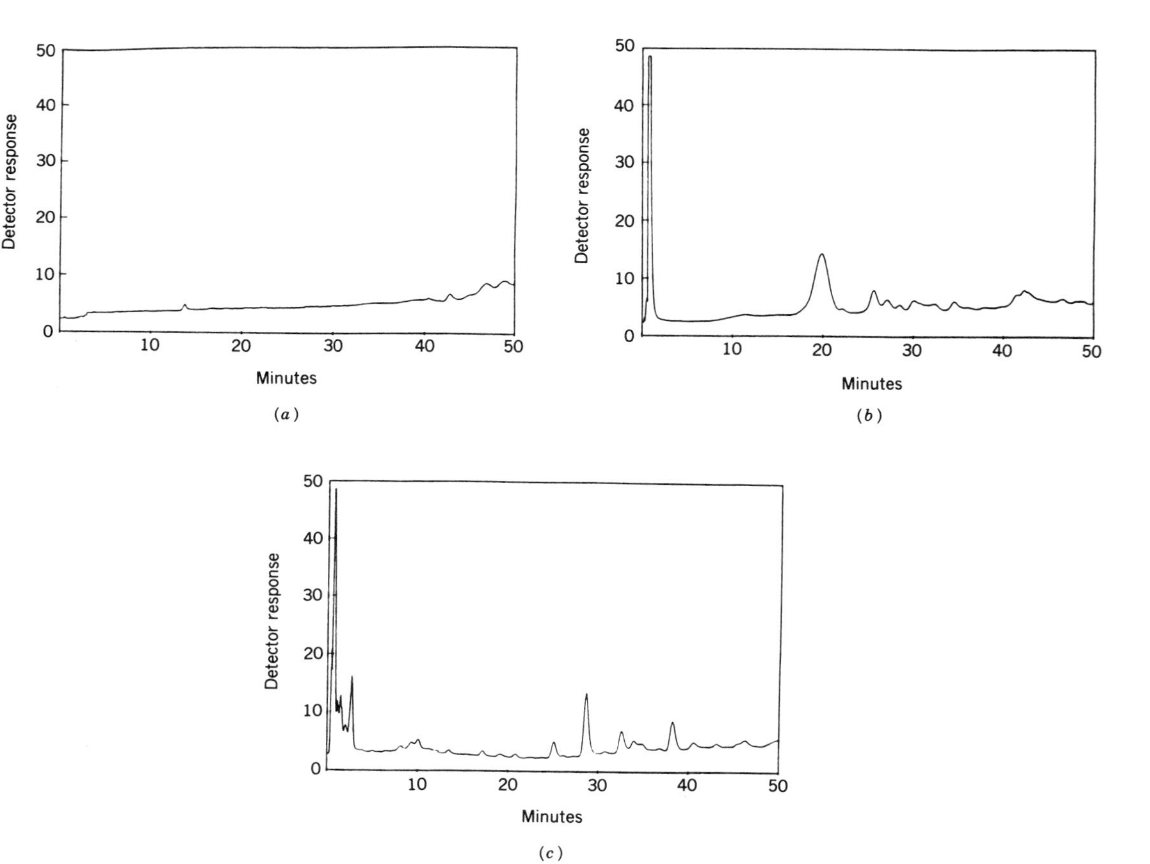

FIGURE 17-5. Example results for a high-efficiency column with the low-pressure gradient former shown in Figure 17-1. Mobile phase: H_2O to MeOH gradient. Column: μBondapak C_{18} (10 μm), 3.9 mm ID × 15 cm. (*a*) Blank gradient. (*b*) Orange peel extract, 25 μL. (*c*) Grapefruit peel extract, 25 μL. (Note: Actual separation will depend upon the quality of the mobile phase and column packing.)

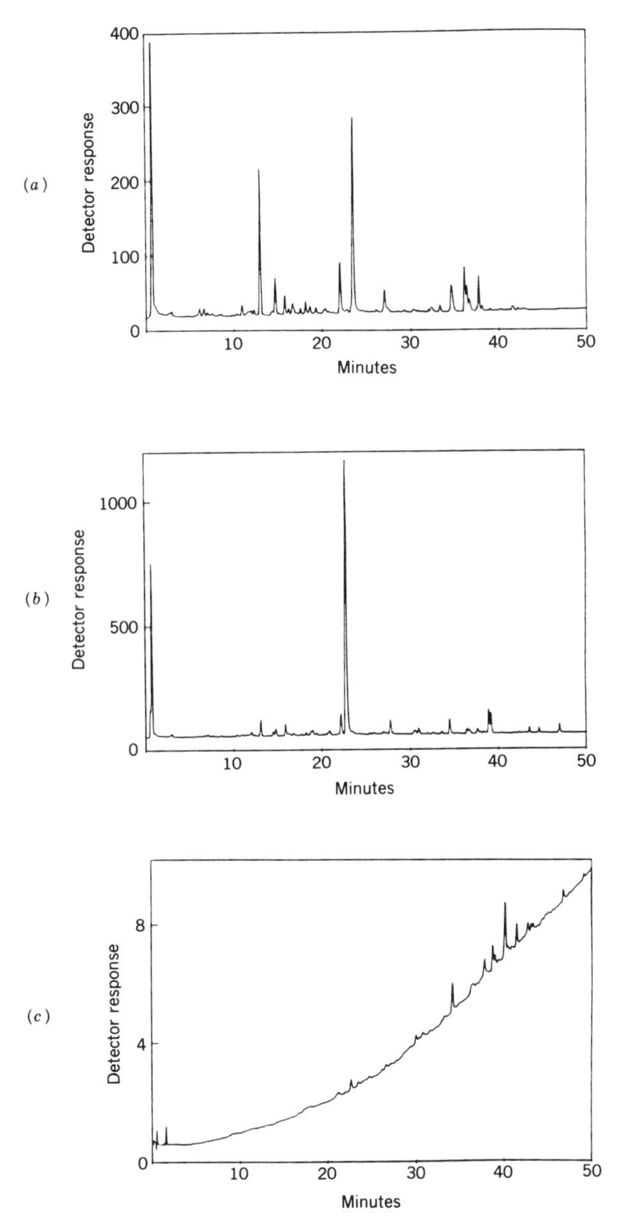

FIGURE 17-6. Example results for a high-efficiency column using an automatic, high-pressure gradient system. Mobile phase: H_2O to MeOH gradient. Column: Resolve C_{18} (5 μm), 3.9 mm ID × 15 cm. (*a*) Orange peel extract, 25 μL. (*b*) Grapefruit peel extract 25 μL. (*c*) Blank gradient. (Note: Actual separation will depend upon the quality of the mobile phase and column packing.)

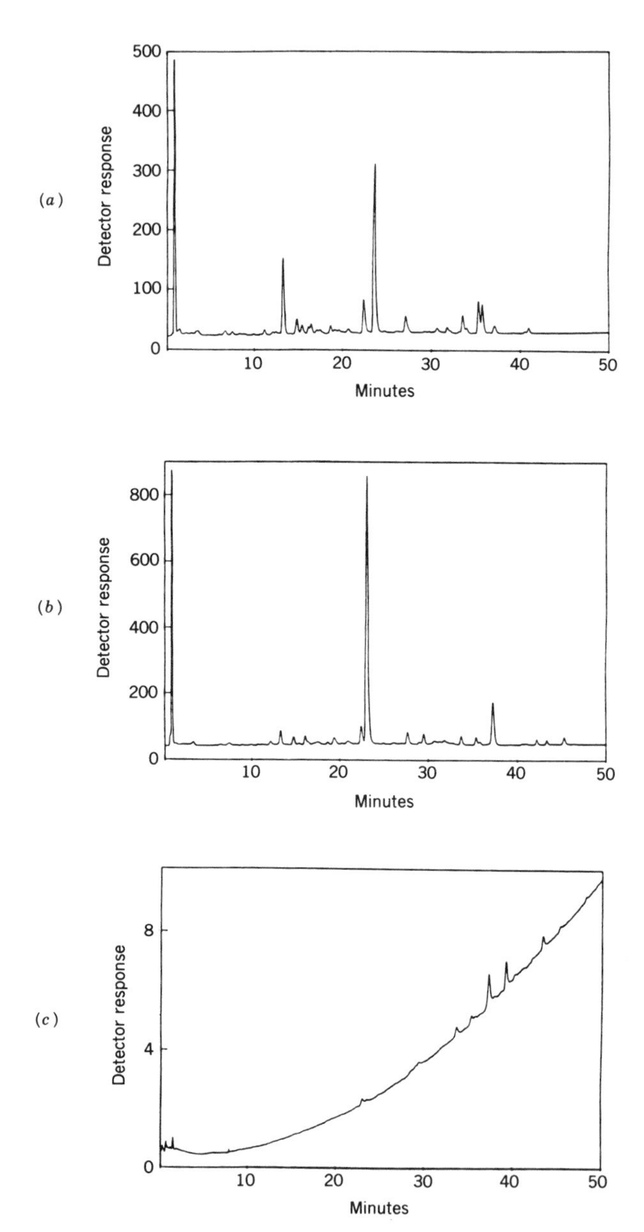

FIGURE 17-7. Example results for a high-efficiency column using an automatic, high-pressure gradient system. Mobile phase: H_2O to MeOH gradient. Column: μBondapak C_{18} (10 μm), 3.9 mm ID × 15 cm. (*a*) Orange peel extract, 25 μL. (*b*) Grapefruit peel extract, 25 μL. (*c*) Blank gradient. (Note: Actual separation will depend upon the quality of the mobile phase and column packing).

DISCUSSION

The peaks observed in the "blank" (gradient) run are real components eluting from the column. Because of the experimental procedure used, water is pumped through the analytical column (e.g., Bondapak C_{18}/Porasil B) column before the gradient is started; therefore, any trace levels of organic compounds in the distilled water will be concentrated on the column and will elute during the gradient. If one wished to obtain a completely flat "blank" gradient baseline, the water could be "cleaned" before making the mobile phase by pumping water through the column. However, if only a qualitative analysis (fingerprint) is desired, subtracting the peaks due to the "blank" from the chromatographic run is sufficient.

After subtracting the peaks from the blank, it is clear that the orange peel extract contains a broad mixture of compound polarities. Very polar compounds elute quickly in pure water and the nonpolar compounds do not elute until the mobile phase is higher in methanol content. If a mobile phase was good for the separation of early eluting peaks, the late eluting peaks would have an extremely long retention time, hence the need for solvent programming.

SELF-HELP QUESTIONS

1. Does the use of a gradient increase resolution or decrease resolution? How can you justify your answer?
2. What are the similarities and differences between the chromatograms obtained for the orange and grapefruit peel extracts?
3. If you did Chapter 12, Experiment 4 (Reverse-Phase Chromatography) how might a gradient be used in that experiment?

SUPPLEMENTAL EXPERIMENT

The following suggested supplemental experiment will increase your understanding of the use and characteristics of high-resolution gradient LC. This may be done under the operating conditions previously established.

To determine a visualization of the concentration of methanol during the gradient, repeat a "blank" run with 0.1% acetone in the methanol flask.

INDEX

Accuracy, 227–229
Adsorption, 18, 96
Air quality, 41
Alpha, 6
Amino acid:
  analysis, 37–40
  sequencing, 37–39
Analyte, 18
Analytical chromatography, 18
Application:
  focus, 9
  gel permeation chromatography, 16
  growth, 11–18
  ion chromatography, 16
  organic synthesis aid, 51–60
  preparative chromatography, 17
  shampoo, 134
  typical uses of HPLC, 13
ASTM:
  committee E-19, 237
AUFS, 18

Band, 18
Band broadening, 18, 88–90, 238–242
Bed, 19
Bonded phase chromatography, 19, 97
  column preparation, 211–213
Buffer, 19, 147–150, 246–248

Capacity factor, 19, 85
Chromatogram, 3, 4, 7, 8, 19, 69, 70
Chromatographic process, 2–6
Column, 19
Column volume, 19
Confidence limit, 232
Correct values, 235
Counter ion, 19

Databases, 106, 107
Decision tree for column choice, 109
Degassing, 249, 313
Detector, 19, 81–85
  wide scope, 82
  high sensitivity, 82
  time constant, 84
Diagnostics, 78–81
Differential Refractometer, 19, 181
Dry column chromatography, 272

Efficieny, 6, 7, 19, 86–90, 214–221
  calculation, 215, 327
  column length, 346, 350
  comparison, 220
  flow rate, 216, 220
Eluent, 19. *See also* Solvent and Mobile phase
Eluent strength, *see* Solvent strength
Elute, 20
Elutropic series, 20, 123, 190
EPA methods for HPLC, 45
Equieluotropic, 138

Error:
  constant, 229
  random, 229, 231
Error curve, 231
Essential oils:
  active ingredient, 429
  experiment, 425–434
Estrogens, 144–150
Exclusion chromatography, *see* Gel permeation chromatography

FD&C dyes, 330
Filter:
  in-line, 71, 74
Fittings, 76–78
Flash chromatography, 272
Flow programming, 20, 314–316
Flow rate, 20, 125, 128, 216–221
  precision, 236
Fronting, 20, 81

Gas Chromatography, 11
Glossary, 18–25
Gel filtration chromatography, *see* Gel permeation chromatography
Gel permeation chromatography, 20, 46–51, 99, 112–114, 177–186, 359, 366
  calibration curve, 178–180
  column types, 114
  efficiency, 183
  exclusion limit, 20
  experiment, 358–369
  preparative capability, 183–186
  small molecule, 177–186, 364, 365, 368
  solvent effects, 180–182
Gene synthesis, 30–33
GPC, *see* Gel permeation chromatography
Gradient elution, 21, 284–314
  applications, 305–309
  considerations, 310–314
  continuous, 286
  experiment, 435–447
  flow rate effect, 301–305
  hardware contributions, 268–293
  high-pressure mixing, 289–292
  ideal system, 288
  low-pressure mixing, 289–292
  method development, 293–305
  mixing devices, 291
  nomenclature, 286
  out-gassing, 313
  precision, 292, 293
  pressure, 312
  rate of change, 294
  refractive index effects, 310–313
  shapes, 294
  step, 286
  viscosity effects, 310–313
Guard column, 71, 74

HETP, 21, 87–90
Height equivalent to a theoretical plate, *see* HETP
Hemoglobin, 34
Historical milestones, 10, 11
Human insulin, 28, 29

Injection, 21
Injector, 21, 73
Instrumentation, 2, 3
  band spreading considerations, 75–78
  diagnostics, 78–81
  influence on performance parameters, 79
  plumbing considerations, 71–75
Interferon isolation, 29
Ion exchange, 21, 98, 122–124, 167–176
  CM packing, 173–175
  DEAE packing, 173–175
  effect of counterion, 171
  effect of pH, 169–171, 174
  effect of pK_a, 169, 174
  proteins, 173–176
  temperature, 171
Ion pair chromatography, 21, 22, 98, 157–164
  retention models, 161
Ion-pair extraction, 157
Ion suppression, 21, 153–157, 170
Isocratic:
  elution, 21, 285
  system, 70–75, 319
Isoelectric point, 174
Isoeluotrophic, 21, 138

k′, *see* Capacity factor
Kinetics, *see* Reaction

Linear velocity, 21
Liquid-liquid chromatography, 21
Liquid-liquid extraction, 22
Liquid-solid chromatography, 22, 96

Matrix, 22
Method development:
  databases, 106, 107
  decision tree for column choice, 109
  general strategy, 105–108
  general tactics, 108–110

next actions after first attempt, 126, 127
thin layer chromatography, 192
Miscibility, 244–246
Mobile phase, 22, 122–124
binary blend, 139
changing, 246
compatibility, 249
considerations for choosing, 243–246, 250
nonaqueous, 139–142
preparation, 254
recirculation, 242–243
refractive index, 312
reservoir, 72
reverse phase, 133–145
selectivity, 135–139
strength, 122, 123
tertiary blend, 139
quality, 251
quaternary blend, 139
viscosity, 312
Mobile phase volume, 22
Molecular weight, 46–51
Molecular weight distribution, 46–51

NIOSH methods for HPLC, 46
N-nitroso compounds, 42
Normal phase, 22, 95–97, 115, 116, 186–204, 332, 333
experiment, 332–344
method development, 191–202
mobile phase, 189–191
stationary phases, 187–189

Oligonucleotide isolation, 30–33
OSHA methods for HPLC, 47

Packing, 22
Paired-ion chromatography, *see* Ion-pair chromatography
Partition chromatography, *see* Liquid-liquid chromatography
Peak, 22
area, 22
height, 22
maximum, 22
width, 23
Pellicular packing, 23, 90
Pesticide analysis, 43–45
Peptide:
analysis, 35
mapping, 36, 37
Plate height, *see* HETP
Plates, 23, 87–90
Plastics, 46–49
Polarity, 23, 122
of the molecule, 191
function group, 193
Polymer analysis, 46–49
Polyaromatic hydrocarbons, 43, 110–113
Porous packing, 23
Precision, 229–231
gradient analysis, 293
LC analysis, 237
Pressure read-out, 78–81
Preparative chromatography, 17, 23, 269–283, 416. *See also* Recycle
approaching the problem, 271
classification, 274
collection, 282
detectors, 280–282
experiment, 415–424
method development, 274–277
monopropionamides of dicyanoheptamethylcobyrinate, 53–60
oligonucleotides, 31–33
recovery, 282
scale-up, 188, 277–280
Protein analysis, 33–35, 173–176

Qualitative analysis, 7, 63–66
Quantitative analysis, 7, 385
experiment, 384–404

R. B. Woodword, 58–60
Reaction:
kinetics, 406
hydrolysis of asprin experiment, 405–414
monitoring using HPLC, 53–57, 346
Recycle, 23, 221–227
experiment, 345–357
shaving, 228, 346, 353
Resolution, 6, 7, 23, 92–94, 328
baseline, 19
influence of parameters upon individual terms, 129
peak purity, 8
relationship of alpha and plates, 93
Retention:
chromatography, 23
equation, 325
relation to phase polarity, 116, 121
relation to sample polarity, 116
time, 23
volume, 23
Reverse phase, 24, 95, 97, 115–121, 132–167, 208–214, 319, 371

Reverse phase (*Continued*)
column choice, 213
differences in columns, 208–213
effect of pH, 153–156
effect of pK_a, 153–155
experiment, 318–331
ionic compounds, 151–167
method development, 133–139, 142–167
method development experiment, 371–383
nonaqueous, 139–142
silica gel, 165–167
Refractive index, 81–83
comparison to UV, 281, 282
detector response, 181, 281
RI, *see* Refractive index

Selectivity, 6, 86, 94, 135–139, 327
Separation, 2, 85
databases, 106, 107
decision tree for column choice, 109
general strategies, 105–108
general tactics, 108–110
mechanisms, 94–102
Separation factor, *see* alpha
Significant figures, 234, 235
Silanols:
contribution to retention, 165
types of, 211
Size exclusion chromatography, 24
Size separation, *see* Gel permeation chromatography
Solid phase extraction, 24, 256–269
conditioning, 256
effect of flow rate, 262–267
eluting, 258
examples, 259–267
guidelines, 268, 270
isolating the analytes, 258
loading, 257, 267
preparation of, 264–267
reconditioning, 259
recovery, 269
Solute, 24
Solvent, 24
binary, 139
boiling point, 245
compatibility, 249
miscibility number, 245
physical properties, 245
polarity index, 245
strength, 122, 123, 190, 198, 200
tertiary, 139
quality, 251
quaternary, 139
viscosity, 245
water saturated, 194, 199
Solvent delivery system, 24, 72
Solvent programming, *see* gradient elution
Sorption, 24
Spices:
major component, 429
steam distillation, 425, 429
Standard deviation, 230, 395
Stationary phase, 25, 115, 187
Statistics:
for chromatographers, 224–238

Tailing, 25, 81, 165
Therapeutic drug monitoring, 60–63
asthma, 61, 62
epilepsy, 62, 63
Thin layer chromatography, 192, 272
Triglycerides, 141–143
Trace analysis, 238–242
Trace enrichment, 25

UV/Visible detector, 25, 81–85

Variance, 230
Van Deemter plot, 25, 88–90
Viscosity, 312
Vitamin B_{12}, 51–60
Void, 25
Void time, 25
Void volume, 25

Water:
purification, 255
specifications, 253
quality, 42–47, 251

Zone, *see* Band
Zone spreading, *see* Band broadening